Renewing Development in Sub-Saharan Africa

After two decades of economic reform programmes, the majority of countries in sub-Saharan Africa still face the challenge of renewing the development progress that they were achieving in the 1960s and early 1970s. *Renewing Development in Sub-Saharan Africa* brings together leading specialists in different aspects of African development, to assess the performance of the sub-Saharan economies over recent decades, the main political, economic and geographical problems and constraints that the continent currently faces, and future strategies and policies which can continue effectively to renew development.

Renewing Development in Sub-Saharan Africa is made up of sections on conflict and power; agriculture and the rural sector; industry and the urban sector; international trade and transport constraints; and gender, health and education. These cover some of the important debates such as the effects of structural adjustment on economic performance, the impact of AIDS, the problems of corruption, the social and economic costs of war and conflict, the role and effectiveness of NGOs, the potential contribution of the private sector, the effect of globalisation on African industrial and agricultural prospects, the problems of the landlocked countries, the strategies for agricultural and rural development, food aid, the options for poverty alleviation in Africa, small-scale enterprise possibilities and education, health and gender issues and policies.

Collectively and individually, the chapters provide succinct statements of the present situation, progress and problems in each of the major sectors, with insights into current issues and debates, as well as a wealth of statistical information. The emphasis of the book is on the identification of appropriate strategies that will enable individual countries to exploit successfully their growth opportunities and to meet poverty-reducing and other key equity objectives, rather than repeating arguments contrasting the general merits of agriculture versus industrial development versus human capital formation.

Deryke Belshaw is Professor Emeritus of Development Studies at the University of East Anglia and Dean of Development Studies at the Oxford Centre for Mission Studies.

The late **Ian Livingstone** was Professor Emeritus of Development Studies at the University of East Anglia and Emeritus Leverhulme Fellow.

D1275873

Renewing Development in Sub-Saharan Africa

Policy, performance and prospects

Edited by
Deryke Belshaw and Ian Livingstone

London and New York

First published 2002
by Routledge
11 New Fetter Lane, London EC4P 4EE

Simultaneously published in the USA and Canada
by Routledge
29 West 35th Street, New York, NY 10001

Routledge is an imprint of the Taylor & Francis Group

Typeset in Galliard by Keystroke, Jacaranda Lodge, Wolverhampton
Printed and bound in Great Britain by T.J. International Ltd, Padstow, Cornwall

British Library Cataloguing in Publication Data
A catalogue record for this book is available from the British Library

Library of Congress Cataloging in Publication Data
Renewing development in sub-Saharan Africa : policy, performance and prospects
/ edited by Deryke Belshaw and Ian Livingstone.
   p. cm.
   Includes bibliographical references and index.
   1. Africa, sub-Saharan—Economic conditions—1960–   2. Africa,
sub-Saharan—Economic policy.   3. Structural adjustment (Economic policy)—
Africa, sub-Saharan.   I. Belshaw, D. G. R. II. Livingstone, Ian.
HC800 .R4567 2001
338.967—dc21   2001019955

ISBN 0–415–25217–2 (hbk)
ISBN 0–415–25218–0 (pbk)

Contents

Figures

Map

Tables

Contributors

Stephen Akroyd is a staff economist with Oxford Policy Management.

Deryke Belshaw is Professor Emeritus of Development Studies at the University of East Anglia and was Visiting Scholar at Wolfson College, University of Oxford, 1998–2000.

Gerald Bloom is a Fellow in the Health and Social Change Programme in the Institute of Development Studies, University of Sussex.

Christopher Colclough is Professorial Fellow in the Institute of Development Studies, University of Sussex.

Christopher Cramer is Lecturer in the School of Oriental and African Studies, University of London.

Andrew Dorward is Senior Lecturer in the Agrarian Development Unit, Imperial College at Wye.

Alex Duncan is a staff economist with Oxford Policy Management and Visiting Professor, Imperial College at Wye.

Oliver Furley is a Research Fellow in the African Studies Centre, University of Coventry.

Reginald Herbold Green was Professorial Fellow in the Institute of Development Studies, University of Sussex. He retired in December 2000.

Carolyn Jenkins is a member of the Centre for the Study of African Economies, Department of Economics, University of Oxford, and an associate of the Centre for Research into Economics and Finance in Southern Africa, London School of Economics.

Jonathan Kydd is Professor of Agricultural Development Economics, Imperial College at Wye.

Peter Lawrence is Senior Lecturer in the Department of Economics, Keele University.

Ian Livingstone was Professor Emeritus of Development Studies, University of East Anglia, and Emeritus Leverhulme Fellow.

Henry Lucas is a Fellow in the Health and Social Change Programme in the Institute of Development Studies, University of Sussex.

Marjorie Mbilinyi is Professor in the Institute of Development Studies, University of Dar es Salaam.

K. Mlambo is a Senior Economist in the African Development Bank, Abidjan.

Oliver Morrissey is Reader in Development Economics in the School of Economics, University of Nottingham.

Michael Mortimore is a partner at Drylands Research, Crewkerne, Somerset.

Paul Mosley is Professor of Economics, University of Sheffield.

Anthony O'Connor is Reader in the Department of Geography, University College London.

T.W. Oshikoya is a Senior Economist in the African Development Bank, Abidjan.

Colin Poulton is a Research Officer in the Agrarian Development Unit, Imperial College at Wye.

Carole Rakodi is Professor in the Department of City and Regional Planning, Cardiff University.

John Shaw was until recently Economic Adviser and Chief of the Policy Affairs Service, World Food Programme.

John Thoburn is Reader in Economics in the School of Development Studies, University of East Anglia.

Lynne Thomas is a Research Officer at the Centre for Research into Economics and Finance in Southern Africa, London School of Economics.

Michael Tribe is Senior Lecturer in the Centre for International Development, University of Bradford.

Martin Upton is Professor Emeritus of Agricultural Economics at the University of Reading.

Tina Wallace is Senior Lecturer in Business Management, Oxford Brookes University.

Steve Wiggins is Lecturer in the Department of Agriculture and Food Economics, University of Reading.

Ian Livingstone
11 October 1933–14 September 2001

A tribute

Ian Livingstone died after a short and unexpected illness on Friday 14 September 2001, a month short of his 68th birthday. Ian had completed his co-editorial duties for this volume and was planning field research on industrial policy in Cambodia, for which he had been awarded an Emeritus Fellowship by the Leverhulme Foundation.

Ian held degrees in economics from Sheffield and Yale universities. He had been Professor of Development Economics at the University of East Anglia since 1978 and was made Professor Emeritus in 1999. He had previously taught and researched at Makerere University, including its preceding Universities of London and East Africa phases; at the Institute of Development Studies at the University of Nairobi; at the University of Dar es Salaam, where he was Director of the Economic Research Bureau; and in the economics department of Newcastle University.

Ian first went to live and work in sub-Saharan Africa in 1975. In the following four and a half decades he researched, advised and published across a wide range of relatively unexplored applied economics issues, mainly in tropical Africa. His economics textbooks for West and Eastern Africa continue to be in extensive demand following their publication in the 1960s, with revised editions in the 1980s. He was instrumental in establishing the Economic Policy Research and Economic Planning Training Centres in Uganda in the 1990s. In the UK Ian was an exceptionally long-serving member of both the Commonwealth Scholarships Commission and the Economic and Social Research Committee of the government's development department. This diverse collection of papers on development in sub-Saharan Africa is a fitting, if unintended, tribute to the life work and professional aspirations of a highly esteemed colleague.

Deryke Belshaw
Norwich
8 October 2001

Preface

The objectives of this volume are: first, to provide for general and more specialised readers a clear introduction to the recent economic experience – primarily the last two decades – of the countries and peoples of sub-Saharan Africa (SSA), providing an overview of the continent as a whole.

The volume aims, second, to set alongside this descriptive baseline the knowledge and experience of specialist researchers working in important sectoral, functional or institutional areas of development activity in Africa. These reviews lead to assessments of the prospects for renewed development on the back of identified improvements in the design of key policies and their execution in specific country contexts.

It aims as a result, third, to provide sufficiently strong indicators to policy makers regarding appropriate policies and directions for strategy that would be of benefit in delineating packages applicable in their own individual country and sub-regional contexts and that would inform their efforts at reversing past setbacks and securing renewed development in the coming decades.

It is anticipated that the comprehensive nature of the review provided here will make this a valuable text for courses in development studies, development economics and geography, as well as various African studies courses, at the undergraduate and postgraduate levels.

No single volume can hope, of course, to cover even partially all the important areas of policy. Such areas include climatic change, global and continental transport systems, governance and political processes – including decentralisation, gender patterns of participation, natural resource management and the integration of environmental issues into development policy.

The multidisciplinary focus and balance of the volume is swayed by the editors' view of the crucial areas calling for action research and pilot testing in specific country and sub-regional contexts. A further underlying intention was to counter single-discipline and single-issue based policy prescriptions, on the one hand, and external donor and activist perspectives about 'Africa', formed in the capital cities of the developed world, on the other.

The initial basis of this volume is a set of papers presented at a conference held at the School of Development Studies in the University of East Anglia, under the auspices of the Standing Committee on University Studies of Africa (SCUSA), in September 1999. In line with the objectives described above, specialists in the various fields from within the UK and Africa were invited to prepare papers to cover each sector or problem area, considering specifically past economic performance, operative constraints on development and desirable policies to be pursued in different sectors.

The volume is structured in five main sections: on conflict and power; agriculture and the rural sector; industry and the urban sector; international trade and transport constraints;

and gender, health and education. These are preceded by the editors' overview of 'Africa's economic problems' and followed by some final observations by the editors.

More specifically, the first substantive section (Part II) addresses the area of conflict as both a major contributor in a number of countries to disappointing or disastrous performance in development and humanitarian terms and as a principal area for taking remedial (conflict resolution) or prophylactic (conflict prevention) action if development is to be renewed. Three chapters address different aspects of this critical subject.

The length of the next section (Part III), which deals with agriculture and the rural sector, reflects the intrinsic complexity and range of issues that need to be considered if effective national development strategies are to be designed for African economies and societies. The ten chapters that have been selected are intended to make accessible important areas of policy and action choice that tend to be relatively neglected in the macroeconomic and international development literature.

Part IV covers both the industrial and urban sectors. Three chapters focus on industry, with its disappointing record, outside South Africa at least, and examine the challenges of restructuring and reinvesting in industrial capacity able to exploit competitive advantages which include 'natural' (transport cost) protection and low-cost access to raw materials and other commodities suitable for pre-export processing.

The next section (Part V) examines some of the central aspects – using the concepts of constraint and opportunity – which international trade, under the changing and evolving patterns of a globalised world market, offers to resuscitated African economies in their varying locations, resource endowments and legacies of past development orientation, experience and international links. Strategic dimensions covered include trade policy reform, capture or recovery of market shares and actions needed to offset the special disadvantages of the numerous landlocked African economies, together with the potential assistance proffered by African common markets and other trading networks.

Part VI consists of three papers where the key areas of welfare policy – gender, health and education – are examined. The impact of globalisation has included trends that increase the economic and political exclusion of women. It is argued that African policy responses should build in capacity for participatory gender interaction; policy formation is reviewed in four key decision areas. In the area of health, recent mortality and morbidity trends in SSA, including those associated with HIV/AIDS, are reviewed, before proceeding to an analysis of public health policies in Africa. The range of impacts specifically on the poor is examined and the requirements of health policies directed towards meeting the needs of the poor outlined. Finally, the main policy issues in the education sector, with particular regard to primary education, are considered. After slowing in the 1980s, primary enrolment growth generally failed to recover in the 1990s, with wide disparities between SSA countries. The importance of raising female child enrolment rates to the male equivalent is emphasised. Finally, in Part VII, an overview chapter identifies aspects of strategy and policy reform that require further attention.

We are grateful to the Rockefeller Foundation in New York for a grant that permitted us to secure substantial participation at the conference by African scholars and to the African Studies Association of the UK and the School of Development Studies at the University of East Anglia (UEA) for additional financial resources. We also wish to acknowledge efficient support from the UEA's Conference Office and the major administrative contribution of Natasha Grist. In particular we wish to record our gratitude to Shona Livingstone for her major input in the production of the index. Finally, we must also thank Shona Livingstone and Barbara Belshaw for their assistance in preparing the manuscript for publication and, at Routledge, Andrew Mould for sound advice and Ann Michael, Audrey Bamber and Belinda Dearbergh for seeing the volume into print.

Acknowledgements

The authors and publishers would like to thank the following for granting permission to reproduce images in this work:

Figures

For Figure 7.2, National Academy Press, Washington, DC, for the figure 'The transition from degradation to intensification in a farming system', from page 66 of *Population and Land Use in Developing Countries*, by C.L. Jolly and B.B. Torrey, 1993.

For Figure 7.3, Taylor & Francis Books Ltd, for figure 2.1 from page 16 of *Working the Sahel*, by M. Mortimore and W. Adams, 1999.

For Figure 9.1, CAB International, Wallingford, Oxon, for figure 5.1 from page 246 of *Smallholder Cash Crop Production under Market Liberalisation*, by A. Dorward *et al*. (editors), 1998.

Tables

For Table 5.2, Club du Sahel/OCDE, Paris, France, for the table 'Les transformations de l'agriculture ouest-africaine', by S. Snrech, Mimeo (Sahl (95) 451), 1995.

For Table 12.5, Institute of Development Studies, University of Sussex, for a table from 'The New Poverty Agenda', by M. Lipton and S. Maxwell, IDS Discussion Paper no. 306, 1992.

Every effort has been made to contact copyright holders for their permission to reprint material in this book. The publishers would be grateful to hear from any copyright holder who is not here acknowledged and will undertake to rectify any errors or omissions in future editions of this book.

Part I

Reviewing development in sub-Saharan Africa

1 Development in sub-Saharan Africa

Progress and problems

1.1 Introduction

In this chapter an initial overview is presented of the pattern of development, broadly defined, to be found in sub-Saharan Africa (SSA) towards the end of the 1990s, together with an idea of trends in key parameters over the two decades from 1980. It should be appreciated that these overall trends mask a great degree of variation between individuals, groups of people and sub-national areas within countries. The information provided should facilitate an understanding, however, of the variation in the situation and performance of the major countries and counter any tendencies to over-simplify or over-generalise the development problems facing Africa at the present time.[1] In the course of this overview the issues and debates discussed in the twenty-five specialist chapters forming the core of the book are indicated. Development as viewed here encompasses progress in providing livelihoods on a sustainable basis, in providing access to education and in basic health provision for the majority of the population. The resources required for these improvements must come from the growth of the economy. Likewise, in assessing economic growth, the quality of growth must be assessed alongside numerical increases in gross domestic product. This quality of growth embraces efficiency, equity and environmental dimensions, as well as the cultural dimension that is normally neglected. Careful assessment of these dimensions is essential in our view in identifying the various appropriate routes to renewed development in SSA.

1.2 What is sub-Saharan Africa?

An apparently clear-cut geographical label – the areas of Africa lying to the south of the Saharan desert – becomes confused when we come to examine the nations whose peoples and institutions comprise and largely account for the variety of activities and systems to be found there. Large parts of certain countries are actually in the Sahara, with strong cultural, religious and historic links to North Africa – the Sudan and Mauritania especially. Most of the statistical tables used here will draw from forty-eight countries officially comprising sub-Saharan Africa in 2001. They usually exclude certain very small countries and those engaged in conflict for which recent statistical data are not available. The map provided on page 4 indicates all forty-eight countries. It allows each country to be located within both the major ecological zones that provide the basis for much of the variation in land use systems and the supranational regional groupings of states.

In 1998 there was an estimated population in sub-Saharan Africa of some 628 million people: almost half in Western Africa and 80 per cent in Western and Eastern Africa combined. Nigeria accounted for approximately 120 million out of 300 million in Western

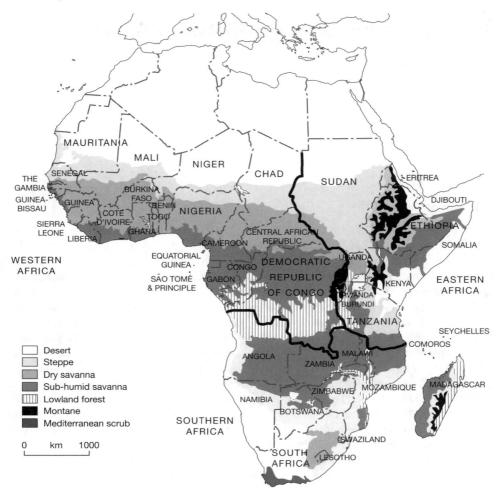

MAURITANIA
MALI
NIGER
CHAD
SUDAN
ERITREA
THE GAMBIA
SENEGAL
DJIBOUTI
GUINEA-BISSAU
GUINEA
BURKINA FASO
BENIN
TOGO
NIGERIA
SIERRA LEONE
COTE D'IVOIRE
GHANA
LIBERIA
CENTRAL AFRICAN REPUBLIC
ETHIOPIA
CAMEROON
SOMALIA
WESTERN AFRICA
EQUATORIAL GUINEA
SÃO TOMÉ & PRINCIPLE
GABON
CONGO
DEMOCRATIC REPUBLIC OF CONGO
UGANDA
KENYA
EASTERN AFRICA
RWANDA
BURUNDI
TANZANIA
SEYCHELLES
COMOROS
ANGOLA
MALAWI
ZAMBIA
MOZAMBIQUE
MADAGASCAR
ZIMBABWE
NAMIBIA
BOTSWANA
SOUTHERN AFRICA
SWAZILAND
SOUTH AFRICA
LESOTHO

Desert
Steppe
Dry savanna
Sub-humid savanna
Lowland forest
Montane
Mediterranean scrub

0   km   1000

* NB: Outside the map area are Mauritius, found east of Madagascar and the Cape Verde Islands, found west of Mauritania and Senegal.

Map   Sub-Saharan Africa: countries, regions and vegetation zones, 2000

Africa, while South Africa accounted for 41 million out of 125 million in Southern Africa. In Eastern Africa countries are more even in population size, although Ethiopia, with nearly 61 million people in 1998, was not far behind South Africa in terms of regional dominance. Nigeria also accounted for 31 per cent of the regional GDP in 1998. However, the economic size of South Africa within the continent is even more striking, accounting for 80 per cent of Southern African GDP and 48 per cent of the whole of SSA.[2]

Average levels of income are not the same across the identified regions, just as they differ markedly between countries within regions and within different parts of each country. Real GDP per capita was equal in 1998 to US $514 in Southern Africa, excluding South Africa, US $332 in Western Africa (but only US $255 in Nigeria, despite its oil, compared with

US $383 on average in the remainder) and just US $209 in Eastern Africa (World Bank, 1999). Incomes, and poverty, are not evenly spread.

1.3 Common and disparate features of SSA countries' experiences

Certain single factors or issues are often popularly held to provide direct explanations for the lack of development progress in SSA countries. On more systematic examination, such narratives fail to explain cases that clearly contradict the suppositions put forward. Examples are:

- *unproductive natural resource bases* In fact, a tropical climate has the advantage that prolific plant growth is prevented, by inadequate solar energy, only at high altitudes: in most locations soil moisture is more likely to be limiting than temperature and rainfall is both adequate and reliable for most of the year or one part-year season at least across the tropical savannah and semi-desert zones. Given reasonable prospects for storing food and for investing in assets, serious famines in the tropical regions can be explained more readily by inaction in the areas of policy and applied research.
- *petroleum resources* The trajectory followed by African petroleum exporting economies is clearly different from the oil importing countries. In addition, the major African oil exporter, Nigeria, is a classic reference for the negative 'Dutch disease' impact on its older agricultural export activities (cocoa, cotton, groundnuts and palm oil), thereby contributing to rural to urban migration and very high rates of urban growth from the 1970s onwards.
- *colonialism* Various adverse effects of colonialism on traditional and emergent African societies are well documented, although the effects of western, Arab and inter-African slave trading were more damaging *prior* to the colonial period proper. Some critical differences in severity of impact can be documented both between the different colonial regimes and within them. For example, Kenya suffered from land alienation to European settlers virtually until independence in 1963, whereas neighbouring Uganda, also under British administration, legislated against this from 1915.
- *neo-colonialism* Francophone countries have retained much closer financial ties with the former colonial powers than has been the case with other ex-colonies. For example, thirteen Francophone African countries in Western Africa share two central banks, whereas Anglophone ex-colonies each possess their own central bank and thus control exchange and monetary policies. Nevertheless, no clear balance of advantage seems to have accrued to those countries with greater freedom of economic policy.

1.4 The economic situation at end-2000: a marginalised continent?

The economic performance of the large majority of countries in sub-Saharan Africa (SSA) over the past two decades is widely seen as having been disastrous and prospects for the future bleak: there is a real danger of the continent being marginalised within the globalisation process and as other parts of the developing world, South and East Asia especially, make great forward strides. The 1980s in particular have been described as the 'lost decade' for Africa. While in the immediate period after independence, from 1965 to 1980, growth overall averaged as high as 4.2 per cent per annum, this rate fell to 1.8 in 1980–90 and just 0.9 per cent in 1990–4 (Table 1.1). While no countries in 1965–80 for which data are available showed negative rates of GDP growth, in 1980–90 there were two such countries and in 1990–94 twelve countries out of thirty-one with relevant data. More hopeful evidence does

Table 1.1 National GDP growth rates in SSA, 1965–98

Average annual growth rate (%)	No. of countries				
	1965–80	*1980–90*	*1990–8*	*1990–4*	*1994–8*
15 and above	—	—	—	—	
10–	1	1	—	—	5
5–	10	—	4	3	9
2.5–	8	14	14	9	4
0–	10	15	10	8	2
(0–)	—	2	1	9	4
(2–)	—	—	3	3	1
(5–)	—	—	—	—	1
(10–)	—	—	—	—	—
(15 and above)	—	—	—	1	—
No. of countries	29	32	32	31	26
SSA (% p.a. rates)	4.2	1.8	2.2	0.9	3.4

Source: World Bank (2000)

Note
In column 1, figures in brackets are negative values.

exist, fortunately, for the later part of the 1990s: in the 1994–8 period the number of countries with GDP growth rates of 5 per cent and above went up from three in 1990–4 to fourteen, for instance, while the overall rate for SSA countries went up from 0.9 to 3.4 per cent. This is a hopeful sign, even though it is not possible to predict its continuation.

With an average population growth rate of 2.9 per cent, significantly higher in many countries, even these GDP growth rates would not produce high rates of per capita growth. The spread of performance among SSA countries in growth rates of GNP per capita is much wider, as well as being around a lower mean (Table 1.2). The pattern over successive time periods is much the same, however: thus the number of countries with rates above 5 per cent went down from ten out of twenty-nine in 1985–90 to just three out of thirty in 1990–4, before increasing again to nine out of thirty in 1994–8. It should be pointed out (based on data given in UNDP, 1999), however, that, tracking the year of highest GDP per capita value within successive time periods from 1975 to 1997, in fourteen out of thirty-eight countries this still fell in the period 1975–9. For these countries the population was on average better off 20–25 years ago. SSA's GDP per capita in 1997 was actually 23 per cent *below* that of 1975, whereas that for South Asia was 7 per cent *above* and that for Southeast Asia and the Pacific 147 per cent above.

Associated with Africa's disappointing growth performance over the past two decades is the performance of exports. Merchandise exports grew in value terms at an average rate of 1.5 per cent over the decade up to 1990 and improved on that to 3.4 per cent over the next period 1990 to 1997–8 (Table 1.3). Over the whole period, however, this still meant a fall in value of exports per capita of 19 per cent and of agricultural exports per capita, especially, of 31 per cent. Performance in exports was, moreover, extremely variable: out of twenty-eight countries, as many as twelve showed decreases in the total *level* of exports, not just per capita, over the full period 1980–98.

Exports continue to be comprised predominantly of primary commodities. Since agricultural export prices fluctuate for reasons outside the control of the countries concerned, physical volumes exported are a better indicator of export effort than value. With respect to volume, the performance since 1980 has been extremely variable between countries in SSA

Table 1.2 National GNP per capita growth rates in SSA, 1963–98

Average annual growth rate (%)	No. of countries				
	1963–83	*1985–90*	*1990–8*	*1990–4*	*1994–8*
15 and above	—	2	—	—	2
10–	—	3	1	—	2
5–	1	5	2	3	5
2–	7	8	3	3	3
0–	14	3	7	2	6
(0–)	5	4	4	5	3
(2–)	2	—	10	7	6
(5–)	—	1	3	8	3
(10–)	—	1	—	1	—
(15 and above)	—	2	—	1	—
No. of countries	29	29	30	30	30

Source: World Bank, *World Development Report*, various years

Note
In column 1, figures in brackets are negative values.

Table 1.3 Total merchandise and agricultural exports, SSA (excluding South Africa), 1980–98

	1980	*1990*[a]	*1998: total exports* *1997: agric. exports*[b]
Exports (US $ m at constant 1995 prices)	42,271	49,202 (1.5)	64,226 (3.4)
Exports per capita (US $)	120	104	109
Exports per capita (US $), 1980=100	100	87	91
Agric. exports (US $ m at constant 1995 prices)	9,695	8,440 (–1.4)	10,904 (3.7)
Agric. exports per capita (US $)	27	18	19
Agric. exports per capita (US $), 1980=100	100	65	69

Source: World Bank (2000) *African Development Indicators, 2000*, New York

Notes
a Growth rate 1980–90 in brackets.
b Growth rate 1990–7/8 in brackets.

and, for individual countries, for different commodities. By and large, countries continue to be dependent on just two or three such exports, in some cases only one.

During the last two decades most SSA countries, with the assistance of the Bretton Woods institutions, adopted multi-year structural adjustment programmes (SAPs) in order to address their macroeconomic problems. Some of the impacts of SAPs are referred to in Chapters 9, 10, 20 and 21 of this volume. Improvements were achieved in the areas of inflation, particularly balance of payments and budgetary balance, but usually at the expense of rising international indebtedness. A number of countries are still grappling with major problems in these areas. Angola and the DRC had particularly high rates of inflation in the period 1990–8. In 1998 half a dozen countries had large government budget deficits and at least thirteen countries showed negative current trade balances in excess of 40 per cent of current exports.

Five SSA countries are significant oil exporters: Nigeria, Gabon, Congo, Cameroon and Angola. Together they account for practically all SSA oil exports, with Nigeria accounting (in 1997) for 70 per cent. The population of these countries is about one-third of the total SSA population. This means, however, that two-thirds of the SSA population are in non-oil economies that are highly vulnerable to oil price shocks that will in crisis periods cause major macroeconomic disruption and impact particularly severely on the poor. For example, in the first half of the year 2000 base petroleum prices rose by 300 per cent. The most severe impacts will be on the landlocked countries, none of which (with the exception of Chad) has oil but all have extended overland routes for both their imports and exports.

Foreign investment into Africa has expanded (Table 1.4): the foreign direct investment (FDI) stock in SSA excluding South Africa more than tripled over the period 1985–98 (UNCTAD, 1999a). It should be noted that a good portion of this investment has been directed towards mineral exploitation, petroleum in particular, in Nigeria, Gabon and Angola. Nevertheless, there were promising signs of increasing investment in eight or nine other countries.

As a percentage of the FDI stock in developing countries (excluding China), however, SSA's share has been falling since 1990 (Table 1.5). If the comparison is made with developing countries *including* China, it has been falling faster still, from 6.9 per cent in 1980 to 3.8 per cent in 1998, due to the size of the inflow into that country. This indicates the likely problem that SSA will face in competing for funds under globalisation.

1.5 Population

As indicated above, high rates of GDP growth are needed in SSA to keep pace with the high population growth rates and to maintain or increase GDP per capita. While women's total fertility rate in South/Central Asia fell from 5.72 in 1970–5 to 3.36 in 1995–2000 and is projected to fall further to 2.25 by 2015–20, that in Eastern Africa fell only from 6.97 to 5.79 in the same years, with a projection to 2015–20 of 3.85; in Middle Africa from 6.31 to 6.17, a minimal difference, with a projection of 4.21; in Southern Africa from a much lower 4.96 to 3.43, with a projection of 2.35; and in Western Africa from 6.99 to 5.47, with a projection

Table 1.4 Inward flow of foreign direct investment (FDI), SSA and other regions compared, 1983–8 and 1998

	1983–8, annual average		1998	
	US $ m	% of total	US $ m	% of total
Developing countries	19,757	100	165,936	100
Developing countries (excl. China)	17,935		120,476	
SSA (excl. S. Africa)	1,065	5.4	5,298	3.2
of which Nigeria, Angola, Gabon	595	3.0	2,196	1.3
Latin America and Caribbean	7,438	37.6	71,652	43.2
Asia	10,042	50.8	84,880	43.3

Source: IMF, *World Investment Reports* (1995, 1999)

Table 1.5 Inward stock of FDI: SSA value as percentage of developing countries, 1980–98

	1980	1990	1998
SSA	6.9	6.0	3.8
Latin America and Caribbean	35.9	30.8	34.1
Asia	52.7	57.7	58.8
of which China	—	5.0	21.4
Developing countries	100	100	100

Source: UNCTAD (1999a, annex B.6)

to 3.69 (data from United Nations, 1999). Thus, while a number of commentators have detected signs of a 'demographic transition' in parts of SSA, the evidence remains unclear, except in some individual countries and regionally in Southern Africa. Out of forty-one countries, five (Angola, Democratic Republic of Congo (DRC), Gabon, Guinea-Bissau and Uganda) showed an actual increase in the fertility rate between the years mentioned and a further eleven (Burundi, Chad, Congo, Equatorial Guinea, Ethiopia, Liberia, Malawi, Mali, Mozambique, Sierra Leone and Togo) a fall of less than 10 per cent. Much of this increase in population has been born into poverty (see also Kirk and Pillet, 1998).

Demographic and health surveys carried out during the 1990s do show positive signs of a transition being under way in Botswana, Zimbabwe and Kenya, as well as in parts of Southern Nigeria (Caldwell *et al.*, 1992; Lockwood, 1995). There, falls in birth rates of the order of 15–25 per cent compare with the fall of 10 per cent which Caldwell *et al.* assert as being conventionally accepted to mark the onset of an irreversible fertility transition. A significant fall in infant mortality rates is a precondition of change, as evidenced by falls below 70 per 1000 births in the first three countries mentioned. As Caldwell *et al.* note (1992: 212), little fertility decline has occurred in those parts of Kenya where infant mortality is higher. The same countries and areas of Nigeria also have particularly high levels of education and relatively high female gross enrolment rates. There is also in each case a high level of contraceptive practice, assisted by effective programmes of contraceptive distribution supported and given legitimation by government. Given a national population policy, the key policy implications to be derived from the above would thus be: widely distributed and accessible public health services; increased female levels of education; and an effective, well organised family planning programme.

1.6 The urban sector

Apart from considering the future of SSA generally, it is evident that it is important to consider the specific prospects of the urban dwellers in Africa. The proportion of the population which is urban-based has increased from 21 per cent in 1975, when one could say that the continent was overwhelmingly rural, to 32 per cent, about a third, in 1997 and is projected to be as high as 43 per cent in 2015 (UNDP, 1999: 200). The changes taking place appear even more striking in Table 1.6, which shows the distribution of forty SSA countries by per cent of urban population over this period. While in 1975 twenty-three of the forty countries had urban percentages below 20, this was down to eight countries in 1997 and a projected two countries

Table 1.6 Trend in urbanisation since 1975: SSA country distribution

% urban population	1975	1997	2015 (projected)
80–	—	—	1
70–	—	—	1
60–	—	2	3
50–	—	3	10
40–	1	7	11
30–	6	13	6
20–	10	7	6
10–	16	6	1
0–	7	2	1
No. of countries	40	40	40

Source: UNDP (1999)

in 2015. Conversely, where only one (small) country was over 40 per cent urbanised in 1975, the number had increased to twelve in 1997 and may be twenty-six out of forty in 2015.

Given the expanding importance of the urban sector, it becomes more and more necessary to identify a clear set of policies for that sector. An initial need, as the basis for such a set, as O'Connor (Chapter 19) points out, is much more accurate statistical description and measurement, because many existing estimates are wildly out. This is required in the first place for the physical planning of urban areas, which should anticipate, rather than follow expansion, and encompass all infrastructural dimensions such as roads, power and water supply. Designs need to be appropriate to the income levels that will obtain within the urban areas and directed towards the avoidance or minimisation of the worst manifestations of urban squalor observable in some other continents.

A challenge will be to raise urban income levels in the likely absence of strong formal sector development: as Rakodi notes (Chapter 18), urban economies have shared in the economic stagnation of the sub-continent. In Nairobi, for example, sample survey data collected during 2000 in connection with the issue of a government Poverty Reduction Strategy Paper indicated that over half of the city's residents lived below the poverty line, 53 per cent of the population occupying 5 per cent of the total residential area and living in informal settlements. According to a statement by the Economic Secretary of the Ministry of Finance and Planning in November 2000 (reported in the *Daily Nation* of 11 November), compared with an estimated 3.7 million poor people living within the city in 1973, there were by then 15 million poor.

Rural–urban migration, rather than being directed towards a high wage formal sector as suggested by the original Todaro model (Todaro, 1969), has in part reflected complementarities between rural and urban sectors. Incomes earned by urban-based family members are a means of risk diversification for the urban–rural family as a whole (Stark, 1991), for instance, while urban remittances constitute a valuable component of rural household incomes and act as a source of social security right across SSA (Schrieder and Knerr, 2000). Careful consideration of advantage produces an equilibrium between poor rural and poor urban income levels, with rural and urban poverty constituting parts of the same problem. Thus, while the proportion of the *total* population below the poverty line in urban areas is invariably much lower, if only the poorest 20 per cent in each sector are considered there is little difference in income level. In practically all countries the bulk of the people in the poorest quintile lives below the poverty line, whether rural or urban (Table 1.7). On its own, this implies that rural and urban incomes, through a variety of means, need to be raised together (see Chapter 14).

1.7 Welfare, health and education

Of the estimated sub-Saharan African population of 580 million in 1995, 220 million fell below the international poverty line (adopted by western donors as the measure of poverty reduction to be monitored over the period 1997–2015) of US $1 per person per day. The number of poor people had risen from an estimated 180 million persons in 1987. In the period 1990–5, 200 million people were stunted as a result of child malnutrition.

While GDP per capita figures were referred to above, a better indicator of wellbeing is now acknowledged to be the Human Development Index (HDI). This combines a measure of income (adjusted per capita income in purchasing power parity dollars) with one health indicator (life expectancy at birth) and indicators of access to knowledge (adult literacy, combined enrolment ratio). In 1998 the HDI stood at 0.464 for SSA, compared with 0.560 for South Asia and 0.642 for all developing countries. The bottom five countries in 1998

Table 1.7 Proportion of population below the international poverty line, rural and urban, selected SSA countries

| | Population below poverty line (%) | | | |
| | All population | | Lowest quintile | |
	Rural	Urban	Rural	Urban
Burkina Faso	65	13	100	67
CAR	77	33	100	100
Côte d'Ivoire	51	29	100	100
Ethiopia	50	21	100	100
Gambia	73	21	100	100
Ghana	44	20	100	100
Guinea	55	18	100	88
Kenya	53	14	100	71
Madagascar	57	21	100	100
Mali	64	8	100	39
Mauritania	55	18	100	94
Niger	55	14	100	69
Nigeria	52	32	100	100
Senegal	66	14	100	69
Sierra Leone	74	36	100	100
South Africa	86	40	100	100
Swaziland	70	36	100	100
Tanzania	51	20	100	97
Uganda	46	16	100	78
Zambia	70	28	100	100

Source: UNDP (2000)

Note
The international poverty line chosen by OECD in the late 1990s is US $1 per person per day, calculated on a purchasing power parity basis.

(Burundi, Ethiopia, Burkina Faso, Niger and, last, Sierra Leone) were all in Africa (UNDP, 2000).

Of the world's countries starting from low human development (HDI below 0.5) in 1975, the three making the slowest progress (Burundi, the Central African Republic (CAR) and Zambia) as measured by the index were all in SSA, with Zambia's HDI actually lower in 1997 (0.431) than in 1975 (0.453). Moreover, while over the period 1980–90 two countries out of 25 showed actual declines in the index (Zambia, Rwanda), during 1990–7 as many as seven (South Africa, Namibia, Botswana, Zimbabwe, Zambia, Kenya and the CAR) showed such declines (UNDP, 1999).

The HDI figures for different countries do not correlate with income per capita due to the fact that other factors besides income affect welfare, including some that would be amenable to policy. This can be gathered from the wide distributions that exist, for instance, in infant and maternal mortality rates and in immunisation rates for different countries. Thus, among forty SSA countries in 1997, three had infant mortality rates below 60 per 1000 live births, while at the other extreme three had rates above 170 per 1000; among thirty-seven countries, four had maternal mortality rates below 500 per 100,000 live births, while five had rates above 1500 per 100,000. In particular, income per capita is an average figure, which does not take account of the distribution of income between individuals, groups and regions within a country, as well the distribution between these of social provision. It has recently been pointed out (Stewart, 2000) that 'horizontal inequality' among groups within a country, inequality in political, economic and social terms, is a prime source of eventual internal conflict (see section 1.11 below).

Overall, improvements in health have been slow; while, for instance, the infant mortality rate in SSA in 1980 was just under that in South Asia, at 115 compared to 119 per 1000 live births, 3 per cent lower, in 1997 the rates were 91 and 77 respectively, with the SSA rate 18 per cent higher.

The quantity, quality and regularity of domestic water supplies are of prime importance in relation to health standards in the population. The provision of treated or protected (from pollution) domestic water supply is one central policy area for which recent inter-country and internal rural–urban sector data are available. Considerable variation in the coverage of provision is indicated in both dimensions. In 1996, in twenty-seven SSA countries, the proportion of the rural population provided with access to improved water supply ranged from 81 per cent (Côte d'Ivoire) down to 8 per cent (Congo Republic). The parallel provision for urban populations ranged from 100 per cent (Botswana) down to 17 per cent or even as low as 14 per cent (Mozambique and Lesotho respectively). In these latter two countries and four others (Ghana, Côte d'Ivoire, Benin and Niger) rural coverage exceeded urban, suggesting the operation of opposite decision processes to those characterising 'urban bias'. On the other hand, particularly low provision for rural populations – less than 25 per cent coverage – is found in the following countries, in addition to the Congo already cited: Madagascar (10 per cent), Angola (15 per cent), Chad (17 per cent), Ethiopia (20 per cent), Sierra Leone (21 per cent) and Cameroon (24 per cent). In five of these seven countries there have been lengthy periods of conflict in the rural areas. In all cases urban coverage is significantly higher than rural, suggesting a causal linkage between conflict in ungovernable rural areas contributing to an uneven pattern of welfare provision, an urban bias process explained here in terms of complex political emergency.

1.8 The HIV/AIDS pandemic

Social and economic welfare in SSA has been further affected immediately and will be increasingly in the coming decades by the devastating effects of the HIV/AIDS pandemic. The total number of HIV victims in SSA amounted at the end of 1999 to some 22 million out of an estimated 33–34 million world wide, that is, two-thirds of the world total. During 1999 there were some 4 million new infections in SSA and in 1998 2.2 million deaths. The effect has been to significantly reduce life expectancies in the most seriously affected countries, reversing decades of steady earlier progress. As Jolly observes (in UNDP, 2000): 'HIV/AIDS is the disaster factor for a dozen or more countries in Africa, pulling down their life expectancy and their ranking in the human development index', widening also the welfare gap between poor and rich countries. The impact on already low life expectancies in SSA (in 1995 a mean of only 47.8 years) can be observed in Table 1.8. Compared with the position 25 years earlier, life expectancies at birth have actually fallen since that time in seven countries out of the twenty-four listed, with only very small decreases in most other countries. The impact of AIDS, affecting adults in particular, can be seen by observing the contrast with the substantial declines that have taken place over a similar period in the infant mortality rates for the same countries.

While the epidemic spread first in East/Central Africa, the most serious impact is now quite evidently in southern Africa, in South Africa especially. Although relevant statistics are obviously uncertain, South Africa accounted for 4.1 million, a significant part of the total for SSA. As a percentage of its population, Botswana was apparently in an even more serious condition, with about 36 per cent adult infections, as are several other Southern African countries. It is evident that the more urbanised countries such as Kenya and Zambia are much more vulnerable to the spread of the disease, as are those countries like Botswana and Lesotho that have workforces with substantial spatial mobility. Rural rates were estimated in 1993 to be running at about 20–30 per cent of urban, apart from those in Zimbabwe, Tanzania and the CAR, which were considered to be about two-thirds (Cornia and Mwambu, 2000: 45).

Inevitably, the impact on the economic and social functioning of SSA countries is and will be immense. Labour forces in the range 30–50 years of age will be depleted and the productivity of both agricultural and industrial workforces and of the public services reduced. These ravages threaten not only the current workforce but probably also those scheduled to join it in the next two decades. In agriculture, with a higher proportion of female-headed households, there will be some switch towards more subsistence-oriented production. Past investments in medium- and high-level trained workers will be lost on a substantial scale: in 1998 Zambia

Table 1.8 Life expectancy, infant mortality and HIV/AIDS adult infection rates: most affected SSA countries, 1997

Country		Life expectancy at birth (LEAB) 1995–2000 (years)	Change in LEAB 1970–5 to 1995–2000 (years)	Infant mortality rate (IMR) 1998 (per 1000 live births)	Change in IMR 1970–98 (per 1000 live births)	HIV/AIDS infection rates 1997 (% adults 15–49)
1	Zimbabwe	44.1	–7.4	59	–27	25.8
2	Botswana	47.4	–5.8	38	–60	25.1
3	Namibia	52.4	3.7	57	–47	19.9
4	Zambia	40.1	–7.2	112	+3	19.1
5	Swaziland	60.2	12.9	64	–76	18.5
6	Malawi	39.3	–1.7	134	–55	14.9
7	Mozambique	45.2	2.7	129	–34	14.2
8	South Africa	54.7	1.1	60	–20	12.9
9	Rwanda	40.5	–4.1	105	–19	12.8
10	Kenya	52.0	1.0	75	–21	11.6
11	CAR	44.9	1.9	113	–36	10.8
12	Côte d'Ivoire	46.7	2.3	90	–70	10.1
13	Uganda	39.6	–6.9	84	–26	9.5
14	Tanzania	47.9	1.4	91	–38	9.4
15	Ethiopia	43.3	2.3	110	–49	9.3
16	Togo	48.8	3.3	81	–47	8.5
17	Lesotho	56.0	6.5	94	–31	8.4
18	Burundi	42.4	–1.6	106	–29	8.3
19	Congo	48.6	1.9	81	–19	7.8
20	Burkina Faso	44.4	3.5	109	–54	7.2
21	Cameroon	54.7	8.9	94	–33	4.9
22	DRC	50.8	4.7	128	–19	4.4
23	Gabon	52.4	7.5	85	–55	4.3
24	Nigeria	50.1	6.6	112	–18	4.1

Source: UNDP (2000: tables 9 and 10).

Notes
1 Countries are ranked by the estimated proportion of adult males infected by HIV/AIDS in 1997.
2 Somalia and Liberia are omitted due to paucity of reliable data; six small countries (less than 1 million population – Cape Verde, Comoros, Djibouti, Equatorial Guinea, St Tomé and Principé and the Seychelles) are also omitted.
3 Sixteen other SSA countries had HIV/AIDS adult infection rates of less than 3.2 per cent.

lost 1300 teachers through AIDS in 10 months. In the CAR primary school closures had to be introduced as a result of substantial falls in the number of teachers. In the Côte d'Ivoire seven out of ten teachers' deaths have been attributed to AIDS (*Independent*, 8 July 2000).

Reduced rates of increase in income will result in reduced rates of private and public saving. The pattern of both private and public spending will be affected: with reduced income and increased medical expenditures on family and relatives, more children will be left out of school, particularly girls. Public health expenditure on AIDS treatment will be diverted from other medical programmes and medical expenditure as a whole from other areas such as education. There is the special problem of the lifetime loss of welfare among orphans, projected to rise to some 13 million in SSA by the end of 2000. Recent research using econometric modelling on the prospective impact of AIDS on the South African economy yields pessimistic outcomes in terms of growth forgone (Arndt and Lewis, 2001), producing GDP per capita in ten years' time 17 per cent lower than it would have been in the absence of the epidemic. Half of this is due to the diversion of government spending towards health, increasing the budget deficit and reducing total investment, and a third due to lower growth in total factor productivity.

An obvious implication is that the failure of most governments in SSA to implement forcefully publicity campaigns at an early stage in the spread of the epidemic has been very costly financially, as well as in terms of welfare. This is evidenced by the relative success achieved in Uganda using a range of preventative approaches; in 2000 the rate of new infections declined for the first time. The need to reinforce campaigns now, targeted to vulnerable groups, is clear.

The negative macroeconomic effects of the pandemic imply more than ever that governments cannot afford costly military conflicts which divert spending and that they will need to prioritise public expenditures still more carefully.

1.9   Education

Provision of education, where access is broadly available, is now widely recognised by economists as an important form of national investment, as well as contributing to welfare. In SSA there was actual retrogression in gross enrolment rates (GERs) during the 1980s and only limited progress in the first part of the 1990s (as discussed here by Colclough in Chapter 26). During 1990–5, only sixteen out of thirty-three SSA countries achieved increases, with fourteen showing falls. This performance, moreover, was in stark contrast to that being achieved at the same time in South Asia. Taking the period 1985–97 (Table 1.9), out of twenty-one SSA countries all but two showed falls in primary enrolment rates and thirteen showed falls in secondary enrolment rates. It is only in the tertiary sector that there have been substantial increases in enrolment rates in most countries.

While the gender balance of school enrolments in SSA has improved, this reflected the greater fall in boys' enrolments rather than an increase in those of girls (Colclough, Chapter 26 in this volume). What his discussion brings out, however, is the wide inter-country variation in achievement and, in respect of the gender gap, that the main cause of discrimination is adverse cultural practice rather than low income levels. Thus it can be calculated (UNDP, 1999) that in 1997 among thirty-four SSA countries the combined GER for females was 90 per cent or more of the male rate in ten countries but in another eleven only in the range 50–67 per cent. Female enrolment rates for twenty-one countries in 1997 averaged 82 per cent at the primary level, 80 per cent at the secondary level and just 52 per cent at the tertiary level. The ranges of values, however, were 70 to 189, 31 to 108, and 12 to 154 per cent respectively.

Table 1.9 Sub-Saharan African countries with negative trends in one or more indices for female educational enrolments in 1997 (1985=100), with female enrolment rates as a proportion of male rates, 1997

Country	Primary enrolment rate (i)	Female rate as % of male (ii)	Secondary enrolment rate (iii)	Female rate as % of male (iv)	Tertiary enrolment rate (v)	Female rate as % of male (vi)
1 Guinea	189	58	73	31	47	12
2 Côte d'Ivoire	114	76	84	53	306	31
3 Namibia	98	106	113	108	n.a.	154
4 Mauritius	97	100	141	106	684	101
5 Burundi	93	96	155	70	136	34
6 Zimbabwe	92	98	111	91	n.a.	n.a.
7 D.R.Congo	91	70	99	63	n.a.	n.a.
8 Lesotho	90	118	93	122	208	115
9 Sierra Leone	88	79	n.a.	n.a.	n.a.	n.a.
10 Botswana	87	106	106	106	349	87
11= Cameroon	86	92	90	77	n.a.	n.a.
11= Kenya	86	105	102	89	139	39
13= Tanzania	85	102	n.a.	n.a.	367	24
13= Comoros	85	83	100	82	n.a.	40
15 Zambia	84	98	104	71	233	39
16 Madagascar	82	102	n.a.	n.a.	58	80
17= Eq. Guinea	80	102	92	90	n.a.	14
17= CAR	80	69	68	50	167	16
19 Congo	77	94	74	79	114	22
20 Mozambique	73	76	74	62	380	31
21 Angola	70	97	73	82	n.a.	n.a.

Source: UNDP (2000: table 28)

Note:
n.a.= data not available.

Both globalisation and structural adjustment programmes have major gender implications, with a variety of potentially negative impacts affecting either men or women through, *inter alia*, changes in the pattern of employment. Such effects in different sectors may call for systematic gender-related analysis and actions across a wide range of areas, economic and social, and at macro, 'meso' and micro levels (see Mbilinyi, Chapter 24 in this volume).

1.10 Aid and debt

While economic and social welfare has been slow to improve, where it has done so the flow of foreign assistance to SSA countries has actually been diminishing during the 1990s, in the post-Cold War period. Between 1992 and 1997 the dollar value of aid received fell by 26 per cent, in per capita terms by 35 per cent, and as a proportion of gross domestic investment by 38 per cent (Table 1.10). At the same time, in 1997 seventeen SSA countries had debt-service ratios (service payments as a proportion of exports of goods and services) in excess of 15 per cent, above 20 per cent in eleven cases and above 25 per cent in four (Ghana, Côte d'Ivoire, Burundi and Madagascar). Accumulated debt as a percentage of GNP was in excess

Table 1.10 Net official development assistance (ODA) to SSA from all donors, 1980–97

	1980	1988	1992	1997	Percentage fall, 1992–7
Value (US $ b at 1995 constant prices)	9.7	16.6	19.6	14.4	26
Value per capita (US $)	21	31	37	24	35
As per cent of GDI	18	46	62	39	38

Source: World Bank (2000), *African Development Indicators 2000*, New York

Notes
Excludes South Africa.
GDI = gross domestic investment

of 100 per cent in sixteen cases and in excess of 200 per cent in six (Congo, DRC, Guinea-Bissau, Mauritania, Angola and Mozambique). The debt situation of the various countries is not closely related to their levels of income and wealth and figures vary widely from country to country (though they are generally much lower in Southern Africa).

Heavy accumulations of debt are related to how effectively previous aid has been put to use. Kanbur (2000) quotes a calculation suggesting that, if the total aid inflow to Zambia between 1961 and 1993 had been invested at a normal rate of return, its current per capita GDP would have been thirty times higher. Attempts at imposing conditionalities have failed to secure good use of aid inflows, for a number of reasons, as Kanbur suggests: the imposition of penalties for failing to direct aid towards assisting the poor could further damage the poor; an element of political clientelism has been quite evident in a number of cases, Mobutu's Zaire for example; donors did not cover the outflow for debt servicing; and donor agencies have a need to maintain 'normal relations', meaning that agency employees themselves have an interest in maintaining their involvement in the aid process. Kanbur goes on to question how far aid to Africa has been useful. He states (2000: 410) that Africa:

> is the last remaining region of the world where official aid inflows outstrip private capital inflows, and they do so by a large margin . . . And, at least for now, this massive quantity of aid does not seem to be helping African development . . . aid has failed in Africa, aid conditionality has failed, and . . . there is very little chance of recovery from this failure under current institutional arrangements.

Kanbur concludes that heavy debt relief as discussed above needs to be combined with institutional reforms directed towards making Africa less aid-dependent and more self-reliant.

A widely quoted econometric study by Burnside and Dollar (1997) was almost as pessimistic, finding no relationship between aid flows and growth in per capita GDP. Aid does assist growth, however, where the macroeconomic environment is a favourable one. One reason why aid and growth are not closely correlated, they suggest, is that aid did not flow especially to those countries that have such environments. Whilst their views have been influential, their analysis is open to counter-argument. First, their approach is ahistorical in failing to give sufficient weight to the destructive effects of the 1970s' oil price shocks, followed by western 'Reaganomics' and international market recession in the 1980s upon the development levels in non-oil-producing and recipient countries. Second, there is no consideration of a counter-factual analytical framework which could test the possibility that aid received prevented an even worse set of outcomes from occurring. In any case, attempts to generalise from the effectiveness of past aid would seem to be less useful than enhancing damaged strategic appraisal and evaluation capacity on a country-by-country basis. The aid debate is now moving on to emphasise the *quality* of aid, particularly with regard to the impact of aid on the poor and to achieving sustainable improvement and change.

1.11 Conflict and governance

One major factor that has affected economic and social development in SSA over the past two decades and longer and continues to affect future prospects has indeed been the extent of armed conflict. In some current situations resolution of ongoing problems must be regarded as essential for any long-term progress. Apart from the countries immediately involved in conflict, other countries in particular sub-regions where there is conflict and in SSA as a whole are affected through influence on outside perceptions of the region or continent and on the willingness of capital to invest in different economic sectors. In addition, influxes of international refugees across their borders impose costs on recipient countries and, in some cases, instability arising from friction between refugees and local nationals. Currently, in 2001, (excluding a number of very small countries) only six out of forty-eight SSA countries can properly be described as in a state of open conflict: four in Western Africa (Guinea-Bissau, DRC, Sierra Leone and Liberia), Sudan in Eastern Africa and Angola in Southern Africa; however, uncertainty remains in a number of other countries.

Cramer (Chapter 3 in this volume) calculates that, categorised as major where they result in at least 5000 deaths, there have probably been as many as twenty-four major conflicts in SSA in the last two decades, with even greater frequency of conflict during the 1990s than in the previous decade. One feature of these conflicts is that much larger numbers of civilian casualties, direct and indirect, are generated, compared with army casualties sustained in battle. Green (Chapter 4 in this volume) estimates the total number of people who would have lived in the absence of such deaths in SSA to be in excess of 10 million; these are predominantly civilian and over half indirect casualties, resulting from lack of access to medical services, food shortages and the like. The GDP losses are estimated as in excess of US $100,000 million.

As the above indicates, the conflicts across the continent have created a secondary problem of large numbers of refugees in adjacent countries as well as of internally displaced persons (Table 1.11). In the thirteen countries listed there were an estimated 2.7 million refugees in 1998 and another 1.5 million displaced persons.

The scale of the economic impact on the development of the continent has attracted development economists to engage in original empirical work on economic effects and to apply their own analytical tools in this new field. Progress in this direction is reviewed here by Cramer (Chapter 3) who, however, points out some limitations of the neoclassical and individualistic models employed. He concludes nevertheless that there is an urgent need for much more research into the origins, dynamics and effects of conflict on the continent, as well as into methods of managing post-conflict situations.

One effect of conflict is to divert public expenditure to military purposes, expenditure that might have been used more productively to promote development. In a number of SSA countries substantial proportions of total public expenditure have been so absorbed (Table 1.12). This does not encourage aid donors, since the fungibility of expenditure means that any additions to the government budget could, in effect, be financing further military spending. All eight of the countries listed here were classified officially in 1998 as 'severely indebted countries'. Calculations for thirty-six SSA countries showed mean values of military spending amounting to 11.4 per cent of central government expenditure in 1992, actually increasing to 13.1 per cent in 1997.

Table 1.11 International refugees and internally displaced people, 1998 (thousands)

Country	Refugees by country of asylum	Internally displaced people	Refugees by country of origin
A: Internal burdens[a]			
1 Sierra Leone	10	670	411
2 Rwanda	33	625	73
3 Tanzania	544	—	—
4 Guinea	414	—	—
5 Sudan	392	—	374
6 Ethiopia	262	—	53
7 DRC	240	—	152
8 Kenya	238	—	5
9 Guinea Bissau	7	196	9
10 Uganda	205	—	9
11 Zambia	169	—	—
12 Côte d'Ivoire	120	—	—
13 Senegal	61	—	—
B: Exodus situations[b]			
1 Burundi	25	—	500
2 Somalia	—	—	481
3 Sierra Leone	----------------	as above	----------------
4 Sudan	----------------	as above	----------------
5 Eritrea	3	—	345
6 Angola	11	—	316
7 DRC	----------------	as above	----------------
8 Liberia	—	—	100
9 Rwanda	----------------	as above	----------------
10 Ethiopia	----------------	as above	----------------
11 Chad	9	—	59
12 Mauritania	23	23	68

Source: UNDP (2000: tables 27 and 32)

Notes
a The first group of countries are ranked by the combined total of refugees granted asylum and internally displaced people. This is taken as a broad measure of additional pressure on resources, including land and relief assistance in each country so affected.
b The second group of countries are ranked by numbers of refugees coming from that country, that is, hosted by asylum providing countries. This is an indication of upheaval and loss of development momentum in the country of origin.
c In the 22 other medium and large SSA countries (population over 1 million people), the internal burden ranges from 0 to 48,000 people, and the exodus size from 0 to 17,000 people; the combined numbers do not exceed 50,000 people in any country.

One question is whether any sensible economic planning is feasible in conflict and immediate post-conflict situations. Green answers this in the affirmative, concluding that even here the adoption of systematic, if modified and specific, post-war rehabilitation planning is both useful and practicable: he provides illustrative guidelines and checklists in support of this contention.

Furley (Chapter 2 in this volume) gives detailed accounts of efforts at conflict prevention, as undertaken both by international agencies and through sub-regional and regional bodies, and does not find the record to date very hopeful. He identifies an alarming change in the nature of conflict, moreover, with multi-country involvement in what started as purely internal, if serious, problems, as in Rwanda and the Democratic Republic of the Congo (DRC), as well as cases of international intrigue and interference related in part to a scramble

Table 1.12 SSA countries with highest levels of military
expenditure as a proportion of central
government expenditure, 1997

Country	%
DRC	41.4
Angola	36.3
Sierra Leone	33.0
Central African Republic	27.7
Burundi	25.8
Uganda	23.9
Rwanda	22.2
Cameroon	17.7

Source: World Bank (2000) *African Development Indicators,
2000*, New York

for rich mineral resources, as in Sierra Leone. These are likely to complicate very significantly the tasks of conflict resolution, as he illustrates with the specific current case of the DRC, involving nine other countries, with different political and financial interests.

A hopeful sign in SSA has been the observable trend in 'democratisation', with abandonment of military dictatorships in a number of countries and the increasing advent of multiparty parliamentary government over the past ten years. Bangura (2000) distinguishes five 'patterns of democratisation' in Africa in 1997: countries with pre-1990 multi-party systems (five countries); countries with post-1990 multi-party parliamentary systems (sixteen in 1997); countries with post-1990 multi-party systems but pre-1990 governments (seventeen countries); countries with post-1990 military regimes (five); and post-1990 populist political systems formed through wars (five countries). Bangura uses two indicators of political pluralism: the proportion of parliamentary seats held by other than the largest party and the number of parties represented in parliament. In 1997 in a list of thirty-six countries, twenty-one had percentages equal to 30 or more, twenty-seven had percentages of 20 or more, while twenty out of the thirty-six countries had five or more parties represented. The average percentage share of other parties in parliament (35) was still less than in South and East Asia (46) and Latin America (57).

Experience in other continents does not show any clear relation between the degree of democratisation in any of these senses and the economic growth achieved. What appears to be more important in this respect is the degree of political stability achieved over a period of time and the quality of general economic management of the economy. Moreover, extended periods of successful development can clearly occur even with significant deficiencies in governance, such as corruption, as exemplified in past decades in Indonesia, if favourable factors are able to override these. Nevertheless, the establishment of democratic government, serving at least as protection against military coups in the short run, could provide an improved and more stable political framework for longer-run economic development.

Apart from macroeconomic reforms, structural adjustment programmes in SSA have targeted poor performance and inefficiency in the public and parastatal sectors (Olowu, 1999). From the early 1980s, civil service reforms involving major retrenchments of staff and the elimination of 'ghost' workers were implemented in many SSA countries. To take one example, in Tanzania, where state enterprises and parastatal institutions were particularly strongly entrenched during the late 1960s and 1970s, reforms have been some of the most thoroughgoing with, for instance, executive powers of regional administrations reduced substantially and their staff reduced to some 20 per cent of their previous level (Government of Tanzania, 1998; see also Therkildsen, 2000). Policies in Tanzania have followed the 'New Public Management' (NPM) paradigm, with a general mandate that ministries should retain core functions such as policy making, regulation and monitoring rather than direct implementation. The latter should be left to managerially autonomous boards operating as service units for the delivery of urban water, health services, roads and so on, with a possible forty-seven such units planned. There have been some resultant benefits, as Therkildsen acknowledges: the main public sector reform, for the civil service (CSR), has been comparatively successful in cost containment at the macroeconomic level, whatever the effect on sectoral programmes and income distribution. However, service delivery, despite the use of performance contracts with private firms, appears not to have improved in most areas, with resource scarcities and continuing problems of implementation affecting the outcome.

Nevertheless, Oshikoya and Mlambo (Chapter 17 in this volume) argue that 'higher and sustained levels of economic growth in Africa cannot be achieved except through the promotion of the private sector', following excessive government regulation in the past, biases against agriculture, and inefficient parastatals and agricultural marketing boards. Despite macroeconomic reforms, Africa 'has remained a high-cost low-investment region', with risk of policy reversal a contributory factor. Measures are needed to improve weak regulatory and institutional frameworks, to provide and upgrade supporting infrastructure and to strengthen weak financial systems.

Related to the problem of 'weak but excessive' regulation is that of corruption, seen by Oshikoya and Mlambo and widely by other authors as a major problem, if certainly not confined to Africa. Ghai (2000: 5) asserts that:

> The cancer of corruption and bribery has spread deep onto the African body politic. Its inevitable consequences have been growing distortion of the economy, with decisions taken without regard to considerations of cost, efficiency, growth or equity. Corruption directs scarce human resources to unproductive channels, imposes massive burdens on the productive sections of the society, violates principles of equity and justice, breeds alienation and eventually justifies rebellions and coups d'etat.

The situation in Kenya in the mid-1990s was described in the following terms:

> Whatever criteria are used, it is no longer an issue in dispute that the growth and extent of corruption in Kenya have reached bizarre proportions. Virtually all our local newspapers, leave alone the foreign ones, have in recent years carried a litany of articles citing occurrences which indicate that corruption has assumed a cancerous state eating into the very fabric of the Kenyan society. Both ordinary Kenyan citizens and prominent decision-makers in the government attest to the dangerous levels of corruption in the country. In 1993, a minister in the government admitted that corruption in Kenya had reached disturbing proportions. 'Although corruption started in 1963, the kickback had reached a level where government officials demand 45 per cent kickback, up from the 10 per cent level in 1963,' he said.
>
> (Kibwana *et al.*, 1996: 123)

In a sample survey of 546 people in Nairobi (urban), Machakos (peri-urban) and Makueni district (rural), the same authors reported that 83.8 per cent of respondents considered corruption to be 'very serious'; 86.1 per cent had come across corruption; and 75.8 per cent blamed the government most, out of a long list of alternatives.

The degree of corruption is but one element within the overall quality of governance, institutions and public service provision that affects economic efficiency and social equity. Different elements have been combined in a recent IMF report as indexes, measured on a scale one (low) to ten (high), to give an overall picture (IMF, 2000: table 6.3). The greatest contrast was between SSA and the newly industrialising economies (NIEs) of Asia. But SSA also compared unfavourably with Asia in general, if less strikingly, the bigger risks in SSA relating to political violence, expropriation and contractual repudiation. Within SSA, countries with markedly poorer scores on all counts are also those in the bottom 20 per cent as regards economic growth.

Collier (2000) examines some of the reasons why corruption in Africa has tended to develop: a massive rise in the opportunities for corruption as a result of over-regulation of private activities, expanded public sector employment (on low pay), and weakened scrutiny. He argues that only a 'big push' in attacking the problem, rather than piecemeal actions, can be expected to be successful.

The quality of governance directly affects the benefits to be secured from foreign aid by recipient countries and the success that donor agencies will be able to claim in providing it. Mosley and Eeckhout (2000) refer to the possible existence of a general 'moral hazard' whereby recipient countries have an incentive to lower performance in order to qualify for more aid, for instance in reducing tax effort, but also to the specific moral hazard of corruption, the opportunities for which may be substantially increased by generous inflows of money.

1.12 Decentralisation and 'participatory' development

A major dimension of many SSA government policies since independence in the early 1960s has been the introduction of diverse forms of decentralised government and planning. The underlying motives have varied widely. In some situations it has been sensible to introduce planning on an area basis covering a particular physical region, as in the case of river basin planning in Nigeria and in southern Tanzania (Rufiji/Usangu) or the Arid and Semi-arid Lands (ASAL) programme in Kenya. These concentrate on technical and physical planning, without replacing routine local government within the districts encompassed. The principal reasons for decentralisation of government itself have been the attempt to achieve 'integrated rural development planning' (IRDP), whereby actions taken in one area simultaneously or sequentially could have a more effective developmental impact; the desire to increase participation in local decision making, including so-called 'bottom-up' planning; and the more overtly political one of giving power to different ethnic groups or regions. Donors have encouraged decentralisation to district and, through participatory civil society organisations, to grassroots levels. This has sometimes been with the objective of evading central-level corruption: in the case of Uganda, however, to take one example, poor management of finance at the district level has restricted substantial devolution of funds.

None of the different types of experience over the past several decades has been particularly successful. The IRDPs were seen and still are being seen – in Uganda, for instance – as useful vehicles for the soliciting of donor funds, donors each being allocated one district or region. These have been useful in the short term but have not been sustainable once donors pull out after a few years, while it may be difficult to secure coverage of all needy districts. In Tanzania, following the Presidential Edict on Decentralisation in 1972, only half the twenty large mainland regions subsequently established and implemented, with single donor assistance, five-year Regional Integrated Development Programmes (RIDEPs).

Decentralisation to a lower government level has been represented in several countries by the preparation of district plans intended to dovetail with national plans. These have tended in practice to take the form of district resource assessment and activity reports, without seriously affecting national development strategies or the financial allocation process, as in the case of Botswana and Kenya, both in the 1980s, and those currently in existence in Ghana and those being prepared in Uganda.

A general feature has been the reluctance by the central ministry of finance and sectoral ministries in each country to relinquish control of recurrent and capital finances. Thus in Kenya, though District Development Funds have been in place since 1978 in each of some forty districts in the country, these have amounted to only about 5 per cent of the budget of central government departments working at district level (Mutahaba, 1989). Devolution of funding has been taken much further in recent years under Uganda's decentralisation programme, however, under which the central government allocates block grants to a large number of district governments, which at the same time are required to use their own tax instruments to raise funds locally. Estimates for the proportion of district revenue raised locally are of the order of 25–35 per cent (Livingstone and Charlton, 2001).

Another example of decentralisation with mixed results is that of Ethiopia. After the fall of the Mengistu regime in 1991, a new regional system with eight regions based on ethnic homogeneity was introduced under the Tigrean and Ethiopian People's Liberation Front (TEPLF) government. Participatory methods are widely used in most regions, whether from the *beito* system at the grass roots in Tigre region or the parish in other regions. At the same time there is a long upward hierarchy of decision levels, through the *woreda* (district) and zone to the region where fiscal budgets are allocated. Consequently the original participatory priorities tend to get submerged in the process and intra-regional inequalities can increase (Young, 1997).

While in Ethiopia decentralisation has been to eight substantial regions, in the very much smaller country of Uganda there has been pressure for additional districts to be created either from parts within a district that considered themselves neglected or from separate tribal groups, leading the number of districts to increase from an initial thirty-nine to forty-five in 1997: by 2000 a further eight were mooted to bring the total to fifty-three. Such big numbers raise questions about the significance of any one district and its needs in national policy making, and the ability of scarce technical and administrative capacity to assist all districts equitably (Belshaw, 2000).

1.13 Rural development

Given the rapid population growth rates observable in the SSA countries and limited and slow-growing employment growth in the formal industrial sector (as described by Livingstone in Chapter 14), the question arises as to how effectively additions to the rural labour force can be absorbed. Bryceson (1996) and others have identified a process of 'de-agrarianisation' whereby people move out of agriculture as such into other occupations within the rural areas or move out of these areas altogether. Despite problems of measurement, evidence exists of a strong trend in this direction. Whether these processes might be associated with diminishing returns in rural areas and with 'over-urbanisation' in the urban areas is considered by Livingstone, examining in particular the possibilities and prospects for rural industry. While informal rural industries play a vital role in providing basic commodities for low-income rural households, as well as being the source of non-farm income and employment, the incomes generated, at the margin especially, are low. In many cases the activities concerned are based on local resources such as wood, the supply of which may be under threat, calling for promotional action. The very optimistic conclusions that have been drawn from surveys of rural industry and rural enterprise in Africa appear to exaggerate their potential contribution to growth and employment as compared, certainly, to urban-based small-scale enterprises. The most important determinant of the level of non-farm incomes, however, including those from rural industries, is the level of demand linkages deriving from farm incomes, and with an agriculture-led development strategy farm and non-farm incomes can be expected to increase together. It is necessary to recognise the complexity of rural livelihoods and to take full account of this complexity in framing policies (see Ellis, 1998, 2000).

In relation to rural development as a poverty reduction strategy, Belshaw (Chapter 12 in this volume) reviews the changing approaches that have been pursued by national and international agencies over successive periods dating from the 1950s, from Community Development to the New Poverty Agenda of the 1990s. He warns against possible limitations both of this agenda and the new community development, favoured by non-governmental organisations (NGOs) especially, with its excessive emphasis on actions at the local and community level to the neglect of sub-national, sectoral and national programmes and policies directed towards pro-poor rural development.

With an increasing role being accorded to NGOs in SSA, Wallace (Chapter 13) assesses the contribution they are making, some of the constraints on their operations and what their proper role might be. Her discussion, with special reference to Uganda, brings out a potential conflict between funders, anxious to ensure cost-effective use of their funding and to make use of NGOs in service delivery, and the NGOs themselves, feeling restricted by the bureaucratic procedures demanded for securing funds and subsequent monitoring and evaluation of their projects. Consequently, many NGOs are tending to become increasingly engaged in broader activities such as advocacy and human rights.

Of relevance here are the results of recent fieldwork carried out in developing countries (Narayan *et al.*, 2000). The relative perceptions by the poor of different institutions' importance and effectiveness were gathered from over 1500 discussion groups, with a membership of over 20,000 poor people in twenty-three countries. NGOs are actually placed below community-based organisations (CBOs) and faith-based organisations (FBOs) for both effectiveness and importance, particularly in rural areas. Results for the reverse question – which are the *least* effective institutions, from the point of view of the poor? – show public sector institutions in the worst light in both rural and urban areas. The authors note (ibid.: 219) that 'in rural areas community-based organisations are most frequently mentioned as both the most important and most effective institutions'. Despite this finding, they nevertheless conclude that:

> NGOs, where present, play important roles in poor people's lives – NGOs have stepped in to fill important gaps created by the breakdown of government – provided basic services . . . poor people . . . would like to be involved in decision-making in programs that NGOs manage. NGOs . . . feature more prominently . . . as effective institutions in urban areas . . . Given the huge scale of the poverty problem and the small scale of most NGOs' activities, it is important to recognise that they still have limited presence, particularly in the communities in Africa.
>
> (Ibid.: 229, 227)

This view from below is instructive but clearly not definitive: it identifies weaknesses in public institutions *and* NGOs compared with community-based organisations, which are, of course, those in most direct contact with the poor.

1.14 The agricultural sector

Given the still predominant role of the agricultural sector in Africa in providing employment and in provisioning the population, its performance remains critical. Wiggins (Chapter 5 in this volume) provides an authoritative review of the situation based on an extensive scanning of the literature. He questions, first, 'whether the nature and extent of the African rural crisis has not been exaggerated'. Certainly, it is important not to confuse the very serious problems and famines of the Sahel and of the Horn of Africa with the situation as regards agricultural production and environmental change in smallholder farming throughout SSA as a whole. He points to a possible 15 per cent fall in per capita farm production over the 30-year period to 1995 as not catastrophic and of course still representing a significant increase in absolute terms. Data on value added in agriculture (Table 1.13), on the other hand, despite their uncertain statistical basis, suggest significant variation in country experience over the period 1980–98, with one group of countries falling behind the population growth rate of around 2.5–2.8 per cent and another succeeding in overtaking this rate. The second decade, to 1998, shows a wider dispersion than the previous one, with negative rates for a number of countries. Fortunately, this appears to be the outcome of events during 1990–4, with results in the more recent period being much more encouraging. Data are usually more reliable for export crops and, where prices are out of the control of the countries concerned, physical outputs are a better measure of performance than market values. Here performance by export crop and by country is quite mixed.

Although performance in food production in SSA was reflected in only a slight decrease in daily per capita supply of calories in 1996 relative to 1970 of 0.9 per cent, this compares with a 14.7 per cent increase in South Asia and a 40.8 per cent increase in East Asia. Again, there was a wide variation between countries over the period from 1970 to 1997, with about as many countries showing falls in supply (in six cases this exceeded 10 per cent, suggesting

Table 1.13 Value added in agriculture, SSA countries: distribution of average annual rates of increase, successive periods, 1980–98

Average annual rate of growth (%)	1980–90	1990–8	1990–4	1994–8
15–	—	—	—	1
10–	—	—	1	2
5–	—	4	4	7
2.5–	12	14	4	18
0–	18	10	12	5
(0–)	—	4	4	1
(2–)	—	3	7	1
(5–)	—	—	—	—
(10–)	—	—	2	—
(15–)	—	—	1	—
No. of countries	30	35	35	35

Source: World Bank (2000) *African Development Indicators, 2000*, New York

quite a serious decline over a long period when an improvement would have been looked for), as showing increases. Though food imports are used mainly to augment urban area supply, they use up scarce foreign exchange (see Wiggins, Chapter 5 in this volume) which countries can scarcely afford and most should not be needing, given the size of their agricultural sectors. Rural households that do not produce sufficient food will often lack income for its purchase and simply do without, culminating in hunger and famine or diversion of resources to relief activity. The many policy issues connected with food aid are discussed comprehensively by Shaw (Chapter 11 in this volume).

From his review, however, Wiggins is correct in stressing the complexity and diversity of African smallholder situations and experience and the key parameters constraining change. A particularly strong finding is the fact that poverty may be related more to factors producing social differentiation, with wide differences in outcomes among households within each area, than to factors underlying the average level of farm production. Accordingly, he sees the challenge as being to understand better how a wider range of variables interact within each complex system.

Changes on the side of livestock production are analysed by Upton (Chapter 6 in this volume). It is encouraging that meat production per head has been increasing in Africa as a whole, as statistics provided here demonstrate. However, SSA (as opposed to North African) countries are lagging behind other developing countries in meat production per head and suffer from a serious shortage of livestock products, particularly in the supply of eggs, compared to other countries. Disease problems are the major constraint. The substantial scope for raising productivity through crossbreeding is indicated, however, by huge differences in productivity between breeds for each category of livestock.

The above findings call for increased research into the factors constraining agricultural production on both the technical and social/economic sides. Agricultural research and extension in many parts of SSA have ground to a halt, however, and are widely considered to have become comparatively ineffective. How far and in what circumstances and areas the state should undertake research and extension and where these could be left for private sector provision are questions explored by Akroyd and Duncan (Chapter 8 in this volume).

It is important to maintain awareness of the diversity in the natural endowment of African countries: many countries are distributed over a range of climatic and agro-ecological zones, with varied agricultural potential. Such zonal variability exists within both large countries such as Nigeria and Ethiopia and smaller ones such as Uganda. Incomes, food supply and overall economic welfare of rural households within these countries will vary greatly according to the particular area and agro-ecological zone in which they find themselves. Market access, including distance from main national and, for export commodities, international markets, affects the available opportunities for local populations and, in the case of the landlocked SSA countries, national populations.

A feature of agriculture in Africa is the relatively limited use of water resources for irrigation. The irrigated proportion of cropland is only about one-eighth of the level found in the other three tropical and sub-tropical macro-regions shown in Table 1.14. Within SSA, the use of irrigation techniques on any scale is limited to a few countries. Madagascar has 35 per of cropland irrigated, irrigated paddy technology having been imported historically by the original settlers from Southeast Asia; otherwise no other SSA country reached a 10 per cent ratio of irrigated cropland in 1995–7. Commercial estates and farms in South Africa and Zimbabwe use irrigation, particularly in sugar production, while in some West African countries swamp and riverine-based irrigation is used, especially for rice. Potential for substantial irrigated areas based on major rivers and lakes exists in a number of countries: it

Table 1.14 Indicators of water use and potential in tropical/sub-tropical macro-regions

Macro-regions	Irrigated land as % of cropland		Fresh water (cu. metres per capita)	% of land area under perennial crops	
	1979–81	*1995–7*	*1998*	*1980*	*1997*
SSA	4.0	4.2	8441	0.7	0.9
Middle East/North Africa	25.8	35.5	1045	0.4	0.7
South Asia	28.7	39.7	4088	1.5	2.1
East Asia/Pacific	37.0	36.3	—	1.5	2.6

Source: World Bank (2001: tables 8 and 9)

has been suggested that the head waters of the Nile Basin, for example, offer significant opportunities to utilise large quantities of water currently being lost to evapo-transpiration (Howell and Allan, 1994).

Of greater relevance, in general, is the potential for exploitation of a wide range of low-cost techniques suitable for small-scale agriculture which could help large numbers of rural people to move away from food insecurity into semi-commercial-cum-subsistence agriculture. These include hillside and valley tanks, low-lift boreholes, gravity-driven sprinkler systems, bunding and other water retention structures, flood recession planting and a variety of lakeside and riverside lifts and pumping schemes. These would be of particular benefit where rains are unreliable.

As Mortimore observes (Chapter 7 in this volume), the arid and semi-arid zones of Africa are not 'a marginal fringe', but account for perhaps 40 per cent of the land area, while the semi-arid zone maintains at least 28 per cent of tropical Africa's population. The pastoralist communities within these countries have been particularly misunderstood and disadvantaged by both colonial and post-colonial governments and, as a result of losing large parts of their past grazing lands, are being forced to adapt in the most difficult circumstances. It is important that their needs and interests are given much fuller attention by national and international authorities and organisations. In the Drylands as a whole, however, local populations have shown considerable resourcefulness, adaptability and capacity to manage their resource bases and positive changes have been achieved.

1.15 Environment

Environmental considerations are becoming particularly serious in the areas of land degradation and forestry. Although statistics published in international reports giving aggregate national rates of deforestation do not show particularly high annual losses, these fail to bring out often serious situations at the local level which affect access to fuelwood and other critical supplies of timber in both rural and urban areas. Statistics that can be used to monitor more closely the diverse situations within different countries are needed. In the area of land degradation there are important gaps in the available knowledge about the various degradation processes and their location, that is, physical erosion of various kinds, soil quality decline, salination and waterlogging and changes in biodiversity. Again, these problems need to be investigated and corrective action taken at the local area level.

In the 1990s several SSA countries drafted National Environment Action Plans (NEAPs). These tend to stress action to protect the environment as against identifying ways in which environmental resources can be used in sustainable ways to reduce widespread poverty. This involves ensuring that resources are available to the poor, rather than simply applying conservation measures that protect resources by prohibiting their use.

Emission problems arising from 'dirty' industries are not yet important locally and are of little global concern, except in South Africa. Sixty per cent of SSA's CO_2 emissions (defined as fossil fuel consumption excluding fuel used in international transport) are contributed by South Africa, 15 per cent by Nigeria and a further 5 per cent or less each from Zimbabwe, Côte d'Ivoire and Kenya (UNDP, 2000: table 21).

1.16 Minerals production

SSA does have a huge, still largely untapped development potential in the form of minerals. Nigeria, Gabon and Angola are major petroleum exporters followed by Congo, Cameroon and, most recently, Sudan and Chad. Other countries with mineral exports in excess of US $100 million in 1997 included in Western Africa Cameroon, DRC, Guinea, Ghana, Mauritania and Niger; and in Southern Africa Botswana, Namibia, Zambia and Zimbabwe, in addition to South Africa. Unfortunately for the Eastern African countries, mineral finds there have so far been quite limited, with the recent exception of Sudan. A number of SSA countries have secured striking increases in their GDPs since the 1970s through mineral exploitation, most obviously in the case of Nigeria. The current contribution of minerals to recorded exports remains quite low overall, however, and the share of value added contributed to GDP has been comparatively stagnant during the 1990s, actually decreasing in a significant number of countries. Allowance must be made, of course, for illegal and unrecorded trade in high value gemstones and metals.

Several constraints have operated. A physical one is the lack of developed physical infrastructure and the investment costs of its provision where such minerals are located deeply inland within the continent, serving as a major disincentive to investment, national or foreign. Second, there have been opportunities for corruption where large cash flows have been involved, with scope for the diversion of vast sums, as in the case of Nigerian and Angolan oil and Sierra Leone diamonds. In many cases the existence of minerals has itself been the original source of intranational or international conflict.

More generally, where revenues have been generated, these have not been put to the most desirable or equitable use and, worse, may have led governments in many cases to adopt inappropriate development strategies, such as inefficient import substitution and poor macroeconomic policies. Thus, comparing per capita growth rates of different categories of countries world-wide, Auty (2000a: 5; see also Auty, 2000b), found that 'the worst performers of all in the last fifteen years were the mineral-driven resource-rich economies', despite the fact that these could have had their capacity to invest and to import enhanced by their mineral exports. He argues that:

> an abundant natural resource endowment provides more scope than resource-paucity does for cumulative policy error. Resource-abundant countries are more likely to engender political states in which vested interests vie to capture resource surpluses (rents) at the expense of policy coherence.

This thesis applies very directly to the SSA sub-continent. An exception here is Botswana, where it is evident that mineral wealth has been translated into sound development policies and comparatively high levels of social provision in the areas of health and education (but unfortunately not into effective HIV/AIDS education programmes, as evidenced in Table 1.8 above).

1.17 The industrial sector

Perhaps the greatest disappointment for African leaders is Africa's record in the development of manufacturing, reviewed in this volume in Chapters 15 and 16. This was seen from the mid-1960s until comparatively recently as the route that Africa needed to follow in order to achieve the economic transformation of its countries. Between 1980 and 1998 the share of manufacturing in SSA's GDP increased only from a roughly estimated 16 per cent to 19 per cent (World Bank, 2000). In 1998 fourteen out of thirty countries had percentages only in the range 5–9 and twenty-one had percentages below 15. The mean increase in the percentage share among twenty-four countries for which estimates are given was just 1.25 per cent over the 18-year period. Taking into account increases in population, it may be observed that value added in industry *per capita* fell by 20 per cent between 1980 and 1998 (Table 1.15).

It is important to remember that internationally published data for 'industry' refers to all industry and not just manufacturing. For some countries the difference will be very significant. Value added in SSA industry grew at the rate of 1.5 per cent over the period 1980–90. It then fell at a rate of –0.6 per cent during 1990–4 (in line with GDP). Since then it has revived, growing rapidly to 1998 at a rate of 9.6 per cent, producing an average rate of increase over the whole period 1990–8 of 4.4 per cent.

Performance, however, has varied widely between countries and for the same countries at different periods of time. Certain countries, Mozambique for instance, have shown very high rates of increase during specific phases of recovery. Nevertheless a significant number of countries, such as Uganda, Côte d'Ivoire, Ghana, Senegal, Namibia, Lesotho and Mali, as well as, of course, Mauritius, attained good rates of increase during the full period of the 1990s up to 1998, as shown by the most recent figures (World Bank, 2000).

Apart from low growth rates overall, however, there has been little structural change in SSA manufacturing, with a high proportion in the food, beverages and tobacco category as well as preliminary processing of primary products. Probably underestimated in the statistics is the substantial amount of value added within the small-scale enterprise sector, directed especially towards low-income consumers.

A number of constraints have held back the development of manufacturing. The major international transport cost disadvantage of the fifteen landlocked countries within SSA, affecting the competitiveness of exports, is discussed in detail elsewhere in this volume (Chapter 22). There has been surprisingly little awareness shown of the implications of 'landlockedness' for development strategy in these countries and little analysis of suitable and practicable policy for manufacturing. A calculation for Uganda, however (Rudaheranwa, 2000), based on 1993–4 data for tariff protection and 1994 data for freight costs, found the average effective rate of tariff protection to be 36 per cent and the rate for surface transport costs to be 49 per cent. The same transport costs will constitute a major obstacle to manufactured exports, of course. While Uganda might wish to emulate other countries in

Table 1.15 Value added in industry, SSA (excluding South Africa), 1980–98

	1980	*1990*	*1994*	*1998*
VA (excl. S. Africa) (US $ m at constant 1995 prices)	39,086	45,582	44,521	52,048
VA per capita (US $)	111	96	85	89
VA per capita, 1980=100	100	87	77	80

Source: World Bank (2000) *African Development Indicators 2000*, New York

developing labour-intensive exports such as textiles, cloth and footwear, the effective rate of protection (handicap in this case) due to surface transport costs for this category was 45 per cent. Further institutional and other costs associated with international transport adds substantially to the latter (see Livingstone, Chapter 22 in this volume).

While the extent of this handicap is not generally realised, it does not, of course, explain progress or lack of it in different coastal countries. Another handicap in many SSA countries is the relatively low population density, which has the effect through transport costs of diluting the domestic market and thus reducing the ease of establishing import-substituting industries. However, not just transport infrastructure but infrastructure generally, including power supply, is in urgent need of rehabilitation and development in most SSA countries.

While the adoption of structural adjustment programmes has not had the anticipated results, it is also the case that misguided general macroeconomic policies have had disastrous effects on domestic industry in particular countries, where these have resulted in serious shortages of foreign exchange and thus in the availability of necessary imported inputs and machinery. While the promotion of resource-based industries should be consistent with SSA's current comparative advantages, Assaf and Hesp (1991), in four country case studies of agro-industry in Angola, Liberia, Tanzania and Zambia, found seriously uneconomic levels of operation in all cases, associated with lack of access to imported inputs. A contrasting case where there has been no such constraint (due to foreign exchange earnings from diamonds and beef) is provided by Botswana, which has been able to develop a successful garment export industry.

The manner in which policies of import-substituting industrialisation were pursued in most SSA countries from the 1960s to the 1980s left a legacy of inefficient industries, as well as missed opportunities. Being applied across the board, rather than selectively, many industries which lacked sufficient local markets exhibited low levels of capacity utilisation or depended on degrees of protection which left them unmotivated to seek out external markets. Related policies that created serious shortages of foreign exchange and thus of the imported inputs needed for production, as indicated above, added to this. It is the case that levels of protection remain high. For nineteen countries for which data were available in the mid-1990s the average level of protection was calculated to be 26.8 per cent compared with 6.1 per cent for OECD countries, while over a third of imports into these countries were also subject to non-tariff barriers (UNIDO, 1998).

As Thoburn argues in this volume (Chapter 16), industries established initially through import substitution can gather strength and go on to serve as a platform for subsequent development of exports, citing the use of protection in the pursuit of 'export-oriented import substitution' in the East Asian countries. As pointed out by UNCTAD (1999b: 89), referring to Southeast Asia, 'in all these countries manufacturing was built up through a fairly prolonged period of import-substituting industrialisation (ISI) which helped to build local capabilities in light and resource-based manufacturing'. In East Africa in the late 1960s Pearson (1969) described manufacturing industries established on the basis of the Kenyan and East African Community markets, referring to these as 'Janus' industries after the two-headed Greek god facing in two directions. Thoburn further suggests that duty-free import of capital goods, a criticised facet of import-substitution policy, may in fact be beneficial by facilitating the acquisition of up-to-date technologies.

For export manufacturing industry, the level of labour costs and productivity are of critical importance, increasingly so in the context of globalisation. A recent report by UNCTAD (1999a) observes that:

A comparison of unit labour costs in African countries and some potential competitors in a number of manufacturing sectors in 1995 shows that in most cases costs in Africa were much higher than in competing countries such as Bangladesh, India and Indonesia . . . Moreover, in general, unit labour costs in Africa actually increased after 1980 relative to those in competing countries, even though in many cases real wages stagnated or even declined.

The data provided (Table 1.16) are not entirely unambiguous but generally support the former contention.

Whatever the reasons for the relatively weak development of manufacturing in past decades, SSA economies now have a major disadvantage as 'latecomers' into world production and sale, where first and then second tiers of NIEs, and increasingly India and China, have established themselves in export markets. Indeed, these have been expanding their shares of manufacturing exports to Africa itself, so that not only does it become more difficult for African firms to compete in export markets but they are subject to more acute competition within their own domestic markets. This competition may be seen as two-pronged, with imports from countries such as Singapore and Korea supplying higher-valued consumer goods such as televisions and refrigerators and China and India competing for the wider low-income market. Thus the share of non-OPEC developing countries in imports of manufactures to Africa has increased dramatically since 1980, while the share of imports from the developed market economies shows some decline (Table 1.17). The large currency depreciations in Indonesia, Thailand and Korea following the Asian Crisis of 1997 have increased even further SSA's competitive disadvantage within and outside the continent.

Table 1.16 Unit labour costs in selected Asian and African countries and industries, 1995

Ratio	Textiles	Clothing	Footwear	Transport equipment
2.00–	—	—	—	*Kenya (2.25)*
1.80–	Bangladesh (1.81)	*South Africa (1.88)*	—	—
1.60–	*Kenya (1.61)*	—	—	India (1.46)
				Indonesia (1.46)
1.40–	*South Africa (1.45)*	Mauritius (1.53)	*South Africa (1.48)*	—
1.20–	—	Zimbabwe (1.30)	—	*South Africa (1.35)*
		Madagascar (1.24)		*Mauritius (1.28)*
				Madagascar (1.28)
1.00–	India (1.09)	—	*Kenya (1.13)*	—
	Ghana (1.05)		Korea Rep. (1.03)	
0.80–	*Mauritius (0.96)*	Indonesia (0.95)	*Zimbabwe (0.97)*	*Zimbabwe (0.98)*
	Korea Rep. (0.81)	Korea Rep. (0.91)	Indonesia (0.85)	Korea Rep. (0.80)
		Bangladesh (0.87)	Bangladesh (0.71)	
0.60–	*Zimbabwe (0.69)*	*Kenya (0.65)*	Turkey (0.60)	Turkey (0.60)
			India (0.60)	
0.40–	*Madagascar (0.49)*	India (0.46)	*Madagascar (0.59)*	—
	Turkey (0.42)		*Mauritius (0.57)*	
	Indonesia (0.32)			
0.20–	—	Turkey (0.39)	—	Bangladesh (0.35)

Source: Derived from UNCTAD (1999a: table 21)

Notes
Ratios are to USA labour costs.
SSA counties are italicised.

Table 1.17 Sources of imports of manufactured goods into
developing Africa, 1980–94

	Share of total manufactured imports	
	Non-OPEC developing countries	*Developed market economies*
1980	8.5	84.4
1990	14.1	80.8
1994	17.3	77.5
	Average annual rates of growth (%)	
1980–90	8.4	–0.7
1990–4	8.2	–1.1

Source: UNCTAD (1995) *Handbook of International Trade and
Development Statistics, 1995*, New York and Geneva

SSA countries may also be latecomers in attracting foreign direct investment. As Pack (1993), for instance, observed, 'given the options they have in Eastern Europe and Asia, multinational companies may be hard to attract'. This relates also to the emergence of world trading blocs in North America, West/Central Europe and East/Southeast Asia, which with globalisation has left SSA marginalised to a significant degree. Apart from movement of financial capital between the three groups, there is a tendency for such capital to think regionally, spilling over to neighbouring countries, as has occurred most strikingly in Southeast Asia. Nor has SSA had the benefit of a 'major player' regionally, such as Japan.

South Africa has obvious potential in this regard. Between 1991 and 1994 it more than tripled its investment stock in the Southern African Development Community (SADC) countries and almost doubled it in the rest of Africa (UNCTAD, 1997). However, apart from facing some of the same difficulties as other investors into the region, its own industries are still emerging from a long period of isolation and import substitution and need to increase their international competitiveness. Unlike the situation in Japan, where investment needed to move out in search of cheaper labour, South Africa has its own supply, supplemented also by in-migration when needed. Manufacturing in neighbouring countries has been limited in the past by the substantial inflow of cheaper South African products.

Difficulty in attracting foreign investment into manufacturing in SSA is increased by the low skill intensity of the labour force compared with Asia, associated with lack of training and education. It is not just a matter of labour costs. Securing foreign investment is important even in the case of labour-intensive goods such as garments and footwear in which design and market tastes and fashions are nevertheless important. This requires organisational and marketing skills, in particular familiarity with overseas markets, which many African entrepreneurs currently lack. The position is somewhat different for sales within African regional and sub-regional markets, where African exporters may find demand patterns similar to those of their own consumers. As stated in a recent international report, however: 'Many African firms which have moved successfully into exports in areas such as textiles and clothing have done so because substantial investment in new equipment and quality control facilities has made it possible to build links with foreign distributors' (UNCTAD, 1999b: 81).

The potential contribution of small-scale enterprises to the development of the manufacturing sector in Africa has been the subject of debate, taken further in Chapter 14 of this volume. In other continents, including North America, small and medium enterprises have played an important role. Many of the small enterprises in Africa are very different from these and fall into the category of micro- or household, often part-time, enterprises. These make an important immediate contribution to supplying goods to low-income consumers locally and to providing supplementary income and employment to the producers involved. At existing levels of income per capita they contribute substantially to import substitution (Elkan, 2000). The vast proportion of microenterprises in Africa do not show enterprise growth and it is questionable whether any significant proportion can be expected to 'graduate' to more substantial operations (Chapter 14). Important opportunities do exist, nevertheless, for their upgrading through product development, the upgrading of skills, and interaction among 'clusters' of enterprises, particularly within the expanding African towns and cities.

Overall, however, the possibility has come under discussion that a process of actual de-industrialisation, rather than simply stagnation or slow growth of industry, is ongoing in SSA. This hypothesis has been the subject of a number of econometric analyses (Noorbakhsh and Paloni, 2000; and Jalilian and Weiss, 2000). Nevertheless, strong conclusions can probably not be drawn, for a number of reasons. In the short run, increasing openness of SSA economies following reforms would be expected to bring, and does appear to have brought about in a number of countries, some retrogression in inefficient import-substituting industries previously operating under protection. This need not necessarily imply a longer-term trend and could even be beneficial to such a trend. Second, significant portions of manufacturing value added (MVA) in SSA are linked to mineral or agro-processing, with MVA value dependent on volumes of underlying natural resource production rather than efficiency in manufacturing as such. In the study by Jolilian and Weiss, results for seven out of sixteen SSA countries were consistent with a hypothesis of de-industrialisation while those for the other nine indicated the opposite trend.

1.18 Trade policy reform

Two chapters in this volume focus on the most important policy changes implemented in SSA during the 1980s and 1990s: those associated with the structural adjustment programmes promoted by the World Bank and IMF. As Morrissey (Chapter 20) points out in his review covering twelve countries, the success of various efforts in this direction has been mixed. While some countries made significant progress in liberalising trade, rationalising tariff structures and removing bias against exports associated with exchange rates and other restrictions, others have been slow and partial in the implementation of reforms. Moreover, where reforms have been implemented, their effects are difficult to isolate from those of other factors, positive or negative, affecting tradables. Lawrence and Belshaw (Chapter 21) attempt to explain the patchy performance among SSA countries in agricultural tradables export described previously and their unclear response to adjustment measures. While proper producer price incentives are clearly essential, non-price factors evidently play a part in explaining countries' differential production responses; these factors need closer analysis than has hitherto been accorded to them.

1.19 Regional integration

The record in SSA of creating and sustaining regional frameworks, as Jenkins and Thomas comment (Chapter 23), has been continuous but generally poor. The ambitious thirteen-country Economic Community of West African States (ECOWAS) has made little progress, being saddled with bureaucratic problems and difficulties of securing agreement among the large number of participants with different kinds of economies and divisions of interest between, for example, poor inland states and better-off and better-situated coastal ones. A further problem here, as in Southern Africa, has been that of overlapping membership between regional groupings, in this case aggravated by the clash between Anglophone and Francophone cultures and systems, in particular between ECOWAS and the West African Community (CEAO). In other cases there have been unrealistic expectations of trade benefits between participating countries, due to ignoring the prohibitive cost of transport of goods between countries, as exists, for instance, within the Preferential Trade Agreement (PTA) between Ethiopia and other East African countries.

Chapter 23, by Jenkins and Thomas, takes as its starting point a very different position from that of most of the early literature on economic integration in SSA, that of the 1960s and 1970s, by examining it against the background of globalisation and mobility of international capital. In contrast the East African Community, which eventually broke up in 1977 after reaching quite an advanced degree of integration, was based on the premise that a larger market would facilitate the process of *import substitution* by allowing the establishment of industries that would not be viable in the absence of the larger combined market. The problem was how to secure simultaneously a degree of equity in the distribution of the new industries that could be established. It was for this reason that a rather sophisticated system of so-called 'transfer taxes' was introduced that would allow only a limited degree of individual domestic market protection within the Community. A major problem, however, was the tendency of industrial investment, local and international, to *polarise* within Kenya, particularly as compared with the landlocked economy of Uganda.

These arguments appear, not necessarily *in toto*, outdated in the context of globalisation, one in which SSA has fallen behind other regions previously at the same stage of development, and in which it has to try to compete in attracting DFI. If the emphasis is on *export*-oriented manufacturing there is less concern, clearly, with the size of a domestic market. Jenkins and Thomas see the attraction of a generally large regional market, but especially one characterised by *openness*, of an area that is attractive to international investment concerned with exporting *from* the region. The distribution of new industries within the region is less of a concern, since it is assumed that all countries are likely to benefit from location within the region, if to varying extents. There is also less concern with the 'problem' of polarisation and dominant (in terms of manufacturing) economies such as South Africa (or Nigeria in Western Africa or Kenya in Eastern Africa). It is considered that smaller regional economies are more likely to benefit from inward investment from, for instance, South Africa, rather than have their own infant industries competed away. Rather than any divergence, Jenkins and Thomas explore the degree to which SADC economies might have achieved or might in the future achieve 'convergence'. They nevertheless suggest that South Africa might need to investigate opportunities for symbiosis with larger economic units outside the continent.

However these arguments are balanced, one advantage of a major regional grouping, as instanced in the Southern African case, may be the pressure on members to conform to stable macroeconomic goals. Even where integration arrangements are not particularly far advanced, as is the case in East Africa in 2000, common adherence to decontrolled exchange rates will be important in facilitating cross-border trade, whether in manufactured goods or agricultural commodities. Thus, with decontrol of the Ugandan currency in 1991 and of the Kenyan and Tanzanian currencies in 1995, the currency black markets at the borders are no longer active, nor are the associated rents being secured. With the easing of border controls generally, there is considerable scope for local regional trade in a wide variety of commodities, including agricultural ones. This can be very much expanded if transport infrastructure is improved through investment and transport efficiency improved through increased management efficiency and reduced bureaucratic control. This is of the greatest importance for the large number of landlocked countries within all the main regions of the continent.

Notes

1   Some important data gaps for making intra-country or inter-country comparisons include, in no particular order:

- information on the human costs of ongoing conflict, especially where displacement of people mainly occurs within a country's borders;
- land degradation processes other than deforestation, especially decline in soil quality due to reduction in organic matter content and other nutrients under unchanged or inappropriate technology;
- changes in the productivity of subsistence agriculture, including changes in vulnerability to climatic and other shocks (pests, diseases) of various kinds;
- shifts in the degree of 'urban bias' in an economy: while rural-to-urban flows can usually be deduced from population censuses, other important information such as the rural:urban burden:benefit ratio and the domestic or intersectoral price terms of trade is often missing;
- the 'parallel economy', the 'grey' and 'black' parts of the informal sector, including incomes generated or transferred by crime, vice, corruption and smuggling, are by their nature difficult to measure, but they may represent a substantial proportion of a country's economic activity;
- remittances from urban workers to rural families or from emigrant relatives to nationals are often not closely measured, especially if they bypass the banking system;
- some major sectoral activities, such as illegally mined precious stones, addictive drugs and small-scale alcoholic beverage and spirit manufacture are often unrecorded or under-recorded;
- quality aspects of education and health service delivery are not widely monitored;
- the economic performance of sub-national regions, farming systems or river basins are rarely monitored or the impact upon these of national development strategies, policies and programmes;
- despite the modelling efforts being made by various agencies, the impact of the HIV/AIDS pandemic means the level and structure of the future populations of the SSA countries remain highly uncertain.

2   The relative economic sizes of the different SSA countries and the large number of very small countries are not always appreciated. While GDP in 1999 was US $133,200 m in South Africa and US $37,900 m in Nigeria, twelve had GDPs of US $2–4,000 m (Rwanda, Zambia, Malawi, Angola, Namibia, Mozambique, Madagascar, Benin, Burkina Faso, Guinea, Mali and Niger) and another eighteen had GDPs below US $2,000 m (Congo Republic, Chad, Mauritania, Sierra Leone, Gambia, Togo, CAR, Equatorial Guinea, Burundi, Eritrea, Djibouti, Swaziland, Lesotho, Guinea-Bissau, Cape Verde Islands, São Tomé and Principe, Seychelles and Comoros). Compared to the latter, Côte d'Ivoire, Kenya, Sudan, Cameroon and Tanzania were large in economic size, in the range US $8–12,000 m. Intermediate between this last group and the twelve-country group listed above were a further eight countries with economies in the mid-size range – from US $4,000 m to US $8,000 m. These were: Ghana, Uganda, Ethiopia, Zimbabwe, Botswana, Senegal, Mauritius and Gabon. Finally, in three countries – the DRC, Liberia and Somalia – conflict and insecurity prevented the construction of reliable national income accounts (World Bank, 2001).

References

Alley, R.B. (2000) *The Two-mile Time Machine: Ice Cores, Abrupt Climate Change and Our Future*, Trenton, NJ: Princeton University Press.

Arndt, C. and Lewis, J.D. (2001) 'The macro implications of HIV/AIDS in South Africa: a preliminary assessment', *Journal of International Development*, forthcoming.

Assaf, G.B. and Hesp, P. (1991) 'Profiles of key branches of agro-industries in sub-Saharan Africa', *Industry and Development*, 30: 1–41.

Auty, R. (2000a) 'The "resource curse" in developing countries can be avoided', *Forum*, 61, Summer (Development Studies Association of the UK).

Auty, R. (2000b) 'How natural resources affect economic development', *Development Policy Review*, 18 (4): 347–64.

Bangura, Y. (2000) 'Democratisation, equity and stability: African politics and societies in the 1990s', in D.P. Ghai (ed.), *Renewing Social and Economic Progress in Africa*, Basingstoke, Hants.: Macmillan.

Belshaw, D. (2000) 'Decentralised governance and poverty reduction: relevant experience in Africa and Asia,' in P. Collins (ed.), *Applying Public Administration in Development: Guideposts to the Future*, Chichester, Sussex: John Wiley.

Bryceson, D. (1996) 'De-agrarianisaton and rural employment in sub-Saharan Africa: a sectoral perspective', *World Development*, 24(1): 97–111.

Burnside, C. and Dollar, D. (1997) 'Aid policies and growth', Policy Research Working Paper 1777, Washington, DC: World Bank.

Caldwell, J.C., Orobuloye, I.O. and Caldwell, P. (1992) 'Fertility decline in Africa: a new type of transition?', *Population and Development Review*, 18 (2): 211–42.

Collier, P. (2000) 'How to reduce corruption', *African Development Review*, 12 (1): 191–205.

Collier, P. and Gunning, J.W. (1999) 'Explaining African economic performance', *Journal of Economic Literature*, 37: 64–111.

Collins, P. (ed.) (2000) *Applying Public Administration in Development: Guideposts to the Future*, Chichester, Sussex: John Wiley.

Cornia, G.A. and Mwambu, G. (2000) 'Health status and policy in sub-Saharan Africa', in D.P. Ghai (ed.), *Renewing Social and Economic Progress in Africa*, Basingstoke, Hants.: Macmillan.

Elkan, W. (2000) 'Manufacturing microenterprises as import-substituting industries', in H. Jalilian, M. Tribe and J. Weiss (eds), *Industrial Development and Policy in Africa*, Cheltenham, Glos.: Edward Elgar.

Ellis, F. (1998) 'Survey article: Household strategies and rural household diversification', *Journal of Development Studies*, 35 (1): 1–38.

Ellis, F. (2000) 'The determinants of rural livelihood diversification in developing countries', *Journal of Agricultural Economics*, 51 (2): 289–302.

Ghai, D.P. (ed.) (2000) *Renewing Social and Economic Progress in Africa*, Basingstoke, Hants.: Macmillan.

Government of Tanzania (1998) *Strategy and Action Plan, 1998–2003*, Public Sector Reform Programme, Civil Service Department, Dar es Salaam: President's Office.

Howell, P.P. and Allen, J.A. (1994) *The Nile: Sharing a Scarce Resource*, Cambridge: Cambridge University Press.

International Monetary Fund (2000) *International Financial Statistics*, LIII (6), New York: IMF.

Jalilian, H. and Weiss, J. (2000) 'De-industrialisation in sub-Saharan Africa: myth or crisis?', in H. Jalilian, M. Tribe and J. Weiss (eds), *Industrial Development and Policy in Africa*, Cheltenham, Glos.: Edward Elgar.

Jalilian, H., Tribe, M. and Weiss, J. (eds) (2000) *Industrial Development and Policy in Africa*, Cheltenham, Glos.: Edward Elgar.

Kanbur, R. (2000) 'Aid, conditionality and debt', in F. Tarp and P. Hjertholm (eds), *Foreign Aid and Development: Lessons Learned and Directions for the Future*, London and New York: Routledge.

Kibwana, K., Wanjala, S. and Okech-Owiti (eds) (1996) *The Anatomy of Corruption in Kenya: Legal, Political and Socio-economic Perspectives*, Nairobi: Claripress.

Kirk, D. and Pillet, B. (1998) 'Fertility in sub-Saharan Africa in the 1980s and 1990s', *Studies in Family Planning*, 29 (1): 1–22.

Livingstone, I. and Charlton, R. (2001) 'Financing decentralised development in a low-income country: raising revenue for local government in Uganda', *Development and Change*, 32: 77–100.

Lockwood, M. (1995) 'Development policy and the African demographic transition: issues and questions', *Journal of International Development*, 7 (1): 1–23.

Mosley, P. and Eeckhout, M.J. (2000) 'From project aid to programme assistance', in F. Tarp and P. Hjertholm (eds), *Foreign Aid and Development: Lessons Learned and Directions for the Future*, London and New York: Routledge.

Mutahaba, G. (1989) *Reforming Public Administration for Development: Experiences from Eastern Africa*, West Hartford, CT: Kumarian.

Narayan, D., Chambers, R., Shah, M.K. and Petesch, P. (2000) *Voices of the Poor: Crying out for Change*, New York: Oxford University Press, for the World Bank.

Noorbakhsh, F. and Paloni, A. (2000) 'The de-industrialisation hypothesis, structural adjustment programmes and the sub-Saharan dimension', in H. Jalilian, M. Tribe and J. Weiss (eds), *Industrial Development and Policy in Africa*, Cheltenham, Glos.: Edward Elgar.

Olowu, B. (1999) 'Redesigning African civil service reforms', *Journal of Modern African Studies*, 37 (1): 1–23.

Pack, H. (1993) 'Productivity and industrial development in Sub-Saharan Africa', *World Development*, 21 (1): 1–16.

Pearson, D.S. (1969) *Industrial Development in East Africa*, Nairobi: Oxford University Press.

Rudaheranwa, N. (2000) 'Transport costs and protection for Ugandan industry', in H. Jalilian, M. Tribe and J. Weiss (eds), *Industrial Development and Policy in Africa*, Cheltenham, Glos.: Edward Elgar.

Schrieder, G. and Knerr, B. (2000) 'Labour migration as a social security mechanism for smallholder households in sub-Saharan Africa', *Oxford Development Studies*, 28 (2): 223–36.

Stark, O. (1991) *The Migration of Labor*, Cambridge, MA: Basil Blackwell.

Stewart, F. (2000) 'Crisis prevention: tackling horizontal inequalities', *Oxford Development Studies*, 28 (3): 245–62.

Tarp, F. and Hjertholm, P. (eds) (2000) *Foreign Aid and Development: Lessons Learned and Directions for the Future*, London and New York: Routledge.

Therkildsen, O. (2000) 'Public sector reform in a poor, aid-dependent country: Tanzania', *Public Administration and Development*, 20: 61–71.

Todaro, M.P. (1969) 'A model of labor migration and urban unemployment in less developed countries', *American Economic Review*, 59 (1): 138–48.

United Nations (1999) *World Population Prospects: The 1998 Revision*, vol. 1: *Comprehensive Tables*, New York: United Nations.

United Nations Conference on Trade and Development (UNCTAD) (1997) *World Investment Report 1997*, New York and Geneva: UNCTAD.

—— (1999a) *World Investment Report 1999: Foreign Direct Investment and the Challenge of Development. Overview*, New York: Geneva: UNCTAD.

—— (1999b) *African Development in Perspective*, Oxford: James Currey, and Trenton, NJ: Africa World Press.

United Nations Development Programme (UNDP) (1999) *Human Development Report 1999*, New York: Oxford University Press.

—— (2000) *Human Development Report 2000*, New York: Oxford University Press.

United Nations Industrial Development Organization (UNIDO) (1998) *Industrial Development Global Report, 1998*, Vienna: UNIDO.

World Bank (2000) *Entering the 21st Century: World Development Report 1999/2000*, New York: Oxford University Press.

World Bank (2001) *Attacking Poverty: World Development Report 2000/2001*, New York: Oxford University press.

Young, J. (1997) 'Development and change in post-revolutionary Tigray', *Journal of Modern African Studies*, 35 (1): 81–99.

Part II
Conflict and power

2 Conflict prevention and conflict resolution

Interventions and results

Oliver Furley

2.1 Introduction

From the period of independence in the 1960s, the continent of Africa has been riven with conflicts and civil wars, and international bodies such as the UN and the Organisation of African Unity (OAU), along with help from regional organisations, have mediated peace agreements and conducted peacekeeping operations with varying degrees of success. There have also been some notorious failures, such as Angola and Somalia. Through a study of more recent events, we can now see significant changes in the nature of African conflicts. In West Africa, conflicts in Liberia, Sierra Leone and Senegal illustrate the blurring of international boundaries through the involvement of neighbouring states or parties in the violence. In Central Africa, the conflict in the Republic of Congo (Congo-Brazzaville) has been severely affected by the massive civil war in the Democratic Republic of Congo (DRC). In East Africa, the Sudanese civil war has dragged in military confrontations with both Uganda and Kenya; while the war in DRC notoriously has involved nine African countries in the fighting, some of them not even neighbouring states. After the failures of the American peace enforcement operation in Somalia and the UN peacekeeping force in Rwanda, the UN and the big powers now show much reluctance to intervene.

2.2 Conflict prevention

Instead, the key aim was to be conflict prevention, and if conflict could not be forestalled by mediation between aggrieved parties before violence broke out (and this was usually the case), then it could be snuffed out before it spread by the timely intervention, or even just the threat of an intervention, by a rapid reaction force. The former Secretary-General of the UN, Boutros Boutros-Ghali, outlined this strategy as early as 1992 in his *Agenda for Peace*, and enlarged upon it in the second edition of 1995, based on Chapter VII of the UN Charter (Boutros-Ghali, 1992: 24–7, and 1995: 12, 18). However, member states have been slow to pledge forces for such an exercise, and when the UN came near to sending a force under Canadian leadership to rescue the plight of Rwandan refugees who were trapped in the DRC, it never transpired, due to the reluctance of member states. Later, when it looked as if a cease-fire agreement was on the point of being signed in the DRC, and a combined UN–OAU peacekeeping force was to be sent, the UN warned that this would take months to organise.[1] The nature and composition of the UN Security Council is one obstacle to making quicker decisions, and proposals for its reform, to include representatives from smaller powers as well as the big powers, are unlikely to materialise until mid-2000. The OAU raised hopes of better conflict prevention when, at the June 1993 Cairo meeting of OAU heads of state, it set up the new Mechanism for Conflict Prevention, Management and Resolution, which 'will have as a primary objective the anticipation and prevention of conflicts' (OAU, 1993), but shortage of funding remains an obstacle.

Meanwhile conflict prevention bodies and attempts to provide 'early warning' systems have proliferated. The US Carter Center, led by ex-President Jimmy Carter, has been prominent in mediation efforts, especially in the Sudan; the Conference on Security and Cooperation in Europe set up a Conflict Prevention Centre in Vienna; and the European Platform for Conflict Management and Transformation is now joined by the German Platform for Peaceful Conflict Management, established in 1998. Many non-government bodies such as the London-based International Alert are active in fact-finding missions and early warning of possible conflicts.

More notable are the attempts by various nations to provide for rapid reaction forces, in the shape of a body of troops readily available for peace enforcement or peacekeeping, and also there are schemes to provide training for the armies of smaller nations to be in a similar state of readiness. Thus in 1999 the British government declared it would contribute up to 15,000 soldiers for such a force; the Scandinavian countries already have a force in readiness. The French, in spite of declaring they were withdrawing the bulk of their forces in Africa, still have 6000 soldiers in various African countries. Some African countries have expressed a willingness to supply troops for peacekeeping or to prevent internal conflict by their presence, for example, Kenya, Nigeria and Ghana. In the present negotiations for a peace agreement in DRC, which stipulates a UN–OAU peacekeeping force, Nigeria and Ghana have already pledged troops for this mission.[2] While the UN lacks its own military and logistical capacity to produce a peacekeeping or a preventative force, member states, including African states, are taking their own measures to arrive at a position of better preparedness to meet threats of conflict and outbreaks of conflict.

The preferred solution is not so much a standing force permanently available, but to supply training in peacekeeping and if necessary peace enforcement to regular troops of those nations willing to participate. The major western nations have usually taken the lead in supplying the funding and training in these schemes. Thus in 1996 the US government launched the African Crisis Response Initiative (ACRI) to enhance the capacity of African states to respond to humanitarian crises and the call for peacekeeping; Canada, the UK, Ireland and Belgium agreed to contribute to the costs. Programmes of 60–70 days' training are provided for contingents of troops, to be followed after six months by sustainment training. The idea is to enable such troops to take action quickly, and all follow a standard curriculum which includes key areas such as inter-operable communication systems. Since its inauguration many African countries have supplied troops to undergo this training (Cleaver, 1998). There are of course some snags in this type of programme. The numbers trained at any one time are quite small – up to battalion strength – and these troops may soon be split up on their return so that the skills they learned may be lost. The institutional memory of units in African armies may be short-lived. Also, as many African states are now actually participants in conflicts, their role as peacekeepers is excluded; neutrality and impartiality are essential ingredients in any peacekeeping force.

Nevertheless, other western countries have followed suit with their own training schemes. The UK has an African Peacekeeping Initiative, with short-term training for African battalions about to deploy in UN peacekeeping operations. There are also peacekeeping modules for police and senior staff, and two military staff colleges in Ghana and Zimbabwe have been assisting. France does the same in Africa itself, with a training centre in Côte d'Ivoire, and it has supplied Senegal with peacekeeping equipment, stored in Dakar. Its scheme is called

Recamp (Reinforcement des Capacités Africaines de Maintien de la Paix); it prepares African battalions for peacekeeping operations under a joint UN and OAU mandate. These are trained to: 'Stabilize a crisis by preventive deployment, react to an open crisis in order to participate in peace restoration, protect the populations and facilitate humanitarian action.'[3] The latter aim sounds more like peace enforcement than peacekeeping, which is often the case anyway when peacekeepers are faced with a peace which is in fact no longer being kept.

Nelson Mandela disliked these western-organised schemes and preferred African regional training schemes. Certainly the Southern African Development Community (SADC) followed his wishes and has a Regional Peacekeeping Centre in Harare, Zimbabwe. Its aim is to train a regional multinational peacekeeping force for either peacekeeping or peace enforcement operations. A long-term aim is to train a cohesive stand-by force 'along lines similar to the NATO concept'. They have laid emphasis on high-level training such as visits by commanders to Denmark and Bosnia, and training seminars for management of missions, preventative diplomacy and peace building in southern Africa.[4] Such programmes might seem merely theoretical but a large multinational training exercise was planned to test all the coordination, communication and logistical skills in an actual operation, called exercise Blue Hungwe. A follow-up to exercise Blue Crane, held in Zimbabwe in 1997, it was to involve 4000 personnel from armies, navies and air forces and a police component.[5] This was in response to the OAU's call for members' armies to be restructured to be able to participate in Africa's own initiatives to resolve conflicts through peacekeeping missions.[6] Britain and France were to back it with funds, but significantly it has had to be postponed from the planned date of November 1998, because so many of the proposed participants were engaged in conflicts or committed to resolving existing conflicts. The South African National Defence Force (SANDF) claimed that it could not afford the costs, due to its involvement in Lesotho; Botswana took the same line; Zimbabwe withdrew because of the economic crisis at home as well as the costs of its intervention in DRC. Other members such as Lesotho and DRC were in no position to participate. It eventually took place in April 1999 in the Kalahari Desert, with Angola, Zimbabwe and Namibia sending only a small number of military observers. In West Africa two large training exercises had already taken place in 1998: Exercise Guidimakha and Cohesion Compiengna, involving 2500 and 4000 troops respectively. Also, in East Africa some 2000 troops from Uganda, Kenya and Tanzania had combined in Exercise Natural Fire.

When or if further conflicts break out, these exercises and initiatives may make little difference to the urgent calls for conflict prevention, resolution and peacekeeping, but they have set a new scene and demonstrate a greater readiness in Africa to cooperate, combine and react on a regional basis whereas for some years before ECOMOG was the only example.

At about the same time, another type of actor appeared on the conflict scene in Africa. This comprised the mercenaries, the private companies and the professional groups, who prided themselves that they had the experience, the expertise and the technical equipment to make it worthwhile for African governments (or rebels) to hire their services. Executive Outcomes is the best known of these, a military consultancy founded in South Africa in 1989. President Jose Eduardo dos Santos of Angola was the first leader to hire them, in his civil war against Savimbi's UNITA party. They put the rebels on the defensive but the US government strongly disapproved of his hiring a band of mercenaries, and pressed Angola and South Africa to expel them. However, the Americans did not disapprove of a company based in the USA, namely the Military Professional Resources Inc. (MPRI), and the US Embassy pressed Luanda to employ them to provide a hundred military trainers in what would effectively be a US military assistance programme.[7] Mercenaries pursued profit rather than deserving causes, and could be found providing security, for instance, in the diamond fields of Angola or Sierra Leone, whether held by governments or rebels. Executive Outcomes at one stage was engaged by Savimbi's UNITA to work with mining corporations in order to provide security on-site.[8] South Africa has been the main base for mercenary companies, such as Combat Force, Investments Surveys, Honey Badger Arms and Ammunition and so on, though others are based in Britain or the USA. The UN disliked their presence in African conflicts: they were unaccountable, not subject to any international or African regional authority and their loyalties were dictated by profit and not by moral or humanitarian imperatives. Their soldiers were of course adventurers, but some of them were experienced sharpshooters and had a not inconsiderable effect in some conflicts. Most notable of these was the effect the British-based Sandline International company had in Sierra Leone, when in a seemingly endless and bloody civil war the British High Commissioner, Peter Penrose, asked Sandline to help train a force to fight the military junta that had ousted the democratically elected President Kabbah, and they played a part in saving Freetown from continued anarchy (Adams, 1999).[9] It should be added that nevertheless the action came in for serious criticism in Britain.

2.3 African regional interventions

Given the readiness of African armies to take part in peace enforcement or peacekeeping in other African countries, it must be said that their military capacity, in terms of weaponry, training, logistics and funding (as well as discipline), varies greatly. The UN peacekeeping mission to Mozambique, involving African troops, was a success, and so was the UN (MINURCA) mission in the Central African Republic, which deployed solely African troops. An earlier example of African armed intervention, for humanitarian purposes (though not sponsored by the UN), was the successful invasion of Uganda by well-trained Tanzanian troops and Ugandan exiles in 1978–9. The ECOMOG forces, consisting of troops from several West African countries but mainly from Nigeria and Ghana, did succeed in securing a peace accord in Liberia, but in Sierra Leone this is proving more difficult. ECOMOG had by no means an unsullied reputation: discipline was lacking and its troops took part in looting and killings so that they were feared rather than welcomed in many areas. Its leaders also had their eyes on the financial prizes to be won by concentrating on the rich diamond or mineral areas. Both these interventions did not at first have authorisation from the UN Security Council, but received it retroactively.

They were instances of regional actors taking their own initiative in response to requests for assistance, and it is clear that President Clinton's Presidential Decision Directive of May 1994 – that African crises should have African solutions (thereby absolving the US from any further disasters like Somalia) – merely confirmed an already established practice. In Guinea-Bissau, President Vieira was faced with a serious mutiny in 1998 which was linked to arms smuggling over the Senegalese border to the Casamance rebels. He requested ECOMOG forces to enforce peace and guarantee the security of the border, which was carried out by troops from Gambia, Benin, Togo, Niger and Mali. Again, the UN Security Council gave it authorisation retrospectively, but the Economic Community of West African States (ECOWAS), the West African international body of which ECOMOG was the military arm, in their Revised Treaty set up a Mechanism for Conflict Prevention, Management, Resolution, Peacekeeping and Security, which became operative in 1998. Clearly this long title echoed the OAU body with a similar (but shorter) title, but more interesting are the stated grounds on which action could be taken: where a humanitarian disaster threatened; where the peace and security of a sub-region were threatened; and where the overthrow of a democratically elected government was attempted or achieved.[10] ECOWAS had gone some way to providing West Africa with its own strategy for security without the need for outside intervention, and further, this had the blessing of the UN. The reference to the defence of democracy is particularly noteworthy, and it was ironic to see the then Nigerian President, General Abacha, who headed a military government established by a coup, sending his troops to Liberia to secure a democratic regime. It could be said that he was anticipating forthcoming events in his own country.

In southern Africa, the economic and military dominance of South Africa could have been a stabilising influence in the region, but President Mandela pursued an extremely cautious policy regarding intervention or peacekeeping operations. Previously, South African forces had sustained some defeats in the Angolan civil war when they intervened, and Mandela wanted to avoid entanglements. Further, there was a question mark over the efficiency of his army in spite of its comparatively plentiful and technologically advanced equipment. Finally, in September 1998, a change in policy came when South Africa backed the forces defending Kabila's regime in DRC; the significance of this was overshadowed when South African forces joined with Botswana troops in intervening in the neighbouring state of Lesotho, where a duly elected government was threatened by a mutiny, amid charges that the 1998 election was rigged. The prime minister asked SADC to intervene, and South Africa showed a further change in policy by deciding to send a force, along with Botswana troops, to assist the prime minister to restore his government. All accounts agree that it was a bungled operation. Their forces suffered casualties and the intervention caused severe rioting and looting in the capital, Maseru. Further, although the danger of a coup and the overthrow of a democratically elected government was forestalled, because doubts were raised whether the mission had proper authorisation from SADC or the OAU, it was 'perceived as illegitimate and biased' (de Coning, 1998: 23). Up to this time, President Mandela had carefully avoided military interventions lest South Africa, with the large forces at its disposal, be perceived as the bully of southern Africa or even of the continent. Lesotho did not provide a good augury for a change in the policy. Yet South Africa joined forces with Angola, Zimbabwe, Namibia and Zambia in sending military assistance to support President Kabila's government in the DRC. Kabila himself had joined SADC in 1997, invited by Mandela. This marked a more aggressive policy to assist regional stability in sustaining what should be South Africa's natural allies and the security of individual states – even though Kabila's government could hardly be regarded as a legitimate or democratically elected one (Marais, 1999). Compared with Angola or Zimbabwe, however, South Africa's contribution has so far been low key, and the new President, Mbeki, prefers the role of a peace-broker through attempts at mediation. He sent his defence and foreign ministers to the Lusaka peace talks, and told Parliament that if a cease-fire was finally signed between Kabila and the rebels, South African troops would be sent to DRC as part of the peacekeeping force.[11]

Turning to the Horn of Africa, conflict prevention appears to be a mirage. After the disasters of UN/US intervention in the civil war in Somalia, in 1995 the country dissolved into anarchy, with various regions dominated by rival warlords, where wars and skirmishes frequently erupted again, and the world has taken little notice. Only when the wars spill over borders do the neighbouring states react, as Kenya does from time to time when refugees from Somalia flood into northern Kenya. The refugees create security problems as well as financial and environmental problems, and minor skirmishes follow new incursions of refugees, some of whom Kenya claims are gunrunners and bandits. A desperate humanitarian situation has prompted NGO activity, and also there is a War-torn Societies Project (WSP) which has started, in north-west Somalia, a Centre for Peace and Development in efforts to mitigate conflict and provide partnership between international and local experts for rehabilitation and development.[12] The north-west was the more promising area, and in fact it had become the self-proclaimed state of Somaliland in what had originally been British Somaliland before the union of Somalia. It is unfortunate that it has remained unrecognised internationally (Woodward, 1998: 153–4) for it could have been a step towards wider conflict prevention in Somalia.

In the Sudan, conflict prevention must seem an impossible dream, though after the 1955–72 civil war the Addis Ababa Accord of February 1972 established the south as a region with a large measure of self-rule in an effort to prevent the return of conflict. The self-rule of the south was never really accorded by the Khartoum government and by 1983 the Sudanese People's Liberation Army (SPLA) resumed the war. The war soon developed its multinational character, with the SPLA in particular obtaining support from allies, as we shall see in the next section. Conflict prevention and conflict resolution in the future will have to take into account the changing nature of conflict itself, as this poses new problems.

2.4 The changing nature of conflict in Africa

The genocidal massacres in Rwanda in April–June 1994 may be taken as a starting-point in discussing how the nature of conflict in Africa has recently undergone an alarming change. The Rwandan crisis began as a largely internal affair, the causes of which have been very thoroughly explored,[13] and the main events need only be summarised here. Foreign intervention, however, quickly became very significant, before the massacres took place. Uganda had been the home for many years of Tutsi exiles from Rwanda, many of whom yearned to return. When President Museveni took power in Uganda in 1986, his guerrilla army, the National Resistance Army (NRA), contained a large proportion of these Tutsis, and Museveni himself is related to Tutsis. For some time the NRA Tutsis had been plotting an invasion of Rwanda, and in October 1990 they attacked, beginning what was partly a civil war against President Habyarimana's regime and partly a foreign invasion. Museveni claimed to be unaware of these plans but there must have been at least some logistical connivance.

The Rwanda government's Forces Armées Rwandaises (FAR) were immediately helped by the French, who had supplied arms to Habyarimana – and who continued to do so to the time of the massacres and beyond. The basic reason for this appeared to be the desire to defend the borders of Francophone Africa against the incursion of Anglophone Africa (Prunier, 1995: 99). A small force of French troops arrived from France and also some from Zaire. Meanwhile attempts at multiparty democracy failed, as Hutu extremists in opposition to the predominantly Tutsi government became more and more violent. The UN and OAU had brokered long-drawn-out peace negotiations in Arusha in 1992–3, in which there was supposed to be power-sharing among the parties and a reconstructed army incorporating the Tutsi/Ugandan force, the Rwanda Patriotic Front (RPF). This was never accepted in the tense situation, and the plane crash that killed Habyarimana was the signal for the already-planned massacre of the Tutsis, by the Hutu 'Interahamwe', the young extremist militia. The small UN peacekeeping force (UNAMIR) failed to act to limit the genocide and actually withdrew most of the mission, a step which proved to be a turning-point in the internationalisation of this crisis as it created a vacuum. UNAMIR II arrived later but only after a crucial delay. France filled the gap, sending troops in 'Opération Turquoise' with UN approval, although French motives were questioned by many who perceived a French desire to continue their strong military presence in Africa (Furley, 1998: 243–6). The RPF swept to victory and took power, causing the former Hutu government, its army (FAR), Interahamwe and hundreds of thousands of Hutu civilians to flee the country, ending up in Zaire, Tanzania, Burundi, Congo-Brazzaville and even further afield.

Unfortunately the huge refugee camps, especially in Zaire, contained not only civilian refugees but ex-FAR soldiers and Interahamwe, who dominated the camps and were determined on nothing other than the re-invasion of Rwanda. Thus the huge relief operations to save the lives of refugees took on the unwelcome aspect of assisting guerrilla forces who used the refugees as pawns to obtain supplies in order to launch attacks on Rwanda and also Uganda, which was now the firm ally of the RPF government. Médecins sans Frontières and other aid agencies knew that 'their resources became the fuel for African war-engines: money, food and medical supplies are what the warlords are fighting for'.[14] Thus, Rwanda had exported its civil war to Zaire; the knock-on effect of this was not only to involve Rwanda and Uganda in hostilities in Zaire, but to add to the causes of a civil war in Zaire itself.

When Laurent Kabila led his guerrilla force in Zaire northwards towards the refugee camps where the Hutus and Interahamwe had their bases, he drove out all the refugees back to Rwanda and destroyed the camps in early 1997; but the Hutu army, some 30,000 strong, remained in the area, a constant threat to Rwanda. Moreover, the forests of north-east Zaire provided ideal concealment for dissident groups from Uganda, especially the Allied Democratic Forces (ADF), who launched attacks on the Uganda border from there. Rwanda and Uganda both therefore declared that they were supporting Kabila in his campaign against President Mobutu's regime in order to safeguard the security of their own borders, as did Burundi. Zaire's civil war was becoming internationalised. Kabila's army was partly a Tutsi force recruited from the Banyamulenge, the Zaire-born Tutsi community in the south, while Museveni of Uganda, who had already assisted the Tutsi RPF government to take power in Rwanda, was accused in some quarters of seeking a 'Tutsi empire' across this part of Africa.

After Mobutu's death in 1997, however, Kabila in his progress to victory proved to be no compliant puppet, and he severed his connection with his former allies and turned to new ones. This was because his victory in setting up his Democratic Republic of Congo (DRC), formerly Zaire, was by no means complete. The various Congolese opposition groups did not accept his rule and fought a second civil war against him, from August 1998, backed this time by his former allies, Rwanda and Uganda. The latter still insisted that their presence in eastern DRC was to protect their own borders – a claim that could be justified in the light of attacks on them by guerrillas based there. Museveni may have had commercial motives as well, and allegedly some of his army officers did. In a report to the UN Security Council, Kabila accused both Uganda and Rwanda of looting the DRC for gold and diamonds. The report alleged that 'men around' Museveni's own brother, General Salim Saleh, were among the main buyers of Congo's gold and diamonds. In Kisangani, Uganda was rumoured to have encouraged a South African mining company to set up operations in 1998, and it first operated in Kisangani and then moved to Gbadolite, near the Yakoma diamond deposits. Such rumours were unproven, but the popular belief remained that Uganda stayed in Congo so long because army officers were doing well out of their business deals.[15] Further, as Bruce Baker suggests with relation to most of these interventionist states there, 'these economic interests on the part of the political elite might well require the prolonged stay of troops to protect them, irrespective of the requirements of the cease-fire agreement to pull out foreign armies'.[16]

It appeared quite likely that this large country would split up, and its great wealth in diamonds, gold, cobalt, copper and other resources became an irresistible temptation for other states to step in for possible gains. The war became, in the words of some commentators, Africa's first continental war. In other African conflicts, one or two other countries had been involved, often under the auspices of an international body such as the UN or the OAU, but the scale of involvement in this Congolese war was something new. The DRC war 'had become an unprecedented regionalised war that threatened the lives of millions of people', as Susan Rice, US Assistant Secretary of State for African Affairs, stated in September 1998.[17] As Kabila approached the capital, Kinshasa, he won the support of Angola and Zimbabwe, quickly followed by Chad (supported by Libya) and Namibia. Sudan joined in, partly to counter Uganda's support for the opposition groups. This meant that Sudan's rebel force, the SPLA, moved troops to the border to stop the Sudanese from establishing a base in DRC.

The countries from southern Africa had the approval of their regional organisation, SADC, for their intervention, and at a meeting of SADC in September 1998, President Mandela of South Africa decided to shift from his neutral stance and hail the intervention of the three southern countries. Altogether, nine countries were involved in the war, for a variety of reasons. President Sam Nujoma of Namibia said that he was in it 'for the spirit of Pan-Africanism, brotherhood and international solidarity' and that African people are each other's keepers,[18] but these lofty reasons were more likely subordinate to the economic motives for most participants, along with the desire, especially in Mugabe's case, to deflect attention from domestic problems. It is significant that the richest diamond mines, at Mbuji-Mayi in East Kabai, have been a prime target first for Kabila's forces and later for the rebel groups.

Then Angola was sucked into the DRC as Congo rebels threatened its border. The rebels had the support of Angola's opposition party, UNITA (having previously supported Mobutu), while the Angolan government supported Kabila; so this was another case of a country exporting its own civil war into the DRC. Zimbabwe was more open than the other interventionists in the DRC, and in 1999 the Ministry of Defence announced a joint venture for dealing with Congolese gold and diamonds between a company in which the Zimbabwean Chief of Staff was the main shareholder and a Congolese company belonging to Kabila and some of his ministers. When this did not prosper as planned, companies from other nations pressed their claims, from France, Japan, USA, Israel and other European and Asian countries,[19] and it would be idle to suppose that such companies from the industrialised world did not see that advantages could be gained from this war.

Turning to West Africa, the civil wars in Liberia and Sierra Leone provide perhaps the clearest example of the way African conflicts have turned into a free-for-all scramble for rich resources and loot, as well as for the fruits of newly won power, whether it be over a nation or over a wealthy region. In August 1990, President Doe of Liberia asked for ECOMOG forces to intervene in his conflict with Charles Taylor's opposition forces. It was an extremely devastating conflict which caused huge numbers of civilian casualties, with hundreds of thousands more civilians becoming internally displaced persons or refugees. At one stage, five different factions were fighting, and Charles Taylor's rival leader Prince Johnson held hostages as human shields in his attacks, including some of ECOMOG's troops.[20] ECOMOG intervened for humanitarian purposes to end the bloodshed and save the country from anarchy. Though they stopped Monrovia from being sacked by Taylor's troops, they could not prevent his victory, as Doe was murdered and Taylor was elected President. Thereafter ECOMOG remained in Liberia as a peacekeeping force, its original intervention receiving the eventual approval of both the OAU and the UN Security Council. Taylor's campaign had been partly an ethnic clash with appalling humanitarian results, and partly a sheer bid for power and riches. He was not content to stop there and turned to threaten Sierra Leone with destabilisation as well: he armed the rebels who were fighting the democratically elected

President Kabbah.[21] The two civil wars bore certain similarities in the violent anarchy and savage crimes inflicted on the civilian population.

ECOMOG, therefore, in intervening in Sierra Leone, could again claim that it did so for humanitarian reasons and in defence of a democratically elected government. In 1996 President Kabbah was elected after five years of civil war, and the Abidjan Accord, backed by the UN, OAU and ECOWAS, should have guaranteed peace. It did not, for Major Johnny Koroma staged a successful military coup to oust Kabbah, and threatened a siege of the capital, Freetown. It was one example of a frequent tendency: that is, an internationally brokered peace accord was never willingly accepted by the opposition groups, who still occupied much of the country – a tendency which will be discussed in the third section of this chapter.

First President Abacha of Nigeria sent a force against Koroma's military junta; then followed the full force of ECOMOG, in a large operation with some 15,000 troops. They had to deal with a state of anarchy in which different rebel groups grabbed whatever they could in the rich diamond-yielding areas or in plundering the towns. Often the fighting was done by youths and even children, drugged and terrorised into fighting and dressed sometimes in women's clothes, wearing wigs, and so on. Freetown was 'being torn apart by mob violence'.[22] At this time President Kabbah was in exile. When the main rebel group, the Revolutionary United Front (RUF), was thwarted in Freetown it took its revenge on the rural civilian population, hacking off limbs in acts of sheer terrorism. Further details of this war are not relevant here, but the Nigerian-led ECOMOG force did succeed in restoring President Kabbah against rebel forces which included not only thousands of Taylor's men from Liberia, but soldiers from Burkina Faso and Ivory Coast (they later switched to support ECOMOG), 300 mercenaries from the Ukraine and other foreign mercenaries, all attracted by the diamond-producing areas in the west. Kabbah in turn had employed the South African-based private company, Executive Outcomes, and later, as we have seen, the London-based company Sandline.[23]

By July 1999, a power-sharing peace deal, under intense pressure from the UN and OAU, was signed after nearly 10 years of war. It was likely to be a very shaky peace, with the RUF rewarded with several ministries including the mineral portfolio, held by Foday Sankoh himself, the leader of the RUF. While the war was called 'the privatisation of conflict', in which so many groups fought for mining profits, Sam Kiley condemned the peace deal as 'the criminalisation of an entire state'.[24] Indeed the war broke out again when Foday Sankoh resumed leadership of the RUF. His capture in May 2000 did not end the war and at one point 500 UN troops were actually captured and held by the rebels.

The Sierra Leone conflict has been called a resource war, and undoubtedly it is fuelled and indeed partly controlled by the greed of various participants for the country's supplies of diamonds. Charles Taylor of Liberia is said to be a 'facilitator' of the RUF's campaign in Sierra Leone as the rebels hold the diamond mining areas and export the gems to world markets via Liberia, in return for supplies of arms. Burkina Faso also is alleged to supply arms to the RUF in return for diamonds. The international community knows this and realises that efforts to end the conflict are almost certain to be frustrated while the rebels enjoy this source of funding. The United Nations Security Council's Resolution 1306 banned the trading in diamonds from the rebel-held areas of Sierra Leone; ECOWAS established an inquiry into the trade, and the Sierra Leone government itself has begun a scheme to issue certificates of origin for the export of legitimately produced diamonds. The big international firms such as De Beers and also Antwerp's Diamond Council professed willingness to collaborate in this and produced their own scheme.[25] But diamonds are easy to smuggle and this trade will continue in spite of such efforts. It is a major hindrance to peacemaking and peacekeeping in the area. Meanwhile, ECOMOG is left with the unenviable task of peacekeeping in both Liberia and Sierra Leone, though in Liberia it is helped by the UN peacekeeping force, UNOMIL, and in Sierra Leone by UN troops under a mandate to use force if necessary.

2.5 Conflict resolution: the ritual cease-fire dance

The changing nature of conflicts in Africa, sucking in so many different participants, national and rebel, for such a variety of reasons, means that conflict resolution has become a very complex and long-drawn out process. Even the most carefully arranged cease-fire as a preliminary to a peace accord seems to be often only a truce, a temporary breathing space, signed reluctantly by some participants and soon broken. When some of the participants still have motives that in their eyes remain valid, they have little interest in maintaining peace. For some, war remains a profitable business, either in political or in financial terms.

The search for peace in the Democratic Republic of Congo (DRC) provides perhaps the best example of this. It involved nine nations, some eight different armies and at least twelve armed groups;[26] the trail of national leaders, statesmen or diplomats who have tried to broker a cease-fire includes Presidents Chiluba of Zambia, Gaddafi of Libya, Chissano of Mozambique, Moi of Kenya, Mandela and later Mbeki of South Africa, Mkapa of Tanzania, Bongo of Gabon, as well as officials from the OAU and SADC, the UN Special Rapporteur Roberto Garreton, the EU Special Envoy Aldo Ajello, and the UK's Special Envoy and deputy minister for foreign affairs, Tony Lloyd. For the belligerents and the would-be peace-makers it has meant an astonishing round of travelling to all parts of Africa and beyond. Venues have included Victoria Falls, Lusaka, Ndola, Addis Ababa, Nairobi, Libreville, Windhoek, Yaoundé, Kampala, Rome, and Syrte in Libya. These travels have been spread over two years; for the exhausted participants the end was not reached until September 1999 when the last of the rebel groups signed the agreement.

The summit meeting at Victoria Falls in September 1998 may be taken as the starting-point of the major effort in peacemaking. Heads of state from Angola, Zimbabwe and Namibia arrived, with President Chiluba of Zambia as chairman; he took on the role of leading mediator and tireless instigator of these meetings. Kabila consented to be present, but he refused to meet the rebel leaders. This was the first opportunity to bring Kabila and his allies, Uganda and Rwanda, to meet the rebels but his refusal meant the rebels went home and the talks flopped.[27] Regional Defence Ministers met in Addis Ababa to discuss renewed efforts, and President Bongo of Gabon organised a summit at Libreville, attended by Kabila and several presidents, while Chiluba and Mkapa went to Rwanda and Uganda as part of SADC's efforts to resolve the conflict. The UN Security Council met to discuss a cease-fire, which they resolved would require the withdrawal of all foreign forces and a dialogue for a peace settlement; there must be proper respect for human rights and international law. At the annual summit of SADC held in Mauritius and attended by fourteen states, no significant progress on peace was made. Curiously, although they recognised the legitimacy of intervention in DRC by Angola, Zimbabwe and Namibia, they did not condemn Rwanda or Uganda. They mandated Chiluba to continue his efforts.[28] Soon other actors appeared on the scene: President Chissano of Mozambique had ambitions to take a role as mediator, and toured the region exploring peace possibilities; Chiluba's supporters were anxious that Chissano might usurp Chiluba's role as chief mediator but Chissano denied any such intention. Tony Lloyd then appeared, appointed as Prime Minister Blair's special envoy to bolster the peace moves: he toured nine countries. At a Lusaka meeting in January 1999, a peace plan was set out, proposing a UN and OAU international force to supervise a cease-fire if it were agreed. The EU offered to help in reconstructing the DRC economy if a cease-fire was established.[29]

A month later Roberto Garreton, UN Special Rapporteur, and Aldo Ajello, EU Special Envoy to the Great Lakes Region, also made their tours and added to the pressure building up. At a meeting in Windhoek some actual progress was made when not only Angola, Zimbabwe and Namibia but also their opponents Rwanda and Uganda agreed they would sign an agreement if a cease-fire materialised – but Wamba dia Wamba, leader of the Rally for Congolese Democracy (RCD), said that only talking directly with Kabila would bring a cease-fire – and this Kabila was adamant he would not do.[30] At Lusaka, the various delegates agreed to split into two teams to try to expedite matters. UN, OAU and SADC representatives would be in both; the first team would comprise Zambia, Rwanda, Uganda, Burundi and DRC representatives and would concentrate on a cease-fire agreement, while the second team, comprising Zambia, Kenya, Mauritius and Botswana, would seek an agreement on border security issues. Thus the peace effort was engaging more and more participants.

Two events at this time, however, affected the peace process. First, the rebels were now two distinct main groups, with tension between Wamba (RCD) and Bemba, leader of the Movement for the Liberation of Congo (MLC). But Wamba himself was ousted by Ilunga. Second, Zimbabwe's contingent in DRC suffered a severe defeat and lost eighty men killed in East Kisai. It meant President Mugabi was under more pressure at home to withdraw from the war – but the split in the rebels meant that obtaining a cease-fire agreement would be more difficult.[31] Certainly, by March the allies on Kabila's side were 'looking for a way out', as *The Economist* headlined, while Uganda's intervention was criticised in the Uganda Parliament, as well as being heavily criticised among the international community,[32] which put pressure on Museveni. Kabila dropped his opposition to negotiating a cease-fire, and an extraordinary meeting took place in Rome between government and rebel representatives, under the auspices of the Sant' Egidio Organisation. An equally unusual meeting was hosted by President Gaddafi at Syrte in Libya, at which a cease-fire document was actually signed by Kabila and four Presidents – but no rebel leaders were present, although Museveni, their supporter, signed it.[33] Gaddafi's attempt to hi-jack the peace process was quickly denounced by the rebels and also by an indignant Chiluba, who stated that the Lusaka process was the only one recognised by the UN, OAU and SADC.[34] In June, after several more meetings at various locations, he declared the final deal would be done 'in Lusaka and nowhere else'.[35]

However, he chaired the important talks in Pretoria following President Mbeki's inauguration, when all the major participants, including Rwanda, agreed that they would sign a cease-fire. Absentees Kabila and the rebel groups were bound to be drawn into this process, even though the rebels were still making ground in DRC nearer the diamond regions. As one commentator put it, 'The warring parties have little incentive to quit': Rwanda and Uganda had captured the area in the east where they could continue to chase Hutu militia threatening their own borders; Kabila still held the diamond and cobalt mines, and the rebels still had hopes of capturing them.[36]

On 16 July 1999 at Lusaka, a cease-fire was duly signed by Kabila and the five main nations involved, setting up a Joint Military Commission to supervise the proposed disarming of militias, but the disputes over rebel leadership meant that neither the MLC, led by Bemba, nor the RCD, with a leadership split between Wamba and Ilunga, would sign. Wamba insisted that he had the right to sign for RCD; Ilunga who had ousted him contested this and refused to sign. This immediately put the peace process in doubt, and Rwanda said that unless the Interahamwe and ex-FAR militias disarmed, Rwanda would have no choice but to continue fighting them.[37] Great pressure was put on the rebels to sign, and Bemba, urged by Chiluba, Mbeki and Museveni, did at last sign it, on 1 August 1999, but he said that if the other rebels did not sign it within seven days, he would continue his push to capture Kinshasa, from which he was just 600 km distant.[38]

The process dragged on and Wamba dia Wamba and Ilunga, leaders of rival groups of the RCD, still refused to sign it. Museveni met them in Kampala to try to persuade them, and Kabila offered a general amnesty to guerilla forces, leading to the surrender to the government of between three and four thousand of them. But violence on all sides continued, including an alleged bombing attack by Sudanese Antonov planes on two fishing towns in rebel territory. The most bizarre episode of all, and one which pointed up the inherent dangers of foreign intervention in a complex civil war, was the outbreak of severe fighting on 14 August between the Ugandan and Rwandan troops themselves, apparently started by a bid to capture control of Kisangani airport. The battle spread through the town and caused severe civilian casualties and damage in four days of fighting. Uganda may have lost up to 200 soldiers, though the ICRC put the figure at thirty-four soldiers killed and 131 soldiers and civilians wounded.[39] The fighting ended with a quick agreement between Museveni and Rwanda Vice-President Paul Kagame (though two further clashes took place), but no doubt some of the trouble sprang from the fact that Uganda backed Wamba dia Wamba's RCD group and Rwanda backed the rival group led by Ilunga. Finally a compromise was reached between the two RCD leaders, in which all fifty-one of the founding members of the party signed together in Lusaka – seven weeks after the original agreement had been signed.[40]

The battle in Kisangani raises the serious question: what are the real purposes of Uganda and Rwanda in DRC? This fight had nothing to do with their avowed aims of defending the security of their own borders. *An Analysis of the Agreement and Prospects for Peace* warns that:

> the conflict seems to be a battle for commercial influence to control diamond, gold and coffee concessions, and for political influence in the region after the war is over . . . The high level of tension between Uganda and Rwanda is likely to affect the geopolitical order of the region; it could lead to further fragmentation and a de facto partition of the DRC.[41]

Very disturbingly, a second battle of Kisangani between Rwanda and Uganda broke out again in June 2000, also with severe civilian and military casualties. Effective control of the two armies is shown to be lacking. Even after the signing of the agreement, new foreign interventionists appear on the scene: North Korean troops have been reported in Lubumbashi, the capital of Katanga in DRC, and they are said to be training Kabila's troops or providing guards for Kabila, but it is equally likely that Kabila may have given them access to the Shinkolobwe uranium mine in return for military assistance against rebels. The uranium is needed for North Korea's plutonium production. Their arrival 'does not augur well for the fragile peace agreement', commented a spokeswoman for the South African Institute for Security Studies.[42]

The agreement provides for a joint military commission to supervise disarmament of the rebel groups, with the aid of UN and OAU military observers; the provision of a UN peacekeeping force; and the setting up of a 'national dialogue' to debate the political future of the country. After such a bitter conflict, baffling in its complexity, they face a huge task. Approaching the end of the year 2000, none of these provisions has been implemented and fighting continues.

The civil war in Burundi provides another example of the difficulties of conflict resolution when some of the participants feel that they can get more advantages by continuing to fight. In this case it is not so much profit that spurs them on as ethnic rivalries and the fear that they could lose out in a peace which imposes democratic elections or power sharing. When Major Pierre Buyoya staged his military coup and set up a Tutsi military government in 1996, Burundi had already had its share of ethnic killings and civil war, and the UN and OAU called for an international peacekeeping force to try to secure peace and to prevent a Tutsi-dominated regime. Buyoya tried to give assurances and promised a government of national unity. No peacekeeping force materialised, however, as many of the would-be participants were wary of becoming entangled in such a chaotic situation, where many groups of militia, Hutu or Tutsi, fought for enclaves of territory. There was an overlap too with the civil war in Zaire, as the exiled Hutu government of Rwanda in Zaire, plotting a re-invasion of Rwanda, linked up with the Hutu militias in Burundi, forming a pan-Hutu alliance.[43] Instead, the East and Central African states agreed to impose trade sanctions on the country, which would not be lifted until 'constitutional legality' was restored, with a democratic system of government. These sanctions were imposed in July 1996 in a very quick reaction to Buyoya's coup, and after a while they caused great economic hardship. Meanwhile former President Julius Nyerere of Tanzania tried to mediate on behalf of the East African states in meetings at Arusha, but it proved very difficult to obtain any agreement from either Buyoya or the opposing groups. In Burundi, the state of continual violence and killings made it impossible. Nyerere was well aware of the cost of these talks where all the travel to and fro was paid for by the donor countries of the international community. He urged a speedy end to the talks, which by January 1999 had cost US $1.1 million, and he warned that donors would not go on paying for endless talks.

The UN Security Council issued the same warning and appealed to the regional leaders to consider lifting the sanctions. The East and Central African states complied, but declared that this was conditional on peace talks making progress. The Uganda government warned that it was losing patience with the way negotiations were dragging on, with 'diversionary obscurantist propaganda by those bent on either delaying or defeating the process'.[44] Meanwhile Hutu rebels continued their attacks in South Makamba and around Bujumbura. In the next round of peace talks in Arusha, the delegates were given sensitising training in negotiating by a team of international mediators who had helped in Bosnia, Northern Ireland and Mozambique – an indication of international concern and even bewilderment at the lack of progress. The peace talks were now organised into four separate commissions to speed things up, but little progress was made on the most important one, dealing with peace and security. Nyerere declared he would be less than honest if he said he was happy with the progress made. The Tanzanian facilitator's representative again warned that donors' patience was running out: 'The fact that we are still at Agenda I, Item I, does show the kind of problems we have to overcome to reach a settlement.'[45]

New clashes followed in Bujumbura Rurale, the rebels being aided by infiltrators from Tanzania. A senior officer said this proved there was a coalition in the Great Lakes Region of the Burundi rebel parties, with ex-FAR soldiers and Interahamwe from Rwanda. While violence was escalating (the peace talks may even have contributed through frustration at the lack of progress), there was stalemate in the talks over whether to include the main rebel group, Forces pour la Défense de la Democratie (FDD).[46]

By May 1999 the donors did indeed lose some of their patience and made the resumption of aid conditional on the signing of a peace agreement. The peace process was reported to be floundering and 'at the risk of disintegrating', as government, parliament and the Hutu FRODEBU party demanded that the Tutsi group of parties withdraw their accusation that the government was colluding with Hutu rebels.[47] In an atmosphere of violence, ethnic hatred and rivalry, it was still impossible to make progress with these long-drawn out peace talks. In July the fifth round of talks ended in Arusha and Nyerere stated that it was a failure: the negotiations had been taken too lightly and obstruction by the Burundi government, on the pretext of violence, had caused the failure. The next round began in September 1999, but the death in October of Nyerere, the renowned elder statesman who was the main inspiration of the peace initiative, removed a key figure. This conflict is proving one of the most difficult to resolve in Africa; meanwhile, since the war began in 1993, 200,000 people have been killed.[48] Ex-President Mandela of South Africa took over from Nyerere as mediator but found the task of persuading all parties to participate equally daunting.

2.6 Conclusion

The recent tendency in African conflicts has been towards multinational conflicts involving several countries, not only government forces and long-standing dissidents but guerrilla forces led by new leaders or parties born out of the conflicts themselves. Quite often these groups are offshoots of parties fighting civil wars in neighbouring states, causing one civil war to spill over into another. Their interventions often develop into quests for profit and wealth in richer neighbouring states, as for instance in DRC, Liberia and Sierra Leone. Few of the conflicts are of the more 'traditional' variety, such as the Ethiopian–Eritrean war, where two states fight for territorial gains or to solve border disputes, with little outside intervention. Instead, in the majority of conflicts, prevention or resolution has become very difficult. Making peace through negotiations and mediation becomes long drawn out, and some of the participants do not want peace, or will not accept it when it is imposed. Those two essential requirements of a peace, agreement and willingness to comply, are difficult to obtain and frequently are reneged upon.

The days of the UN-sponsored multinational peacekeeping missions may be numbered, for two reasons. First, the major member states are less willing to become embroiled in these complex African conflicts and decline from making such commitments when the prospects of successful and permanent completion are far from certain. Second, their place is being taken by more directly involved regional organisations which are more acceptable to the country concerned and are more capable of speedy action. They are in fact gearing themselves to quicker reaction and response, with new strategies, as we have seen, which have at least some success. Acknowledged African leaders such as Chiluba and Mbeki, and the pressures they are able to bring to bear on warring parties, may be the best hope for the future. Even then, funds to bear the not inconsiderable costs of negotiation and maintenance of a peacekeeping presence must come from the world powers.[49]

Notes

1 *IRIN Weekly Update* (hereafter *IRIN*), 3–9 July 1999.
2 *IRIN*, 10–16 July 1999.
3 UN Department of Peacekeeping Operations Training Unit, December 1998; and *Independent*, 25 June 1999.
4 Ibid.

5 *Weekly Mail and Guardian*, 9 October 1998.
6 Panafrican News Agency, 23 March 1999. It was declared an outstanding success, and 'the international community has recognised the African states are willing to resolve African issues themselves and the international community is prepared to assist in providing the means to this end' (address by the Minister of Defence, the Hon. Mr Joe Modise at the Exercise Blue Crane Luncheon, 26 April 1999).
7 *US News and World Report*, 8 February 1997.
8 'Diamond Mercenaries of Africa': Background Briefing, Radio National, Australian Broadcasting Corporation Transcript, 4 August 1996. Executive Outcomes was apparently closed down in January 1999 because the South African government passed anti-mercenary legislation and was concerned at the number of apartheid-era soldiers fighting elsewhere in Africa (*News and Views*, 11 December 1998).
9 I am indebted to my colleague Gerry Cleaver for this reference and for his advice on this section.
10 The ECOWAS Mechanism for Conflict Prevention, Management, Peacekeeping and Security, Section II Article 46, 31 October 1998. I am indebted to Jeremy Levitt for this reference and for some of these comments.
11 Ray Kennedy, 'South Africa takes a gamble on peacekeeping forces in Congo', *The Times*, 1 July 1999.
12 *Ethnic Studies Network Bulletin*, no. 16, April 1999.
13 See in particular, Prunier (1995), and his chapter on 'The Rwanda Patriotic Front', in Clapham (1998).
14 Sam Kiley, *The Times*, 25 April 1996.
15 Uganda newspaper, *The Monitor*, 2 October 2000.
16 Bruce Baker, 'Going to war democratically: the case of the second Congo War, 1998–2000', *Contemporary Politics*, 6 (3) (2000): 75.
17 *IRIN*, 12–18 September 1998.
18 Panafrican News Agency, 14 September 1998.
19 *Africa Confidential*, 40 (22), 5 November 1999; 41 (9), 28 April 2000; 41 (17), 1 September 2000; BBC *Online News*, 5 September 2000 and *IRIN*, 8 September 2000. See also A. Alao, 'The role of African regional and sub-regional organisations in conflict prevention and conflict resolution', UNHCR Working Paper no. 23, July 2000, p. 17.
20 *Guardian*, 17 April 1996, and *The Economist*, 13 April 1996.
21 *West Africa*, 1–14 February 1999. ECOMOG itself did not come out of this war with its reputation unscathed: it was frequently accused by other West African countries of not being a neutral force and of being an agency for expanding Nigeria's influence. Also, its troops were guilty of widespread looting and some Nigerian officers were accused of regarding service with ECOMOG as a lucrative opportunity to enhance their personal wealth. Gerry Cleaver, 'Liberia: lessons for the future, from the experience of ECOMOG', in Furley and May (1998: 232).
22 *The Times*, 1 June 1997.
23 *The Times*, 22 January 1999 and 11 February 1999.
24 Sam Kiley, *The Times*, 15 July 1999, and Mark Doyle, 'Rebels may profit from a reign of terror', *Guardian Weekly*, 8–14 July 1999.
25 *Africa Confidential*, 41 (16), 4 August 2000.
26 *IRIN*, 19–25 February 1999.
27 Panafrican News Agency, 14 September 1998.
28 Ibid., and *IRIN* 18–24 September 1998.
29 Panafrican News Agency, 22 and 23 February 1999, *IRIN*, 19–25 February 1999.
30 *IRIN*, 16–21 January 1999.
31 *IRIN*, 30 January – 4 February 1999, and BBC *Online Network*, 16 March 1999.
32 *The Economist*, 20 March 1999.
33 *Africa News Online*, 14 April 1999 and BBC *News Online Network*, 18 April 1999.
34 Panafrican News Agency, 24 April 1999.
35 Panafrican News Agency, 15 June 1999.
36 'The Congo cease-fire begins to drag', *International Herald Tribune*, 1 July 1999.
37 *IRIN*, 3–9 July 1999.
38 *Uganda Newsline*, 4 August 1999.
39 *IRIN*, 14–20 August 1999, and 21–27 August 1999.

40   *The Times*, 1 September 1999.
41   *The Agreement on a Cease-fire in the Democratic Republic of Congo: An Analysis of the Agreement and Prospects for Peace*, 22 September 1999, p.7; http://www.intl-crisis-group.org/
42   *The Times*, 12 October 1999.
43   See W.M. Cyrus Reed, 'Guerrillas in the midst', in Clapham (1998: 142); and Shashi Tharoor, 'The future of peacekeeping', in *After Rwanda: The Coordination of United Nations Humanitarian Assistance*, London: Macmillan, 1996, p. 23.
44   Uganda Newspaper, *New Vision*, 2 February 1999, and *IRIN*, 16–21 January 1999.
45   *IRIN*, 5–12 March and 17–23 April 1999.
46   *IRIN*, 24–30 April 1999. For ties between the rebel groups in Burundi and Rwanda, see African Rights, *Rwanda: The Insurgency in the Northwest*, London, September 1998, pp. 70–4.
47   *IRIN*, 1–7 May and 10–16 July 1999.
48   Panafrican News Agency, 17 July 1999.
49   Since the time of writing, in the DRC, President Laurent Kabila has been assassinated and his son Joseph has been appointed President. Joseph has shown a much more positive attitude to the peace process; also Rwanda and Uganda have withdrawn some of their troops so that the outlook has improved.

References

Adams, T. (1999) 'The new mercenaries and the privatisation of conflict', *Parameters* (US Army War College Quarterly), summer: 103–16.

Boutros-Ghali, B. (1992) *An Agenda for Peace*, New York: United Nations; 2nd edn 1995.

Clapham, C. (ed.) (1998) *African Guerillas*, Oxford: James Currey.

Cleaver, G. (1998) 'African crisis response initiative', paper presented to the African Studies Association of the UK, September.

de Coning, C. (1998), 'Conditions for intervention: DRC and Lesotho', *Conflict Trends*, October: 23.

Furley, O. (1998) 'Rwanda and Burundi: peacekeeping amidst massacres', in O. Furley and R. May (eds), *Peacekeeping in Africa*, Aldershot, Hants.: Ashgate.

Furley, O. and May, R. (eds) (1998) *Peacekeeping in Africa*, Aldershot, Hants.: Ashgate.

Marais, H. (1999) 'Diplomacy discarded for intervention: South Africa carries a big stick', *Le Monde diplomatique*, March.

Organisation of African Unity (OAU) (1993) *Resolving Conflicts in Africa: Implementation Options*, OAU Information Services Publication, Series III, Addis Ababa: OAU.

Prunier, G. (1995) *The Rwanda Crisis, 1959–94: History of a Genocide*, London: Hurst.

Woodward, P. (1998) 'Somalia', in O. Furley and R. May (eds), *Peacekeeping in Africa*, Aldershot, Hants.: Ashgate.

3 The economics and political economy of conflict in sub-Saharan Africa

Christopher Cramer

3.1 Introduction

Until the end of the 1980s development economists showed little interest in the problems of civil war. Yet within the past decade this has changed dramatically, with a rapidly expanding literature on the economics or political economy of conflict, civil war, or so-called complex humanitarian emergencies. Major research programmes have included: UNRISD's War-Torn Societies Project (www.unrisd.org/wsp); Queen Elizabeth House at Oxford (Stewart *et al.*, 1997); a WIDER project on complex humanitarian emergencies (for example, Nafziger and Auvinen, 1997); the ILO (for example, Cramer and Weeks, 1997); the World Bank (for example, Collett *et al.*, 1996), which in mid-1999 signalled its commitment by opening a dedicated web-site (www.worldbank.org/research/conflict); and the OECD (OECD, 1997). There have also been useful surveys of aspects of the literature, including Carbonnier (1998) and Luckham *et al.* (1999).

This is not just an academic response to empirical developments, since outbreaks of armed conflict of various kinds have long been a common feature of developing countries. Two plausible, and probably only partial, explanations may be as follows. First, until the end of the Cold War many economists considered war to be purely exogenous, a given of global politics, something that occasionally disrupts from time to time more normal conditions under which economic laws apply (Fischer and Schwartz, 1992: 239). However, a rash of post-Cold War conflicts has encouraged or freed up economists to consider more intimate relations between such conflicts and other variables within their models. Second, this has coincided with a growing confidence within the mainstream economics profession that the precepts of economic orthodoxy can be applied to an increasingly wide range of social phenomena. This theme is taken up further in the section below on the causes of conflict.

Whatever the actual mix of reasons, there has been a shift within the economics of conflict. The interest of development economists was at first concentrated on the costs of war (UNICEF, 1989; Green, 1991; Stewart, 1993; and more recently Stewart *et al.*, 1997; Brück, 1997). During the 1990s, however, there has been increasing economic analysis of the dynamics of wartime (Azam *et al.*, 1994; Chingono, 1995) and, even more, of the economic or political economy origins of conflict (for example, Nafziger *et al.*, 1999; Collier and Hoeffler, 1996). Throughout, there has been an interest in the post-conflict role of economics. Post-war rehabilitation analyses can draw on cost-of-war exercises and also on analyses of the causes of war and of wartime economic and social change (for example, Boyce *et al.*, 1996). Despite the expansion of the literature on these subjects, there remain areas of significant analytical and empirical weakness, and the subject, as part of development economics, is still in its infancy.

As well as a shift from focusing on the consequences of conflicts towards analysing their 'root causes', the shift in perspective within the economics of war in poor countries has taken other forms. From treating these conflicts as exogenous events, whose economic effects are then traced through the mechanisms of economic models, economists have increasingly brought the determination of conflicts within the interaction of model variables. At the same time, there has been something of a shift in intellectual confidence as development economists have taken more eagerly to analysing wars. Early exercises might have spoken of the 'vacuum' in the political economy of war (Green, 1991) or might have asked modestly whether economic analysis can help to mitigate the costs of conflict (Stewart, 1993). More recently, this tone has shifted to incorporate a more assertive sense that economics is the best-placed discipline for understanding the origins, dynamics and consequences of conflicts in poor countries.

After briefly reviewing the spread of conflicts in SSA and emphasising their diversity, this chapter focuses chiefly on the literature on the costs of conflict and on the causes of conflict. The main argument of the chapter is that conflict directly expresses social, political and economic relations and that therefore it requires analytical tools directly geared towards understanding relations. However, the majority of the literature only addresses conflict via concepts with an indirect grasp of relations at best. So, the majority of the literature is dominated either by structuralist explanations and concepts of stratification (such as ethnicity, income inequality, poverty) or by the non-relational methodological individualism of neoclassical economics. The two are sometimes combined, for example in the expanding amount of cross-sectional large sample statistical or econometric analysis. This methodological distinction is yet to be adequately highlighted and explored in the literature. Research into direct social relations in individual contexts, and their expression through armed conflict in SSA, has so far chiefly been confined to case-study material of considerable value.

The literature on conflict in SSA and other parts of the world has also become a site of reductionist colonisation by mainstream economics. The chapter discusses this, for example, in the light of implicit theories of violence and war assumed by much of the literature. The two most commonly accepted theories are: that violence and conflict are functions of scarcity and that conflict is a function of difference, expressed through measures of inequality or ethnic fragmentation, for example. This chapter suggests that, at least, more subtle discussion of underlying theories of violence needs to be made explicit, and introduces one theory of human violence with suggestive links to the problems of late capitalist development. The chapter also seeks to escape the constraints of the classical liberal interpretation of war, that it is exclusively and always negative in its consequences; instead, economists should be prepared to explore the conditions under which conflicts, however terrible and destructive, can also play more constructive roles.

3.2 The prevalence of conflict in sub-Saharan Africa

If we take a major armed conflict to be one in which at least 5,000 people died, then at a rough estimate there have been perhaps 24 major armed conflicts in sub-Saharan African countries during the past two decades (Figure 3.1). Of the countries affected, seventeen have had earlier major conflicts this century too and some eight of these conflicts very obviously have roots going back to before independence. Another nine or so countries that have experienced conflicts in the past 20 years or so have had fewer casualties. Conflicts have been especially frequent perhaps during the 1990s. By another measure, in which an armed conflict is defined as a war when more than 1000 people die per year (Wallensteen and Sollenberg, 1998), between 1989 and 1997 there were fourteen wars in SSA.

	1900–	1960–	1970–	1980–	1990–
Angola		1961–75		1976–95	1998–
Burundi			1972–80	1988–95	
Cameroon	1955–60				
CAR					1996–7
Chad				1980–94	
DRC		1960–5		1993	1996
Congo				1993–	
Djibouti				1990–6	
Eritrea				1974–91	1998–2000
Ethiopia	1935	1941	1974–9,	1976–83	1998–2000
Guinea-Bissau		1962–74			1998–
Kenya	1954–6			1991–6	
Liberia			1985–8		1990–7
Madagascar	1947–8				
Mali				1988–94	
Mozambique		1965–75		1976–92	
Niger				1991–6	
Nigeria		1967–70		1980	1989	1991
Rwanda	1956–65			1990–7	
Senegal			1982		
Sierra Leone				1991–	
Somalia				1988–95	
Sudan		1963–72		1983–	
South Africa	1899–06		1976	1983–94	
Uganda		1966	1970–8	1980–7	1992–
Zambia		1964			
Zimbabwe			1972–9	1983–4	

Notes: Togo, Ghana, Gabon, Comoros, Lesotho, Mauritania, Namibia, Western Sahara and Zambia have all experienced minor conflicts with less than 5,000 casualties since independence. Data are unreliable estimates in many countries, especially for Angola, DRC, Eritrea, Ethiopia, Rwanda, Sierra Leone, Sudan and South Africa.

Legend:

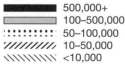

■■■■■■■ 500,000+
▨▨▨▨▨ 100–500,000
┊:::::::: 50–100,000
///////// 10–50,000
\\\\\\\\ <10,000

Figure 3.1 War casualties in selected SSA countries
Source: Luckham *et al.* (1999)

For the most part, these conflicts are what are normally defined as 'internal' wars, that is, not wars between countries, with the exceptions including recent conflict between Eritrea and Ethiopia. However, in virtually every one of these wars there has been external involvement of one kind or another, ranging from international support to both sides of the MPLA–UNITA conflict in Angola to French military and political support for the Habyarimana regime in Rwanda during the war against the invading Rwandan Patriotic Front (RPF) forces, in the build-up to the genocide of 1994. Other wars have had foreign involvement of a more private, commercial nature, including the contribution of multinationals to political and military feuding in, for example, Sierra Leone and Liberia (Reno, 1998). Traditional notions of a 'civil war' are less relevant in a world where the ways in which wars are fought reflect international economic integration. Thus, the use of international arms markets, external loans or foreign exchange earnings, especially mineral based, to purchase arms, the integration of cross-border weapons and criminal networks and foreign military 'aid', are characteristic of many modern 'civil' wars. Furthermore, these wars commonly have spillover effects, humanitarian, economic and even military: thus, countries like Tanzania and Côte d'Ivoire have been affected by war in neighbouring countries. Wars in SSA during the past two decades reflect widespread concerns that the nature of typical armed conflicts has changed (de Waal, 1997). These wars generate huge numbers of civilian casualties rather than being confined to traditional battles between well-defined armies. Clear-cut distinctions between military forces and civilians are often missing in these wars, in which, sometimes, a person may slide between civilian behaviour and militia activity or in which, as in Sierra Leone (Richards, 1996), there are so-called 'sobels', government soldiers by day, rebels by night.

One of the major difficulties that compromises the precision of economic analysis – particularly at the level of cross-country comparisons – is that of defining wars. There are two main parts to this difficulty. One is that it is often hard accurately to distinguish the beginning, and indeed sometimes the end also, of a period to be defined as a war. The other is that not all wars are alike. This goes for their causes,[1] and it goes too for their intensity, their geographical spread, their duration, their military characteristics, and so on. In a country like Angola, warfare is written into the whole fabric of social relations and the majority of the population have lived their whole lives in a context of war; in others, such as the Sudan, war has also been protracted, but it has arguably been more intermittent and more geographically concentrated, though its indirect economic, political and social effects are widespread if not always so direct. In still others, such as Burundi or Nigeria, armed conflict has been briefer. Yet these examples reveal another problem for analysis, in that the distinction between war and other forms of violence or other forms of social conflict is not always hard and fast. In South Africa there was a form of civil war that principally affected what is now KwaZulu/Natal Province and that had a satellite conflict in the townships of the Pretoria–Witwatersrand–Vereeniging (PWV) industrial heartland. However, in another sense, South Africa could be described as a war economy for a much longer period, given the political conflict over the apartheid system, a prolonged if not immensely effective guerrilla campaign by the armed wing of the ANC (*Umkhonto we Sizwe* – Spear of the Nation), and the substantial military expenditure and efforts to destabilise much of the Southern African region by the apartheid regime. Further, to the extent that violence and political conflict undermine economic performance through, for example, the disincentive to investment, arguably post-apartheid violence has weakened the distinction between war and violence.[2]

Another major problem is the unreliability of data. Socio-economic data for much of SSA are notoriously unreliable even in non-conflict periods. This point can be made for different sectors (see Yeats, 1989, on trade statistics; Riddell, 1990, on manufacturing data; de Haan and Kochhaier, 1997, on labour market data; and Svedberg, 1990, on nutrition and agricultural data). War exacerbates this problem in a variety of ways. Even if data collection agencies remain well staffed, the collection of data in many areas can become too perilous. It is likely that there is a significant increase in most internal wars in the degree to which economic activity is unenumerated, partly because enumeration efforts decline and partly because normal marketing channels either break down or are evaded. This can make it difficult, for example, to quantify with any confidence the effect of conflict on agricultural output (FAO, 2000). Brown *et al.*, (1992) suggest that in the Horn of Africa the share of subsistence agriculture in total output has increased during war, and growing informal or underground activities have pushed much economic activity outside the coverage of national accounts. Furthermore, it is difficult at times to know when to attribute observed outcomes in the form of economic data to war or to other, separate or inter-related developments. That unreliable data weaken the economic analysis of war-affected countries is widely acknowledged. None the less, this has not stopped many observers attempting to quantify the costs of conflict or to make confident conclusions from cross-country statistical analysis of war economies. At the case-study level, a relevant example is the international response to conflict and genocide in Rwanda. In the immediate aftermath of the genocide, sweeping and misleading judgements were made about the imminence of famine and the impact of conflict on the country's seed stock. These estimates were generated partly by external assumptions and partly by the new government in Rwanda, most of whose members had little recent knowledge or experience of the agriculture sector in the country (Pottier, 1996).

3.3 'Mortgag'd states, wasted nations':[3] assessing the costs of conflict

This section first highlights in broad brushstrokes the developmental impact of internally fought wars in Africa, then discusses formal analytical and quantitative models of the economic impact of war and, lastly, focuses on a number of problematic issues opened up by the recent literature. The discussion highlights the role of Sen's entitlements approach in recent literature and draws attention to the influence of the classical liberal interpretation of the economic impact of war, that is, the idea that war entails exclusively a loss in economic terms.

3.3.1 Pervasive effects

The economic effects of wars in Africa are felt throughout society; the costs are particularly high as a result of wars being internally fought and because of their integration into international economic (and political) relations. Resources are usually diverted from social and economic investment and recurrent spending towards the war effort.[4] Wars are frequently financed by inflation, since in a civil war in a low-income country the scope for paying for a war out of increased wartime production and employment (*à la* Keynes's *How to Pay for the War*) are negligible. Productive investment is deflected by wartime conditions (Azam *et al.*, 1994). Wartime insecurity reduces competition and, for example in trading, creates monopoly rents or a war tax with negative consequences for consumers.[5] The trading infrastructure is commonly disrupted partly by direct damage to roads, bridges, buildings, irrigation networks and power lines, and partly by the diversion of vehicles to the military.

Agricultural production is affected in a number of ways: marketed output is typically reduced by the impossibility or soaring cost of transport; foreign exchange shortage and military absorption reduce the availability of agricultural marketed inputs; migrant labour sources dry up; the agricultural labour force shrinks due to casualties, flight and military recruitment; military action and the laying of landmines have a direct effect on cultivation. In Rwanda, for example, a 'year's tea and coffee harvests had been lost, and vandals had left all the tea factories and about seventy per cent of the country's coffee-depulping machines inoperable' (Gourevitch, 1999: 229). Over 28 years, 1970–97, conflict-related losses to agricultural production in SSA amounted to a total of some US $52 billion (in 1995 dollars), according to one estimate (FAO, 2000). This represented some 30 per cent of the agricultural output of affected countries during conflict years. In extreme cases such as Angola, with virtually continuous warfare since the early 1960s, empirical estimates suggest that agricultural output at the end of the 1990s was well below half what it would have been in the absence of war. Other countries where conflict effects on agricultural production were almost as severe are Ethiopia, Mozambique and Sudan.

According to a World Bank report: 'In Central and Eastern Africa alone, about 22 million people are displaced . . . Natural disasters and famine used to be the root of human suffering and displacement in Africa; now, internal conflict is the leading cause of emergencies in many countries' (World Bank, 1996: v). The effects of conflict on the poor are potentially considerable. For Messer *et al.* (1998: 1), armed conflicts constitute a significant cause of deteriorating food scenarios in developing countries. Luckham *et al.* (1999: 4) argue more forthrightly that armed conflict has 'become the single most important determinant of poverty in Africa'.

3.3.2 *Modelling the costs of conflict*

A number of sectoral, macroeconomic and comprehensive economic analyses have attempted to formalise and quantify the economic costs of war in Africa and elsewhere. Many of these share a distinction made between direct war damage and indirect developmental costs of war, including an argument (Stewart, 1993) that the indirect costs are larger and longer lasting than the direct costs. These exercises in assessing the costs of war began with efforts by the Southern Africa Development Coordinating Conference (SADCC) (1986), UNICEF (1989), and Green (1991) to establish the developmental effects of war in Africa, particularly focusing on Southern Africa and the effects of the South African apartheid regime's destabilising activities in the region. UNICEF (1989), for example, argued that the cost of destabilisation in Angola and Mozambique between 1980 and 1988 amounted to close to US $47 billion while losses for the remaining seven (at the time) SADCC member states amounted to about US $17 billion, though these figures were based on somewhat nonchalant counterfactual assumptions.

During the 1990s Stewart (starting with Stewart, 1993) has been at the forefront in formalising this analytical framework. Stewart *et al.* (1997) analyse a sample of sixteen countries, including African ones, in which more than 0.5 per cent of the population died due to direct and indirect war effects during the 1970s and 1980s. This analysis concluded that:

- indirect effects have more significant long-run developmental effects than immediate, direct effects. For example, social, cultural and institutional degeneration have consequences for economic production and transactions and generate lasting negative multiplier effects on investment;
- war costs are more severe where conflict is geographically pervasive and where state collapse is manifest in the loss of core revenue collecting and public good provision capacities;
- international wars are economically less harmful than civil wars;
- the considerable variations in the economic consequences of war may be affected by variations in the duration, intensity and geographical spread of conflict, and whether or not international trade embargoes are imposed, the public sector collapses, and non-state social support networks are available;

• the costs of war also are affected by pre-war features of the economic structure, including the import dependence ratio, the level of subsistence production and the incidence of poverty.

The methodology adopted in Stewart *et al.* (1997) is to compare social and economic performance indicators in conflict-affected countries with both the pre-war period and with non-conflict countries in the same region as conflict-affected countries; and to combine this empirical analysis with an application of Sen's entitlements conceptual framework to guide a disaggregated assessment of which groups are affected in different ways by the effects of war.[6] The original idea of entitlements based in legal production and market exchange is broadened to consider public entitlements, civil entitlements and non-legal acquisition or loss of assets and income during war. Luckham *et al.* (1999) offer a matrix of the mechanisms through which wars affect economies and societies, distinguishing between entitlements versus needs, and direct and indirect (macro, meso and micro level) channels. Within this framework, it is then possible to consider the macro and disaggregated effects of, for example, the tendency in wartime to shift towards shorter-term and more consumption-based economic activities (Collier and Gunning, 1995; Brück, 1997; Addison, 1998) as well as to trace through the impact of conflict on health and education service delivery, market segmentation and regional price differentials and the disruption of rural livelihoods.

Besides these approaches there have been formal attempts to model the impact of war on food production and total agricultural output. Messer *et al.* (1998: 16–21) conduct with–without investigation of actual and 'peace-adjusted' food production in SSA. First, the difference between mean annual food production per capita in war and non-war years is calculated, as a percentage of non-war production averages. Second, the same procedure is followed for mean rates of growth of food production. Then, using both the mean production and rate of growth approaches, estimates are made of what food output might have been in the absence of war and contrasted with actual trends. The findings suggest that in thirteen out of fourteen countries food production was lower in war years, with falls ranging from 3.4 per cent in Kenya to 44.5 per cent in Angola, around a mean annual wartime decline in food production of 12.3 per cent. Food production growth rates in the 1970–93 period declined by as little as 0.3 per cent in Chad and as much as 8.1 per cent in Liberia, with a mean decline of 2.9 per cent per annum. Chad and Uganda present anomalies since food production appears to rise during war years in both.

The FAO (2000) estimates direct agricultural output losses in SSA due to war, with no attempt to include capital losses or indirect effects, for major conflict-affected regions. This is done by specifying a simple model in which the level of agricultural output in the absence of war is determined by relative prices and a trend variable capturing, respectively, short-run farmer decisions and the growth of the labour force together with technical change. Both the level and trend of estimated losses rose after the early 1980s, despite the return to peace in the early 1990s of Ethiopia (a qualified peace) and Mozambique. For the 24 years for which this estimation method and that applied to food production by Messer *et al.* (1998) overlap, the food production approach gives an estimated average loss of total food production of 2.8 per cent per annum, while the FAO estimates losses averaging 4.0 per cent of total agricultural production across all sub-Saharan African countries. Given the difference between food and total agricultural production, these two estimates are fairly consistent in the overall pattern observed. Both, however, are cautious not to claim more than a very rough indication of losses due to conflict in agriculture, due to extreme data unreliability, the difficulty in separating conflict effects from those of weather, health trends (for example,

HIV), and volatile commodity price trends, and the difficulties of separating the effects of conflict on agriculture from the effects of earlier agricultural decline or crisis on the origins of conflict.

Data weaknesses in these models are of more than marginal significance. Brown *et al.* (1992: 200) point out that 'the figures we usually use in analysing and measuring the severity of the crisis . . . are so hopelessly inadequate that they cannot provide a full account of the actual situation and in some cases give a totally wrong impression'. Indeed, the most sensible approach is that taken by Luckham *et al.* (1999: 15), who suggest that they do not intend 'to portray what we know already, which is far too little, but to highlight some of the relationships that deserve more investigation if we are to fill the gaps in our understanding'.

3.3.3 *Critical concerns about war-costing*

It seems reasonable to expect rapid growth in research attempting to fill out with greater and more disaggregated empirical detail the analytical framework developed in these war-costing exercises. But there remain some significant analytical questions regarding this approach. A more critical alternative in making cost-of-war assessments is provided by recollecting the history of their precursors in the nineteenth and early twentieth centuries. These earlier war-costing exercises have been characterised as the classical liberal interpretation, viewing war as exclusively negative in its social and economic consequences: 'Since war was a loss, the best way its effects on the economy could be finally expressed was to calculate that loss as accurately as possible' (Milward, 1984: 11). However, these analyses came under a critical spotlight. Whenever any lingering assumption of full employment is dropped, it is very difficult to assess the economic cost of an individual war casualty. The effect of war on growth is complicated further by the reality of changing technology during a war, as well as changing institutions and shifting social relations. Both direct and indirect costs proved, for example during the First World War, exceedingly difficult to measure. Increasingly, it became clear that the social and institutional changes wrought during war weakened the depiction of war as simply a loss. And there was evidence that some part at least of the extra expenditure (and the diversion of expenditure) on the war effort could be beneficial in generating income by stimulating new employment and production. An example was agriculture in the UK during the First World War, in which both farmers and farm labourers – to different degrees – benefited from wartime price guarantees and other interventions designed to boost domestic production.

Present-day poor countries fighting civil or internal wars differ considerably from industrialised countries fighting international conflicts. Expenditures are diverted to war, war is typically financed by borrowing and by inflation, but there is less productive effort geared to the war – in terms of provisioning the military or replacing imported food by domestically produced food. The differences are greater where internal conflict takes place in a context of state collapse, as for example in Somalia or Liberia. The import and foreign aid dependence of the economy are likely to rise in developing countries at war. Therefore, there are some grounds for thinking that the classical liberal interpretation of the economic impact of war is more appropriate in these cases.

However, first, there remain the difficulties of quantifying accurately the costs of conflict. There are complications from assuming average life expectancies reflecting the social composition of war casualties, and in measuring unemployment and/or underemployment in poor countries. Second, at this stage little research has been done on key aspects of the economics of war in LDCs. There is little empirical detail on the role of war in stimulating production increases and changes in technology and/or organisation of production in some sectors. Indeed, even in a civil war certain rural areas, with military protection, may undergo productive stimulus just as others fall idle. There is little knowledge about the role of war in the emergence of a national capitalist class. Third, little is known about how civil wars affect social relations. This effect is of course variable, between two extremes: one characterised by Sorokin's idea that disaster, including war, was a favourable ground 'for the emergence of radically different social forms' (cited in Milward, 1984: 44); and the other by the impulse to a 'return to normalcy' (Form and Loomis, cited in ibid.). The former effect is commonly associated with the experience of conflict and post-conflict change in, for example, South Korea. The latter is more in line with the assumption in much of the recent literature on conflict in SSA that conflict becomes central to socio-economic path dependence, and hence the inclusion of 'history of conflict' variables in some models (for example, Nafziger and Auvinen, 1997; Collier and Hoeffler, 1996). Fourth, what is the appropriate counterfactual non-conflict scenario against which war losses are to be measured? For example, it might not always be clear that economic losses due to war are greater than in the absence of a war where there may yet be militarisation or poor governance of the society and economy. And what presumptions should be made about the direction and effectiveness of policy in the absence of war?

The main difficulties in cost-of-war exercises are that they presume implicitly a static world in which one can clearly identify benchmarks for measuring loss, and that their accounting framework neglects the economic implications of complex shifts in social relations during a conflict. There is also a problem with choosing the appropriate time scale for analysis. Despite Stewart's (1993) suggestion that the long-run costs of conflict endure after the end of a war, it is also possible for socio-economic changes during war – however unpalatable at the time – to have longer-run progressive consequences. An example might concern the so-called 'war tax'. After a war, when some monopoly trading rent has dissipated, wartime speculators and entrepreneurs may provide a source of productive investment.[7] Evidence suggests, for example, that the post-war national bourgeoisie in Mozambique bought privatised enterprises from the state through the successful exploitation of war-rent (personal interviews). It is also possible that post-conflict infrastructure rehabilitation creates a faster and more comprehensive replacement and upgrading of infrastructure stock than would have occurred otherwise. Further, the liberal interpretation of the economic impact of war implicitly assumes that many war costs derive from market distortions. However, here as in other economic areas it is unwise to presume that divergence from an idealised benchmark of market perfection has exclusively negative implications. Amsden (1997) points out that, contrary to the common assumption that development is a movement closer and closer towards market perfection, development is often driven partly by the emergence of market imperfections where these are value enhancing and favourable to investment and innovation. Wartime 'distortions' in the form of a war tax and monopoly rents to entrepreneurs may, within this perspective, be part of a process that, if channelled effectively after the end of conflict, contributes to development.

Towards the end of the 1990s, the literature on the economic consequences and costs of conflict in Africa and elsewhere exhibited a shift. On the one hand, some analyses have remained under the influence of the basic assumptions of the classical liberal interpretation of war. On the other hand, elements of the critique of this interpretation have begun to be addressed. The insight that some people do well out of a war has been incorporated into analysis of the impact of war. One exploration of this is Collier and Gunning's (1995) discussion of the portfolio preferences of those with wartime savings and on the post-conflict policy options for influencing the reallocation of such savings towards productive investment in, for example, export-oriented manufacturing. Similarly, the application of an entitlements framework allows for exploration of the uneven consequences of conflict. This acknowledgement of wartime accumulation has been influenced by work on the political economy of 'asset transfers' during conflicts and of the role of war in creating famines (Duffield, 1991, 1994; de Waal, 1998; Keen, 1995).

To sum up, there are analytical and methodological tensions in the literature on the economic consequences of conflict in Africa. Assumptions rooted in the classical liberal interpretation of the economic impact of war do not fit comfortably alongside acknowledgement of the economic beneficiaries of conflict and the possibility of long-run positive economic change caused by the experience of conflict. Applying the entitlements framework may help in that it enables a disaggregation of the economic consequences of conflict (Stewart *et al.*, 1997; Luckham *et al.*, 1999). However, the entitlements framework itself is compromised by a tension between neoclassical assumptions of methodological individualism and greater openness to the primacy of social relations (Fine, 1997).

3.4 Theories of violence and the origins of conflict in economics and political economy

In much of the recent literature on war in Africa there is little discussion of underlying theories of violence and conflict. This section briefly reviews earlier strands in development theory regarding conflict and then locates recent analyses, with their focus on scarcity and difference as propagators of conflict, within this theoretical context. It is argued that the majority of recent work on the subject is either 'structuralist' or is rooted in the individualist, rational choice presumptions of neoclassical economics. Both traditions support cross-section, multicountry statistical analysis of conflict and its correlates in which, it is argued, the significance of context and social relations is left out of the analysis. After reviewing briefly some major contributions to a search for empirical regularities, the section goes on to discuss the emerging emphasis on the role of economic agendas in the origin of conflicts in Africa. Next, there is a discussion of the treatment in the literature of major manifestations of difference, that is, ethnicity and inequality. The discussion then summarises alternative ideas about the origins of violence and conflict, and considers their relevance to the problems of understanding war in Africa.

3.4.1 'Development' and conflict

Within the broad schools of development theory since the 1950s, there have been a number of ideas about the sources of social violence, conflict and civil war. These mostly share a view of conflict arising from friction in social relations, and of this friction being generated by processes of development, problems and contradictions in development, and/or the social upheavals brought about by capitalist development. Individual economic motivations – greed for some, survival for others – are typically understood within these theories as bound by, embedded in or flowing from larger social processes and tensions.

Commentators such as Olson (1963) and Huntington (1968) came to see the process of 'modernisation' as itself almost inevitably accompanied by conflict, because of the social upheavals that development created.[8] For example, social grievances arose from the disruption to traditional ways of life that were characterised, for example, by commercialisation of consumption goods and labour. Dependency theorists explained conflicts in terms of arrested development that was imposed on peripheral societies by the core capitalist nations, and in terms of the 'proxy' conflict between core powers, fought out on the terrain of the periphery. Another kind of explanation of conflict was influenced particularly by Tilly's (1975) ideas about the tensions that typically arise from the efforts of central authorities to mobilise resources towards the ends of constructing nation-states. If there is always a reaction against such mobilisation, whether or not this generates armed conflict will depend on the ways in which the central state manages the mobilisation of resources (in an exclusive or a more reciprocal manner) and the way that it manages the reactions provoked.[9] Within this framework, the relationship between democratic institutions and the propensity for conflict is interesting but never straightforward (cf. Stewart and O'Sullivan, 1997). Finally, 'moral economy' analyses of peasant societies disrupted by the introduction and spread of capitalism have generated explanations of peasant radicalism, some focusing more on the internal factors generating crisis and conflict, others more on external provocations (Scott, 1976, 1985; Wolf, 1969; Moore, 1966). Within this framework, Wolf (1969), for example, stressed the role of the group economic interests of middle peasantries in conflict, since they had most to gain from capturing material benefits previously denied them within a particular social structure, but most to lose (compared with poorer peasants) from the costs of conflict.

3.4.2 *Scarcity, difference and shock*

More recent analyses of conflict in developing countries, including those in sub-Saharan Africa, have tended to stress two underlying causes of conflict – scarcity and difference – and their combination. Scarcity is revealed either in stagnation and/or poverty, or in direct resource scarcity, for example, food production, environmental resource degradation and scarcity (Homer-Dixon and Blitt, 1998), or water shortages (Starr, 1991). Difference is typically analysed through either or both of ethnic (or more broadly identity) difference or inequality in the distribution of assets and income. These theories owe much to Gurr's theory of relative deprivation (Carbonnier, 1998). These underlying theories of conflict are structuralist, in that they focus only indirectly on relations between groups, via structural representations or attributes of differentiation and relative deprivation. These structural representations include, for example, use of the Gini coefficient to capture inequality and use of an ethno-linguistic index of collective identity fragmentation, cohesion or polarisation. The more these structural concepts of the social background to conflict are appealed to, the less the real content of specific social relations can be central to the analysis. A consequence is that it becomes easier to lift such concepts as inequality or ethnicity out of their actual context, out of 'effective reality', and to append them, as embodying the 'social', to non-social methodologies.

With a picture of susceptibility to conflict, or 'latent social cleavages' apparently revealed by structural measures, it is then possible to consider the effect on such a system of a major change, for example, an external shock or a shift in policy direction. In the case of Rwanda, for instance, the most obvious external shock was the collapse of international prices for the main export commodities (coffee, tea and tin) during the second half of the 1980s. Rodrik (1998) expands on this idea, analysing the differential impact of external shocks on economic growth under variable measures of latent social conflict and of institutional capacity for conflict management. He finds that external shocks do generate conflict where underlying proclivity for conflict is strong at the same time as the institutions of conflict management are weak.

Alternatively, policy shocks, including structural adjustment, might negatively affect 'latent social cleavages'. There have been various attempts to explore whether structural adjustment policies have an inbuilt tendency to raise the likelihood of conflict (Morrison *et al.*, 1994; Nafziger and Auvinen, 1997; Cramer and Weeks, 1998; Woodward, 1996; Storey, 1999). Their findings vary but it is clear that there is no clear direct causal effect of structural adjustment leading to conflict. Arguably, none the less, there are ways in which structural adjustment policies might raise the potential for conflict. This could happen if sudden and dramatic effects, including devaluation and capital account liberalisation, plunged an already frail society sharply into economic crisis. It could also happen if the effects of stabilisation and adjustment included a shift of material power between either classes or collective identity groups. Hence, there is a common argument that privatisation has the potential to increase political instability by shifting resources from a public sector with a particular ethnic character to a private sector dominated by a different ethnic mix (Mkandawire, 1994).

3.4.3 *The search for regular patterns across countries*

A number of attempts have been made in recent years to find regular patterns relating the behaviour of certain variables derived from such measures to outcomes of conflict or complex humanitarian emergencies. Some of these are relatively eclectic in their modelling approach (for example, Nafziger and Auvinen, 1997), while others (for example, Collier and Hoeffler, 1996) are built on tighter explanatory models. In a quest for causal uniformity, these exercises deprive their subjects of context. None the less, they have generated influential findings, at times at odds with each other. For example, Nafziger and Auvinen focus chiefly on the links between their dependent variables (measures of conflict and complex humanitarian emergencies) and low income (poverty), low growth and food output per capita (stagnation and scarcity), the Gini coefficient (inequality), high inflation, high military expenditure and a history of conflict variable. These are the main variables that do seem to predispose – by correlation – a country to conflict. However, they also stress that together these variables only account for a small proportion of the variance observed in the data. Collier and Hoeffler (1996) conduct econometric analysis in the hope of isolating variables that will predict civil war. Their results confirm the expectations of their underlying model. This model states that civil war is the outcome of a cost-benefit analysis conducted by would-be rebels against a government. Utility is maximised by starting a rebellion if the gains from victory outweigh the costs of coordinating a rebellion and the likelihood that the government will be able to sustain a military effort large enough to contain or put down rebellion. These deciding factors are proxied by strictly economic measurements, except for the cost of coordination, which is measured by means of an index of ethnic fragmentation.

Like most models of civil war this one finds that poverty is significant in the background to conflict, but there are certain outstanding features of this model. In treating difference, via ethnicity, as a predictor of conflict, Collier and Hoeffler argue that a high degree of ethnic fragmentation, by raising coordination costs, is likely to represent a low risk of civil war in a country; on the other hand, a sharp polarisation, say between two ethnic groups, clearly predisposes a country to civil war.[10] The other treatment of difference, or relative deprivation, concerns the role that inequality plays in this model. Contrary to assumptions made in much of the literature on conflict in developing countries, Collier and Hoeffler propose that greater inequality reduces the risk of war. The reason for this does not lie in any social theory of differentiation and conflict, but in a public policy framework. Greater inequality indicates the presence of an elite that would allow the government to raise taxes on its income for a war effort which, if successful, would also preserve the status quo on which the elite thrives.

A number of critical quibbles could be raised about this model and its empirical basis; not least the questionable assumption that armed conflict in developing countries begins with rebellion against the government. This over-simplification derives from the common idea that violence and conflict involve an aggressor and a victim, the two being more or less separate, rather than being tied together in a mutual and escalating process of violence. However, there are two striking features of this approach; first, that the model is built on the foundations of methodological individualism and rational choice that drive most mainstream economic analysis; the methodological implications of this are considered further below. The second is that the authors are resolutely unmoved by common traditional assumptions that conflict is a function of relative deprivation, grievance or inequity. Instead, they propose that conflict is a function of the same basic human motivation that dominates mainstream economic notions of human behaviour, that is, individual greed or self-interest.

3.4.4 Economic agendas in the origins of conflict in Africa

Indeed, there has been an increasing emphasis in the recent literature upon the leading role of economic factors in the origins of conflict – such as poverty and inequality – and also on the role of economic motivations as the prime cause of war. This theme, however, has been treated in slightly different ways, not all relying so heavily on neoclassical economics. There is a widespread argument that war in Africa has become increasingly apolitical (PRIO, 1999) and that politics, ideology, ethnicity and so on have been replaced by straightforward power contests motivated chiefly by the pursuit of economic agendas. Some writers, such as Keen (1995), Duffield (1994) and Reno (1998), have stressed the role of civil war in enabling 'asset transfers' and monopoly over mineral resources, as well as in enabling basic survival for many, but have done so within specific historical, political and socio-economic contexts. These tend analytically to focus on specific and shifting contexts of social relations between identity groups, political groups, and classes. From this perspective it is possible to assess why qualitative breaks occur in the historical record, that is, why wars have broken out at a particular moment in certain countries.

A related analysis is that of Collier (1999), which again ties the insight that there are economic factors involved in the origins of and motivations for conflict to a rigorous neo-classical framework. Arguably, what this analysis gains in the elegance of its formal model it loses in social understanding. Collier states that wars in developing countries have become less ideological and are principally fuelled not by grievance but by greed, the root individualist assumption of all orthodox economics. This is contrary to arguments such as those of Wolf (1969) or Homer-Dixon (1991, 1995) that stress grievance arising from the mix of resource scarcity with social inequality and human rights abuses, but that argue that 'violent struggles arise as much from *perceptions* of unfairness as from absolute shortages' (Messer *et al.*, 1998, my emphasis).[11] Collier's (1999: 1) finding is that 'group grievances beneath which inter-group hatreds lurk, often traced back through history' are not good predictors of conflicts. Instead, economic agendas are important and economic opportunities are far more likely than social or group grievances to cause wars.

The model designed to capture this distinction in causality proxies economic agendas through measures of the share of primary commodities in GDP, the proportion of young men in the population, and average years of education. Primary commodities signal economic opportunities chiefly because they are readily 'lootable' or liable to predatory taxation. The more prevalent are young men in a society, the lower the cost of recruiting rebels. And this cost is lowered also if there are few peaceful income-earning opportunities for these young men, which may be captured by the educational attainment proxy. Grievances may be captured – similarly to Rodrik's (1998) model of the effects of external shocks on growth in the presence of latent social conflicts – by indices of ethnic and religious fragmentation, inequality in land ownership, an index of political rights, and the rate of growth of per capita income. Collier's statistical results suggest that some societies are much more prone to conflict 'simply because they offer more inviting economic prospects for rebellion' (1999: 5). On the other hand, inequality does not matter at all, and political repression and a high degree of ethnic and religious fragmentation have the opposite of their predicted effects.

Collier argues in support of his findings that since justice, revenge and relief from grievance are public goods, they are subject to free-rider problems that are a disincentive to rebellion. Further, people are unwilling to fight for a cause unless they are convinced that the rebellion will succeed, hence initially rebellions face a coordination problem. There is also a time-consistency problem in that potential recruits can recognise that a leader who promises to assuage grievances may, once in power, turn out not to deliver. Yet a greed-motivated rebellion, by restricting the benefits to the participants, avoids free-rider problems; by enabling the sporadic or territorially restricted predatory taxation of primary commodities, such rebellions avoid the need to assume final victory; and for the same reason, rebel recruits can be paid off from primary commodity rents instantly, so escaping the time-consistency problem (Collier 1997: 7). This approach explains the social purely on the basis of market theory, explicitly in the case, for example, of free-rider problems, and implicitly in the assumption that wars emerge from the aggregation of individuals' rational utility maximising choices.

There is little sense that conflicts may feed on both self-interest and concern for the public good at the same time, nor an acknowledgement that the complexity of cause and motivation cannot adequately be captured by the 'grievance' variables, which, as empirically questionable measures of attributes of difference or stratification, are crude tools for capturing social relations in their diversity. Furthermore, as Keen (1998) suggests, some economic motivations for participating in conflict, and indeed for perpetuating conflict, may only become significant once a conflict has already begun rather than being the prior deciding factor in starting a rebellion. Nor is it easy to separate motivations or short-term interests that pull people together behind a programme of violence: money, food or clothing represent one form of interest, but are probably tied closely to coercion and fear (Stewart, 1998: 29). Further, in contrast both to Collier (1999) and to Reno (1998) and others, it can be argued that politics has not been abandoned. Atkinson (1996), for example, argues that it is necessary when analysing Liberia to distinguish between different time periods of conflict and also between different warring factions. Further, Atkinson suggests that for some groups, in particular Taylor's NPFL, combined successful exploitation of war economy conditions with a genuine political agenda and pursuit of legitimacy. Indeed, not even the organisation of the war economy has been particularly different from pre-conflict informal and formal trading activities, in many cases. There is, therefore, a danger that the insight into economic motivations becomes too reductionist, and simply replaces one simplification – the tribal or cultural explanation, as for example, in Kaplan (1994) – with another, the economic opportunism explanation.

3.4.5 *Accommodating the social*

There is a tension in much of the literature between the rational and the irrational, or the rational and the social, centred on the role of ethnicity or, more broadly, collective identities.

On the one hand, there is a notion that modern conflicts manifest in Africa pre-modern, ancient ethnic rivalries – a view commonly held of the Tutsi–Hutu relationship in Rwanda, for example. On the other hand, ethnic differences are brought into 'rationalist' economic models, typically on the assumption that ethnic fragmentation or polarisation is a structurally functional sign of animosity, rivalry and weak social trust or social capital. For Stewart, for example, structural attributes of difference in a society do matter in the origin of conflicts, but less so the horizontal differences captured by measures such as the Gini coefficient, while more important are vertical inequalities such as religious or ethnic distinction (Stewart, 1998).[12] The tension partly reveals a methodological problem. Most orthodox analysis of the political economy of conflict in Africa is based on methodological individualism, but at the same time assumes the significance of collective groupings such as ethnic identities, without fully working through the ways in which this might be incorporated into individualist rationality.[13] To put this another way, most of the economic and political economy literature suggests that war in Africa makes sense, but it is not clear exactly what form of rationality, or making sense, is at work. In some of these models, notably in Collier and Hoeffler (1996) and Collier (1999), the social – manifest in forms of collective identity – is introduced into the model but excluded from the underlying assumptions of the model, which are neoclassical and individualist. What emerges is a combination of neoclassical utility maximisation assumptions with social concepts, for example, ethnicity, stripped altogether of their history and context. This analytical development is common to much recent orthodox work in economics, for example in the explosion of work on 'social capital', and is arguably a facet of the increasing colonisation by mainstream economics of the social sciences (Fine, 1999a and b).

Given the analytical and methodological tensions in this approach to the economics of conflict, future work may take two different directions. One would be signposted by the asymmetric information paradigm that has guided much of the recent literature on social capital, institutions and growth theory puzzles, which owes much to the work of Stiglitz (1994). If this route is taken, future studies of the economics of conflict in Africa and elsewhere may focus theories of ethnicity and collective identity that are derived from individual rationality in the presence of multiple information asymmetries and gaps. Arguably, work in this vein might also provide a means of investigating the aggregation of different kinds of motivation into an active military force. For one of the weaknesses of much of the literature on war in Africa is the lack of understanding of relations between those organising a rebel or state military effort and those, with highly variable degrees of 'choice', joining the war effort.

The alternative direction for economic or political economy analysis of conflict would be to abandon analytical frameworks that begin with the individualist, non-social assumptions of neoclassical economics and that consequently consider the economic and the social as separable spheres. Instead, there could be greater commitment to the premise that economic relations, behaviour and performance are organically embedded in the social and the political.[14] From this perspective, scarcity, poverty and economic or environmental crisis are themselves to be understood as social events. In this vein, Messer *et al.* (1998: 10–11) review succinctly a variety of resource conflicts embedded in ideological contests in different regions of the world. On the case-study level, Keen (1995), for example, argues against the economism of Sen's entitlements approach to understanding famine and shows how power relations and war generated and exploited famine in Southern Sudan.[15]

What the majority of mainstream economic analyses of conflict in Africa have in common is a functionalist framework. Given structural attributes of difference such as a particular ethno-linguistic index score, or a particularly high Gini coefficient, a propensity to conflict can be more or less read off, especially when combined with the effects of measures of institutional development and trends in economic performance variables such as GDP growth, terms of trade shocks and so on. Such an approach fits closely the work done in recent years that has tried to link the so-called new political economy with endogenous growth theory, generating cross-country sample-based arguments about the relationship between political instability and growth (for example, Alesina and Perotti, 1993) or income inequality and growth (Alesina and Rodrik, 1994). The most obvious problem with such an approach stems from the neglect of relations pertaining or evolving in specific historical contexts. Thus the descriptive measure of difference is taken to have causal powers irrespective of the actual social relations that generate and sustain such outward observations of difference. Ethnicity in Africa, for example, has very different social and historical meanings, and precise definitions of ethnic difference, with comparable value across the region, are elusive (McIlwham, 1998). Similarly, aside from the descriptive difficulties in measuring inequality precisely in SSA, little in the literature has been influenced by Sen's (1992) argument that a descriptive measure of a category like inequality can have very different social significance in different contexts. As a result, analyses stressing the role of such difference-attributes in the origin of conflict may pick up something, but only indirectly, and cannot, for example, account for where in a continuum of stratification difference begins to produce a qualitative break in terms of causal impact (Wood, 1995: 76).

This approach, inferring the potential for war from functional attributes of difference or indicators of scarcity, is particularly suitable for cross-country comparison and the methodological predilection in economics for probabilistic prediction. However, this analytical framework suggests that wars in Africa, or in developing countries generally, have the same causes. Yet a cursory return to one of the themes of section 3.2 – the diversity of war experiences and types in Africa – suggests that it is not necessarily sensible to assume that we are comparing like events.

3.4.6 Rethinking violence and conflict

It is possible that further research into the origins of violence and armed conflict in the region – and elsewhere – will open up new avenues of economic analysis suitable, perhaps, to methodologies less dependent on the axioms of neoclassical economics. Here, rather than claiming a superior or complete theory of violence and war, it is worth at least introducing ideas that lead in a slightly different direction from the simple assumptions that scarcity and difference are functional generators of conflict. The work of Girard (1977, 1996) is particularly suggestive and, though based on the history of early societies and the role, particularly, of religion in containing violent relations between people within a society, it may have some bearing on studies of development and conflict. In particular, Girard challenges the conventional assumptions of violence rooted in deprivation and difference, as well as challenging the 'myth of the individual' (in his case with reference to pyschoanalytical theory).

Briefly, at the core of Girard's work is the argument that the mimetic impulse, the need to copy a model other, is fundamental to human behaviour. This impulse leads to the need to own or appropriate those attributes and objects associated with the model, setting up an inherent potential for rivalry and conflict and generating complex reciprocal relations and the scope for mimetic escalation that can, of course, become violent (Girard, 1996). Conflict then arises not from the chance clash of two or more people with independent needs for a given object, but directly out of social relationships; and scarcity itself is then a product of the way that these relationships are framed. Social institutions – for example, religion or modern legal systems, and social hierarchical distinctions – work to contain the scope for violent conflict within this framework: partly by restricting the scope for appropriation and partly by displacement of violence, often on to scapegoating rituals that act as a legitimising vent for violence and a unifying institution. Ehrenreich (1997) complements this idea with her argument that the social significance of war lies in the roots of human fears of predation, and in the historically entrenched ability of war or organised, legitimised violence to unite societies. These ideas are efforts to acknowledge underlying tendencies that may commonly be at work in human societies, but they do not generate laws in the sense of predictable event-regularities. Rather the degree and forms in which they emerge in observable outcomes will clearly vary with particular historical contexts and developments. A further point, in Girard, is that social differences and distinctions themselves do not inevitably produce violent conflict but that, to the contrary, the collapse of established differences, distinctions or hierarchies is a more common spur to violent conflict. Within Girard's analytical framework this is because the erosion of differences releases the scope for uncontained mimetic rivalry and conflict.

Transferred to modern societies and the development literature, this may have some relevance. Mimesis is central nationally and internationally to Gerschenkron's (1962) argument about the significance of lateness in industrialisation, since in late industrialisation there is a tension between the observable benefits of industrialisation and the obstacles to securing these benefits. Mimesis is equally central to Anderson's (1983) argument about how the nation-state became a model for political legitimacy internationally. Furthermore, Girard's focus on the erasure of difference and not difference itself as producing violent conflict fits the observation that the spread of capitalist relations, and of social change generally, is associated with the breakdown of prevailing social structures and hierarchies and, indeed, with conflict.[16] This is closer to the arguments reviewed above in the earlier development theory literature, suggesting that conflict may be integrally tied to the processes of capitalist development and their implications for social upheaval and change, whether or not that change is progressive.

However unpalatable, and against the grain of late twentieth century notions of harmonious, inclusive development (Sen, 1997; Wolfensohn, 1999) and the idea that instability is bad for development, it may be that conflict commonly accompanies progressive change, that much that human societies hold dear, as Girard puts it, is founded in violence. The most renowned version of this idea is Heraclitus's remark that war is the father of all things. Most modern societies have emerged from violent conflict, too. The literature on convergence of growth rates, for example (Pritchett, 1997; Jones, 1997; UNCTAD, 1997), tends to start with data on the first industrialising nations from the 1870s and tends to neglect the precursor of conflict to such take-offs – for example, the Prussian wars and the process of German unification, and the American civil war. As Gourevitch notes (1999: 331), Yoweri Museveni is 'a student of how the great democracies emerged from political turmoil, and he recognised that it did not happen quickly, or elegantly, or without staggering setbacks and agonising contradictions along the way . . .'. More recently still, countries that, since 1950, have managed rapid industrialisation combined with reasonably low economic inequality have been characterised by violent conflicts unleashed by the possibility of changing highly unequal social structures; South Korea, Taiwan and Greece, for example (Bowman, 1997).

To put this another way, in contrast to institutional functionalism that presumes a given form of institution will work across all contexts, it could be argued that effective conflict-management institutions are unlikely to emerge in any society without the experience of conflict of one kind or another.[17] As Hirschman (1995) argues, social conflict is inevitable, every change throws up new conflicts, and the mark of successful societies is their management of conflicts rather than the lack of them. Of course, conflict does not always generate progressive institutional and political developments, but it is equally clear that having a history of conflict does not inevitably consign a country to perpetual repetition of violence. Hirschman reviews the intellectual history – from Hesiod and Heraclitus through to twentieth century sociologists and political scientists – of the idea that conflict can have progressive social consequences. In particular, he stresses the argument that democracy (as one set of conflict-management or conflict-tending institutions) does not often appear from a prior universal acceptance of 'basic norms' but only after lengthy periods during which different groups that have been at each others' throats eventually recognise that unilateral dominance cannot be secured by either. In SSA, if the recent history of Angola and Rwanda is one of face-off and the belief in total domination, there are none the less signs of this emergence of greater democracy from the experience of conflict. One example is Mozambique, where there was no democratic tradition but where parliamentary democracy has fared reasonably successfully since the end of war in 1992 and elections in 1994. Amongst the greatest challenges for further research into the political economy of conflict in SSA is this need to investigate the diverse experiences of institutional change in the wake of war. Now that the classical liberal interpretation of war – with its assumptions of war as exogenous and wholly negative in its socio-economic effects – has been overtaken by more nuanced analyses of the dynamics of wartime economies and societies, this area of research could expand productively in coming years.

3.5 Conclusions and policy considerations

The literature on the economics and political economy of conflict in Africa has bloomed during the past decade. This literature has shown that the disciplines of economics and political economy are highly relevant to efforts to understand, respond to and even possibly prevent outbreaks of armed conflict that are costly in humanitarian and economic terms. There have been considerable advances in our understanding of the ways in which conflict pervades economy and society. There have been advances in our understanding of the differential experience of internal war, including the increasing focus on those who are the economic (and not just political) beneficiaries of conflict. This chapter, however, has argued that there is not a consensus in the literature, and that there is a vital need for more research and analytical work. There are two aspects to this. First, we still do not know enough about the origins and dynamics of conflict and, even more importantly, about the diversity of conflict origins, experiences and effects across SSA. The single greatest need here is for further individual and comparative case-study research. Second, this chapter has argued that the literature under review abounds with methodological tensions that are not sufficiently exposed and confronted. There is, therefore, a need for more work at the methodological level in terms of framing coherent analyses, generating alternative frameworks and so on.

The different methodologies, analytical assumptions and research emphases throughout the literature feed into the wide range of policy implications considered in the literature. Basically, what goes into a conflict analysis by way of assumptions and theory tends to result in particular policy implications. For example, Stewart (1998) is concerned to show that horizontal inequality is especially important in the causes of war and, as a result, recommends that the IMF and the World Bank pay more attention than they do to the implications of their reform programmes for horizontal inequality. In particular, she argues that IMF/World Bank programmes should not obsessively focus on cutting back state resources and spending, since these are a critical means of increasing horizontal equality. On the other hand, Collier (1999) builds on his assumptions about the nefarious effects of market distortions in providing benefits for wartime speculators and recommends that an effective way to prevent wars and in their aftermath to weaken the hand of those with a stake in perpetual conflict is to reduce monopoly profits, through market deregulation. More broadly, the literature is often divided between those who consider standard IMF/World Bank policy reforms theoretically sound and those who consider them fundamentally flawed and especially inappropriate to conflict-prone or post-conflict scenarios. None the less, there are areas where analyses constructed on different assumptions generate more or less equivalent policy recommendations. For example, many conflict analyses focus on poverty or under-development as a cause of conflict and therefore seek to prevent conflict by accelerating structural change and broad or inclusive economic growth. This would include diversification from traditional concentration on raw primary commodity exports. For those approaching the subject from a different angle, much the same conclusion might be made. Thus, for example, Collier (1999) argues that conflicts are especially associated with a high proportion of primary commodities in total exports and that, as a result, conflict prevention should include policies to advance diversification of exports.

Most analyses distinguish in policy terms between:

- policies aimed at prevention of conflict by tackling the 'underlying causes' of conflicts;
- policies reducing the vulnerability to conflict by avoiding pressure on potential 'trigger' or 'permissive' elements or policies;
- policies focused on post-conflict recovery, including methods for reabsorption of refugees and demobilised soldiers and the question of whether or not these should include 'targeting' policies or more non-discriminatory, inclusive support programmes (Cramer and Weeks, 1997);
- policies to reduce the individual incentives to lead or participate in a rebellion;
- policies at the international level to provide political weight to peace settlements, to reduce the resourcing of wars, or to reduce the scope for international aid to fuel conflict.

One of the most common concerns of policy recommendations in this literature is with the need for early-warning systems capable of increasing awareness of the potential for imminent conflict. There are difficulties here though. An early-warning system presumes understanding of the causes of conflict. To the extent that our knowledge of the causes of conflict remains fuzzy, such prediction systems may be rather unrefined and imperfect. Further, to the extent that the causes of conflict vary enormously, there is a danger in building international, institutionalised early-warning systems that have a tendency to favour a particular set of causes as more or less common to all conflicts. Further, there may be a danger of a 'cry wolf' syndrome emerging: since, arguably, countries in internal conflict are not massively different from others but are exaggerated reflections of the tensions and social conflicts common to virtually all developing countries, characterised by late industrialisation and fragile 'nation-ness', then a conflict early-warning system might be triggered like an inner-city car alarm at the slightest rumble or whiff of wind. There is also an international governance or institutional issue here: for the tangle of prediction failure, standard external agendas and international politics that helped allow the Rwandan genocide to happen makes it clear that international sensitivity to the potential for conflict needs to be managed in as open and politically independent a manner as possible.[18] Finally, there is a tension here that is more broadly relevant throughout the literature and its policy implications. This is between the quest for generalised statements and quantitative event-regularities and the preference for what Luckham *et al.* (1999) call more differentiated and contextual conflict analysis.

Most of the literature favours, for obvious reasons, fairly general political or institutional reforms to make SSA countries more democratic. For some, the emphasis needs to be on the role of external guarantors of peace settlements, to help entrench democratic compromise (for example, Collier, 1999, or Boyce, 1996, on 'peace conditionalities'). For others, it is important to recognise that representative majority democracy may not be the most effective institution for 'inclusive' politics, and that varieties of proportional representation and power-sharing might be more effective in countries where the institutionalised exclusion or repression of minorities has been important to the origins of conflict (Stewart, 1998). Perhaps it could also be argued that international bodies should resist the temptation to foist their own preferred institutional settlement on a country, and instead should acknowledge that the most effective conflict-management institutions will be those that are designed precisely by parties that have been in conflict and that are relevant to the particular context.

The literature has focused to an increasing extent on the economic agendas that support or perpetuate conflict and on the way that conflict can become socially and institutionally entrenched. Consequently, there is considerable attention to policies that might help prevent this or bring it to a close. These policies range from rehabilitation of the judiciary, to power-sharing and pay-offs for demobilised soldiers, to the promotion of more competitive markets that offer less monopoly profit attraction to would-be warlords, and recommendations that international measures be taken to stop the flow of resources that fund war efforts, including diamonds (Angola and Sierra Leone), from illegitimate trading networks into the more centralised, legitimate networks. More awkwardly, there may be a need to accept that monopoly is not always developmentally unsound, and that the critical issue after a conflict, in which wartime accumulation via monopoly has taken place, is not necessarily to evaporate monopoly profits but to manage the state–private sector relationship so as to harness, in some cases, the socially progressive potential latent in monopolies.

There are three points that should be made to conclude this review. First, for analytical and policy purposes, there is an urgent need for far more research, and research conducted through the application of different methodologies, into the origins, dynamics and effects of conflict in SSA. In particular, it is important to focus on institutional change arising from social tension or conflict in Africa: research should explore this in the context of the management of conflicts before they become armed warfare, as well as where there have been relatively successful transitions to peace and where there appear to have been failures of institutional development. Second, this chapter has suggested that it is not just difference or differentiation *per se* that is significant in the build-up to conflict, but the particular economic and social relations underpinning outward attributes of difference and also the end or threat-ened end of a given social structure of differentiation. Finally, again at the international level, helping to prevent wars in Africa may involve extraordinarily difficult political judgements, given the implications for virtually any aid programme for the balance of political and material interests in a country. These judgements should not be shirked; rather, they should be acknowledged and thought through in as open a manner as possible.

Notes

1 Stewart (1998: 2) argues that there can be no simple generalisations and it should no longer be possible 'to state as many do that conflict is inevitable because of primordial ethnic divisions, nor that it is the outcome of underdevelopment and that policies to combat low incomes and poverty will also automatically reduce the risk of conflict'.

2 Violent crime cannot always be clearly distinguished from political violence either, as, again, in the case of South Africa and the involvement of political groups in apparently 'criminal' violence (Ellis, 1999). On the other hand, some forms of investment may be relatively immune to political violence, as in the oil sector in Angola (see also, on Nigeria, Frynas, 1998).

3 See Samuel Johnson's imitation of Juvenal's tenth Satire:

> Yet Reason frowns on War's unequal Game,
> Where Wasted Nations raise a single Name,
> And mortgag'd States their Grandsires' Wreaths regret,
> From Age to Age in everlasting Debt.
>   (In Milward, 1984: 10)

4 Stewart (1993) makes it clear that such 'meso-level' spending diversions are not always evident: for example, in some countries health spending is stable or rises slowly while military expenditure increases sharply.

5 On the war tax in Mozambique see, for example, Castel-Branco and Cramer (forthcoming).

6 For a case study approach, see O'Sullivan (1997) on Sri Lanka.

7 See also Collier and Gunning (1995) on the portfolio preferences of wartime accumulators.

8 A lengthier review of the ideas of development theory on conflict is in PRIO (1999).

9 For an analysis of the origins of war in Angola and Mozambique drawing on this approach, see Cramer (1994).

10 One of the more notorious outings for this ethno-linguistic fragmentation index is in Easterly and Levine (1997), where the ethno-linguistic variable, ETHN, is strongly correlated with public policy variables including the black market premium, poor financial development, poor infrastructure and low education, and combined with a small direct ethnicity effect; combined, the ethnic-related variables account for half the difference in growth rates between Africa and East Asia. Besides a range of analytical problems with this model, it could be pointed out that among the nine SSA countries that count as monolingual (on the judgment that 90 per cent plus of the population speaks one language) are countries typically associated with ethnically fuelled war – including Somalia, Burundi and Rwanda (McIlwham, 1998).

11 On perceptions of inequity as opposed to just objective measures of inequity, see also Stewart (1998: 28–9).

12 The same idea has been put forward by the Ugandan president, Yoweri Museveni, as reported in Gourevitch (1999: 330).

13 More generally, as Storey (1999: 43–4) argues, there is little consensus about the relationship between economic and ethnic factors. Horowitz (1998) argues that ethnic division of labour may, by reducing direct ethnic competition, work as a 'shield' against ethnic conflict rather than a 'sword' sharpening such conflict. Others suggest that ethnic distinctions in the division of labour, for example, the distribution of employment in the public and private sectors, may make a society vulnerable to conflict when a shock or policy change (such as privatisation) has sectorally differentiated consequences ('Bayo Adekanye, 1995, in Storey, 1999; Mkandawire, 1994: 209–10).

14 On the contrast between external and internal relations or between atomism and organicism in conceptualising the nature of social reality, see Lawson (1997: 166). According to external/ atomistic concepts, reality consists of externally related entities so that all things exist and act in ways that are quite independent of any relationships in which they stand; while concepts of internally or organically related social reality have in mind that the essential aspects of any particular entity can only be determined from a knowledge of the relationships in which it stands.

15 See also Richards (1996) and Reno (1998) on Sierra Leone.

16 Violent conflict can also offer a means of escaping prevailing hierarchies – for example, see Kriger (1992) on the way that joining the liberation struggle in Zimbabwe was often an attractive means of escaping the oppressive stagnation and rule by elders in rural society, rather than some straightforward motivation of greed or the expectation of the gains from victory.

17 For a critique of functionalism in the new institutional economics, see Khan (1995) and, for example, on the difficulties in defining and measuring corruption, Khan (1999).
18 See the Joint Evaluation of Emergency Assistance to Rwanda (1996).

References

Addison, T. (1998) 'Underdevelopment, transition and reconstruction in Sub-Saharan Africa', *Research for Action*, no. 45, Helsinki: UNU/WIDER.

Alesina, A. and Perotti, R. (1993) 'Income distribution, political instability and investment', NBER Working Paper no. 4486, Cambridge, MA: National Bureau of Economic Research.

Alesina, A. and Rodrik, D. (1994), 'Distributive politics and economic growth', *Quarterly Journal of Economics*, CIX (436): 465–90.

Amsden, A. (1997) 'Bringing production back in: understanding government's economic role in late industrialization', *World Development*, 25 (4): 469–80.

Anderson, B. (1983) *Imagined Communities: Reflections on the Origin and Spread of Nationalism*, London: Verso.

Atkinson, P. (1996) 'Liberia: war economy or political economy?', mimeo, London: London School of Economics.

Austin, G. (1996) *The Effects of Government Policy on the Ethnic Distribution of Income and Wealth in Rwanda: A Review of Published Sources*, London: Department of Economic History, London School of Economics.

Azam, J.-P., Bevan, D., Collier, P., Dercon, S., Gunning, J. and Pradhan, S. (1994), 'Some economic consequences of the transition from civil war to peace', Policy Research Working Paper no.1392, Washington, DC: Policy Research Department, World Bank.

Bowman, K. (1997) 'Should the Kuznets effect be relied on to induce equalizing growth? Evidence from post-1950 development', *World Development*, 25 (1): 127–43.

Boyce, J. (ed.) (1996) *Economic Policy for Building Peace: The Lessons of El Salvador*, Boulder, CO and London: Lynne Rienner Publishers.

Brown, S., Schraub, J. and Kimber, M. (eds) (1992) *Resolving Third World Conflict: Challenges for a New Era*, Washington, DC: United States Institute of Peace Press.

Brück, T. (1997) 'Macroeconomic effects of the war in Mozambique', QEH Working Paper, QEHWPS11, Oxford: Queen Elizabeth House.

Carbonnier, G. (1998) 'Conflict, postwar rebuilding and the economy: a critical review of the literature', War-torn Societies Project (WSP) Occasional Paper no. 2, Geneva: UNRISD.

Chingono, M. (1995) *The State, Violence and Development: The Political Economy of War in Mozambique*, Aldershot, Hants.: Avebury.

Colletta, N., Kostner, N. and Wiederhofer, I. (1996) *The Transition from War to Peace in Sub-Saharan Africa*, Washington, DC: World Bank.

Collier, P. (1999) 'Doing well out of war', paper prepared for Conference on Economic Agendas in Civil Wars, London, 26–27 April.

Collier, P. and Gunning, J. (1995) 'War, peace and private portfolios', *World Development*, 23 (2): 233–41.

Collier, P. and Hoeffler, A. (1996) *On Economic Causes of Civil War*, Oxford: Centre for the Study of African Economies.

Cramer, C. (1994) *A Luta Continua? The Political Economy of War in Angola and Mozambique*, PhD thesis, Faculty of Economics and Politics, University of Cambridge.

Cramer, C. (1997) '"Civil war is not a stupid thing": exploring growth, distribution and conflict linkages', Department of Economics Working Paper no.73, London: School of Oriental and African Studies, University of London.

Cramer, C. and Castel-Branco, C. (forthcoming) 'Privatisation and the challenges of underdevelopment, transition and reconstruction in Mozambique', paper for UNU/WIDER project on 'Underdevelopment, Transition and Reconstruction in Africa', Helsinki: UNU/WIDER.

Cramer, C. and Weeks, J. (1997) *Analytical Foundations of Employment and Training programmes in Conflict-affected Countries*, ILO Action Programme on Skills and Entrepreneurship Training for Countries Emerging from Armed Conflict, Geneva: ILO.

Cramer, C. and Weeks, J. (1998) 'Adjusting adjustment for complex humanitarian emergencies', paper for UNU/WIDER project on 'War, Hunger and Displacement: the Economics and Politics of the Prevention of Humanitarian Emergencies', Helsinki: WIDER.

De Haan, A. and Koch Laier, J. (1997) 'Employment and poverty monitoring', Issues in Development Discussion Paper no.19, Geneva: ILO.

De Waal, A. (1997) *Famine Crimes: Politics and the Disaster Relief Industry in Africa*, London: James Currey.

De Waal, A. (1998), 'Contemporary warfare in Africa', in M. Kaldor and B. Vashee (eds), *New Wars*, London: UNU/WIDER.

Duffield, M. (1991) *War and Famine in Africa*, Oxford: Oxfam GB.

Duffield, M. (1994) 'The political economy of internal war: asset transfer, complex emergencies and international aid', in J. Macrae and A. Zwi (eds), *War and Hunger: Rethinking International Responses to Complex Emergencies*, London: Zed Books.

Easterly, W. and Levine, R. (1997) 'Africa's growth tragedy: policies and ethnic divisions', World Bank Research Working Paper, Washington, DC: World Bank.

Ehrenreich, B. (1997) *Blood Rites: Origins and History of the Passions of War*, London: Virago Press.

Ellis, S. (1999) 'The new frontiers of crime in South Africa', in J-F. Bayart, S. Ellis and B. Hibou *The Criminalization of the State in Africa*, International African Institute with Oxford: James Currey, and Bloomington: Indiana University Press.

FAO (2000) 'Conflicts, agriculture and food security', in *The State of Food and Agriculture: Lessons from the Past 50 Years*, Rome: FAO.

Fine, B. (1997) 'Entitlement failure?', *Development and Change*, 28 (4): 617–47.

Fine, B. (1999a) 'The World Bank and social capital: a critical skinning', mimeo, London: Department of Economics, School of Oriental and African Studies, University of London.

Fine, B. (1999b) 'A question of economics: is it colonizing the social sciences?', *Economy and Society*, 28 (3): 403–25.

Fischer, D. and Schwartz, R. (1992) 'Economists and the development of peace', in M. Chatterji and L. Forcey (eds), *Disarmament, Economic Conversion and the Management of Peace*, New York: Praeger.

Frynas, J.G. (1998) 'Is political instability harmful to business? The case of Shell in Nigeria', University of Leipzig Papers on Africa, Politics and Economics no.14, Leipzig: Institut für Afrikanistik.

Furley, O. (ed.) (1995) *Conflict in Africa*, New York and London: Tauris Publishers.

Gerschenkron, A. (1962) *Economic Backwardness in Historical Perspective*, Cambridge, MA: Harvard University Press.

Girard, R. (1977) *Violence and the Sacred*, Baltimore, MA: Johns Hopkins University Press.

Girard, R. (1996) *The Girard Reader*, ed. J.G. Williams, New York: Crossroad Publishing Company.

Gourevitch, P. (1999) *We Wish to Inform You that Tomorrow We Will Be Killed with Our Families: Stories from Rwanda*, London: Picador.

Green, R.H. (1991) 'Neo-liberalism and the political economy of war: sub-Saharan Africa as a case study of a vacuum', in C. Colclough and J. Manor (eds) *States or Markets? Neo-liberalism and the Development Policy Debate*, Oxford: Clarendon Press,.

Green, R.H. (1994) 'The course of the four horsemen: costs of war and its aftermath in sub-Saharan Africa', in J. Macrae and A. Zwi, with M. Duffield and H. Slim (eds), *War and Hunger: Rethinking International Responses to Conflict*, London: Zed Books.

Gurr, T. (1970) *Why Men Rebel*, Princeton, NJ: Princeton University Press.

Hirschman, A.O. (1995) *A Propensity to Self-Subversion*, Cambridge, MA: Harvard University Press.

Homer-Dixon, T. (1991) 'On the threshold: environmental changes as causes of conflict', *International Security*, 16 (2): 76–116.

Homer-Dixon, T. (1995) 'The ingenuity gap: can poor countries adapt to resource scarcity?', *Population and Development Review*, 21: 587–612.

Homer-Dixon, T. and Blitt, J. (eds) (1998) *Ecoviolence: Links Among Environment, Population and Security*, Oxford: Rowman and Littlefield.

Horowitz, D. (1998) 'Structure and strategy in ethnic conflict: a few steps towards synthesis', paper presented at the Annual World Bank Conference on Development Economics, Washington, DC.

Huntington, S. (1968) *Political Order and Changing Societies*, New Haven, CT: Yale University Press.

Jones, C.I. (1997) 'On the evolution of the world income distribution', *Journal of Economic Perspectives*, 11 (3): 19–36.

Kaldor, M. and Vashee, B. (eds) (1997) *New Wars*, London: UNU/WIDER.

Kaplan, R.D. (1994) 'The coming anarchy: how scarcity, crime, overpopulation, and disease are rapidly destroying the social fabric of our planet', *Atlantic Monthly*, February: 44–76.

Keen, D. (1995) *The Benefits of Famine: The Political Economy of Famine and Relief in Southwestern Sudan, 1983–89*, Princeton, NJ: Princeton University Press.

Keen, D. (1998) 'The economic functions of violence in civil war', Adelphi Paper 320, Oxford: Oxford University Press.

Khan, M. (1995) 'State failure in weak states: a critique of new institutionalist explanations', in J. Hunter, J. Harriss and C. Lewis (eds), *The New Institutional Economics and Third World Development*, London: Routledge.

Khan, M. (1999), 'The new political economy of corruption', paper for SOAS series of critiques of the Post-Washington Consensus, London: Department of Economics, School of Oriental and African Studies, University of London.

Kriger, N.J. (1992) *Zimbabwe's Guerrilla War: Peasant Voices*, Cambridge: Cambridge University Press.

Lawson, T. (1997) *Economics and Reality*, London: Routledge.

Luckham, R., Ahmed, I. and Muggah, R. (1999) 'The impact of conflict on poverty in sub-saharan Africa', background paper for World Bank Poverty Status Assessment for Sub-Saharan Africa, Sussex: Institute of Development Studies.

McIlwham, F. (1998), '"Africa's growth tragedy" reconsidered', MSc dissertation, Department of Economics, School of Oriental and African Studies, University of London.

Messer, E., Cohen, M.J. and D'Costa, J. (1998) 'Food from peace: breaking the links between conflict and hunger', Food, Agriculture and the Environment Discussion Paper 24, Washington: International Food Policy Research Institute (IFPRI).

Milward, A. (1984) *The Economic Effects of the Two World Wars on Britain*, 2nd edn, London: Macmillan Press.

Mkandawire, T. (1994) 'The political economy of privatisation in Africa', in G.A. Cornia and G. Helleiner (eds), *From Adjustment to Development in Africa: Conflict, Controversy, Convergence, Consensus?*, London: Macmillan.

Moore, B. (1966) *Social Origins of Dictatorship and Democracy: Lord and Peasant in Making the Modern World*, Boston, MA: Beacon Press.

Morrison, C., Lafay, J.-D. and Dessus, S. (1994) 'Adjustment programmes and politico-economic interactions in developing countries: lessons from an empirical analysis of Africa in the 1980s', in G.A. Cornia and G. Helleiner (eds), *From Adjustment to Development in Africa: Conflict, Controversy, Convergence, Consensus?*, Basingstoke, Hants.: Macmillan, and New York: St. Martin's Press.

Nafziger, E.W. and Auvinen, J. (1997) 'War, hunger and displacement: an econometric investigation into the sources of humanitarian emergencies', WIDER Working Paper no.142, Helsinki: WIDER.

OECD (1997) *DAC Guidelines on Conflict, Peace and Development Cooperation*, DAC Taskforce on Conflict, Peace and Development Cooperation, Paris: OECD.

Olson, M. (1963) 'Rapid growth as a destabilizing force', *Journal of Economic History*, 23 (1): 529–52.

O'Sullivan, M. (1997) 'Household entitlements during wartime: the experience of Sri Lanka', *Oxford Development Studies*, 25 (1), Special Issue on War, Economy and Society.

Pottier, J. (1996) 'Agricultural rehabilitation and food insecurity in post-war Rwanda: assessing needs, designing solutions', *IDS Bulletin*, 27 (3): 56–75.

PRIO (1999) *To Cultivate Peace: Agriculture in a World of Conflict*, Report 1/99, Oslo: International Peace Research Institute (PRIO).

Pritchett, L. (1997) 'Divergence, big time', *Journal of Economic Perspectives*, 11 (3): 3–17.

Prunier, G. (1995) *The Rwanda Crisis*, New York: Columbia University Press.

Reno, W. (1998) 'Humanitarian emergencies and warlord economies in Liberia and Sierra Leone', paper presented to conference on 'War, Hunger and Displacement: the Economics and Politics of the Prevention of Humanitarian Emergencies', Stockholm, 15–16 June, UNU/WIDER.

Richards, P. (1996) 'Fighting for the rain forest: war, youth and resources in Sierra Leone', in *African Issues*, International African Institute with Oxford: James Currey and Portsmouth, NH: Heinemann.

Riddell, R. (1990) *Manufacturing Africa*, London: James Currey.

Rodrik, D. (1998) 'External shocks, social conflicts, and growth collapses', Cambridge, MA: John. F. Kennedy School of Government.

Scott, J. (1976) *The Moral Economy of the Peasantry: Subsistence and Rebellion in Southeast Asia*, New Haven, CT: Yale University Press.

Scott, J. (1985) *Weapons of the Weak: Everyday Forms of Peasant Resistance*, New Haven, CT: Yale University Press.

Sen, A. (1992) *Inequality Re-examined*, New York: Russell Sage Foundation and Oxford: Clarendon Press.

Sen, A. (1997) 'Development thinking at the beginning of the 21st century', Development Economics Research Programme no.2, Suntory and Toyota International Centres for Economics and Related Disciplines, London School of Economics, paper presented to conference on 'Development Thinking and Practice' at the Inter-American Bank, September 1996, Washington, DC.

Southern Africa Development Coordinating Conference (SADCC) (1985) 'The cost of destabilisation: memorandum presented by SADCC to the 1985 summit of the Organisation of African Unity', in J. Hanlon (1986) *Beggar Your Neighbours: Apartheid Power in Southern Africa*, Bloomington: Indiana University Press and London: CIIR/James Currey.

Starr, J.R. (1991) 'Water wars', *Foreign Policy*, 82: 17–36.

Stewart, F. (1993) 'War and underdevelopment: can economic analysis help reduce the costs?', *Journal of International Development*, 5 (4): 357–80.

Stewart, F. (1998) 'The root causes of conflict: evidence and policy implications', paper prepared for conference on 'War, Hunger and Displacement: the Economics and Politics of the Prevention of Humanitarian Emergencies', Stockholm, 15–16 June, UNU/WIDER.

Stewart, F. and M. O'Sullivan (1997) 'Democracy, conflict and development: three cases', QEHWPS15, Oxford: Queen Elizabeth House.

Stewart, F. *et al.* (1997) 'Civil conflict in developing countries over the last quarter of a century: an empirical overview of economic and social consequences', *Oxford Development Studies*, special issue, 25 (1): 11–41.

Stiglitz, J. (1994) *Whither Socialism?*, Cambridge, MA: MIT Press.

Storey, A. (1999) 'Economics and ethnic conflict: structural adjustment in Rwanda', *Development Policy Review*, 17: 43–63.

Svedberg, P. (1990) 'Undernutrition in sub-Saharan Africa: a critical assessment of the evidence', in J. Dreze and A. Sen (eds), *The Political Economy of Hunger*, Oxford: Oxford University Press.

Tilly, C. (ed.) (1975) *The Formation of National States in Western Europe*, Princeton, NJ: Princeton University Press.

Turton, D. (1997) 'War and ethnicity: global connections and local violence in North East Africa and former Yugoslavia', *Oxford Development Studies*, 25 (1): 77–94.

UNCTAD (1997) *Trade and Development Report, 1997*, Geneva: UNCTAD.

UNICEF (1989) *Children on the Frontline: The Impact of Apartheid, Destabilization and Warfare on Children in Southern and South Africa*, New York: UNICEF.

Uvin, P. (1996) 'Development, aid and conflict: reflections from the case of Rwanda', Research for Action 24, Helsinki: UNU/WIDER.

Wallensteen, P. and Sollenberg, M. (1998) 'Armed conflict and regional conflict complexes, 1989–97', *Journal of Peace Research*, 35 (5): 621–34.

Wolf, E. (1969) *Peasant Wars of the Twentieth Century*, New York: Harper and Row.

Wolfensohn, J. (1999) 'A proposal for a comprehensive development framework', discussion draft, Washington, DC: World Bank.

Wood, Ellen Meiksins (1995) *Democracy Against Capitalism: Renewing Historical Materialism*, Cambridge: Cambridge University Press.

Woodward, D. (1996) 'The IMF, the World Bank and economic policy in Rwanda: economic, social and political implications', report for Oxfam, Oxford: Oxfam UK.

World Bank (1996) *A Framework for World Bank Involvement in Post-conflict Reconstruction*, Washington, DC: World Bank.

Yeats, A. (1989) 'On the accuracy of economic observations: do sub-Saharan trade statistics mean anything?', Policy, Planning and Research Working Paper WPS 307, Washington, DC: International Economics Department, World Bank.

4 Planning for post-conflict rehabilitation

Reginald Herbold Green

To plan is to choose. Choose to go forward.

(Mwalimu J.K. Nyerere)

4.1 Introduction

As they enter the next decade(s), a significant proportion of sub-Saharan countries find themselves in post-conflict, or in other cases actual conflict situations, in which economic planning is extremely difficult and, some would say, impossible. It is argued here that planning for post-conflict rehabilitation and reconstruction is in fact both necessary and practicable, although its form needs to be rethought, restructured and contextualised.

At least fifty countries in the world either presently have national or substantial regional conflicts or have recently achieved cessation of these conflicts without achieving a state which can be described as full rehabilitation. The population of the countries (or afflicted areas, in ones in which the conflict is zonal) is of the order of 500 million. It is possible to challenge a few of the inclusions but rather easier to suggest plausible additions and to identify countries likely to tip into conflict.

The list is not limited to Africa: indeed, a majority are non-African. Nor is it confined to very small states, the average population size being about 10 million, with several over 20 million and some near or over 50 million:

> Papua New Guinea (Bouganville), East Timor, Cambodia, Sri Lanka, Afghanistan, two or more CIS Asian Republics, Turkey (Kurdish zone), Iraq, Yemen, Lebanon, Palestine, Georgia, Armenia, Azerbaijan, Chechnya, Albania, Croatia, Bosnia, Kosovo, Serbia, Mexico (Chiapas plus), Haiti, Colombia, Peru (Andean Plateau), El Salvador, Nicaragua, Guatemala, Honduras, Algeria, Western Sahara, Mali (Saharan Zone), Guinea Bissau, Sierra Leone, Liberia, Congo (Brazzaville), Tchad, Central African Republic, Congo (Democratic Republic), Angola, Rwanda, Burundi, Uganda, Mozambique, Namibia, Comoros, Ethiopia, Eritrea, Somalia, Somaliland, Sudan.

To begin with, some working definitions of conflict and rehabilitation are provided. As the further starting point for analysis of rehabilitation/reconstruction needs, the costs and impacts of war are subsequently described in some detail and itemised. Possible guidelines for and checklists of the components of post-war rehabilitation planning and programmes are then suggested, ending with sketched programmes for Somaliland and Mozambique.

4.2 Conflict

Conflict is defined for purposes of this chapter as widespread armed conflict (perhaps, as a rule of thumb, directly afflicting at least half the territory and 40 per cent or more of the population). Conflict of interest or non-violently managed/mediated tensions, even if there are local outbreaks of violence, are not included. Neither are military coups if not associated with widespread, continuing violence. Borderline cases include regionalised violence (for example, the eastern half of Ghana's Northern Region) which at regional, but not national, level have analogous results and an identical need for reconstruction, rehabilitation and reconciliation (R-R-R). Other borderline cases arise where an oppressive regime which was widely perceived as illegitimate and which was facing rising, if still largely suppressed, opposition has been removed rapidly, by basically non-violent means. Nigeria is such an example. In Mali the so-called Touareg insurgency led to the relatively peaceful, if military, overthrow of the regime and to major strategic policy shifts. In these cases there are certainly differences of degree and usually of kind in respect to the robustness of state administration and service delivery capacity, of the economy and of the financial system and a basic absence of the return home (and need for R-R-R) of large numbers of internally displaced/ international refugee households.[1]

It may be argued that 'failed states' – in the sense of entities which have lost the capacity to administer, to enforce law and reasonably just order, to deliver either basic services or to create a setting consistent with growth or even stability in production, thus losing legitimacy in the eyes of most of their subjects, produce for a new government what is very similar to a post-conflict situation. In practice the distinction rarely arises since total state failure (as in Sierra Leone from the 1970s, the then United Republic of Somalia from the mid 1980s and then Zaire, starting in the early 1980s or before) almost always leads either to civil war (whether largely domestically or, as is usual, partly externally manipulated) or to a highly oppressive regime, the latter somewhat increasing state capacity but with a high risk of future armed conflict. Some would cite Burkina Faso as an example, although its characterisation as a 'failed state' under President Sankara is more questionable. State failure can precede and lead into or be caused by civil war; in practice it is usually associated with it.[2]

It is not always clear when we are in a post-conflict situation. In some cases a significant level of widespread armed conflict remains and threatens to erupt into a new war period, for example, Rwanda, Burundi, post-Mobutu Congo before the 1998 invasion/insurgency, Sierra Leone, probably Liberia. No turnpike theorem is appropriate to describing the path towards established peace. Initial post-conflict situations usually incorporate pockets of violence and real possibilities of renewed war. The opposite side of that coin is that, even in obvious civil war conditions, there are likely to be areas (geographic and programmatic) in which even an enfeebled state can carry on significant basic service delivery, law and order and administration functions, as was the case in Mozambique in main towns and some rural areas.

How a conflict ends, as well as its prior causes, has a major impact on priorities for rehabilitation as well as on the probabilities of success. In general it can be said that rehabilitation will be faster and less difficult the shorter the conflict; the less the cost (especially in lives, livelihoods, human capital and social capital, including state legitimacy); the greater the degree to which the war is perceived[3] as either a liberation struggle (for example, Somaliland) or as one instigated and manipulated by external actors (for example, as was perceived in Mozambique); and the less deep and systemically entrenched the divisive or corrosive forces that led to the violence (for example, Rwanda and Burundi). Physical damage seems to matter less as, surprisingly, does brutality – short of systemic genocide. Renamo in Mozambique behaved brutally as a strategy, as did Barre forces in Somaliland, but both are cases of relatively rapid, sustained and rooted rehabilitation.

Also crucial are the reality and the perception of permanence with respect to the cessation of conflict; absence of general opposition to its renewal (including opposition by the elite and the presence or absence of external bases/allies/sources of supply for continued attempts to destabilise/vanquish the new regime. Even in pure material resource terms, planning is hampered if allocations purely for security must be massive relative to resources available for state revival and for improving the condition of the people. Here contrasts can be made between Mozambique vs. Rwanda, Somaliland vs. Sierra Leone and Mali vs. Liberia.[4]

4.3 Rehabilitation

Rehabilitation after conflict (or return–reconstruction–rehabilitation–reconciliation) is a process of restoring livelihoods, infrastructure, institutions, human capital, basic services, social capital (especially across groups on opposite sides of the clash) and a political culture of pre-conflict resolution of differences. Not all aspects are equally suitable for formal planning techniques but, up to a point, resource costings and probable levels of direct and indirect benefits can be estimated.[5]

Four general points can be made here. First, rehabilitation does not imply return to the prewar *status quo* (or *modus vivendi*), for at least the following reasons:

- the causes of the conflict are likely, in large part, to lie in the previously existing structures and relationships;
- in many cases a radical political restructuring has occurred which necessarily informs all or almost all aspects of rehabilitation;
- very often the nature of relationships, as well as orders of magnitude, will require alteration, for example, the state farm–plantation–smallholdings balance or the purposes and patterns of economic intervention;[6]
- the costs and destructions of war may create analogues to 'greenfield sites', economically, socially and politically. Restoration of production may be most efficient if the products are altered. To take an example, it is doubtful whether Mozambique would be well advised to rebuild its heavily damaged sugar plantation sector, given current global price patterns and trends for sugar, as well as the plantations' historic need for subsidies.[7]

Second, post-conflict rehabilitation execution as such cannot take place nationally until the war has ceased (or at least formally halted in those areas that incorporate the bulk of the population, physical capital and national resources) and is not expected to reignite.[8]

Evidently there is a blurred area as to the extent and probable tenacity of peace, or at least non-war. Somaliland has been in the process of rehabilitation since 1991 but occasional insurgency has occurred in particular places in some years, so that even today peace is neither total nor fully secure. Over time the trend in breadth and depth of peace and the falling number, scope and durations of interruptions in a country allow a positive judgement to be made. By this standard Sierra Leone is not yet past conflict and Liberia is very much a borderline case.

In such cases – as in all public or private policy – risk needs both to be taken into account and to be managed. The cost of waiting for certain, stable peace before beginning rehabilitation is likely to be never attaining the starting point.[9] The cost of ignoring risk is likely to be a large volume of wasted capital spending in the form of destroyed, damaged or unusable constructions and, less uniformly, serious loss of personnel. In high-risk situations/areas, an economic case exists for concentrating on visible, quick, priority need-related programmes with low capital costs. Primary and educational/preventative health and some aspects of agricultural extension meet that test, as do seed and tool distribution, vocational training and some rural and small towns' water supply rehabilitation. Main highways, bridges and dams do not, though ports may.

Some discussions of rehabilitation, however, include countries like the Sudan which are very clearly not in any sense post-conflict, despite occasional broader or narrower truces. These are cases in which only survival support can be provided, except in some cases (not the Sudan) for isolated, stable peaceful areas. In the Sudan even forward planning toward post-conflict rehabilitation (beyond a broad list of strategic objectives and targets) is very difficult, since it is unclear whether one should posit two states, or a confederation, with what types of state and at what dates.

Third, how survival assistance is organised and, to a lesser extent, its content will affect initial potential for rehabilitation. Keeping people alive turns on access to food, medical services (particularly health, education, vaccination and first aid) and water. Except in secure refugee settlements with good fortune as to funding, little more is likely to be possible and sometimes, as in the Sudan, much less.

With whom external actors work is important, and how seriously they seek to build up recipient participation in programme design and in operation (including operational management), because these represent an investment in sustaining or rebuilding social capital. Similarly, attempting to support productive activity (whether in places of refuge or in the field) matters not only because own output enhances survival chances, but because a return to sustainable livelihoods after the conflict is likely to be faster and more effective. How much is possible, other than providing seeds and hand tools outside especially favourable camps or for those living with kin in peaceful transborder contexts is, however, doubtful.[10]

Fourth, neither rehabilitation nor rehabilitation planning is likely to be effective if it is perceived as no more than a marginal social service or safety net add-on outside core macroeconomic planning and programming. There are two very simple but very compelling reasons for this:

- in the absence of real opportunities to regain livelihoods and service access, displaced persons, refugees and demobilised combatants will lose hope and quite probably go back to conflict (the old war, a new insurgency or banditry);
- the output, export and food supply payoffs of rehabilitation can often be as high and rapid as in those sectors on which macroeconomic programming usually focuses. Risking a return to violence and overlooking/underinvesting in high, early growth potential opportunities do not constitute good macroeconomic planning.

4.4 The costs of war

The costs of war – and therefore the starting point for reconstruction – are high in terms of lives lost; physical condition of survivors (their ability to rebuild); GDP; livelihoods; physical damage; institutions; human capital; social capital; and residual instability. The relative weights of these vary widely by country.

The total number of people who would have lived in the absence of war deaths in SSA almost certainly exceeds 10 million.[11] In Mozambique it was of the order of 2 million over 1980–92, while the Sudanese death toll may be higher. In Angola, and probably Rwanda, it is 1 million or more, as it is in Ethiopia if one assumes that the 1984–5 famine would have been contained at low mortality levels in the absence of war.[12] Sierra Leone, Somalia, Burundi, Liberia, Nigeria (during the Biafra war) and Congo (Democratic Republic) have probably each lost at least 200,000 lives each to war, 500,000 in the last two cases.

The loss of life is predominantly civilian (probably 95 per cent), generally over half indirect, resulting from lack of access to medical services, food and water, and to exhaustion from freed flight. Probably over two-fifths are infants and the young, who are the most vulnerable to these shocks.[13] The old, who are almost equally vulnerable, may account for only 2 per cent, because in most of SSA only about 5 per cent of populations are over 60 years of age. Famine deaths account for up to a third.[14] Genocide (narrowly defined, so that only Rwanda and Burundi are included)[15] accounts for over a tenth, while other civilian deaths at the hands of combatants – predominantly by massacres[16] not incidental to combat – about a tenth.

The impact on survivors is also often desperate and in part irreversible. Sustained malnutrition, with lack of access to health services, is permanently damaging mentally, emotionally and physically, even if less evident than mine and terrorist victims who lack limbs, eyes, ears and/or noses.[17] Equally hard to estimate without in-depth studies are cases of traumatisation and desocialisation as a result of being combatants outside normal social structures or mores or as a result of repeated forced migration and harassment. In Mozambique UNICEF/SCF estimates suggest a figure of 250,000 traumatised children.

GDP losses in SSA probably exceed US $100,000 million,[18] about half of this in Southern Africa (excluding South Africa)[19] and the rest concentrated in Central Africa, the Sudan and the Horn, although current and cumulative losses relative to territorial output are also very high in Liberia and Sierra Leone and were so during much of 1979–81 and 1981–6 in Uganda. In Mozambique, 1992 output was less than half what it would have been in the absence of war, a situation probably pertaining in the Sudan, Liberia and Sierra Leone today.

The largest source of output loss – and one continuing after the cessation of war – emanates from lack of maintenance and investment exacerbated by the destruction of existing capital stock. Military expenditure is damaging in two main ways: it eats up resources otherwise available for directly productive, infrastructural and human investment capital,[20] along with destruction or immobilisation of existing physical capital (and bills for repair and replacement).

Loss of livelihoods is frequently higher (as a proportion of prewar livelihoods significantly damaged or wiped out by war) than that of GDP. This is hardly surprising because almost all SSA wars have been fought predominantly in rural areas. In extreme cases (Mozambique, Angola, Southern Sudan) over half of all rural households have become refugees or internally displaced persons with low to negligible new livelihoods. Serious livelihood attrition is consistent with lower than average (Angola) or very low (Namibia) GDP loss if most rural households are – even initially – very poor and relatively small (under 10 per cent in Angola and perhaps a quarter of that in Namibia) contributors to GDP.

Physical damage – especially to transport infrastructure, all rural and small town infrastructure and housing, and power transmission – is probably of the same order of magnitude as GDP but is hard to calculate because the present value of older portions of the capital stock is exceedingly hard to estimate and the cost of repairing or catching up with neglected maintenance, in contrast to replacement, are equally subject to widely varying estimates.

In some cases war has severely intensified pre-existing failure of (especially governmental) institutions, while in others it has eroded ones which were becoming stronger in the prewar period (for example, Sierra Leone, Zaire/Congo in the first case and Mozambique in the second). In either case the cost has been substantial. Lengthening the period from breakdown to reconstruction increases the time and resources needed for recovery, and restoring institutional capacity requires much more than replacing capital stock, budgetary flows and personnel.

Human capital loss results from loss of life, gaps in or reductions of flows of newly qualified personnel and forced or semi-voluntary emigration. For example, there has been no functioning university in Somalia or Somaliland for a decade and few who obtained higher education abroad since then are now in Somalia or Somaliland. As a result most senior public servants in Somaliland are aged 55–65, with a very limited number of fully qualified 30–40 year old professionals to serve as their successors, let alone to expand the service as necessary to achieve at least moderately adequate capacity for basic service delivery. Similarly, Rwanda lost over half of its central and over three-quarters of its local government servants – dead or fled.

Social capital (including political culture) costs are widespread and high, albeit unequal both as to present impact and post/present/future potential for rebuilding. Trust – especially among civilian supporters on opposite sides and among the very large numbers (sometimes majorities) who were disenchanted with both (or all, in the case of multi-sided conflicts) – has frequently been eroded toward the point of non-existence. It is much harder to evaluate how lasting that situation is, and may be highly contextual. Both mediation and reconciliation have positive – and surprisingly fast – track records in Mozambique and Somaliland, but not in Angola or Somalia. Similarly, social networks have frequently become narrower. Thus mixed villages and neighbourhoods were common in prewar Rwanda and Burundi, but no longer are since 1994.

The last major cost category is residual/consequential violence. Demobilised combatants – especially in the absence of well-designed, adequately resourced programmes for reabsorption into livelihoods above the absolute poverty level – are likely in numerous cases to return to the only skill and tools they know: those of violence and weaponry. Residual pockets of violence – whether of banditry or of fighters to the bitter end – are common. The two greatest potential costs, however, are:

- reignition of war (whether on the old or on new lines) because of an inadequate peace settlement (including a very differently interpreted one) or inadequate rehabilitation;
- continuation of the war from nearby external bases by the vanquished party (Interahamwe against the Rwanda government and its analogue against the Burundi government).

Perhaps surprisingly, revenge killings after the cessation of war have not been a substantial factor. In some cases – Rwanda for instance – this has required sustained government pressure, including draconian action against some offenders. In others, such as Mozambique, while the government has certainly opposed retaliation, there appears to have been limited need for action to enforce that opposition. It is argued that revenge killings in Sierra Leone are likely to be a serious problem – a belief apparently shared by many ex-army and Revolutionary United Front cadres – but logic as well as pre-settlement speculation in Mozambique ran on the same lines.[21]

Relative levels of loss – relative to each other and relative to prewar – vary widely. No generalisation, other than that they will be appallingly high, appears likely to be sustainable.

This would suggest that entry points and priorities for rehabilitation planning also may need to be substantially different in different countries.

Calculated at the global level, it is just possible to argue that the epidemic of wars has only marginal costs. The countries afflicted probably account for less than 2 per cent of global GDP (even in the absence of war, probably at most 5 per cent) and have very limited military capacity relative to large economy armed forces[22] and, except in the Balkans, are not next door to OECD or EU club members. But that would hardly be a decent human approach, given that up to one human being in ten is afflicted by war or the consequences of recent war in her/his country. In any event it is less than accurate, because of very real regional spillover effects. These are perhaps most severe in SSA but clearly apply also to the Balkans, the Middle East, the Caucasus and the Himalayan periphery. Spread and spillover costs include:

- deterrent (or border repellent) military spending;[23]
- military spending and participation in conflict in support of neighbours;[24]
- involvement in regional peace maintaining, restoring or enforcing involvement;[25]
- loss of sources and markets;
- costs of hosting refugees;
- external image regionally and globally, not least in relation to external official and enterprise sources of finance.

Tanzania and Zimbabwe have had military budgets primarily related to external threats and been successful in keeping conflict on their own soil to a minimum. Similarly, both have deployed troops into combat in support of neighbours (as has Angola), neighbours whom they perceived to be in danger of being overwhelmed by external or externally backed aggression, with consequential threats to their own security. Nigeria has incurred heavy costs in seeking to avert, reduce or end war in Liberia and Sierra Leone.[26]

The Central African conflict triggered by the RPF invasion/Interahamwe genocide in Rwanda has spread to involve at least eight states (Rwanda, Burundi, Congo (Democratic Republic), Congo (Brazzaville), Chad, Angola, Zimbabwe, Namibia and, perhaps, the Sudan) in combat and has imposed heavy deterrence costs on Tanzania. It has led to two externally catalysed civil wars in Congo and is – despite the Lusaka Peace Accord – by no means in sight of resolution, either in its transborder or, especially, its internal aspects. Nor has it resolved the initial problem – the red thread of Interahamwe (IH) genocide (together with pre-emptive anti-Hutu violence in Burundi), which has woven itself into both Congo (Democratic Republic) wars as well as those in Congo (Brazzaville) and the CAR (in both cases IH have fought as mercenaries). IH remains in the north-east quadrant of Congo (Democratic) and now in the Kabila army and, while crippled through base point constriction, in the analogous Burundian external armed opposition.[27]

Trade losses, while not comparable to military costs, can be significant, especially zonally. In Tanzania, Ngara District's largest markets (for beans and coffee) before 1994 were in Rwanda and Burundi and the latter were in turn important manufactured goods suppliers (for example, beer, textiles, cigarettes). Transit traffic revenues were arguably the district's most dynamic sector. That ended abruptly in 1994, since when impacted sectors have recovered slowly and partially.

Refugee costs are a contentious issue. A 1994 study in Ngara suggests that on current household account Ngara may have been in a break-even position, with commercial sector and banana gains offsetting bean grower losses. But the ecological impact, including demands on women and girls' time in having to go further to secure water and fuel, were very significant, as were the security costs incurred by the Tanzanian government on the border and in/around the camps.[28]

Countries in what are perceived as war zones, even if they are not significantly unstable and involved in neighbours' wars, pay a high price because both tourists and investors are often vague on distances, actual locations of conflict, and how rapidly conflicts are likely to spread. They avoid risk and stay out of the whole zone. For a country – however at peace domestically – to be in a zone in which wars exist is to live in a high-cost and dangerous neighbourhood.

4.5 Guidelines and checklists for post-war rehabilitation planning

A number of propositions can be put forward as guidelines for planning in relation to the context we are considering:

- the type of planning which can be effective depends on the context: the weaker the data base (and the less suitable for projections), the greater the degree of structural change sought, the more limited existing analytical capacity, and the greater the uncertainty as to resource availability and/or the impact of external shocks, the greater the case for strategic planning focused on a limited number of priority targets, and focusing on achieving movement in the right direction rather than the literal attainment of quantified targets by specified dates;[29]
- in SSA, strategic and sectoral planning linked directly to budgets, via policies, programmes and projects, is usually easier to formulate than comprehensive plans: though a drawback here is that this approach does make it harder to estimate linkages and side-effects to understand trade-offs among objectives;
- there is a case, therefore, for a comprehensive one-year policy–budget package and a strategic overarching formulation covering inter-relationships and trade-offs;[30]
- in these circumstances especially, the uses of a planning process are not limited to the economic (or even the readily quantifiable material) aspects of decision taking: all goals and targets require allocation of scarce resources and almost all meaningful targets can be compared with results, at least with respect to the direction and frequently to the order of magnitude of change;
- wars have particular, identifiable costs and leave behind contexts distinctly different from those of economic, public service and governance decline without war, calling for different overall planning and budgetary processes;
- whenever a sectoral approach is used, this must feed into an overarching analysis in relation to resource allocation: any sector not treated as of macroeconomic significance and linked in with macroeconomic analysis and programming will, in practice, be marginalised and under-allocated.[31]

We can move from these very general propositions to more specific checklists of plan content. As an example of a very simple checklist, one of the clearest formulations is that identified by a very senior, very old, very peace- and order-oriented (albeit willing to fight for those ends) Somali elder at Baidoa in 1995. He set out three goals within an overarch or foundation:[32]

- water, without which people, livestock and crops cannot survive;
- food, also essential to survival and ability to produce;

- health and education, without which ability to learn and to produce are stunted and full human life, rather than mere survival, is unattainable.

The overarch stated was a framework of peace and order within which people could set about achieving the other goals/targets.

At first glance this formulation may seem notably non-economic. In part, that is typical of non-economists' (whether ministers' or village women's) formulations: financial magnitudes and production are to them vital means or intermediate goals to other ends. In fact the focus is on production of water, food, health and education services and on a law/order context in which that can be done. The goal set is adequate to carry out a strategic planning exercise and, in principle, can be elaborated for targets and then iteratively programmed and budgeted. More generally, a schematic checklist might include, by cluster:

Security–law–order
- border defence, to keep war out or repel it;
- insurgency defence;
- day-to-day security from violent interference in people's normal lives from any quarter (including army, police, politicians, officials, or 'big men') by a user-friendly civilian police force;
- intelligible laws and operational, accessible magisterial courts operating with consistency, impartiality, speed and honesty.

Basic service access
- water (human and livestock/crops);
- health services (especially preventative, educational, primary curative);
- education (especially primary and vocational/adult).

Livelihood rehabilitation/strengthening
- facilitating return home/re-establishing farm (or trade);
- ensuring access to inputs/knowledge (for example, extension services/markets);
- avoidance of needless/complex regulation.

Provision of safety nets
- initial return home /demobilisation;
- calamity (for example, drought) response;
- absolute poverty alleviation (aged, disabled, mother and child, epidemic, for example, HIV).

Provision of infrastructure
- physical (for example, rural roads, buildings and equipment), as needed by all the above;
- institutional and procedural, as needed by the above;
- market (ensuring there are commercial, transport, financial operators – preferably not governmental).

This would not be dissimilar to a strategic checklist for absolute poverty reduction planning, programming and budgeting for any African and indeed any low-income country. The differences are contextual, relating to the starting point. Priorities and sequences relating to action turn on initial gaps and on which elements of each priority are crucial in allowing progress on other aspects of that priority and, especially, on other priority clusters.

In particular, law–order–security have higher priority – in resource allocations and timing – than in non-war contexts. It is quite true that law and order are never enough. It is, however, equally true that without some achieved levels of law, order and security there cannot be much else.[33]

'Resettlement' does not appear as a separate cluster. It is an initial stage of livelihood establishment and rehabilitation; it usually means 'return home'; and it is frequently so rapid that 'planned' support is overtaken by events. However, one particular aspect, demobilisation with economic/social reintegration, is usually under-prioritised – at least in terms of resources deployed in articulated, relevant programmes. If young men equipped with and trained in the use of guns (and often little else), with limited social/human capital, are demobilised into absolute poverty, with neither hope for themselves nor respect from others, there will very likely be insecurity and violence. This applies in the case of all ex-combatants, not only 'victors' or formal government troops.

The list is specifically intended to apply to R-R-R planning as sectoral planning. That is not to deny that overall fiscal and financial institutions and flows, arterial infrastructure, higher education, large nodal or growth pole projects/enterprises and so on are crucial. They are. But it is probably more practicable to pursue them as part of one or more than one sectoral planning process with systematic attention to linkages/feedbacks within a simple, overarching, macro-strategic plan frame and iterative incorporation and reconciliation at programming and budgetary process stages.

The perception of all causes and conditions leading to violence – of how to formulate these in ways which help us to find specific means to reducing/reversing them – is probably important to any results and certainly to efficient resource allocation.

To argue that the removal of a lid (in the African case, colonialism) allows unresolved tensions to re-emerge and to intensify (the 'Balkan Model') is not wholly invalid. Nor is the statement that 'ethnicity matters', albeit ethnicity/nationalism are frequently manipulated and used to deflect wrath[34] and can be (and are) used to build states and reduce tensions as well as the reverse. A somewhat different set of propositions, serving as points to study for operational implications, may be more useful for R-R-R planning:

- lawless and violent governments lead to lawless and violent people;
- inclusion (or gaining power) at the centre, not secession, is the usual central demand of insurgency supporters;
- cross border violence is usually the result, not the cause, of domestic war;
- territorial identity/loyalty neither requires a central historical 'national' culture and societal pattern nor automatically flows from it. Botswana and Tanzania, and particularly Somaliland, are nation states in the territorial loyalty and social capital sense but Tanzania is very diverse in 'historic nations' and none has irridentist ambitions in respect of the numbers of its 'historic nationals' living across present frontiers;
- if a territorial state ceases to function (for example, Sierra Leone, Zaire) territorial loyalty may persist but violence is likely to emerge – especially if the failed state makes spasmodic violent attempts to reassert its authority;
- perceived inequity – by exclusion or drawing off of resources with few, if any, flowing back – is much more likely to lead to intense alienation and violence than poverty and relatively low inward resource allocations. Bayelsa, as the largest contributor to Nigeria's exports and fiscal flows, believes it is being robbed; Ngara in Tanzania is poor, at least until recently under-allocated, but sees some central concern and results (most notably 1994 on preservation of law, order and security) and knows it is not being robbed to support Dar es Salaam.

Each proposition does lead to priority actions. For example, in Nigeria the automatic allocation of oil revenue shares to producing and (at lower levels) other states, not just the federal government, was a unifying means. Its partial reinstatement is a move in the right direction but, given the intervening period, no longer enough by itself.

Similarly, malgovernance can be broken down into components facilitating identification of actions to reverse them and for testing packages of measures for joint feasibility and (ordinal) efficiency:

- the most pervasive form of malgovernance is non-governance, that is, failure to provide acceptable and increasingly broad levels of services in the five clusters cited (people cannot live by participation, competitive elections and prudential financial sector regulation alone);
- next is general lawlessness and aspects of malgovernance promoting resort to violence including systemic theft at upper and middle levels;[35]
- waste matters because resources are scarce and because, when gross, it brings the government – both political figures and public servants – into contempt;
- non-participation matters by reducing the perception of belonging, of having control over a range of meaningful 'local' decisions and of being able to hold leaders accountable. The forms of participation matter less than that they are seen as real and, to a degree at least, effective.[36]

With respect to perceptions of participation and of territorial national identity, symbols and mixed symbolic/functional unifying means do matter – on occasion negatively as well as positively. Here a planning technician can usually do little more than inform on costs and potential negative feedbacks.

Somaliland has invested very heavily in decision taker time for reconciliation conferences of guurti (elders who, in Somaliland, are close to and legitimate for most people); in electing a House of Elders (strategic politics and political economy) and an Assembly (to oversee government operations); and a draft constitution with a pre-finalisation/referendum consultation/dialogue process; as well as in re-establishing a functioning magisterial to supreme court system, a civil police force and a national army. All have functional roles, but in addition (apart from the courts and the army) they create a perception of participation, even if indirect, from household to country and of a state which stands as, but is also more than, a federation of sub-clans.

Indigenous national – in the sense of non-European – languages are an example of instruments that have different impacts in different contexts. The ideal is an indigenous language which is fairly broadly understood (even if imperfectly), easy to learn and the home language of a relatively weak group. In such a case – notably Swahili in Tanzania – the gains are real and the costs can be reduced by developing the language and by wide teaching of a secondary international language. Unfortunately these conditions are rarely met. If one region or elite would gain, rather than others, the 'national' language can prove more divisive than a move towards universal access to the colonial one: hence Mozambique's option for Portuguese over Swahili (or English), which it was feared would disadvantage the Centre and North versus the South (or vice versa).[37]

4.5.1 *Case study: Somaliland*

What can be said about elaborating from the guidelines and checklists in particular countries at particular times? This is more easily done at case level: generalisations are more difficult and likely to result in too many sets of alternatives. Two sketches may illustrate what is likely to be practicable: one starting from a long-collapsed state and severe damage to livelihoods and commerce as well as initial near total breakdown of law and order – Somaliland. The other is in the context of a functioning state and institutions with order and livelihood existing (even before the end of the war), however attenuated, for perhaps half the population – Mozambique. Differences in professional personnel and data availability are almost as marked.[38]

In the first case the starting point for a planning process is to identify goals, results to date, and a limited number of key actions/programmes which would solidify what has been achieved, and fairly rapidly create the basis, over two to four years, for setting a broader range of targets.

Somaliland has law, order, security and a broadly shared sense of being a state. Basic institutions, central and local, including a public service, are in place, but are very weak because of exiguous resources, including limited external official, personnel and investment flows, resulting from non-recognition.

Three ongoing priorities are identifiable:

- restoration of revenue collection by central and local government, from the order of US $25 million to US $150 million (about an eighth of GNP), through import (not export and remittance) taxes for the central government and urban rates and service charges for the municipal ones. Devising simple systems and standard procedures, plus training of staff (requiring perhaps up to a dozen expatriate specialist professionals), should be attainable over 18 to 36 months, with build-up to a target ratio to GDP at, say, 60 months;
- restoration of the key Berbera–Hargeisa–Djijiga–Addis Ababa trade and transport backbone, building from present UNCTAD channelled work at Berbera via an EU Horn Regional programme with Ethiopia. The US $50–75 million cost (and related specialist personnel) over 3–5 years is not out of line with EU major regional project funding and require only Ethiopian involvement and EU acceptance of Somaliland as a 'Region' in an EU sense, not formal recognition;
- demobilisation and reintegration into civilian livelihoods of the 12,500 'surplus' army personnel, together with livelihood training for the 5000 (out of 20,000) 1991 demobilisees who have not yet successfully built civilian livelihoods. Because skilled workers and artisans are in demand – in civil engineering and building construction – an 18-month programme of training and labour-intensive reconstruction would be feasible, for example, under ILO auspices. As it could be classed as humanitarian – even though underpinning security, social stability, poverty reduction and economic growth (and lightening the budget burden of food, shelter, clothing and so on for the 'extra' troops) – foreign assistance should be attainable. Phased over five years, the annual external cost would be of the order of US $20 million, less than US $10 per Somalilander.

If these three strategic programmes could be carried out, the domestic revenue, the trunk infrastructure rehabilitation and the reconstruction of secondary infrastructure and boosting of human capital and livelihoods would facilitate progress towards basic services provision, sustained domestic growth, and return of some diaspora professionals.

For Somaliland rehabilitation 1999–2000 was a year of success, slightly impeded by drought in some provinces and continued problems of access to Saudi livestock markets. Politically the progress toward an agreed constitution and 2001–2 multiparty elections continued. Radical institutional reform – plus increases in civil service salaries from, say, a quarter to three-quarters of the household absolute poverty line – led to tax collection doubling from about US $25 million to US $50 million, with a possible US $75 million in 2000–1. The main uses are water, education, veterinary services and health. At the same time quiet diplomacy has led to more countries, including the EU, accepting that Somaliland is a regional governance unit with which they can work. This trend has been accelerated by the drought crisis in Ethiopia's Somali region, because the direct route is Berbera–Hargeisa–Djijiga (not Djibouti–Addis–Djijiga) and Somaliland evidently does provide a law and order climate conducive to rapid highway repair.

Demobilisation remains largely stalled – warehousing up to 15,000 ex-militiamen is a better buy than demobilising them into grinding poverty. But, as the failure of the (rather ludicrous exile/Barre war criminal dominated) 'regime' emerging from the Djibouti Conference to restore decent governance in Somalia becomes clearer and Somaliland both increases government capacity and holds an election, creeping *de facto* recognition appears likely, with entry of external firms (not least commercial banks) as well as external assistance at 'normal' poor SSA levels.

4.5.2 Case study: Mozambique

Mozambique's planning process could be and was more formal. The suggested rehabilitation programme was prepared for/presented to Somaliland decision takers *ad hoc* and could be internalised or not, or used in part; whereas the Mozambican proposals for *Reconstrucao* were part of an ongoing domestic planning process.

Because state institutions and decision takers existed during the war, with substantial pre- to post-war continuity, a pre-peace strategic plan could usefully be set out in 1992, drawing on experience with survival support during the war. It was explicitly a strategic sectoral plan, intended to be iteratively integrated into the comprehensive macroeconomic planning process.

The components were: livelihood rehabilitation (including supporting activities such as rural trade and transport); return of basic services to war-ravaged areas; and local level infrastructure (including municipal and provincial decentralised governance and market access institutions). Security (except for the demobilisation of veterans into livelihood rehabilitation) was a separate exercise. Linkages, multipliers (estimated at about 1, that is, total GDP impact twice the value of direct household production) and impact on external and fiscal balances were estimated, as were progress toward food security (including household production for own use) and absolute poverty reduction.

Positive results were projected: increase in output over 5 years to 50 per cent above early 1990s' GDP; basic food balance in non-drought years, except for wheat and vegetable oil; a self-sustaining position on fiscal and external account (for the results of the strategic planning exercise, not the economy as a whole) by year six.

It was proposed to meet the programme's external cost of perhaps US $1500 million over 5 years (US $20 per Mozambican per year) by shifting survival support funding (Emergencia was US $350–500 million a year), UNHCR support of refugees (US $150 million a year) and the committed demobilisation/reintegration component of UNOMOZ (US $200 million) to the reconstruction sector, phasing this down over 5 years as domestic feedback payoff rose.

In the event, the planning process – except in cases such as food production and to a lesser extent rural household – was a failure. The proposed funding sources were transferred to other countries (or were lost to donor budget tightening). External official financial sources – especially the World Bank and the IMF – did not so much criticise *Reconstrucao* as treat it as peripheral (perhaps as a component of safety nets) and not integral to overall macroeconomic strategy. In particular, basic services restoration was constrained even when initial-period non-inflationary external finance was to hand, while main roads were overfinanced, and so absorbed almost all rural capital expenditure.

The Mozambican experience calls attention to the external actors' role. R-R-R is more national and contextual than, say, dams or the prudential regulation of banks. Therefore the necessary level of national parameter setting should be higher and the involvement in specific permutations greater. This is especially true because external financial sources appear to give much lower priority to R-R-R than to either survival support or 'standard development' and also to focus on those aspects of governance of special relevance to macroeconomic frameworks and to foreign enterprises, with much less attention to the aspects which are of direct concern to the vast majority of households.

Unfortunately countries in which R-R-R is a central strategic priority are typified by low levels of domestic resource availability and weak human, physical and institutional infrastructures, as well as low governance capabilities. They cannot proceed very rapidly with the more financial resource-intensive aspects of R-R-R without significant external resources, human as well as financial.

The apparent sectoral and territorial exceptions appear to be special cases. Mozambique has met the target of being basically food self-sufficient (except for wheat and vegetable oil) in non-drought years. Peace, return to homes and removal of state barriers to trade were sufficient to achieve that goal, which was seen as laughable when formulated in 1992. But donor conditionality (even more than resource commitments) has blocked the extension of basic services and 'low' level infrastructure.

In Somaliland massive remittances (of the order of 75 per cent of GDP and over 40 per cent of GNP) have facilitated rapid advance in some aspects of R-R-R but not fiscal rehabilitation, so that basic services and infrastructural rehabilitation lag badly. The paucity of technical assistance/conditionality flowing from non-recognition has enabled, in fact has forced, Somalilanders to set their own strategic priorities, but the parallel absence of specialised professional inputs and revenue for large programmes limits implementation.

Similarly with external NGOs: New Government Options to donors; New Governance Overarches for themselves; National Growth Obstacles to many recipients. Integration into national strategies is practicable together with a perception-priority-programming impact which becomes internalised, as demonstrated by UNICEF, which is widely seen as akin to an ideal NGO, even if it clearly is a UN family agency. So is cooperation within local governmental and community frames, as with ActionAid in Mozambique. Even basically anti-state bodies, if competent, may be ideal for survival support channelling during war and in other non-governance settings: for example, Médecins sans Frontières in contested rural areas of Mozambique to 1993. In contrast, the generalised use of semi-anarchically autonomous NGOs, accountable to external donors and to their own lenders' consciences and visions, to deliver R-R-R (and basic services) both fragments and erodes governmental and civil society capacity and rehabilitation. Domestically designed and driven, partly donor funded and programmed is a fairly easy overarching goal to prescribe but one much harder to get official donors, NGOs and many African governments to internalise and to act upon consistently.

4.6 Conclusion: ways forward for post-war rehabilitation planning

The case for the usefulness and practicability of post-war rehabilitation planning appears to be fairly robust, as does actual or potential interest on the part of several African governments. The examples of Somaliland and Mozambique suggest that adaptation to different contexts/starting points is feasible. The most immediate constraint on post-war rehabilitation planning is its marginalisation outside main resource allocation processes. The first step towards loosening that constraint is to enter into wider dialogue and debate on potential and observed positive and negative factors. Assuming that the process has something to offer, being ignored or treated to minor handouts as a form of safety net or survival support is much more damaging than serious open controversy.

The analytical aspects of the dialogue and debate would serve a second function: exploring certain key relationships/aspects which have not been adequately incorporated (except by assumption) into rehabilitation planning – notably gender.[39]

In parallel, both because of their immediate value and in order to broaden the base of available experience for analysis and dialogue, rehabilitation planning processes should be continued as, where and to the extent possible. This is probably useful even if in the end particular components are removed and fitted *ad hoc* into a standard, more comprehensive and macroeconomic planning exercise. At least in that case their relative importance and linkages would be more fully understood.

To plan is to choose and to choose necessarily includes to choose not to do or to give actions very low priority (which with very scarce resources, as in most post-war contexts, comes to the same thing). In the case of post-war rehabilitation planning, however, the evidence supports choosing to go forward on the analytical, operational and feedback/evaluation fronts.

Notes

1 This definition does not imply that high levels of tension and/or coup d'états are benign, nor that transitions like that in Nigeria are non-problematic and benign (especially as a civil war in two or three key oil producing states is a real risk). It simply argues that they are different in kind to the aftermaths of war in, for example, Rwanda, Burundi, Mozambique, Somaliland, Sierra Leone or Liberia (assuming in the last two cases that war is ended and not just temporarily suspended).

2 Not all civil wars result in state failure as sketched. The Mozambican war certainly limited the geographic area in which the state was effective, and its operational capability even within that area, but a perfectly recognisable state did exist. The same has characterised, to date, Angola and the post-genocide Rwanda state.

3 Perception, as linked to and as independent from reality, matters. Somaliland – then independent – voted against union with Somalia and was annexed by force: hence the perception that 1985–91 was a liberation struggle, even though during the pre-Barre civilian era Somalilanders had become more integrated into the then United Republic.

4 The external context factor is one that affects the likelihood of the conflict spreading. Had the Interahamwe (IH) not been given the time and resources to consolidate, reorganise and strike back from Eastern Zaire, not only would rehabilitation in Rwanda have had a far more propitious context but the expansion of the war to both Congos, Uganda, the Central African Republic and Zimbabwe would have been averted (possibly at the cost of delaying the implosion of the Mobutu regime by a few years). This incidentally illustrates bad international humanitarian and peacekeeping planning. The failure to allocate resources to prevent IH from using the Zaire camps as rehabilitation centres and bases has been costly (even to the Security Council member states), out of proportion to money/lives 'saved'. That the exercise would have been possible is illustrated by the fact that Tanzania – which planned and prioritised more effectively, with fewer resources – largely succeeded in preventing any such consolidation and resurgence of IH in its territory.

5　In general this is not a one instrument per target case. For example, re-establishment (or establishment) of a competent, user-friendly civilian police force contributes to social reconstruction by creating peace/law/order and to enhanced production by reducing risk, damage or theft and transport costs. The police force needs to be complemented by a just, intelligible legal code and an impartial, honest and accessible primary court system.

6　The last point is consistent with IMF and World Bank positions. They are not opponents of all regulation or market intervention but against types they believe place serious constraints on production and do little to further level playing field competition or to prevent enterprise or public servant abuses. For example, they – like many large financial enterprises – advocate more prudential scrutiny and regulation of banks.

7　This question is independent of the public–private one. Private sugar industry rebuilding would require infrastructural reconstruction, probably domestic guaranteed prices/purchase levels and other open or camouflaged subsidies.

8　A good test may be the proportion of refugees and internally displaced persons who choose to 'return home'. In Mozambique (as in Kosovo) it has been very high and rapid while in Angola (as in Sri Lanka) it has been limited and slow.

9　The needs for livelihood rehabilitation, access to basic services and to infrastructure (including markets) and civil law, order and policing are immediate. Failure to act towards at least most of them will rapidly erode support for the new governmental structures and processes. External donors in this context are very risk-averse, in a perverse way: they make substantial rehabilitation support to bankrupt governments in devastated economies conditional upon prior achievement of the very targets for which the resources are needed.

10　Most refugees in camps do want to take part in education, basic service provision and/or production. However, in few instances have a majority been able to do so (SWAPO of Namibia's Angolan settlement camps were exceptions). A higher proportion is able to do so when living 'unofficially' with kinsfolk, as in the case of Mozambican refugees in Northern Malawi and south-eastern Tanzania.

11　The method used is that of UNICEF's 1987 and 1989 *Children on the Front Line* studies of Southern Africa. In the case of Mozambique, population projection data for 1994 exceeded actual population – after the vast majority of refugees had come home – by about 2,500,000.

12　Arguably this is to overestimate. The first drought/dearth that did not lead to mass famine in Ethiopia was in 1994–5 under the present government. In 1974 massive famine deaths were a cause, not a consequence, of the overthrow of the Emperor Hailie Selassie's regime.

13　In Iraq proper, even with limited violence on the ground, but with inefficient distribution and lack of access to adequate food, pure water, preventative educational and basic curative health services, under-five mortality has more than doubled and risen absolutely by about 80 per 1000.

14　In SSA since 1960 (and to an only somewhat lesser extent since 1920) droughts and dearths (major losses to food crop harvests) have not resulted in famines with substantial loss of life, with the known exception of Ethiopia (and perhaps Italian Somaliland). Motorised transport and access to imported food make that possible, but war prevents distribution even in the absence of any 'starve out the enemy' strategy.

15　Genocide is defined here as a systematic strategy of eliminating an ethnic group or sub-group, not as Jaquerie-type uprisings, even when manipulated (as in Nigeria in 1965 and Rwanda in 1959–60). Clear cases are Burundi, in the early 1970s and since 1994, and Rwanda from 1994. (Interahamwe still exists, is still committed to genocide and still kills an average of at least a score of human beings a day in Rwanda.)

16　Massacres have been predominantly in Liberia, Sierra Leone, Uganda (under Amin), Angola, Mozambique, Sudan and Ethiopia (including then-occupied Eritrea).

17　Mass wilful amputation is largely concentrated in Sierra Leone and Liberia, ear and nose cropping was a typical Renamo tactic in Mozambique.

18　Also broadly on the UNICEF estimation basis of reasonably projected GDP in the absence of war versus actual.

19　This GDP cost applies also to states (for example, Zimbabwe, Tanzania) that have incurred large additional military expenditure, either to successfully deter/repulse spill-over wars or invasions (neither has had major war or insurgency on its own soil) or to provide military combat support to other states in its region in order to prevent their collapse or to try to make peace (Zimbabwe, Tanzania, Nigeria).

20 This cost continues even once growth is restored, because it resumes from a lower base.

21 If the relatives of the massacred and the mutilated (especially those in the *de facto* National Guard, the Kamajors) do seek to exact retribution (as many Watutsi, within and without the army, were inclined to do in Rwanda), it is doubtful whether the government will (unlike the RPA government in Rwanda) be strong enough to block them. However, the most probable short-term prospect in Sierra Leone – recurrent outbreaks of violence and quite probably near full-scale war because of an inherently unsound 'peace' agreement – renders such speculation at the least premature.

22 Assuming of course that full-scale civil war does not break out in Russia, that the Kashmiri civil war does not lead to another full scale India–Pakistan war, that neither Iran nor Iraq implodes or returns to war with its neighbour and that Indonesia does not dissolve into a series of overlapping civil wars.

23 On occasion 'forward deterrence' has been practised. Angola's participation in the removal of the Lissouba regime in Congo (Brazzaville) and the defence of President Kabila's in Congo (Democratic) are basically forward deterrence to deny UNITA bases and links to the outside world. Ethiopia's support for Somalian forces inimical to Sergeant Aideed is consequential on seeking to destroy Ogadeni ravanchist terrorist bases in the Lugh area.

24 The cost is largely independent of the motivation. Tanzania has acted primarily because it views violence as contagious and out of solidarity and only secondarily for self-protection, whereas in its Mozambican involvement keeping the Beira Corridor open and a credible Mozambican state in being were vital Zimbabwean interests.

25 Integration oriented groups of states have political, social, humanitarian as well as economic reasons to try to render their neighbourhoods safe by mediating, containing, ending civil wars in member states and bordering ones. The Southern Africa Development Coordinating Conference (SADCC), the Economic Community of West African States (ECOWAS) and the Intergovernmental Authority on Development (IGAD) have recognised and sought to act on this principle.

26 Again the cost is largely independent of the motivation. Nigeria's early intervention in Liberia may in large part have been intended to defend Generals Babangida's and Abacha's friend, Sergeant Samuel Doe, and throughout has included a will to at least regional prestige as well a genuine concern that state collapse and civil war are contagious and that there is an imperative to limit and end mass barbarism where possible.

27 The external wa Hutu opposition to the wa Tutsi military establishment in power in Burundi is not uniformly genocidal. However, some transcripts of its Radio Democracy read as virtual paraphrases of the former Radio Milles Collines – the voice that called Rwanda to genocide – with only the names of places and people changed.

28 Two studies were done at the instigation of the Tanzanian government using substantial UN agency personnel inputs. Both reached similar conclusions, albeit the second (via UNICEF) examined distributional, workload and ecological issues in more depth.

29 Targets do matter but, up to a point, sustained movement in the intended direction is an acceptable second best. Whether universal child vaccination and mother and child care is achieved in year 10 or year 15 is probably less important (*not* unimportant) than whether a set of institutions, policies and allocations steadily driving up the achieved percentages are put and kept in place.

30 The overarch could be called a model but – for historic reasons – this carries overtones of data input levels, mathematical programming and linearity not usually appropriate to Africa.

31 There are exceptions to that generalisation but they appear to be in contexts in which one or two committed domestic or external actors can raise initial resources, demonstrate results and win approval of main macro economic decision takers for continuation, for example, universal vaccination/immunisation as spearheaded by UNICEF with a growing number of African health ministers and officials. In practice that approach is practicable only if resources needed are relatively small (compared to total uses) and are to a significant degree non-fungible and additional (carrot conditionality).

32 Set out in dialogue with author, Baidoa, September 1995. Just how high the degree of uncertainty is in many rehabilitation cases is illustrated by the fact that he was already engaged in the district's defence and Baidoa fell a week later with significant 1993–5 water, food security and (to a lesser extent) health, education and local governance gains wiped out (or at least badly damaged and put on hold) literally overnight.

33 The criticism of the 'night watchman' state as inadequate of the 1950s through 1970s and, slightly later, of the deformation of 'law' and order as oppression were valid, if perhaps overdone. They are

not inconsistent with accepting the need for law, order and security (in senses meaningful and acceptable to ordinary people/producers) as vital.

34 In Nigeria in 1965 the basic grievances of Northern and Middle Belt peoples were against the ruling Northern (Hausa/Fulani) elites. The elites skilfully refocused and manipulated their wrath against the middle tier outsiders (the Ibo) who served and profited from the elite but were not the decision takers. The result was pogroms against the Ibo and the catalyst to the Biafran war (initially of secession, later towards conquest of Southern Nigeria).

35 The 'corruption of need' engulfing lower and middle level public servants – including the two-thirds who are primary school teachers, nurses, constables, agricultural extensions and rural water officers – who cannot possibly live on their pay is damaging but less so. Their need is usually comprehended and their privatised, generalised, decentralised user fees grumblingly accepted if they do provide services. That it would be much more efficient for the state to pay them living emoluments remains valid.

36 'Traditional', *ad hoc* community-based or more formally elected bodies from village to national level can be successful if listened to and seen as legitimate by those for whom they speak or work and also perceived to produce some useful product (not just words and aspirations).

37 Thus the regular extension of the co-official language role of English in India where not only the south but also Bengal and Bombay perceive Hindi alone as disadvantageous to them and the *de facto* (probably increasing) role of English parallel to the official Tagalog in the Philippines which may end bilingual. The opposite case is Sri Lanka where Sinhala only was – correctly – perceived as anti-Tamil chauvinism and proved deeply devisive (as well as functionally costly).

38 For simplicity, recurrence of armed conflict, as in Mozambique from 1992 and Somaliland from 1995, is excluded.

39 Given initial unequal access, generalisation of basic services should benefit women. Similarly because water collection weighs heavily on women and girls, improved provision should allow reallocation of time – especially to schooling for girl children. But such premises – as in planning in respect to gender in general – require examination both as to validity and specification and as to what complementary measures (not least in sectors that, on the face of it, have little direct gender content). This was done to some extent in the Mozambican case but not in that of Somaliland.

Part III

Agriculture and the rural sector

5 Smallholder farming in Africa
Stasis and dynamics

Steve Wiggins[1]

5.1 Introduction

This review examines change in smallholder farming in Africa South of Sahara, as seen in case studies, and discusses some implications. To try to be both brief and clear, what follows is necessarily selective.

The chapter will argue the following points. First, although there may be an African rural crisis, it is easy to exaggerate its seriousness as well as to misunderstand its nature. Second, as a gross generalisation, evidence from micro studies suggests that successful agricultural development depends first and foremost on market access and conditions on the demand side, and in lesser part on supply-side factors. Third, smallholder farming is nothing if not dynamic, at different times and places experiencing booms, at other times more modest growth, stagnation or decline. Although we can identify a limited number of variables that explain a great deal of the changes, these interact at different levels in ways that provide a bewildering complexity. This mocks efforts at building strong theories of change capable of providing detailed policy guidelines that may be applied widely. Instead, we may derive some broad principles, but their application in concrete cases requires detailed understanding of local specificity. Fifth, the linkages between success in farming and more generalised improvements in rural welfare (and reductions in poverty) are inadequately understood. Rural inequality is often ignored and under appreciated.

5.2 The context: crisis in rural Africa

Few would deny that events in rural Africa over the last quarter century or so have produced more disappointments than successes. As Michael Mortimore (1989: 1) has put it, 'The problem is not whether there is a crisis in rural Africa, but what its nature really is.' The problems faced may be divided, to take the clinical analogy, into the acute and the chronic. The latter stems from the observation that agricultural production per head in Africa has apparently been falling since the early 1970s. In some accounts it is accompanied by reports of (serious) environmental decline. In the acute version, these problems are so serious as to give rise to famines.

The acute crisis of famine cannot be denied, but it has been localised and sporadic. Mass death owing to famine (or 'famines-that-kill') has mercifully not been general throughout the continent, having been almost entirely restricted to the Horn of Africa. Horrific as these famines have been, only a small fraction of rural Africans have suffered them. For most Africans the chances of death in famine are, in the words of Seaman (1993), 'vanishingly small'. Even famines as dearth and destitution have largely been confined to the drylands north of the Equator, the Sahel broadly defined.

Evidence for the chronic crisis of production and the environment is less compelling than first meets the eye. Three comments are worth making here. First, for only a few environmental problems is there substantial and compelling evidence. Desiccation in the Sahel and the frequency of drought in African drylands would be examples. Evidence of the extent and seriousness of other, often more widely reported degradations, desertification, loss of forests, overgrazing, and so on is less clear. Against the official story of serious damage from these problems are set critical questions and some 'counter-narratives' (see, for example, the collection of essays in Leach and Mearns, 1996). These question both the incidence of damage as well as conventional accounts of the processes at work. Of course, dispute does not prove the argument either way, but it should cause observers to reflect on what is known about environmental change with any certainty.

Second, although the evidence of declining farm output per head at first sight looks impressive and alarming – (see Figure 5.1 for data published by FAO), several points need to be borne in mind: the database is weak; most of the data reported to the FAO by national ministries of agriculture are at best informed guesses, since few countries have been able to carry out sample surveys regularly; production may well be underestimated.

Moreover, aggregate figures for sub-Saharan Africa as a whole are not necessarily a fair reflection of the diversity of the region's experience. Aggregate statistics are dominated by data from a handful of countries. Looking at food crops, Nigeria alone produces almost one-third of the calorific value of cereals and roots and tubers. The largest five farm sectors produce 59 per cent of the total and the largest ten more than three-quarters.

For the thirty years between 1965 and 1995, farm production per capita fell by all of 15 per cent: hardly a catastrophic decline. Indeed, the entire thesis of decline depends on being sure of the numbers to within 15 per cent accuracy, a figure well within the range of errors in the data. Furthermore, FAO indices aggregate the output of different products by valuing them, so the index may fall if prices fall regardless of any change in the physical quantities produced. Hence some of the decline seen during the 1970s may reflect little more than falling world prices for export crops after the boom times for primary commodities of 1972–4. Similarly, during the 1980s, devaluations of currencies lowered the dollar value of estimated national agricultural output. Block (1995) recalculated the indices to eliminate the effect of price movements, and promptly found much larger increases in farm output during the 1980s than appear when looking at the usual indices.

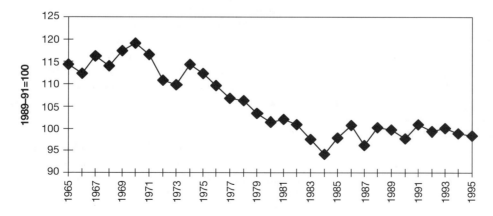

Figure 5.1 SSA: index of agricultural production per head

Third, the decline has been relative to a rapidly growing population of just under 3 per cent a year from the mid-1970s to the mid-1990s. Farm output has not declined absolutely. Indeed, compared to the slower growth of the rural population (around 2.2 per cent a year mid-1960s to early 1990s), recorded farm output has, by and large, kept pace. The failure to produce food in Africa can be seen, above all, as a failure to provision the cities (Jaeger, 1992; Morgan and Solarz, 1994). But this has not necessarily meant food scarcity in the towns and still less urban hunger: what has usually happened is that imports have filled the gap. Thus, Africa south of the Sahara imported 2.5 million tons of cereals in the mid-1960s, but almost 12 million tons by 1993–5. This may seem alarming, but it is only 22 kg or so per person a year, or two sacks of grain for every household. The problem is less one of food shortage than of using scarce foreign exchange (US $2889 million in 1993–5) to import what it should be possible to produce in Africa: just 270 kg more grain from each hectare of cereals currently planted would deal with the shortfall.

If so far some scepticism is expressed about the degree and nature of the chronic crisis as seen in physical terms of the environment and farm production, fewer doubts apply to the social calamity of widespread rural poverty (Heyer, 1996). But is this a reflection of a declining farm sector, or is it a consequence of the social and economic problems of inequality and disadvantage? If the data on national farm production and cereals imports come anywhere close to reality, then sufficient food is available in Africa. Poverty is not necessarily a supply-side problem, but one of effective demand. It is by no means clear that boosting farm output in the early 1990s by the 15 to 16 per cent necessary to have maintained farm output a head at the levels of the early 1960s would have significantly reduced rural poverty.

5.3 The evidence from micro studies

If the evidence from national statistics is in doubt, then we must turn to studies of farming at the household, village and district level in order to appreciate the nature of smallholder farming and ongoing change within it. Case studies have been collected to investigate changes taking place in the apparently most blighted decade for African farming in recent times: that between the mid-1970s and the mid-1980s. These have been collected from databases of published material, principally that of CAB International, focusing initially on ten countries and later widening the search to include all the countries of West Africa. This produced a surprisingly small number of usable cases, just twenty-six in all. Table 5.1 lists them. In comparing the studies, use was made of a framework employed by Snrech (1995) who, reviewing cases in West Africa, saw two factors as affecting change in that region: market access and pressure on natural resources. From this he produced the schema shown in Table 5.2. This captures variables of prime interest to both natural and social scientists. Pressure on resources invites considerations of agro-ecological potential and carrying capacity, while market access brings in the demand for farm produce and the influence of economic policy on such demand. Theoretically it comes close to the ideas of Boserup (1965, 1987), as reflected in the top left and bottom right-hand quadrants of the Snrech chart. That the chart has two other quadrants recognises that some situations do not easily enter Boserup's account (although she was well aware of these possibilities).

This scheme has been used to classify the case studies. In doing so, the weakness of such frameworks readily becomes apparent. The good access/poor access and high pressure/low pressure dichotomies of what are in reality continuous variables may allow us to think through stereotypical cases, but fail us empirically, since some cases inevitably lie in the middle of the range and cannot easily be assigned to one category or another. Accordingly, in the absence

Table 5.1 Case studies of change in African farming

Case	Main sources
Nigeria	
Awka-Nnewi, Anambra State	Okafor, 1993; Okorji and Obiechina, 1990
Dagaceri, Birniwa District, Kano	Mortimore, 1989
Ibarapa District, Oyo State	Guyer, 1989, 1992; Guyer and Lambin, 1993
Imo State	Goldman, 1993; Martin, 1993
Kano close-settled zone	Mortimore, 1993
Lower Sokoto Valley	Adams, 1987, 1988, 1993; Swindell, 1986; Swindell and Mamman, 1990; Swindell and Iliya, 1992
Namu District, Benue Valley	Netting *et al.*, 1989; Stone *et al.*, 1990
Ghana	
Kusasi	Webber, 1996; Cleveland, 1991
Manya Krobo District	Amanor, 1994
Yensiso, Amanase, Sekesua	Gyasi *et al.*, 1995; Gyasi and Uitto, 1997
Guinea Bissau	
Kandjadja village, Mansaba sector	Rudebeck, 1988, 1990
Senegal	
Bignona and Oussouye, lower Casamance	Linares, 1992
Middle Senegal valley	Bloch, 1993; Diemer *et al.*, 1991; Jamin and Tourrand, 1986; Lecomte, 1992; Niasse 1990, 1991; Park, 1993; Sow, 1983; Tourrand and Ndiaye, 1988; Woodhouse and Ndiaye, 1991 a and b
The Gambia	Baker, 1992, 1995; DeCosse, 1992; Haswell, 1991; Mills *et al.*, 1988; Osborn, 1990; Posner and Gilbert, 1989; Webb, 1992
Burkina Faso	
Boromo, Guinea zone	Matlon, 1991; Reardon, 1991; Vierich and Stoop, 1990
Djibo, Sahelian zone	Matlon, 1991; Reardon, 1991; Vierich and Stoop, 1990
Yako, Sudanian zone	Matlon, 1991; Reardon 1991; Vierich and Stoop, 1990
Kenya	
Lower Machakos District	Tiffen *et al.*, 1994
Mid Machakos District	Tiffen *et al.*, 1994
Upper Machakos District	Tiffen *et al.*, 1994
South Nyanza District	Kennedy, 1989; Kennedy and Cogill, 1987
Upper Embu District	Haugerud, 1983, 1988, 1989
Tanzania – Iringa Region	Birch-Thomsen, 1990; Boesen and Ravnborg, 1992; Friis-Hansen, 1988; Nindi, 1988; Rasmussen, 1985
Mbozi District	Bantje, 1986; FSG 1992
Zambia – Northern Province	Allen, 1987; Bolt and Holdsworth, 1987; Francis, 1988; Francis and Rawlins-Branan, 1987; Holden, 1993; Moore and Vaughan, 1987; Ndiaye and Sofranko, 1988; Pottier, 1990; Sano, 1990; Seur, 1990; Sichone, 1990
Zimbabwe – Chivi	Scoones 1996

Note
To these can be added other insights from studies at the village level which were not complete enough to add to the list.

Table 5.2 Probable direction of agricultural and rural development in West Africa, according to market access and pressure on natural resources

	Poor or risky access to market	Strong access to market
Weak pressure on natural resources.	Extensive crop production, or emigration, livestock important	Large-scale mechanised farming, capital-intensive (for example, cotton pioneer zones)
Strong pressure on natural resources:	Subsistence production, labour-intensive, low returns, strong emigration	Intensification by labour or capital, depending on the pre-existing conditions.

Source: Snrech (1995)

of reliable guidance to define degrees of access and resource pressure corresponding to one category or another, a further category in the middle was created to accommodate the in-between cases.

'Market access' is here taken to represent the levels of farm-gate prices for output received by farmers. These prices reflect both (a) physical access to the market and the costs of transport to it, and (b) the prices prevailing in the markets to which they ship produce: largely a function of the size of the centre, the incomes of the populace, and the availability of competing supplies, these moderated by the existence of market imperfections and the effects of government policy on any of the variables mentioned.

Pressure on natural resources is seen as a combination of rural population density, the terms of access to such resources (land tenure), and the quality of the resources in terms of rainfall and soils.

Making qualitative judgements on the case studies gives Table 5.3. The evidence suggests the importance of access to markets, physically and economically. When there is access and farmgate prices are high enough, farmers will usually expand production. Where population pressure is low and land is available, the most likely response will be to bring new land into production or, where there is less land, to reduce fallow periods. This will be facilitated by use of animal draught or occasionally by tractors. Notwithstanding mechanisation of land preparation, other farm operations such as weeding and harvesting typically remain manual, so labour demand rises. Only when the land frontier is closed do we usually see intensification of land use: through fertiliser and manure use, irrigation, bunding, terracing, and so on, and increasing yields.

Market access is thus a necessary, but not sufficient condition for expanded smallholder production. Beyond that lies a trinity of supply-side challenges: access to capital, inputs and the ability of farmers to bear risk. These may limit response, particularly by individual farmers within communities, but rarely prevent at least some farmers taking advantage of market opportunities. For policy makers, overcoming such obstacles requires careful analysis, a detailed understating of particular cases, and a willingness to try unfamiliar policies, some of them politically awkward.

Technical innovations, it seems, rarely lead or impede the growth of farm production. In the cases seen, farmers have usually been able to draw on either indigenous knowledge or some pre-existing techniques known to other groups of farmers or extensionists, researchers or input suppliers to support extensification or intensification. This is not to belittle the role of agricultural technology generation. For example, the experience of Kenyan smallholders in the 1950s and 1960s adopting coffee and tea cultivation would hardly have been possible without decades of previous tropical research on these crops. Intensive dairying in the African

Table 5.3 Market access, pressure on resources and agricultural response

	Low pressure on natural resources	*Medium pressure on natural resources*	*High pressure on natural resources*
Good access to markets	*Successful extensive development:* • Northern Province, Zambia • Iringa, Tanzania • Mbozi, Tanzania • Ibarapa, Nigeria	*Successful intensification and extensification:* • S.Nyanza, Kenya • Boromo, B. Faso • Namu, Nigeria *Successful intensification:* • mid-Machakos, Kenya *Problematic intensification:* • Yensiso, Amanase, Sekesua, Ghana	*Successful intensification:* • Upper Machakos, Kenya • Upper Embu, Kenya • Kano CSZ, Nigeria *Problematic intensification:* • Lower Sokoto, Nigeria *Failing intensification:* • Imo State, Nigeria • Awka-Nnewi, Nigeria
Medium access to markets	*Coping with problematic farming:* • Lower Machakos, Kenya	*Coping with problematic farming:* • Lower Casamance, Senegal • Yako, B. Faso • Chivi, Zimbabwe *Failing or highly problematic agriculture:* • The Gambia • Middle Valley, Senegal • Upper Manya Krobo, Ghana	
Poor access to markets	*Struggling for subsistence:* • Djibo, B. Faso	*Struggling for subsistence:* • Dagaceri, Nigeria • Kandjadja, G. Bissau	*Failing intensification:* • Kusasi, Ghana

highlands benefits greatly from knowledge of artificial insemination, nutrition and disease control. The point, however, is that efforts to develop smallholder farming through the supply side, by the promotion of apparently more productive techniques, only prosper when there is market opportunity.

The public role in these cases is instructive. Governments can be powerful promoters of smallholder farming when they build roads, clear bush of tsetse fly and control the black fly of river blindness. Although not as easy to demonstrate, smallholder farming probably benefits from the provision of schooling and health care. The much-maligned parastatal crop and livestock authorities have also had similarly potent effects when, for example, they have boosted output prices in remote areas through the payment of pan-territorial prices, subsidised the price of fertiliser and other inputs (and maintained pan-territorial price levels), and provided input credits. On the other hand, in the cases reviewed, there is much less evidence of specific agricultural projects having been a primary motor in the developments seen.

The environment in these cases is clearly an important, sometimes critical, conditioning factor. In the semi-arid margins, low and unreliable rainfall makes crop farming risky, with half or more of harvests hit by drought, one or more in three severely, so there are few stories of successful farming. At best there are systems where farmers cope, by diversifying their crops and livestock to reduce risks, by building up assets in the good years to be set against the bad, and by undertaking diverse occupations to reduce dependence on farming.

Environmental change was less apparent as a primary factor in these cases except for desiccation in the Sahel. In other cases, accounts of reduced soil fertility, weed invasions and so on were set within the context of stagnant or failing agriculture resulting primarily from lack of demand, where farmers had little incentive to remedy local environmental problems. Conversely, situations are described where drives to increase farm production have been accompanied by soil and water conservation measures and tree planting: Machakos being the best-known example, the Kano close-settled zone being another.

5.4 Stasis and dynamics

The analysis of field-level studies has so far been largely cross-sectional. It thus sees how particular cases exhibit different characteristics for a particular period, usually one in which change remains in the same direction, the main causal factors being constant. It thus tends to omit the longer-term dynamics as causal factors themselves change.

At the risk of rehearsing the obvious, these causal factors will be explored. Smallholder farming fits into a pattern of peasant livelihoods in which households try to maximise their material welfare, subject to the demanding limitations of doing so within acceptable risk (and otherwise ensuring survival for the future), avoiding excessive drudgery, and otherwise respecting social norms regarding work, community obligations and so on. This may be modelled by focusing on the allocation of household labour between competing opportunities, within a set of limitations, some of them physical. Key parameters include the farm-gate prices of outputs and purchased inputs, and off-farm wage rates that determine both the price of hired help as well as the returns to using household labour off the farm. To this list may be added the opportunity cost of capital, although this rarely seems to be as critical a parameter as those previously mentioned.

These parameters change frequently, provoking changes in farming systems and indeed rural economies as a whole. Most obviously this occurs through changes exogenous to the local economy, as for example when the international price of an exported crop shifts, or when the government subsidises an input such as fertiliser.

But there are also changes arising through processes within local and wider systems, endogenous processes where the combined effect of the actions of individual households feed back, usually with a time lag. These include the following:

- within the farm and village, land use will have longer-run consequences on the costs of production, as soil fertility is either run down or improved, soil is eroded or conserved, water is retained or allowed to run off rapidly. Removal or planting of trees may also affect soil fertility and erosion, as well as micro-climates;
- within the village and local rural economy, population increase will exert pressure on the land and alter the relative prices of labour and land. It will also, following Boserup (1965), make it increasingly difficult to maintain soil fertility through fallowing, forcing farmers to work harder on their land, manuring and composting to maintain fertility, or to accept lower yields and lower returns to their labour;
- in the local rural economy, increased farm output may generate multipliers through both production and consumption, creating jobs and raising local wage rates;
- regionally, population growth is likely to create towns, expand urban markets, reduce the average distance of farm to market, and lower transport costs as better tracks and roads are built. This process, proposed by Boserup (1965), leads to the farm-gate terms

of trade improving (as output prices rise and the cost of bought inputs falls), thus providing farmers with a positive incentive to intensify production; and

- nationally and internationally, increased supply of particular commodities will at some point impose downward pressure on prices, whilst increased use of purchased inputs will tend to drive up their prices. In the first case, given relatively inelastic demand for farm produce, quite modest increments in marketed deliveries may depress prices. In the second case, the supply curve for manufactured farm inputs may have a gentle upward slope over a large range of output, so that input prices are little affected by the level of demand.

These endogenous processes overlap and interact, while time lags will differ from process to process, within a hierarchy of economies from farm to global level. The hierarchy may be summarised as shown in Table 5.4.

Thus, while we may start with a simple model of smallholder actions, once this is specified in any detail, we rapidly have a complex model, in which in particular cases a wide variety of processes may play a critical role. Moreover the variables are endogenous to different systems, giving different rates of feedback. And this complexity arises before any consideration is given to random variations and stochastic processes inherent in any complex system.

Hence the first conclusion to be derived is that comparatively few basic variables can generate considerable complexity. It is not surprising, then, that the experiences of smallholders in Africa are so diverse. It also means, and this is frustrating, that one can find an example of almost any variable that might reasonably be imagined driving a system at some place or time. Selective use of the evidence can be used to support a wide range of propositions about smallholder development! Small wonder that there are so many competing hypotheses to explain the disappointments of African agriculture from the 1970s onwards.

Table 5.4 Effects possible on an African smallholding through the economic hierarchy

Level of economy	Processes
Farm and village	• Population change and demand for land • Use of soil and soil fertility and erosion • Weather risks
Regional, local rural	• Population change affects supply of labour and wage rates; as well as demand for land and fallowing possibilities • Farm output has multiplier effects in stimulating the off-farm rural economy and raising wage rates
National	• Population increases and economic growth lead to urbanisation and rising demand for farm produce • Policies affect domestic prices for farm produce and farm inputs • Economic performance affects the non-farm economy and thus demand for off-farm labour and thus urban and non-farm wage rates
International	• Changes in world market prices for the main exported products – cotton, groundnuts, cocoa, coffee, tea, rubber, palm oil, sugar, tobacco, timber, fruit (bananas, plantain, mango, pineapple, citrus, etc.), vegetables (okra, beans, peas, peppers, etc.), and beef • Changes in world market prices of competing imports of grains and other foodstuffs, moderated by trade policy • Possibilities of international migration • Flows of aid and foreign investment

Policy-wise, this should make one wary of accepting advice based on the logical propositions derived from a much-reduced scenario (for example, that farmers will increase output as the price paid rises). Similarly, arguments for the replication of a success in one context need to be treated with scepticism.

To return to the empirical evidence, what is observed when the key parameters, for whatever reason, change? When costs of production fall or output prices rise, and returns to farming change – often dramatically so, since a 10 per cent change in price can change a gross margin by 50 per cent or more – then we can see transformation in smallholder farming. Booms can occur suddenly, and their effects can be rapid: within a decade, farming systems are frequently transformed and the landscape with it. Dramatic though these developments may be, most change in African farming occurs within the organisational form of smallholdings, without major changes in land tenure, massive infusions of capital or dramatic changes in technique. Instead the following are typically seen when farming prospers:

- switches in land use to the more profitable crops;
- the application of more labour, sometimes accompanied by in-migration;
- small increases in the use of capital (fertiliser, pesticides, purchased seeds, irrigation pumps, animal draught and occasionally tractors); and
- in some cases, especially with export crops, building of infrastructure such as storage and processing facilities and access roads by government or a monopsonistic trading company.

At other times, the fortunes of farming may decline as output prices fall, or costs of production rise. Under decline, farmers tend to adapt within the peasant household, rather than abandoning farming or their villages. Commonly observed reactions include:

- temporary migration out;
- the adoption of petty enterprise, requiring almost no capital and little new skill; and
- some reduction in the intensity of production, with sharp reductions in the use of purchased inputs such as chemical fertiliser or sprays, as well as the application of less labour on tasks such as weeding or pruning.

In both cases, experience shows the importance of the peasant household as the unit to organise production and consumption and to assure livelihoods. It also shows how important labour use is.

If the fortunes of farming systems depend on the (complex) interplay of a restricted range of variables, then this raises the policy question of which parameters to act on, and by how much, in order to get a favourable outcome, preferably a sustained boom in farming with positive feedback to the rest of the rural economy though multipliers. Such is believed to have underwritten economic success in Asia over the last quarter-century. There is no simple answer to this. Even the simplest and most reduced propositions, for example, estimating the elasticity of response of cotton output to its price, is less than simple. There seems little alternative to careful scrutiny of particular farming systems and rural economies to arrive at qualified synthetic judgements, as seen, for example, in Ruthenberg's (1980) great work on tropical farming.

On the other hand, a particular and recurring interest among policy makers is what to do when a fall in the international price drives down returns to an export crop. Neo-liberals would advocate letting the system adapt through the market. But, as can be seen, that might well involve a widespread retreat into subsistence farming, reduced export crop production, with sharply reduced export earnings, job losses in the export marketing chain, further downward multipliers and a reduced GDP. In theory this would free resources to be used in higher-return activities. But this supposes a much greater flexibility and mobility of factors of production than applies. Instead, this scenario seems likely to give us an African rural variant of a low-capacity trap, as expected by Keynes in a very different context.

5.5 Inequality and poverty

The discussion so far traces some familiar ground, but there is another consideration which has been out of fashion with most observers since the early 1980s: that of inequality and social differentiation in the countryside. As Heyer (1996) points out, the rural poor in Africa are often not as clearly disadvantaged structurally as the poor may be in other parts of the developing world. They usually have access to land, are able to migrate and to undertake non-farm work, just as their more prosperous neighbours. The apparent similarity of circumstance may also tend to obscure the extent and depth of their poverty. It certainly makes it less simple to be clear about the causes of poverty.

One view of poverty is geographical. Even Heyer appears to subscribe to the 'well-supported view [but no sources cited!] that poverty in sub-Saharan Africa is more a matter of the existence of poor rural areas than of inequality within them' (1996: 282). If only poverty could be neatly sketched on the map, so that efforts to alleviate it could be geographically focused. Instead, there are worrying signs that the converse may apply: that poverty is pervasive and occurs even where farming is relatively successful.

For example, Tiffen *et al.*'s (1994) much-admired study of Machakos District, Kenya, demonstrates much success in agricultural development and improved rural welfare over the last 40 years. But there is precious little in that work about differentiation. Instead, we have had to wait for Rocheleau *et al.* (1995) for reports of increasing social differentiation in the District. This has been reinforced by the detail in Murton's recently published work (1999), based on surveys in one village in upper Machakos, supported by rapid appraisal for other villages. Landlessness in this community is uncommon, but 55 per cent of the land is held by the upper quintile of farmers, and this has increased since the 1960s. Forty per cent or more of households have not had the capital to invest in cash crops of coffee and French beans, most of these cannot produce enough staple food to feed themselves, and depend on seasonal work on their neighbour's fields for their subsistence. When such work is scarce, these households go hungry. Thus Mbooni Location, although one of the most naturally favoured in Makueni District, has higher levels of malnutrition than other parts.

Similarly, Scoones *et al.* (1996) write that in Chivi, southern Zimbabwe, fully 75 per cent of marketed farm output comes from just 25 per cent of farms. Table 1.1 in that book shows wide variations within the community in access to livestock, area tilled, grain output and access to remittances. Moreover, the differences co-vary: thus those with more cattle have larger fields and also more access to remittances. In the same country, data from household surveys carried out in 1989–90 by Muchena (1993), in two contrasting parts of the communal lands show similar patterns of differences to those reported by Scoones. These data have been used by Piesse *et al.* (1999) to estimate Gini coefficients for household incomes of 0.45 and 0.48 for the two areas – remarkably large values for rural communities where differences between households are inconspicuous.

These and other surveys reveal surprisingly large variations in the fortunes of small-holder households within communities, even when almost all have access to land. Thus, whilst in some parts (upper Machakos, for example) there may be many signs of successful small-scale farming – well-kept fields, glossy-coated cows, tin-roofed houses, piped water, bustling market centres with well-stocked shops and so on – a large number of households, perhaps the majority, may be subsisting at survival level, no more. The same must surely apply in less prosperous zones. The poor, it seems, eke out their livelihoods by being prepared to undertake miserably petty jobs, to migrate to earn pittances whilst enduring much hardship, and by cutting consumption to the bone. And yet they seem not that different from their neighbours who, although not rich by any measure, enjoy some comfort in their livelihoods.

The rural poor, it seems, are crippled by lack of access to (often small amounts of) resources other than land, unable to invest in anything that might enhance their earnings, either for lack of capital or because they cannot afford to bear any risk. They often depend mightily on the fortunes of their neighbours for jobs (Murton, 1999), loans and gifts. That said, comparatively little is known about differentiation at village level. Much of the 1970s debates on differentiation, along Marxian lines, offered clear and cogent hypotheses of how capitalism might create classes amongst the peasantry, but were weaker on providing detailed evidence of the processes and degree of differentiation.

Despite considerable work on sub-components of the rural economy, too little is known about how changes in the economy result in changing absolute and relative standards of living for different actors. In particular, it is only in the last 10 years that researchers have turned their attention to the multipliers that act between farming and other sectors of the rural economy (see, for example, Delgado *et al.*, 1994). Moreover, the relations between economic variables and fundamental aspects of welfare such as nutrition and health are known in outline at best (Sahn, 1994).

5.6 Conclusions

Whatever crises are seen in the African countryside, the peasantry and their smallholder farms persist. Relatively few variables seem responsible for most of the changes seen in different areas and times. Yet those few variables operate within systems defined at various levels, replete with much feedback, producing considerable complexity in particular cases.

This implies that, although policy fundamentals may be fairly clear, policy fundamentalism is dangerous, since at the margin the particular system may be driven by a variable that is generally insignificant. Thus while we can define necessary conditions for the successful development of smallholder farming, a list of those which are sufficient is likely to be lengthy and hedged by many caveats. Policy making thus demands a great deal of understanding of particular circumstances, rather than the general application of ready-made recipes. In making policy, therefore, it may be that the ability to monitor and adapt is more valuable than to be able to select the 'best' policy in advance.

Academically, the challenge is to understand better how variables interact in complex systems. In particular, the linkages between increased farm output, the growth of the rural off-farm and non-farm economy, and the impacts on poverty and nutrition merit special attention. Some thought about the impact of chance might add a useful dimension to understanding.

Note

1 The thinking in this chapter has benefited from work with Millie Gadbois, conversations with Hugh Bunting, and the studies done by doctoral supervisees Mary Muchena, Ntengua Mdoe, Kofi Marfo, Dickson Nyariki and Kwasi Ohene-Yankyera. They do not necessarily share the ideas and opinions, still less the responsibility for any errors and omissions in this piece.

References and bibliography of case studies consulted

Adams, W.M. (1987) 'Approaches to water resource development, Sokoto Valley, Nigeria: the problem of sustainability', in D. Anderson and R. Grove (eds), *Conservation in Africa: People, Policies and Practice*, Cambridge: Cambridge University Press.

Adams, W.M. (1988) 'Irrigation and innovation: small farmers in the Sokoto Valley, Nigeria', in J. Hirst, J. Overton, B. Allen and Y. Byron (eds), *Small-scale Agriculture*, Canberra: Commonwealth Geographical Foundation and Department of Human Geography.

Adams, W.M. (1991) 'Large-scale irrigation in northern Nigeria: performance and ideology', *Transactions of the Institute of British Geographers*, NS 16: 287–300.

Adams, W.M. (1993) 'Development's deaf ear: downstream users and water releases from the Bakolori Dam, Nigeria', *World Development*, 21 (9): 1405–16.

Adams, W.M. and Mortimore, M.J. (1997) 'Agricultural intensification and flexibility in the Nigerian Sahel', *Geographical Journal*, 163 (2): 150–60.

al Hassan, Ramatu, Kydd, J. and Warner, M. (1996) 'Review of critical issues for natural resources in Ghana's Northern Region', Briefing Paper 4, research project: 'The Dynamics of Smallholder Agriculture in the Guinea Savannah Zone of West Africa, with Particular Reference to Northern Region Ghana' (NR Policy Research Continuum, R5584CA), Wye College, UK.

Allen, M. (1987) 'Questioning the need for seasonal farm credit: cases from Northern Zambia', *Agricultural Administration and Extension*, 25: 25–36

Amanor, Kojo Sebastian (1994) *The New Frontier: Farmer Responses to Land Degradation. A West African Study*, London and New Jersey: Zed Books, in association with Geneva: UNRISD, Geneva and Netherlands Technical Centre for Agricultural and Rural Cooperation (CTA).

Andersson, Jens A. (1996) 'Potato cultivation in the Uporoto mountains, Tanzania: an analysis of the social nature of agro-technological change', *African Affairs*, 95: 85–106.

Ariyo, J.A. and Ogbonna, D.O. (1992) 'The effects of land speculation on agricultural production among peasants in Kachia local government area of Kaduna State, Nigeria', *Applied Geography*, 12: 31–46.

Awumbila, Mariama and Henshall Momsen, J. (1995) 'Gender and the environment: women's time use as a measure of environmental change', *Global Environmental Change*, 5 (4): 337–46.

Baker, K.M. (1992) 'Traditional farming practices and environmental decline, with special reference to The Gambia', in K. Hoggart (ed.), *Agricultural Change, Environment, and Economy*, London: Mansell.

Baker, K.M. (1995), 'Drought agriculture and the environment: a case study from The Gambia, West Africa', *African Affairs*, 94: 67–86.

Bantje, Han (1986) 'Household differentiation and productivity: a study of smallholder agriculture in Mbozi District', Research Paper no. 14, Institute of Resource Assessment, University of Dar es Salaam, Tanzania.

Bassett, T.J. (1988) 'Development theory and reality: the World Bank in northern Ivory Coast', *Review of African Political Economy*, 41: 45–59.

Bassett, T.J. (1993) 'Introduction: the land question and agricultural transformation in sub-Saharan Africa', in T.J. Bassett and D.E. Crummey (eds), *Land in African Agrarian Systems*, Madison, WI: University of Wisconsin Press.

Bates, R. (1981) *Markets and States in Tropical Africa*, Berkeley, CA: University of California Press.

Becker, Laurence C. (1990) 'The collapse of the family farm in West Africa? Evidence from Mali', *Geographical Journal*, 156 (3): 313–22.

Bemstein, Henry (1979) 'African peasantries: a theoretical framework', *Journal of Peasant Studies*, 6 (4): 421ff.

Berry, S. (1993) *No Condition is Permanent: The Social Dynamics of Agrarian Change in Sub-Saharan Africa*, Madison, WI: University of Wisconsin Press.

Berry, S. (1997) 'Tomatoes, land and hearsay: property and history in Asante in the time of structural adjustment', *World Development*, 25 (80): 1225–41.

Bingen, J., Carney, D. and Dembelé, E. (1995), 'The Malian union of cotton and food crop producers: its current and potential role in technology development and transfer', Agricultural Research and Extension Network Paper, London: Overseas Development Institute.

Binns, Tony (ed.) (1996) 'People, environment and development in West Africa', *Geography*, 81 (4): 361–407.

Binswanger, H. and Pingali, P. (1989) 'Technological priorities for farming in sub-Saharan Africa', *Journal of International Development*, 1 (1): 46–65.

Birch-Thomsen, T. (1990) 'Agricultural change: prospects for the future: A case study from the Southern Highlands of Tanzania', *Quarterly Journal of International Agriculture*, 29 (2): 146–60.

Block, S.A. (1995) 'The recovery of agricultural productivity in sub-Saharan Africa', *Food Policy*, 20 (5): 385–405.

Bloch, P. (1993) 'An egalitarian development project in a stratified society: who ends up with the land?', in T.J. Bassett and D.E. Crummey (eds), *Land in African Agrarian Systems*, Madison, WI: University of Wisconsin Press.

Boesen, J. and Ravnborg, H.M. (1992) 'Peasant production in Iringa District, Tanzania', CDR Project Paper 93 1, Copenhagen: Centre for Development Research.

Bolt, R. and Holdsworth, I. (1987) 'Farming systems economy and agricultural commercialization in the south-eastern plateau of Northern Province, Zambia', ARPT Economic Study no. 1, Kasama, NP, Zambia: Adaptive Research and Planning Team.

Borton, J. and Clay, E. (1988) 'The African food crisis of 1982–86: a provisional review', in D. Rimmer (ed.), *Rural Transformation in Tropical Africa*, London: Belhaven.

Boserup, E. (1965) *The Conditions of Agricultural Growth: The Economics of Agrarian Change under Population Pressure*, London: Earthscan.

Boserup, E. (1987) 'Agricultural growth and population change', in J. Eatwell, M. Milgate and P. Newman (eds), *The New Palgrave Dictionary of Economics*, London and Basingstoke, Hants.: Macmillan.

Bryceson, D. (1996) 'De-agrarianization and rural employment in sub-Saharan Africa: a sectoral perspective', *World Development*, 24 (1): 97–111.

Bryceson, D. (1998) 'African farm labour: where to?', paper given to conference on 'Africa and Globalisation: Towards the Millennium', University of Central Lancashire, April.

Bush, Ray (1997) 'Africa's environmental crisis: challenging the orthodoxies', *Review of African Political Economy*, 74: 503–13.

Carney, J. and Watts, M. (1990) 'Manufacturing dissent: work, gender, and the politics of meaning in a peasant society', *Africa*, 60 (2): 207–41.

Carter, Michael R. (1997) 'Environment, technology, and the social articulation of risk in West African agriculture', *Economic Development and Cultural Change*, 45: 557–90.

Chaléard, Jean-Louis (1989) 'Risque et agriculture de plantation: l'exemple des cultures comerciales developpées dans le departement d'Agboville (Côte d'Ivoire)', in M. Eldin and P. Milleville (eds), *Le Risque en agriculture*, Paris: Editions de l'ORSTOM, Institut Français de Recherche Scientifique pour le Developpement en Coopération.

Chaléard, Jean-Louis (1996) 'Les mutations de l'agriculture commercials en Afrique de l'Ouest', *Annales de géographie*, 105 (592): 563–83.

Cleveland, David (1991) 'Migration in West Africa: a savanna village perspective', *Africa*, 61 (2): 222–45.

Cliffe, Lionel (1977), 'Rural class formation in East Africa', *Journal of Peasant Studies*, 4 (2): 195ff.

Collinson, M.P. (1989) 'Small farmers and technology in Eastern and Southern Africa', *Journal of International Development*, 1 (1): 66–82.

Collion, Marie-Hélène and Rondot, Pierre (1998) 'Partnerships between agricultural services institutions and producers' organisations: myth or reality?', Agricultural Research and Extension Network Paper no. 80, London: Overseas Development Institute.

DeCosse, P. (1992) 'Structural change in Gambian agriculture: stagnation or silent transformation?', Banjul, The Gambia: USAID, mimeo.

Dei, George J.S. (1991) 'The re-integration and rehabilitation of migrant workers into a local domestic economy: lessons for "endogenous" development', *Human Organization*, 50 (4): 327–36.

Delgado, Christopher L. (1994) 'Africa's changing agricultural development strategies: past and present paradigms as a guide to the future', Food, Agriculture and the Environment, Discussion Paper 3, Washington, DC: International Food Policy Research Institute.

Delgado, Christopher L., Hopkins, J.C. and Kelly, Valerie A. (with Peter B.R. Hazell, Anna Alfano, Peter Gruhn, Behjat Hojjati and Jayashree Sil) (1994) *Agricultural Growth Linkages in Sub-Saharan Africa*, Washington, DC: International Food Policy Research Institute.

Dercon, Stefan (1994) 'Peasant supply response and macroeconomics policies: cotton in Tanzania', *Journal of African Economies*, 2 (2): 157–93.

de Waal, Alex (1989) *Famine that Kills: Darfur, Sudan, 1984–1985*, Oxford: Oxford University Press.

Diemer, G., Fall, G.B. and Huibers, F.P (1991) 'Promoting a smallholder-centred approach to irrigation: lessons from village irrigation schemes in the Senegal river valley', Irrigation Management Network Paper no. 6, London: Overseas Development Institute.

Eyoh, Dickson (1990) 'National policy versus local power: the Lafia Project in Nigeria', *Canadian Journal of African Studies*, 24 (2): 216–34.

Eyoh, Dickson L. (1992) 'Reforming peasant production in Africa: power and technological change in two Nigerian villages', *Development and Change*, 23 (2): 37–66.

Fairhead, J. and Leach, M. (1996) 'Relics of colonial science: rethinking West Africa's forest-savanna mosaic', from the draft of *Challenging Received Wisdom on the African Environment*, ed. Melissa Leach, and Robin Mearns, Oxford: James Currey.

Fairhead, J. and Leach, M. (1996) 'Enriching the landscape: social history and the management of transition ecology in the forest-savanna mosaic of the Republic of Guinea', *Africa*, 66 (1): 14–36.

Faussey-Domalain, Catherine and Vimard, Patrice (1991) 'Agriculture de rente et démographic dans le sud-est Ivoirien: une économie villageoise assisté en milieu forestier péri-urban', *Revue tiers monde*, 32 (125): 53–114

Food Studies Group, Oxford, UK and Department of Rural Economy, Sokoine University, Morogoro, Tanzania (1992) 'Agricultural diversification and intensification study: final report', vol. II [on farming systems in Tanzania], Oxford: Food Studies Group.

Francis, P. (1988) 'Ox draught power and agricultural transformation in Northern Zambia', *Agricultural Systems*, 27: 35–49.

Francis, P. and Rawlins-Branan, M.J. (1987) 'The extension system and small-scale farmers: a case study from Northern Zambia', *Agricultural Administration and Extension*, 26: 185–96.

Friis-Hansen, E., (1988) 'Villagization and changes in Tanzania farming systems: a case study', in *EADL* (European Association of Development Research and Training Institutes) *Bulletin*, 1 (1): 71–99.

Funna, S.M., (1987) 'Sierra Leone: economic structure and recent performance', in A. Jones and P.K. Mitchell (eds), *Sierra Leone Studies at Birmingham 1985*, University of Birmingham.

Gilbert, E. (1990) 'Non-governmental organisations and agricultural research: the experience of The Gambia', Agricultural Administration (Research and Extension) Network Paper 12, London: Overseas Development Institute.

Goldman, Abe (1993) 'Population growth and agricultural change in Imo State, southeastern Nigeria', in B.L. Turner II, Goran Hyden and Robert W. Kates (eds), *Population Growth and Agricultural Change in Africa*, Gainesville, FL: University Press of Florida.

Goldman, Abe and Smith, Joyotee (1995) 'Agricultural transformations in India and Northern Nigeria: exploring the nature of green revolutions', *World Development*, 23 (2): 243–63.

Guyer, J. (1984) 'Women's work and production systems: a review of two reports on the agricultural crisis', *Review of African Political Economy*, 27/8: 186–90.

Guyer, Jane I. (1989) 'Women's farming and present ethnography: thoughts on a Nigerian re-study', 19th Annual Hans Wolff Memorial Lecture, Indiana University.

Guyer, Jane I. (1992) 'Small change: individual farm work and collective life in a western Nigerian savannah town, 1969–88', *Africa*, 62 (4): 465–89.

Guyer, Jane I. and Lambin, Eric F. (1993) 'Land use in an urban hinterland: ethnography and remote sensing in the study of African intensification', *American Anthropologist*, 95 (4): 839–59.

Gyasi, Edwin (1994) 'The adaptability of African communal land tenure to economic opportunity: the example of land acquisition for oil palm farming in Ghana', *Africa*, 64 (3): 391–405.

Gyasi, Edwin A. and Uitto, Juha I. (1997) *Environment, Bio-diversity and Agricultural Change in West Africa*, Tokyo, New York and Paris: United Nations University Press.

Gyasi, Edwin, Agyepong, G.T., Ardayfio-Scandorf, E.L., Enu-Kwesi, Nabila J.S. and Owuso-Bennoah, E. (1995) 'Production pressure and environmental change in the forest-savanna zone of southern Ghana', *Global Environmental Change*, 5 (4):, 355–66.

Haswell, M. (1991) 'Population and change in a Gambian rural community, 1947–1987', in M. Haswell and D. Hunt (eds), *Rural Households in Emerging Societies: Technology and Change in Sub-Saharan Africa*, Oxford: Berg.

Haugerud, A. (1983) 'The consequences of land tenure reform among smallholders in the Kenya highlands', *Rural Africana*, 15–16: 65–89.

Haugerud, A. (1988) 'Food surplus production, wealth, and farmers' strategies in Kenya', in R. Cohen (ed.), *Satisfying Africa's Food Needs*, Boulder, CO: Lynne Reiner.

Haugerud, A. (1989) 'Land tenure and agrarian change in Kenya', *Africa*, 59 (1): 61–90.

Heyer, Judith (1996) 'The complexities of rural poverty in sub-Saharan Africa', *Oxford Development Studies*, 24 (3): 281–97.

Holden, Stein T. (1993) 'Peasant household modelling: farming systems evolution and sustainability in northern Zambia', *Agricultural Economics*, 9: 241–67.

Hyden, Goran (1980) *Beyond Ujamaa in Tanzania: Underdevelopment and an Uncaptured Peasantry*, London: Heinemann.

Jabbar, Mohammad (1993) 'Evolving crop-livestock farming systems in the humid zone of West Africa: potential and research needs', *Outlook on Agriculture*, 22 (1): 13–21.

Jaeger, William K. (1992) 'The causes of Africa's food crisis', *World Development*, 20 (11): 1631–45.

Jamin, P.Y. and Tourrand, J.F. (1986) 'Évolution de l'agriculture et de l'élevage dans une zone de grands aménagements: le delta du fleuve Sénégal', *Le Cahiers de la recherche développement*, no. 12: 21–34.

Jayne, T.S. and Jones, Stephen (1997) 'Food marketing and pricing policy in Eastern and Southern Africa: a survey', *World Development*, 25 (9): 1505–27.

Jones, Samantha (1996) 'Farming systems and nutrient flows. A case of degradation?', *Geography*, 81 (4): 289–300.

Kelly, V., Reardon, T., Diagana, Bocar and Abdoulaye Fall, Amadou (1995) 'Impacts of devaluation on Senegalese households: policy implications', *Food Policy*, 20 (4): 299–313.

Kennedy, E.T. (1989) 'The effects of sugarcane production on food security, health, and nutrition in Kenya: a longitudinal analysis', IFPRI Research Report no. 78, Washington, DC: International Food Policy Research Institute.

Kennedy, E.T. and Cogill, B. (1987) 'Income and nutritional effects of the commercialization of agriculture in south-western Kenya', IFPRI Research Report no. 63, Washington, DC: International Food Policy Research Institute.

Kennedy, E. and Cogill, B. (1988) 'The commercialisation of agriculture and household-level food security: the case of south-western Kenya', *World Development*, 16 (9): 1075–81.

Kimmage, K. (1991) 'The evolution of the "wheat trap": the Nigerian wheat boom', *Africa*, 60 (4): 471–501.

Kimmage, K. and Adams, W.M. (1992) 'Wetland agricultural production and river basin development in the Hadejia-Jama'are valley, Nigeria', *Geographical Journal*, 158 (1): 1–12.

Leach, Melissa and Mearns, Robin (eds) (1996) *The Lie of the Land: Challenging Received Wisdom on the African Environment*, Oxford: James Currey.

Lecomte, B. (1992) 'Senegal: the young farmers of Walo and the new agricultural policy', *Review of African Political Economy*, 55: 87–94.

Le Roy, Xavier (1989) 'Fragilisation de systèmes de production par l'introduction de cultures de rapport, Nord Côte d'Ivoiré', in M. Eldin and P. Millevill, (eds), *Le Risque en agriculture*, Paris: Editions de l'ORSTOM, Institut Français de Recherche Scientifique pour le Développement en Coopération.

Linares, Olga (1992) *Power, Production and Prayer: The Jola of Casamance, Senegal*, Cambridge: Cambridge University Press.

Linares, Olga F. (1996) 'Cultivating biological and cultural diversity: urban farming in Casamance, Senegal', *Africa*, 66 (1): 104–21.

Low, Allan, R.C. (1994) 'Environmental and economic dilemmas for farm-households in Africa: when "low-input sustainable agriculture" translates to "high-cost unsustainable livelihoods"', *Environmental Conservation*, 21 (3): 220–4.

McClintock, John (1993) 'Trees and economics in the Sahel: a case study of two Senegalese villages', Report of McNamara Fellowships Program, Economic Development Institute of the World Bank.

Mace, Ruth (1989) 'Gambling with goats: variability in herd growth among restocked pastoralists in Kenya', Pastoral Development Network Paper 28a, London: Overseas Development Institute.

McIntire, J., Bourzat, D. and Pingali, P. (1992) *Crop–Livestock Interaction in Sub-Saharan Africa*, Washington, DC: World Bank.

Mackintosh, M. (1989) *Gender, Class and Rural Transition: Agribusiness and the Food Crisis in Senegal*, London: Zed Books.

McMillan, Della E. and Meltzer, Martin I. (1996) 'Vector-borne disease control in sub-Saharan Africa: a necessary but partial vision of development', *World Development*, 24 (3): 569–88.

McMillan, Della E., Sanders, John H., Koenig, Dolores, Akwabi-Ameyaw, Kofi and Painter, Thomas M. (1998), 'New land is not enough: agricultural performance of new lands settlement in West Africa', *World Development*, 26 (2): 187–211.

Martin, Susan (1993) 'From agricultural growth to stagnation: the case of the Ngwa, Nigeria, 1900–1980', in B.L. Turner II, Goran Hyden and Robert W. Kates (eds), *Population Growth and Agricultural Change in Africa*, Gainesville, FL: University Press of Florida.

Matlon, Peter (1991) 'Farmer risk management strategies: the case of the West African semi-arid tropics', in D. Holden, P. Hazell and A. Pritchard (eds) *Risk in Agriculture: Proceedings of the Tenth Agriculture Sector Symposium*, Washington, DC: World Bank.

Mills B.F., Kebay, M.B. and Boughton, D. (1988) 'Soil fertility management strategies in three villages of eastern Gambia', Research Paper 2, Department of Agricultural Research, Ministry of Agriculture, The Gambia.

Milner-Gulland, E.J., Mace, Ruth and Scoones, Ian (1996) 'A model of household decisions on dryland agro-pastoral systems', *Agricultural Systems*, 51 (4): 407–30.

Moore, H. and Vaughan, M. (1987) 'Cutting down trees: women, nutrition and agricultural change in the Northern Province of Zambia, 1920–1986', *African Affairs*, 86 (345): 523–40.

Morgan, William B. and Solarz, Jerzy A. (1994) 'Agricultural crisis in sub-Saharan Africa: development constraints and policy problems', *Geographical Journal*, 160 (1): 57–73.

Mortimore, M. (1989) *Adapting to Drought: Farmers, Famines, Desertification in West Africa*, Cambridge: Cambridge University Press.

Mortimore, Michael (1993) 'The intensification of the peri-urban agriculture: the Kano close-settled zone, 1964–1986', in B.L. Turner II, Goran Hyden and Robert W. Kates (eds), *Population Growth and Agricultural Change in Africa*, Gainesville, FL: University Press of Florida.

Mosley, P. (1993) 'Policy and capital market constraints to the African Green Revolution: a study of maize and sorghum yields in Kenya, Malawi and Zimbabwe, 1960–1991', Discussion Papers in

Development Economics, Series G, vol. 1 (6), Departments of Economics and of Agricultural Economics and Management, University of Reading, UK.

Muchena, Mary (1993) *Cattle in Mixed Farming Systems of Zimbabwe: An Economic Analysis*, PhD Thesis, University of Reading, UK.

Murton, J. (1999) 'Population growth and poverty in Machakos District, Kenya', *Geographical Journal*, 165 (1): 37–46.

Ndiaye, S. and Sofranko, A.J. (1988) 'Importance of labor in adoption of a modern farm input', *Rural Sociology*, 53 (4): 421–32.

Netting, R.M., Stone, M.P. and Stone, G.D. (1989) 'Kofyar cash-cropping: choice and change in indigenous agricultural development', *Human Ecology*, 17 (3): 299–319.

Niasse, M. (1990) 'Village irrigation perimeters at Doumga Rindiaw, Senegal', *Development Anthropology Network*, 8 (1): 6–11.

Niasse, M. (1991) 'Production systems in the Senegal Valley in a no-flood context, Senegal', *Development Anthropology Network*, 9 (2): 12–20.

Nindi, B.C. (1988) 'Issues in agricultural change: case study from Ismani, Iringa Region, Tanzania', in D.W. Brokensha and P.C. Little (eds), *Anthropology of Development and Change in East Africa*, Boulder, CO: Westview Press.

Nyerges, A. Endre (1996) 'Ethnography in the reconstruction of African land use histories: a Sierra Leone example', *Africa*, 66 (1): 123–44.

Okafor, Francis C. (1993) 'Agricultural stagnation and economic diversification: Awka-Nnewi Region, Nigeria, 1930–1980', in B.L. Turner, G. Hyden and R.W. Kates (eds), *Population Growth and Agricultural Change in Africa*, Gainesville, FL: University Press of Florida.

Okorji, E.C. and Obiechina, C.O.B. (1990) 'Implications of overdependence on women's labor for food production in traditional farming systems of Anambra State, Nigeria', *Culture and Agriculture*, 40: 7–11.

Okoth-Ogendo, H.W.O. (1993) 'Agrarian reform in sub-Saharan Africa: an assessment of state responses to the African agrarian crisis and their implications for agricultural development', in T.J. Bassett and D.E. Crummey (eds), *Land in African Agrarian Systems*, Madison, WI: University of Wisconsin Press.

Osborn, T. (1990) 'Multi-institutional approaches to participation technology development: a case study from Senegal', Agricultural Administration (Research and Extension) Network Paper 13, London: Overseas Development Institute.

Ouedraogó, Robert S., Sawadogo, Jean-Pierre, Stamm, Volker and Thiombiano, Taladia (1996) 'Tenure, agricultural practices and land productivity in Burkina Faso: some recent empirical results', *Food Policy*, 13 (3): 229–32.

Painter, Thomas, Sumberg, James and Price, Thomas (1994) 'Your terroir and my "action space": implications of differentiation, mobility and diversification for the *approche terroir* in Sahelian West Africa', *Africa*, 64 (4): 447–64.

Park, T.K. (ed.) (1993) *Risk and Tenure in Arid Lands: The Political Ecology of Development in the Senegal River Basin*, Tucson and London: University of Arizona Press.

Piesse, J., Simister, J., Thirtle, C. and Wiggins, S. (1999) 'Modernisation, multiple income sources and equity in the communal lands in Zimbabwe', *Agrekon*, 38: 243–58.

Pingali, P., Bigot, Y. and Binswanger, H. (1987) *Agricultural Mechanisation and the Evolution of Farming Systems in Sub-Saharan Africa*, Baltimore, MD: Johns Hopkins University Press for the World Bank.

Platteau, Jean-Phillippe (1996) 'Physical infrastructure as a constraint on agricultural growth: the case of sub-Saharan Africa', *Oxford Development Studies*, 24 (3): 189–219.

Porter, Gina (1994) 'Food marketing and urban food supply on the Jos Plateau, Nigeria: a comparison of large and small producer strategies under "SAP"', *Journal of Developing Areas*, 29: 91–111.

Porter, Gina (1996) 'SAPs and road transport deterioration in West Africa', *Geography*, 81 (4).

Porter, Gina and Phillips-Howard, Kevin (1997) 'Comparing contracts: an evaluation of contract farming schemes in Africa', *World Development*, 25 (2): 227–38.

Posner, J.L. and Gilbert, E. (1989) 'District agricultural profile of Central Baddibu, North Bank Division', Gambia Agricultural Research Papers, Working Paper 2, Ministry of Agriculture, The Gambia.

Posner, J.L., Kamuanga, M. and Sall, S. (1988) 'Production systems in the lower Casamance and farmer strategies in response to rainfall deficits', International Development Paper 20, East Lansing, MI: Michigan State University.

Pottier, J. (1990) 'Village responses to food marketing alternatives in Northern Zambia: the case of the Mambwe economy', paper delivered to workshop held at African Studies Centre, Cambridge, March.

Poulton, Colin, Dorward, Andrew and Kydd, Jonathan (1998) 'The revival of smallholder cash crops in Africa: public and private roles in the provision of finance', *Journal of International Development*, 10 (1): 85–103.

Raikes, P. (1988) *Modernising Hunger. Famine, Food Surplus and Farm Policy in the EEC and Africa*, London: Catholic Institute of International Relations and James Currey.

Rasmussen, T. (1985) 'The Green Revolution in the Southern Highlands of Tanzania', in CDR Project Paper, no. A.85.7, Copenhagen: Centre for Development Research.

Rasmussen, T. (1986) 'The Green Revolution in the Southern Highlands', in J. Bosen, K. Havnevik, J. Koponen and R. Odgaard (eds), *Tanzania: Crisis and Struggle for Survival*, Uppsala: Scandinavian Institute for African Studies.

Reardon, Thomas (1991) 'Income diversification of rural households in the Sahel', mimeo.

Reardon, Thomas (1995) 'Sustainability issues for agricultural research strategies in the semi-arid tropics: focus on the Sahel', *Agricultural Systems*, 48: 345–59.

Reardon, T., Kelly, V., Crawford, E., Diagana, B., Dioné, J., Savadogo, K. and Boughton, D. (1997) 'Promoting sustainable intensification and productivity growth in Sahel agriculture after macroeconomic policy reform', *Food Policy*, 22 (4): 317–27.

Reardon, Thomas, Edward Taylor, J., Stamoulis, Kostas, Lanjouw, Peter and Balisacan, Arsenio (1998) 'Effects of nonfarm employment on rural income inequality in developing countries: an investment perspective', symposium paper for the Agricultural Economics Society Annual Conference, Reading, UK, March.

Rocheleau, E., Steinberg, E. and Benjamin, A. (1995) 'Environment, development, crisis, and crusade: Ukambani, Kenya, 1980–1990', *World Development*, 23 (6): 1037–51.

Rudebeck, Lars (1988) 'Kandjadja, Guinea-Bissau 1976–1986: observations on the political economy of an African village', *Review of African Political Economy*, 41: 17–29.

Rudebeck, Lars (1990) 'The effects of structural adjustment in Kandjadja, Guinea-Bissau', *Review of African Political Economy*, 49: 34–51.

Ruthenberg, Hans (1980) *Farming Systems in the Tropics*, 3rd edn, Oxford: Oxford Scientific Publications.

Sahn, David E. (1994) 'On economic reform, poverty and nutrition in Africa', *American Economic Review*, 84 (2): 285–90.

Samuels, F. and Leplaideur, A. with the collaboration of Harriss, B. (1991) 'Changing agrarian structure and petty commodity production in the Northern Region of Ghana', DCV/Labo *Agro-Economie no. 24*, Programme Riz, Institut de recherches agronomiques tropicales et des cultures vivrières, Montpellier: CIRAD.

Sandford, S. (1983) *Management of Pastoral Development in the Third World*, Chichester, Sussex: John Wiley.

Sano, H.-O. (1990) 'Political economy and agrarian change in Zambia: the case of Northern Province', paper delivered to workshop held at African Studies Centre, Cambridge, March.

Savadogo, K., Reardon, T. and Pietola, K. (1994) 'Mechanization and agricultural supply response in the Sahel: a farm-level profit function analysis', *Journal of African Economies*, 4 (3): 336–77.

Savadogo, Kimsey, Larivière, S. and Martin, Frédéric (1995) 'Stratégies des ménages ruraux en matière de sécurité alimentaire dans un contexte d'ajustement structural: le cas de la province du Passoré au Burkina Faso', *Economies et sociétés*, 22 (3–4): 145–66.

Schatz, S.P. (1986) 'African food imports and food production: an erroneous interpretation', *Journal of Modern African Studies*, 24 (1): 177–8.

Schroeder, Richard A. (1993) 'Shady practice: gender and the political ecology of resource stabilization in Gambian garden/orchards', *Economic Geography*, 69 (4): 349–65.

Schroeder, Richard A. (1994), 'Contradictions along the commodity road to environmental stabilization: foresting Gambian gardens', *Antipode*, 27 (4): 325–42.

Scoones, I. with Chibudu, C., Chikura, S., Jeranyama, P., Machaka, D., Machanja, W., Mavedzenge, B., Mombeshora, B., Mudhara, M., Mudziwo, C., Murimbarimba, F. and Zirereza, B. (1996) *Hazards and Opportunities: Farming Livelihoods in Dry Land Africa. Lessons from Zimbabwe*, London: Zed Books.

Seaman, John (1993) 'Famine mortality in Africa', *IDS Bulletin*, 24b (4): 27–32.

Seur, H. (1990) 'The introduction and diffusion of agricultural innovations in Chibale chiefdom, Serenje District, Zambia: peasants, policy, and the plough', paper delivered to workshop held at African Studies Centre, Cambridge, March.

Sichone, O.B. (1990) 'The motive forces of development: views from an Isoka village', paper delivered to workshop held at African Studies Centre, Cambridge, March.

Sinclair, A.R.E. and Fryxell, J.M. (1985) 'The Sahel of Africa: ecology of disaster', *Canadian Journal of Zoology*, 63: 987–94.

Smith, J., Barau, A.D., Goldman, A. and Mareck, J.H. (1993) 'The role of technology in agricultural intensification: the evolution of maize production in the Northern Guinea Savannah of Nigeria', *Economic Development and Cultural Change*, 42 (3): 537–54.

Snrech, Serge (1995) 'Les transformations de l'agriculture ouest-africaine: evolutions 1960–1990. Défis pour l'avenir: implications pour les pays saheliens', mimeo (Sahl (95) 451), Paris: Club du Sahel, December.

Sow, A. (1983) 'Les contraintes au développement de la petite exploitation agricole dans la moyenne vallée du fleuve Sénégal', Étude No RRD. 19, Division de l'étude du développement, Paris: UNESCO.

Stiles, Daniel (1995) 'Desertification is not a myth', *Desertification Control Bulletin*, no. 26: 29–36.

Stone, G.D., Netting, R.McC. and Stone, M.P. (1990) 'Seasonality, labor scheduling, and agricultural intensification in the Nigerian Savanna', *American Anthropologist*, 92: 7–23.

Swift, Jeremy (1996) 'Desertification: narratives, winners and losers', in Melissa Leach and Robin Meams (eds), *The Lie of the Land: Challenging Received Wisdom on the African Environment*, Oxford: James Currey.

Swindell, K. (1986) 'Population and agriculture in the Sokoto-Rima basin of north-west Nigeria', *Cahiers d'études africaines*, 101–2 (XXVI–1–2): 75–111.

Swindell, K. and Iliya, M.A. (1992) 'Accumulation, consolidation and survival: non-farm incomes and agrarian change in north-west Nigeria', Working Paper no. 6 of the Project on African Agriculture, New York: Social Science Research Council.

Swindell, K. and Mamman, A.B. (1990) 'Land expropriation and accumulation in the Sokoto periphery, north-west Nigeria 1976–86', *Africa*, 60 (2): 173–87.

Tellegen, Nina (1998) 'Survival or growth? Rural enterprises in Malawi 1983–1993', paper presented to Conference on 'Africa and Globalisation: Towards the Millennium', University of Central Lancashire, April.

Thomas, D. (1997) 'Desertification: the uneasy interface between science, people and environmental issues in Africa', *Review of African Political Economy*, 74: 583–9.

Tiffen, M. (1992) 'Environment, population growth and productivity in Kenya: a case study of Machakos District', paper presented to the Annual Conference of ESRC Development Economics Study Group, University of Leicester, March.

Tiffen, M., Mortimore, M. and Gichuki, F. (1994) *More People, Less Erosion: Environmental Recovery in Kenya*, Chichester, Sussex: John Wiley.

Tosh, J. (1980) 'The cash-crop revolution in tropical Africa: an agricultural reappraisal', *African Affairs*, 79 (314): 79–94.

Tourrand, J.F. and Ndiaye, M. (1988) 'Innovations techniques en milieu paysan dans le delta du fleuve Sénégal pour l'alimentation du cheptel', *Cahiers de la recherche développement*, 17: 47–53.

Turner, B.L., Hyden, G. and Kates, R. (eds) (1993) *Population Growth and Agricultural Change in Africa*, Gainesville, FL: University of Florida Press.

van der Ploeg, J.D. (1991) 'Autarky and technical change in rice production in Guinea Bissau: on the importance of commoditisation and de-commoditisation as interrelated processes', in M. Haswell and D. Hunt (eds), *Rural Households in Emerging Societies: Technology and Change in Sub-Saharan Africa*, Oxford: Berg.

Vierich, H.I.D. and Stoop, W.A. (1990) 'Changes in West African savanna agriculture in response to growing population and continuing low rainfall', *Agriculture, Ecosystems and Environment*, 31: 115–32.

Watts, M. (1983) *Silent Violence: Food Famine, And Peasantry in Northern Nigeria*, Berkeley: University of California Press.

Webb, J.L.A. (1992) 'Ecological and economic change along the middle reaches of the Gambia river, 1945–1985', *African Affairs*, 91 (1108): 543–65.

Webber, Paul (1996) 'Agrarian change in Kusasi, North-East Ghana', *Africa*, 66 (3): 437–57.

Wiggins, Steve (1995) 'Change in African farming systems between the mid-1970s and the mid-1980s', *Journal of International Development*, 7 (6): 807–46.

Wiggins, Steve (1998) 'African farming seen from village studies: changes from the 1970s to the 1990s', Paper for the African Studies Association supported conference on 'Africa and Globalisation: Towards the Millennium', University of Central Lancashire at Preston, UK, April.

Wiggins, Steve (1999) 'Setting the scene: recent change in West African farming and natural resource management', in Roger Blench (ed.), *Natural Resource Management in Ghana and its Socio-economic Context*, London: Overseas Development Institute.

Wiggins, Steve and Gadbois, Millie (1995) 'What do we know about change in the agricultural and rural economies of West Africa? A review of changes since the early 1970s', paper for Conference of the African Studies Association, 1–3 September, University of Central Lancashire.

Woodhouse, P. and Ndiaye, L. (1991a) 'Structural adjustment and irrigated agriculture in Senegal', Irrigation Management Network Paper no. 7, London: Overseas Development Institute.

Woodhouse, P. and Ndiaye, L. (1991b) 'Structural adjustment and irrigated food farming in Africa: the "disengagement" of the state in the Senegal River Valley', Working Paper no. 5 of the Project on African Agriculture, New York: Social Science Research Council.

Woodhouse, P., Chenevix-Trench, P. and Tessougué, M. (1997)'After the Flood: local initiative in using a new wetland resource in the Sourou Valley, Mali', *Geographical Journal*, 163 (2): 170–9.

6 The intensification of small-scale livestock enterprises

Progress and prospects

Martin Upton

6.1 Introduction

In this chapter, an attempt is first made to clarify the concept of intensification of livestock enterprises. The process is viewed in terms of increasing stocking rates, a switch to more intensive types of livestock (pigs and poultry) and increasing variable inputs per animal. Crude, 'ballpark' statistics published by FAO are then used to indicate that all these forms of livestock intensification have occurred in Africa over recent decades. Associated changes in livestock systems include the shift from purely grassland-based systems to forms of agro-pastoralism and thus to mixed farming, and the change from mixed farming to the use of purchased feeds and a shift to landless animal production systems. Components of the intensification process include improvements in animal health, nutrition, management, marketing, housing and breeding. These components are discussed in detail to stress the need for them to be integrated in a package or for their introduction to be properly phased. The discussion is illustrated with examples from small-scale dairy development in East Africa and attempts at intensifying small ruminant production.

6.2 The concept of livestock intensification

We shall attempt here first to clarify the concept of intensification of livestock enterprises. Crude, 'ballpark' statistics published by FAO are then used to give an indication of progress in livestock intensification over recent decades. Finally the micro-level components of the intensification process are analysed and conclusions drawn on future prospects.

There has long been debate about the precise meaning of 'intensification' or the process of increasing the 'intensity of production' although most would agree that, in broad terms, it means an increase in the levels of variable inputs, and therefore of product output, per hectare of land. However, it has been suggested that in the case of pure cropping systems, a distinction should be drawn between (a) an increase in the cropping frequency or intensity (for example, by reducing the length of fallows, or by sequential cropping) and (b) an increase in the quantity of variable inputs used per hectare of cropped land.

Assessment of the intensity of livestock production is more difficult because plant material, which is the direct product of the land, is processed through the livestock to produce 'second-stage products' such as meat, milk and eggs. For grassland-based systems, with ruminant cattle, sheep or goats, the stocking rate provides a rough indication of 'first-stage' intensity. However, at the second stage, intensity also depends upon the level of variable inputs, such as concentrate feeds or veterinary treatments and the output achieved, per animal.

A problem then arises in that input requirements and product outputs differ substantially between different types and species of animals and even between age cohorts within a specific system. Thus stocking rates, inputs and outputs are often related to the number of (tropical, grazing) livestock units, the standard for which is usually a milking cow. Conversion factors are necessarily crude but are generally based on the metabolic weight of the animal, which reflects the nutritional energy requirement. For example a sheep or goat is usually assumed to represent ⅛ or 0.125 of a livestock unit (LU), while a cow or buffalo represents 1 LU.

In many cases, particularly for the mono-gastric pigs and poultry, there is no direct link between the inputs used per hectare of land and the quantity of meat, milk or eggs produced. Indeed such livestock production systems are sometimes described as 'landless'. Estimates of metabolic weights, or digestible energy requirements, have been used to arrive at livestock unit conversion factors for these species. Since they are fed on grains and other 'concentrate feeds', more feed energy is produced per hectare of land than under grazing systems, so the ratio of livestock units per hectare of 'feed' is likely to be higher, but so too will be the cost per unit of feed and per livestock unit. For both these reasons pig and poultry enterprises are generally assumed to be more intensive forms of land use than grazing livestock.

A minor digression is needed at this point, to consider the classification of livestock systems. One such classification of livestock production systems worldwide has been produced by the FAO (Seré and Steinfeld, 1996). After consideration of a range of criteria, including integration with crops, relation to land, agro-ecological zone, intensity of production, and type of product, the study arrived at three main types of system, with the middle group divided into two main sub-categories, as follows:

- grassland-based systems, solely based on livestock, in which more than 90 per cent of the dry matter fed to animals comes from rangelands, pastures or home-grown annual forages and in which annual average stocking rates are less than 10 LU per hectare of agricultural land;
- mixed farming systems, in which more than 10 per cent of the dry matter fed to animals comes from crop by-products or more than 10 per cent of the total value of production comes from non-livestock farming activities. This category is subdivided into:

    (a) rainfed mixed farming systems; and
    (b) irrigated mixed farming systems;

- landless systems, based solely on livestock, in which less than 10 per cent of the dry-matter fed to animals is farm produced and in which annual average stocking rates are above 10 LU per hectare of agricultural land (Seré and Steinfeld, 1996).

Different systems are associated with different livestock species or types. Grassland-based systems are entirely dependent on ruminant livestock, such as cattle, sheep and goats, which can readily digest green fodder. At the other extreme, most landless systems in the developing world are based on pig or poultry production. Incidentally, the above classification of systems, like this chapter, is concerned only with cattle, buffalo, sheep, goats, pigs and chicken. By neglecting a series of smaller animal species and game animals, the analysis underestimates the availability of animal protein, particularly in the more rurally based developing countries.

These three main types of system may be ranked in order of increasing intensity of production. Grassland based systems are the least intensive (or most extensive), mixed farming systems are intermediate, while the landless systems are the most intensive. Thus intensification may occur either by a change in technique, involving an increase in stocking rate or of variable inputs per livestock unit, within a given system or by a change in technology and a switch to a different, more intensive system. The latter may involve substitution of a different species or type of animal or bird.

6.3 Evidence available from published statistics

Some broad indications of trends in livestock production may be derived from the statistics on agricultural production published by the Food and Agriculture Organisation (FAOSTAT, 1999). These statistics must be viewed with scepticism, as the methods of estimation are rather crude. None the less they may be used to provide approximate measures of directions of change.

For the purposes of this analysis, data on areas of permanent pasture, livestock numbers and production are extracted for the main sub-continental groupings of Africa: namely north-western (Maghreb), west, central, east, southern and near-east in Africa. From these statistics are estimated changes in stocking rates, in livestock species and systems and in production per head. The changes are estimated over the 10 years from 1988 to 1998. However, comparisons with the rates of change over the previous 20 years from 1968 to 1988 show a fairly consistent process. First we review types and quantities of meat production in the developing countries of Africa.

6.3.1 Livestock production and supply in Africa

Data on per caput consumption of the main livestock products in 1997 are summarised in Table 6.1. In comparison with the rest of the developing world, African countries suffer from a shortage of livestock products. Only in the southern countries of Botswana, Lesotho, Namibia and Swaziland do supplies of meat and milk per caput exceed the developing world average. However, the supply of eggs even in this sub-region is well below the world average. Central Africa appears to have the lowest per caput supplies of livestock products, of any sub-region.

The main meat sources in Africa are cattle (41 per cent), sheep and goats (22 per cent) and poultry (25 per cent). Dependence on the ruminants, cattle, buffaloes (only in Egypt), sheep and goats (66 per cent) is much higher than the average for the developing world (30 per cent). Pigmeat is, as yet, relatively unimportant in Africa (9 per cent) while, on average for all developing countries, pigs provide 42 per cent of all meat supplies.

Table 6.1 The supply of livestock products, 1997
          (kg per capita per year)

	Meat	*Milk*	*Eggs*
All developing countries	24.6	42.4	6.8
All developing africa	13.1	35.4	1.8
North-Western	18.6	66.8	4.7
West	11.1	16.5	2.0
Central	9.7	8.6	0.3
East	11.0	35.1	1.2
Southern	24.8	52.4	1.0
Near East in Africa	20.7	77.8	2.2

6.3.2 *Grazing livestock units per hectare of permanent pasture*

This part of the analysis is particularly unsatisfactory, since permanent pasture is not the only form of grazing available. Furthermore, ruminant diets are often supplemented from crop by products and other sources. None the less, some indications of increasing intensity are provided by the growth in numbers of grazing livestock units: that is cattle, buffalo, sheep and goats expressed in terms of livestock units.

The results, given in Table 6.2, show increasing numbers of grazing livestock units over the last 10 years in all regions except Southern Africa. The growth has been fastest in West Africa, and the near-east in Africa (Egypt and Libya). For all developing countries of Africa the increase is over 2 per cent annually. There is relatively little change in the area of permanent pasture, other than in East Africa, where the area has fallen and the Near East in Africa where the area has increased by 0.44 of 1 per cent annually. The overall picture for Africa as a whole is generally one of increasing stocking rates as measured by livestock units per hectare of permanent pasture. These results indicate that, in most parts of Africa, production per hectare of land devoted to livestock is tending to rise; in short, intensity is increasing.

6.3.3 *Changes in livestock production systems*

Although grazing livestock numbers have been increasing for the developing countries of Africa as a whole, the numbers of pigs and poultry have grown faster. This mirrors a trend in the rest of the developing world (see Figure 6.1). However, there are discrepancies between sub-regions within Africa. The high growth rate of pig numbers, of 5 per cent annually for all developing countries of Africa, is heavily influenced by a growth rate of 7.5 per cent recorded for West Africa. Numbers of other livestock species have grown faster than pigs in all other sub-regions. Chicken numbers have increased quite rapidly in all sub-regions except Central Africa, where sheep and goat numbers have grown faster. None the less the overall trend appears to be one of intensification through increased emphasis on pigs and poultry, the intensive livestock enterprises.

In comparison the populations of ruminant livestock are growing at a slower rate, or even declining in some cases. Within the small-ruminant sub-group, goat numbers are increasing relative to sheep numbers in most sub-regions. It may be argued that the shift from sheep to goat production is a form of intensification, since goats are generally more prolific and productive than sheep.

The general picture that emerges shows Africa lagging behind other parts of the developing world in terms of meat supply per capita. This appears to be associated with much greater dependence on pigs and poultry in other continents. However, pig and poultry production is expanding rapidly in Africa in common with the rest of the developing world.

Table 6.2 Growth in number of livestock units per hectare of permanent pasture, 1988–97 (% per annum)

Change in	All Africa	North-West	West	Central	East	Southern	Near East
LSUs	2.06	0.44	2.64	2.20	0.74	−0.12	5.48
Perm. pasture	−0.31	0.22	0.16	0	−0.99	0.01	0.44
LSUs/ha	2.38	0.22	2.48	2.20	1.75	−0.13	5.02

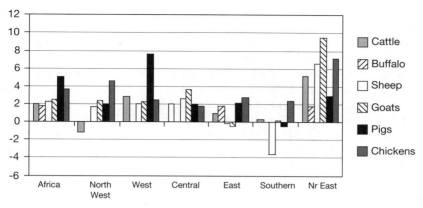

Figure 6.1 Growth in livestock numbers by species

6.3.4 Changes in production per head of livestock

A comparison of the expansion of production of meat and other livestock products, with the growth in animal numbers, provides a further measure of intensification. An increase in product output per head of a particular type of livestock represents an increase in intensity and productivity.

For the developing countries of Africa as a whole, productivity per head has increased over the last 10 years for most livestock products (see Table 6.3). Beef and veal production per head of cattle has declined slightly, but meat production per head of all the other main species has grown significantly. Milk production per head has increased for cattle, buffaloes and goats, although production of milk per sheep has fallen. This last trend is entirely due to a decline in the North and near-eastern African countries, probably due to the increase in the price of imported feed grains following structural adjustment. Egg production per chicken has declined in all sub-regions except West and Southern Africa. This trend may also be the result of increasing feed prices in recent years although the question then arises as to why poultry numbers and production of poultry meat are increasing.

6.4 Stages and components of the intensification process

From the above broad analysis it appears that intensification is occurring in three ways. First, grazing livestock numbers are increasing in relation to a fairly static area of grazing land; second, there is a shift in emphasis towards the more intensive pig and poultry enterprises; and third, output per animal is increasing in most cases, presumably as a result of increased input use. There is also a spatial dimension to this process, with a gradient of increasing intensity from the more remote, extensively grazed areas to the landless, commercial

Table 6.3 Growth in production per head of livestock for all Africa, 1988–98 (% per annum)

Growth in:	Cattle	Buffaloes	Sheep	Goats	Pigs	Chickens
Numbers	1.98	1.79	2.20	2.46	4.97	3.62
Meat	1.93	4.93	2.88	3.50	5.84	4.36
Meat per head	−0.05	3.08	0.67	1.02	0.83	0.71
Milk or eggs	2.14	4.54	1.01	3.33	—	2.24
Milk or eggs/head	0.16	2.70	−1.16	0.85	—	−1.33

production systems of the so-called 'peri-urban zones' in the vicinity of large urban markets (see Upton, 1997).

These observed trends are associated with ongoing processes of evolution of the production systems outlined above. Extensive grassland-based systems evolve towards forms of agro-pastoralism under the pressures of increasing livestock numbers on limited and possibly shrinking areas of grazing land. Thus grassland-based systems develop into mixed farming systems (for example, see Tiffen *et al.*, 1993). At higher levels of intensity, confinement of animals and hand feeding become financially attractive. Ultimately the use of purchased feeds may become economically viable and mixed farming systems evolve into 'landless' livestock production systems. In terms of crop–livestock interactions the process follows an inverted U form 'as (*human*) population density increases: integration is very weak at the beginning (*in grassland based systems*), increases (*in mixed-farming systems*), and then decreases (*in landless production systems*)' (McIntire *et al.*, 1992).

Within this process of intensification the following elements are of importance: (a) nutrition; (b) health; (c) management; (d) breeding; and (e) marketing (Devendra and McLeroy, 1982). These authors suggest that, at least in the case of sheep, improvements should be introduced in this sequence. However, there are situations where health measures, such as endemic disease eradication, should have first priority or where the simultaneous introduction of a package of innovations would be more appropriate. Furthermore 'animal confinement', by the construction of fences or livestock housing, represents a sixth important input.

A suggested sequential framework for productivity improvement, or intensification, is given in Figure 6.2. Again this relates particularly to small ruminants. Health measures may be introduced for free-range, scavenging flocks. Improved management and marketing then allow a modest increase in numbers and off-take. However, to take full advantage of improved forage production animals must be permanently housed or tethered.

The rapid growth in pig and poultry production is largely dependent on the development of large-scale commercial units by urban business people and civil servants. Small-scale pig and poultry production is also expanding (see Esrony *et al.*, 1996; Kitalyi 1998; Munya *et al.*, 1991; Rendel, 1996; Sonaiya, 1995), but in view of the large-scale activity, these enterprises are omitted from the remainder of this chapter. The remaining discussion focuses particularly on smallholder production of ruminant cattle, goats and sheep and improvements in health, nutrition, management, marketing, housing and breeding.

6.5 Animal health improvement

Animal health services, with their related breeding or genetic material inputs, are largely exogenous to the farm as a system; they are provided from external sources. (Husbandry practices for disease prevention are internal to the farm system but are treated separately in section 6.7.) The same is true of purchased concentrate feeds for landless production systems but, apart from this special case, the other inputs are endogenous or generated from within the farm household system. Advice and assistance may be provided from outside the system, but the service itself is provided from within the farm household. Thus improvements in animal health are generally dependent on the development of an effective institutional framework for their delivery. Frequently this framework is deficient or lacking, along with poor infrastructure, markets and access to information.

Veterinary disease control services have traditionally been provided by the state, but public finance constraints have limited the availability and effectiveness of public services. Over

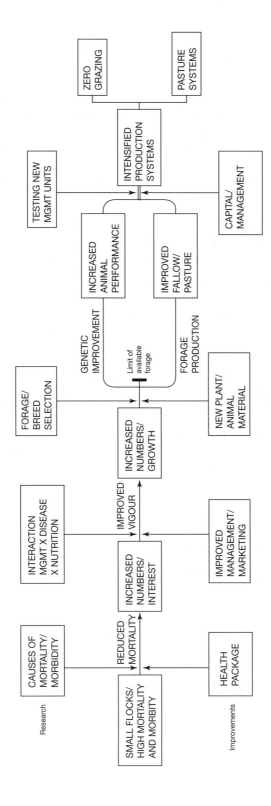

Figure 6.2 Generalised development path for small ruminant production
Sources: ILCA (1980); Upton (1988)

recent decades, pressures for government disengagement and greater reliance on the private sector have been applied to the provision of veterinary services. Many African states have adopted policies of cost-recovery and privatisation of animal health provision, but with limited success (see Holden, 1999; Oruko and Upton, 1998).

One critical problem is the high fixed cost of establishing and operating a veterinary practice. There are substantial economies of scale in the provision of such services. In addition transaction costs of delivery are generally high. As a result, the operation of a professional veterinary practice only becomes economically viable above a certain threshold level of livestock intensity, with a high stocking density and emphasis on intensive production systems. Thus veterinary practices are unlikely to be economically viable in pastoral livestock production regions (Umali *et al.*, 1992). A corollary is that where a veterinary practice is viable, it is likely to form a natural monopoly, so some state regulation of prices and quality of services is highly desirable.

'By far the largest number of practitioners in Africa are not veterinarians but para-professionals with non-degree training ranging from a few months to three years' (Leonard *et al.*, 1999). Most of these are in government service, although some are employed by non-government organisations (for example, Intermediate Technology in Kenya). Under the pressures for privatisation, an appropriate institutional framework is needed for the employment and supervision of the paraprofessionals previously in government service. Delivery of animal health services by these personnel may be economically justified in remote areas and for simple treatments, where the services of professional veterinarians are not available.

Even in extensive livestock production areas, there may be a case for vaccination campaigns or other measures to control major epidemic diseases, such as *rinderpest* or *peste des petits ruminants*, or contagious bovine pleuropneumonia (CBPP). Similarly programmes for the control or eradication of endemic diseases such as *trypanosomiasis* or its vector the *tsetse fly* may be launched. Such control programmes are essentially public goods (with non-exclusion of, and non-rivalry between, beneficiaries). There is limited scope for independent action by the individual livestock producer, beyond the use of protective or curative drugs or the use of traps and targets for tsetse control. Even then, communal action is more likely to be effective.

Thus such programmes are usually provided by the government or even by international aid (for example, the Pan African Rinderpest Campaign, PARC). Cost recovery may be possible, by means of a levy on all livestock producers, but organisation and collection of the levy may itself be a costly exercise.

With intensification, made possible by control of major epidemic and endemic diseases, other endemic and production-related diseases, such as *helminthiasis* (parasitic worms) or mastitis, may become more important. It is suggested that animal health risks are lower under intensive systems, because of greater scope for managerial control (McDermott *et al.*, 1999). However, a counter-argument suggests that risks of rapid disease spread are increased with increasing stocking density, while the introduction of exotic breeds of animals leads to loss of genetic disease resistance found in many indigenous breeds.

Thus the level of animal health provision influences, and is influenced by, the intensity of livestock production. However, the relationship is neither simple nor direct. The incidence of disease, and hence the need for animal health services, is uncertain and a source of production risk. The benefits of control of epidemic diseases include reduced risk of disastrous loss as well as improvements in expected productivity.

6.6 Animal nutrition

There is a close relationship between nutritional inputs and output represented by liveweight growth, reproduction rates and milk production. Animal scientists express this relationship in terms of metabolisable energy (ME) and protein requirements for maintenance, growth and production (Chesworth, 1992; Preston, 1995). The maintenance ration, necessary to keep the animal alive and in reasonable condition, is an overhead cost which declines per unit of output as production is increased. Faster growth to maturity ensures that the maintenance 'cost' accrues over a shorter period with a consequent saving in total. Thus improved nutrition, in both quantity and quality, should lead to increased feed conversion efficiency.

However, the costs of animal feeds vary according to the source and the season. Generally speaking, improvements in feed quality (that is, with a higher protein content) are linked with increased cost. Feed costs generally rise in the dry season, as suitable plant material becomes scarcer. Grazing or browsing of communal rangeland provides feed that is free to the individual herder (although social costs may be incurred through overgrazing). Similarly crop by-products are free or low-cost goods, providing one of the beneficial complementary interactions, along with animal draught power and manure used on crops, associated with mixed farming systems. Additional costs may be incurred in forage conservation (for example, hay making or fodder banks) for dry season use, or in processing crop by-products to improve the quality (for example, straw treatment).

Natural grazing and untreated crop by-products are generally of low feed quality. Intensification, through improved nutrition, may necessitate the growing of fodder crops, such as Napier grass, berseem (lucerne) or fodder trees. This introduces competition with human food and cash crops, for cultivated land and other resources. There is an obvious opportunity cost involved.

The competition with human food supplies is even more direct where cereal- and legume-based concentrate feeds are used. In many cases concentrate feeds have been imported, so that greater self-sufficiency in intensive livestock products has been bought at the expense of large increases in imports of coarse grains (see Durning and Brough, 1991; Cunningham, 1992). For example, several near-eastern and North African countries relied on overvalued currencies and direct subsidies to provide sheep producers with imported barley and other concentrate feeds at a fraction of the border-parity price. This encouraged rapid increases in animal numbers and the establishment of feedlot fattening units. Following structural adjustment programmes, and the consequent rise in feed prices, these intensive production systems have collapsed, while there is an urgent need to find alternative, lower-cost sources of feed.

In summary, improved nutrition can lead directly to increased intensity of production, but for economic viability and sustainability the costs must be carefully weighed against the benefits.

6.7 Management

All production systems must be controlled and managed. Many relatively extensive, traditional pastoral and mixed farming systems are well managed using indigenous knowledge (for example, ethno-veterinary methods) and technology. It has proved very difficult to design improvements other than by a major change of production system. Movement of livestock is one response to seasonal variation in food supplies and disease risks. Intensification invariably necessitates tighter control and increased management inputs.

With tighter managerial control, disease risks may be reduced by rotational or zero grazing, by better hygiene, by regular prophylactic treatments, and by isolation or culling of diseased animals. Nutrition may be improved since feed inputs can be more readily adjusted to the needs of the individual animal. Benefits may be gained by matching the seasonal pattern of reproduction to available food supplies; that is, by arranging for the whole breeding flock or herd to give birth near the beginning of the rainy season. In these ways feed conversion efficiency can be raised without increasing the feed cost.

Fertility management is an important factor in increasing intensity of production. Reproductive performance is affected by:

- age at first calving, kidding or lambing;
- service period, from parturition to first service;
- pregnancy or conception rate per service;
- birth (calving, kidding or lambing) rate;
- interval between births;
- numbers born per parturition; litter size.

All these parameters are controllable to some extent and can be improved with careful management.

The economic benefits of improved fertility management include:

- increased numbers of offspring;
- increased efficiency of milk production;
- increased lifetime production, due to prolonged lactations and shorter parturition intervals;
- increased feed conversion efficiency, since fewer unproductive animals are carried;
- reduced replacement costs;
- reduced breeding and veterinary costs.

A major component of the cost of improved management is that of careful recording and monitoring of individual animals and their production costs. Without improved management, intensification of livestock production is likely to fail.

6.8 Marketing

Although limited numbers of small ruminants, and even cattle, may be slaughtered for home consumption and small quantities of milk are used within the household, production of a marketable surplus necessitates the existence of an effective marketing system. In traditional areas of extensive rangeland grazing and mixed farming, a market usually exists for live animals, prior to their movement to urban slaughter slabs, and for small quantities of fresh or soured milk or ghee.

Existing markets for live animals often function effectively and many ambitious projects for increasing intensity of production, by constructing stock routes, watering points and marshalling yards, have had little impact. However, improved slaughterhouse and meat chilling facilities may be needed to comply with meat hygiene and health requirements, especially if meat exports are intended.

Intensification of milk production, on the other hand, necessitates the establishment of cooling and processing facilities. As there are economies of scale in these activities, the individual smallholder cannot afford the establishment and operating costs of a viable plant. In many cases, producer cooperatives have been formed to establish dairy facilities. Even then the supplies of milk are often insufficient to justify the capacity of the processing plants, while many have suffered from poor and ineffective management. The Kagera District Livestock Development Project (KALIDEP) in Tanzania is a typical example, where smallholder dairying with grade cows has been established over wide areas of the District. Milk marketing problems are such that the whole project would probably founder without external assistance in the provision of milk processing facilities (Upton *et al.*, 1997). It is concluded that the establishment of economically viable processing facilities is essential for the success and sustainability of dairy development and intensification, but that this is not easily achieved.

6.9 Confinement/housing

Even under extensive rangeland systems animals are often confined in enclosures (bomas) at night to avoid injury and loss. Suckling calves and their mothers may remain during daytime in the pre-weaning period. Free-ranging and scavenging sheep and goats may be housed at night also. However, as intensity increases, under mixed farming systems confinement of animals, at least in fenced enclosures, becomes essential to avoid crop damage by the animals. Under alley farming systems in Nigeria, fodder shrubs have been used as living fences to confine goats and sheep. If animals are yarded or housed in a purpose-built structure, then hand feeding or zero-grazing becomes necessary.

Confinement of animals is important in facilitating improved management and better control of production. Animals can be afforded individual attention and appropriately controlled feeding regimes. Risks of becoming infected with diseases transmitted by contact or by vectors, such as ticks, may be reduced. Health treatments are more easily administered. Regular observation of the animals is possible although identification of oestrus, when a breeding female is ready for service, may be more difficult. Overall the advantages in improved management made possible by confinement of animals outweigh the disadvantages but this is at the cost of establishing and maintaining the housing and providing the greatly increased labour required.

6.10 Breed improvement

Large observed differences in productivity between different breeds of the same livestock species, or even between individual animals in the same herd or flock, offer scope for intensifying production by improving the genetic make-up through selection and controlled breeding. However, the actual productive performance of an animal depends upon the interaction between genetic potential and the environment. In order to benefit from the genetic potential of an improved breed, nutritional levels and health measures must also be improved. Without a package of improved inputs, the newly bred animal may perform less well than the traditional 'unimproved' members of the species (Richardson and Hahn, 1994).

There are two main alternative approaches to genetic improvement: (a) selection of breeding stock from within the flock or herd; and (b) the introduction of exotic genetic material by crossbreeding or breed substitution. The former is a slow process which requires careful recording and monitoring of individual animals to identify the better performers, which will then be used for the breeding programme. Against this may be set the advantage that the animals are already well adapted to the existing environment and production system.

Improvement is slow, because heritability (the likelihood of passing on a characteristic from parents to offspring) is low for certain key characteristics, such as fertility; the generation interval (age of parents at first parturition) is long, especially for cattle (about 4 years); and for a small-scale producer with few animals there is little scope for culling unsuitable members of the breeding herd. The scope for improvement by selection is better for small ruminants as the numerical size of the breeding flock is likely to be larger, with more opportunities for culling, and the generation interval is only about 1 year. None the less it may take several generations of selection for breeding to achieve a significant improvement in productive performance.

The other alternative, which may produce faster results, is the introduction of a new, possibly exotic, breed as a substitute for local stock or for crossbreeding. An exotic breed, such as the Friesian dairy cow, in its home environment produces a much higher yield than indigenous African cows. However, introduction of pure-bred Friesian cows is fraught with danger, since they are not adapted to the local environment, may lack the tolerance to endemic diseases which is inherent in local cattle, and require a high standard of management to survive and approach their potential level of production. The challenge facing animal breeders in sub-Saharan Africa is to improve the productivity of livestock without losing traits that are essential to survival.

The offspring of a first cross between an exotic sire and an indigenous dam is often the preferred choice for development programmes. Many dairy development schemes in East Africa are based on the distribution to farmers of first-cross heifers. Crossbred animals retain desirable characteristics from both parents and benefit from *heterosis* or hybrid vigour. Substantial costs are involved in establishing the foundation stock of crossbred heifers for distribution and maintaining animals (usually males) of the 'superior' breed to continue the crossbreeding programme.

There are problems in maintaining the genetic balance in the second and future generations of crossbred heifers, since the emphasis will shift towards the type of sire used, local or improved. Ideally a system of criss-cross breeding should be used, with alternation between sire types at each generation, but this may be difficult to arrange (Matthewman, 1993). Frequently the choice of sire is based simply on what is currently available at the time of oestrus.

Artificial insemination (AI) service provision can be an effective means of breed improvement. There are substantial costs involved in establishing the centre, selecting and keeping improved bulls, bull testing and delivery of the semen to farms as and when needed. Economies of scale mean that the service must be provided by the public sector or by communal group activity, even though cost recovery from individual producers is feasible. If the service becomes unreliable farmers are likely to revert to using local bulls all the time, and genetic regression to the original local breed will occur.

Small-ruminant breed improvement programmes differ in that the distribution of improved males (bucks or rams) to run with village flocks may be more feasible. Also, because of the shorter generation interval, faster progress can be made. In the case of the Small Ruminant (CRSP) programme, in Kenya (and other countries), crossbreeding was carried to the stage where a true-breeding intermediate type of dual-purpose goat was developed before does were distributed to farmers (Semenye *et al.*, 1989).

6.11 Summary and conclusions

Analysis of the available statistics suggests that intensification of livestock production is occurring in much of Africa. Numbers of all species are increasing, particularly those of the more intensive pig and poultry enterprises. Furthermore, output per head is increasing for all species of farm livestock. These increases in intensity have occurred despite the fact that intensification generally necessitates a change in technology, with improvements in disease control, nutrition, management, marketing, housing and breeding of livestock.

References

Chesworth, J. (1992) *Ruminant Nutrition*, Basingstoke, Hants.: Macmillan.

Cunningham, E.P. (1992) *Selected Issues in Livestock Development*, Economic Development Institute Technical Materials, Washington, DC: World Bank.

Devendra, C. and McLeroy, G.B. (1982) *Goat and Sheep Production in the Tropics*, London: Longman.

Durning, A.B. and Brough, H.B. (1991) 'Taking stock: animal farming and the environment', Worldwatch Paper no. 103, Washington, DC: Worldwatch Institute.

Esrony, K., Kambarage, D.M., Mtambo, M.A., Muhairwa, A.P. and Kusiluka, L.J.M. (1996) 'Intestinal protozoan parasites of pigs reared under different management systems in Morogoro, Tanzania', *Journal of Applied Animal Research*, 10 (1): 25–31.

FAOSTAT (1999) FAO computerised database, <http://apps.fao.org> (accessed August 1999).

Holden, S. (1999) 'The economics of delivery of veterinary services', *OIE Scientific and Technical Review*, 18 (2): 425–33.

International Livestock Centre for Africa (ILCA) (1980) *ILCA: The First Years*, Addis Ababa.

Kitalyi, A.J. (1998) 'Village chicken production systems in rural Africa: household food security and gender issues', FAO Animal Production and Health Paper 142, Rome: FAO.

Leonard, D.K., Koma, L.P.M.K., Ly, C. and Woods, P.S.A. (1999) 'The new institutional economics of privatising veterinary services in Africa', *OIE Scientific and Technical Review*, 18 (2): 544–61.

McDermott, J.J., Randolph, T.F. and Staal, S.J. (1999) 'The economics of optimal health and productivity in smallholder livestock systems in developing countries', *OIE Scientific and Technical Review*, 18 (2): 399–418.

McIntire. J., Bourzat, D. and Pingali, P. (1992) *Crop-Livestock Interaction in Sub-Saharan Africa*, Washington, DC: World Bank.

Matthewman, R.W. (1993) *Dairying: The Tropical Agriculturalist*, London: Macmillan/Centre for Tropical Veterinary Medicine (CTVM).

Munya, S.J.M., Agumbah, G.J.O., Njenga, M.J., Kuria, K.J.N. and Kamau, J. (1991) 'Causes of pre-weaning mortality in small-scale and medium-scale intensive piggeries in Central Kenya', *Indian Journal of Animal Sciences*, 61 (2): 126–8.

Oruko, L.O. and Upton, M. (1998) 'Reforms in Africa's agricultural sector: the case of input markets in Kenya', in *Proceedings of Conference on Africa and Globalisation*, University of Central Lancashire, Preston, 24 April, mimeograph.

Preston, T.R. (1995) *Tropical Animal Feeding: A Manual for Research Workers*, Animal Production and Health Paper 126, Rome: FAO.

Rendel, J. (1996) 'Sustainable use and development of national animal and feed resources in the production of human food in Africa', in R. Lindberg (ed.), *Veterinary Medicine: Impacts on Human Health and Nutrition in Africa*, Proceedings, Addis Ababa, Ethiopia, 27–31 August. Uppsala: Sveriges Lantbruksuniversitet (Swedish University of Agricultural Sciences).

Richardson, F.D. and Hahn, B.D. (1994) 'Models for the selection of cow types for extensive meat and milk production in developing areas', Pastoral Development Network, Paper 36d, London: Overseas Development Institute.

Semenye, P.P., Onim, J.F.M., Conelly, W.T. and Fitzhugh, H.A. (1989) 'On-farm evaluation of dual-purpose goat production systems in Kenya', *Journal of Animal Science*, 67 (1): 3096–102.

Seré, C. and Steinfeld, H. (1996). 'World livestock production systems: current status, issues and trends', FAO Animal Production and Health Paper 127, Rome: FAO.

Sonaiya, E.B. (1995) 'Feed resources for smallholder poultry in Nigeria', *World Animal Review*, 82: 25–33.

Tiffen, M., Mortimore, M. and Ackello-Ogutu, A.C. (1993) 'From-agro pastoralism to mixed farming: the evolution of farming systems in Machakos, Kenya, 1930–1990', Network Paper no. 45, Agricultural Administration (Research and Extension) Network, London: Overseas Development Institute.

Umali, D.L., Feder, G. and de Haan, C. (1992) 'The balance between public and private sector activities in the delivery of livestock services', Discussion Paper no. 163, Washington, DC: World Bank.

Upton, M. (1988) 'Goat production in the humid tropics: actual and potential contribution to agricultural development', in O.B. Smith and H.G. Bosman (eds), *Goat Production in the Humid Tropics*, Wageningen: Pudoc.

Upton, M. (1997) 'Intensification or extensification: which has the lowest environmental burden?', *World Animal Review*, 88 (1): 21–9.

Upton, M., van Weperen, W., Mwazyunga, P. and van Munster, B. (1997) *Kagera Livestock Development Programme (KALIDEP). Phase II: Report of the Mid-term Review Mission*, Dar es Salaam: Royal Netherlands Embassy, and RDP Livestock Services B.V.

7 Development and change in Sahelian dryland agriculture

Michael Mortimore[1]

7.1 Features of the dryland areas

The arid and semi-arid agro-ecological zones of Africa are not a marginal fringe but contain from 36 to 43 per cent of its area, depending on the definition used (Figure 7.1). The semi-arid zone – where crop and livestock production are about equally important – accommodates at least 28 per cent of the population of tropical Africa on 18 per cent of its surface, with 27 per cent of its ruminant livestock (Jahnke, 1982). As vacant land was, until recently, still abundant, the drylands have witnessed unprecedented surges in their human populations. The conservation of natural resources and achievement of sustainable livelihoods in the drylands therefore deserve a prominent place in discussions of African development policy.

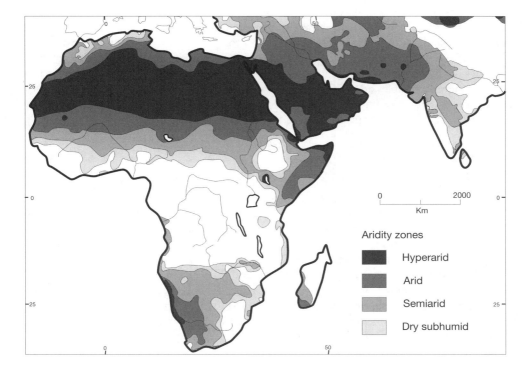

Figure 7.1 The tropical dryland in Africa
Source: After UNEP (1992: 4)

The aridity index, on which this map is based, is defined as annual precipitation (P) over potential evapotranspiration (PET). This scheme recognises the three zones: arid (P/PET = 0.05–0.20); semi-arid (P/PET = 0.20–0.50); and dry subhumid (P/PET = 0.50–0.65).

Four major properties of the drylands may be identified.

7.1.1 Aridity

Drylands under a unimodal rainfall regime have little or no rain for 5–8 months each year or, under a bimodal regime, for 2–3 months twice each year. Ecosystems are adapted to drought stress, the production of biomass is sharply concentrated into the wet seasons and there are long quiescent periods when annual plants die and perennials produce little growth. Domestic animals undergo weight loss and (in drier years) starvation. They may have to be watered from wells, consuming much labour. Rainfed farming is episodic, with periods of intense and exhausting work separated by periods of relative inactivity. The monthly distribution of rainfall may be profoundly influenced by latitude (as in West Africa) or by altitude and rain-shadowing effects (as in eastern Africa).

The exceptions are the wetlands, which offer dry season grazing, flood-recession farming, or irrigation opportunities. The value of these wetlands is a function of their scarcity (generally less than 10 per cent of the rainfed areas).

7.1.2 Rainfall variability

Rainfall averages from 250 to 1000 mm per year. It is not only sharply seasonal but it is also variable: both between years and during seasons. For both animals (wild or domestic) and humans, variability introduces an element of risk into almost all life-supporting activity. The impact of variability increases with aridity. The greater part of the rainfall occurs in short episodes of high intensity. Variability can be expressed in probabilities. Unfortunately, while such a statement enables an assessment of average risk and returns, it provides no basis for predicting the timing or intensity of drought events. The incorporation of rainfall variability into development policy or project design is extremely difficult.

7.1.3 Natural diversity

Drylands are ecologically diverse. An experimental classification based on major ecological parameters identified over eighty environmental units in the semi-arid zone of sub-Saharan Africa (Mortimore, 1991). The dimensions of this diversity in the West African francophone Sahel – in soils, water, and natural vegetation, as well as rainfall distribution – have been effectively shown by Raynaut and his colleagues (1997). The semi-arid and sub-humid zones of sub-Saharan Africa are associated predominantly with savannas, the density and size of trees diminishing with increasing aridity. In the arid zone steppes, the dominant species of the grasslands are quite variable. Owing to burning, clearance for cultivation and grazing, the natural communities have been transformed into mosaics, consisting of farmland, fallow, rangeland, residual woodland and eroded or degraded land. Topographical wetlands interrupt these patterns. Local communities and households have adapted in intricate ways to the micro-diversity of their environments.

7.1.4 Demographic diversity

While high rural population densities are rare in the arid zone, surprisingly high densities are found in certain semi-arid areas. A range varying from >300 to <5 persons/km^2 has implications both for theory and for development policy (English *et al.*, 1993; Tiffen *et al.*, 1994; Raynaut *et al.*, 1997; Mortimore and Adams, 1999). The relationship between agro-ecological potential and population density is weak, confounding any attempt to establish a generalised relationship between density and degradation.

The variation in rural population densities demands a diversity of responses in the systems of natural resource management (NRM), from the one extreme where labour scarcity is a major driving factor, to the opposite, where land scarcity has provoked various forms of agricultural intensification. On these differences in rural population density, a variegated ethnic spatial distribution is superimposed, along with significant (though little understood) variations in human fertility.

It should be noted also that the drylands are not, and in general never have been, closed economic systems. Trading networks and diasporas were a feature of their pre-colonial histories. Political linkages between arid and sub-humid regions gave expression to fundamental complementarities. Mining and urbanisation have intensified these linkages through offering new employment opportunities to migrants. A useful report, the *West Africa Long Term Perspective Study* (Snrech *et al.*, 1994) has highlighted the north–south and city–hinterland patterns of dependency that are critical in the economy of the western Sahel. At the household level, income opportunities in distant places provide a vital supplement to those from agriculture, especially when drought causes cereal production failures or increases livestock mortality, also providing, therefore, insurance against risks.

7.2 Drought and desertification

The great Sahel drought that culminated in widespread famine in 1972–4 has had a pivotal significance in the evolution of dryland development policy. It was perceived at the time, in official circles, as a fivefold crisis of:

- *drought*: technically viable indigenous systems of production quite suddenly came to be seen as maladaptive. New technical or management solutions appeared to be necessary (Gorse and Steeds, 1987);
- *food scarcity*: persistent dependency on food aid continued through the 1980s in the francophone Sahelian countries (Somerville, 1986), casting doubt on the region's ability ever to feed itself;
- *overstocking*: the massive mortality among livestock holdings convinced some that the populations exceeded the levels supportable, and must be reduced (Western, 1982);
- *degradation*: the effects of drought were easily confused with 'desertification': 'overcultivation' (or 'soil mining'), 'overgrazing' (or pasture degradation), 'deforestation' (or removal of woodland) and – it was assumed – 'overpopulation' (Cloudsley-Thompson, 1984);
- *'coping'*: a disaster that affected, first and foremost, the crop and livestock sectors, and made them inherently insecure; this was believed to be the underlying cause of increased poverty, asset losses and extensive out-migration (Garcia and Spitz, 1982).

A strongly interventionist philosophy of government, and a theory of development based on economic 'take-off' through public investments and export agriculture, had earlier given support to an assumption that the solution to poverty would have to be imposed from the top down. What many saw as a collapse of livelihoods across the region, therefore, *necessarily* called for massive interventions by governments and donors. Saving the natural resources from further destruction became a primary target. The scenario of an 'advancing Sahara' (resuscitated from before the Second World War) appeared to call for an emergency international effort. Scenarios for the East African drylands had been pessimistic since the 1940s or earlier (see Tiffen *et al.*, 1994).

Even before the Sahelian drought, pessimistic assessments were made of degradation in Africa (Dregne, 1970), but it was that event which popularised the issue (Eckholm and Brown, 1977). It provoked a veritable sandstorm of literature surrounding the United Nations Conference on Desertification in 1977. The human dimensions were the primary focus of interest (for example, Johnson, 1977; Biswas and Biswas, 1980), and human agency was uppermost in the official understanding of desertification. The Conference concluded by approving a Plan of Action to Combat Desertification (the PACD), which was coordinated by the Desertification Branch of the United Nations Environment Programme (UNEP), as the basis for national plans. Although global in its scope, the sheer size of the African drylands ensured the continent prominence in the implementation of the PACD.

Desertification is used to justify strong planning, policies or interventions in smallholders' NRM; more resources for centralised forestry activities; spending by aid donors and development agencies; and politically appropriate responses to drought and famine crises. Yet the PACD, after its approval in 1977, persistently failed to receive the support expected: 'Many governments did not recognise the enormity of the desertification threat, or appreciate the costs or complex processes of desertification' (Buonajuti, 1991: 31; see Odingo, 1992). Africa, 'the bedeviled continent' (Kassas *et al.*, 1991: 20), continued to attract the greatest attention, though not the necessary funds. Time-hallowed associations between population growth, over-exploitation and degradation were repeated in the literature; and assessments continued to make extensive use of approximations and assumptions (UNEP, 1990; Dregne *et al.*, 1991). The struggles of the Desertification Branch of UNEP to mobilise funds for the PACD, the controversies surrounding the history, definition and operation of the term 'desertification', and the institutional dynamic which perpetuated it in the face of much scepticism are documented in successive issues of *Desertification Control Bulletin* (Nairobi: UNEP) and other literature (see Mortimore, 1989, 1998; Swift, 1996).

The UNEP redefined desertification for the Earth Summit in 1992 (UNEP, 1993; Cardy, 1993) as follows: 'Desertification means land degradation in arid, semi-arid and sub-humid areas resulting from various factors, including climatic variations and human activities.' This became the basis of the United Nations Convention to Combat Desertification (CCD), which came into force in December 1996. The activity which preceded and followed the signing of this Convention has continued to be influenced by the same conventional wisdom as before, though participation and institutional issues have been elevated in the rhetoric. The focus of the Convention is the preparation of national plans for furtherance of the 'combat'. It was not long before the Intergovernmental Negotiating Committee was complaining of difficulties in attracting the necessary finance (UNEP, 1997).

7.3 Development efforts in the drylands

With the exception of the irrigation sector, which only affects directly a small proportion of the drylands and their populations, returns to investment (whether public or private) are low and uncertain. Resulting directly from the properties of aridity and rainfall variability, this problem has always constrained donor enthusiasm and created a conflict between policy recommendations to privilege high potential areas (Lele and Stone, 1989) and those advocating a broader approach to rainfed systems of production (Gorse and Steeds, 1987; English *et al.*, 1993). In a previous era the colonial governments tried to improve the profitability of dryland agriculture by developing export commodity production, and this agenda was pursued unchallenged after independence in such policies as the *Programme National Agricole* of Senegal:

> Past development efforts have largely focused on promoting productivity improvements in a single sector – crops or livestock or forestry – without paying much attention to the contexts in which traditional production systems developed. While this cutting-edge approach has produced results in wetter, more fertile areas, it has proven inadequate in the SSZ [Sahelian and Sudanian Zones of West Africa]. The new techniques were not much more productive than existing practices, nor were they designed to fit into production systems based on local rainfall and soils. While some successes have been achieved, overall results have been disappointing.
>
> (Gorse and Steeds, 1987: iii–iv)

While this analysis may not do full justice to the impact on dryland production systems of improved groundnuts, Allen cotton or hybrid maize, for example, scepticism about reliance on physical productivity advances to secure improved livelihoods in the drylands is often justified.

The same authors went on to advocate a new approach which had three elements: holistic design which overcomes traditional disciplinary specialisms and recognises multiple and diverse land use; an approach which takes account of variability; and participatory design, which searches for possible improvements within existing systems (ibid.:18–21). Although the study made some use of the then-popular concept of carrying capacity, now largely discredited (Behnke *et al.*, 1993), such a systems approach has subsequently been prioritised by researchers and funding agencies (for example, NRSP/DfID, 1999), and applied to the design of interventions elsewhere, such as the Indo-British rainfed agricultural development projects in East and West India.

Notwithstanding some acceptance of systems approaches, scenarios of downward spirals in bioproductivity, of negative interactions between population growth and the degradation of natural resources, of persistent and unmanageable drought, out-migration and urbanisation continued to be reflected in agency thinking during the 1990s, as, for example, in reports prepared on environmental policy in Nigeria by the FAO and the World Bank. In the literature, negative interpretations of system linkages were more influential than positive ones (Cleaver and Schreiber, 1994; Stiles, 1997). Arresting such trends was seen as a justification for the application of external knowledge, capital and pressure; a fundamentally unnatural pathway was advocated for the indigenous systems of natural resource and livelihood management to follow without assistance. In macroeconomic management, losses of natural capital (for example, 'soil mining') and the costs of rectification (for example, tree planting, erosion control) were uppermost; the possibility that the drylands could be a source of capital generation was not considered.[2]

7.4 Investing in dryland agriculture

7.4.1 Public versus private investment

In practice, the development agencies have focused on intervention capital rather than on mobilising indigenous resources. In fact, an obsession with credit has characterised many projects, and every government has operated credit schemes for small producers. The underlying assumption is that capital is scarce and therefore investment for technical modernisation, enhanced productivity and resource conservation depends on assistance from outside the household. Yet credit is risky for the smallholder given the aridity (= low productivity) and variability (= uncertainty) of arid or semi-arid ecosystems.

A study of long-term environmental and economic transformation in Machakos District, Kenya, from 1930 to 1990 showed that, notwithstanding the prevailing orthodoxy about credit, the greater part of the capital invested in new farm technologies, new management methods, soil and water conservation structures and other projects was generated by the Akamba households from their own incomes and labour, or by cooperating in groups (Tiffen *et al.*, 1994). For example, even during the peak of soil conservation work promoted by the EC-funded Machakos Integrated Development Programme (MIDP) (1978–82), as many kilometres of terraces were constructed without project assistance as with it. At other times, much less (or no) assistance was available. Notwithstanding the importance of investment in neoclassical economic theory of development, many of the studies of farming systems which were carried out (by various authors and agencies) in the District either underestimated or ignored autonomous investment.

Nor were the achievements of Akamba smallholders (for details, see Tiffen *et al.*, 1994) the result of disproportionate public investment in infrastructure or government services. The data show that government expenditures per capita were approximately the same, on average, as in other parts of Kenya, even including the periods of intense development activity in Machakos under the Land Development Board, the Swynnerton Plan and the MIDP.

7.4.2 The variety of smallholders' investments

Since the creation of capital is integral to improving household welfare, and its absence is a prime indicator of poverty, it is worth asking why the process has sometimes been neglected. In a context of development interventions, such as the *PNA* of Senegal, or the *Projet Développement du Maradi* in Niger, a conventional view of investment restricts the concept to capital equipment such as ploughs, seeders, carts and so on and inputs of improved seeds and agrochemicals. Such investments were seen as essential for the transformation from 'traditional' to 'modern' modes.

Given such a view, one is likely to conclude that (a) investment is low or non-existent in 'traditional' systems, (b) capital scarcity is the main factor, and (c) credit is the only available solution. In order to accelerate the technical transformation, which is seen as a necessary condition of progress, credit is supplied for the purchase of additional, preferably innovative, capital equipment and inputs, and repayment is secured through the control of markets and of payments to producers. Over-capitalisation and chronic credit dependency are possible outcomes: in the Bassin Arachidier of Senegal, under deteriorating economic conditions, the disintegration of these technical assets is now called *décapitalisation*. Some achievements of the MIDP seem to have been similarly dismantled. This raises the question whether credit does not frequently distort the agricultural economy of drylands.

A broader view of investment, however, could include the following categories, which are often ignored and yet are important in the circumstances of poor smallholders:

- *investments created by labour, family or group*: field boundaries and enclosures; tree planting and protection; manure making and composting systems; field distribution of organic materials (which have residual effects after the target crop has grown); field ridging and other micro-conservation practices;
- *investments combining labour with purchased inputs of skills or components*: storage structures; stables, bomas (covered livestock pens) and chicken houses; houses and other domestic capital; stone walls or terraces; drainage structures and so on;
- *low-cost investments*: seeds purchased in markets or from other farmers; salt licks or medicines for animals; metal tools or plough components made by local blacksmiths; wooden components; animal harnesses, yokes; fencing; granary roofs and so on;
- *livestock produced by breeding* (often enhanced by owner selection);
- *purchased livestock*: commonly draught or transport animals; animals for fattening; small ruminants for savings, including womens' and childrens' animals;
- *off-farm productive investments*: tools, transport, marketing facilities and so on;
- *expenditures on education or placing migrants in employment* in order to increase household incomes;
- *gifts or 'social investments'* which may be regarded, among their other purposes, as a form of risk insurance.

Only a long-term frame of analysis can identify and estimate the significance of such small, irregular or episodic investments. It is no wonder that they are underestimated in 'one-off' studies and in short-term development projects whose purpose is to disburse assistance from outside. Yet some scepticism is warranted as to whether the constraints of dryland environments permit a significant accumulation to occur. The conventional view recognises only an exploitative management of natural resources, which has to be supplemented to an increasing extent by incomes from other sources.[3] This essentially Malthusian view appears to confirm the numerous claims of degradation. An alternative view, however, which is consistent with a 'Boserupian' interpretation of change, recognises the possibility of an ongoing improvement in land use and a rising curve of productivity (Figure 7.2). Machakos is by no means the only supporting case. The Kano Close-settled Zone in Nigeria has a longer history under pre-colonial, colonial and post-colonial economic systems, including the impact of structural adjustment (Mortimore, 1993). In the Sahel as a whole there is evidence of analogous transformations in population densities, market relations and agricultural capitalisation (Snrech *et al.*, 1994).

If such an interpretation is sound, then the objectives of a policy intended to alleviate poverty in the drylands should be: (a) to facilitate a transition to more intensive and productive land use through mobilising indigenous resources; (b) to support sustainable practices; and (c) to protect investments through drought episodes or other periods of production failure, when asset loss can all too easily result in impoverishment.

7.4.3 Investment in diversification

Income diversification is neither new nor unusual in African households (Haggblade *et al.*, 1989). It is frequently associated with travel, either permanent migration or (more commonly) seasonal or shorter term circulation (Chapman and Prothero, 1984). In the

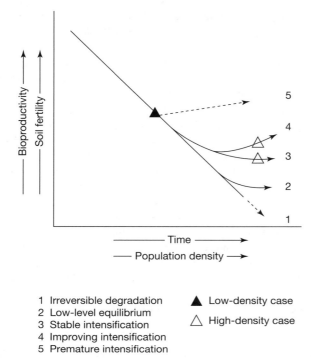

1 Irreversible degradation   ▲ Low-density case
2 Low-level equilibrium
3 Stable intensification   △ High-density case
4 Improving intensification
5 Premature intensification

Figure 7.2 The transition from degradation to intensification in a farming system
Source: Jolley and Torrey (1993)

drylands, and in particular during the search for scapegoats after the Sahel drought, population mobility was more often understood within a deterministic frame of reference (Amin, 1974; Colville, 1982) rather than an opportunistic one (Mabogunje, 1972). That is, the Sahel (in particular) was seen as a net exporter of capital and labour to the privileged coastal regions and cities, rather than there being a view of these places as a constituent part of the resource portfolio of Sahelian communities (Mortimore, 1989).

This issue, which varies with the subtleties of time and place, impinges on the management of natural resources in the following way. Every household must manage its labour resources, as well as its capital, prudently, as they are determined by its demographic endowment and are generally scarce. The opportunity costs of agricultural labour do not fall to zero during the dry season, as is sometimes supposed, because there are essential tasks connected with livestock, field preparation, marketing, maintenance and other activities. However, in some conditions, labour may be withdrawn from such work if the benefits justify its transference to other ways of earning income, including those in other places. Labour management is therefore the pivot of household livelihoods. In balancing their resources of labour (and skills) in sequential, everyday decisions, households move away from exclusive dependence on crop production towards increasingly complex income structures with their logistical and geographical implications. Eventually off-farm activity may supplant agricultural work, even (for some individuals) during the growing season. But diversification options have their entry costs – investment needs – of which travel is but one.

Policy which fails to recognise the multisectoral basis of household livelihoods – more especially in the drylands, where survival depends on it – may be rewarded with failed agricultural interventions as people vote with their feet. Certainly in eastern and southern Africa the mining and urban sector has distorted some smallholder systems to an extent which is rather rare in western Africa. What policies can best assist households whose farming is done by women, in the absence of their husbands, and whose savings are invested in animals?

It may seem paradoxical that, notwithstanding their often remote locations, dryland households are highly dependent on access to markets, not merely for evacuating produce but in the more general sense of access to labour markets, trading and other income opportunity markets at the national and even international level (for example, the migration of Senegalese to France). Official support of many such activities would raise sensitive issues. Yet to focus dryland development policies exclusively on natural resource management looks increasingly myopic.

It has been suggested here that public investment and/or credit have received an emphasis in development practice that underestimates the potential for private investment in NRM, even in drylands subject to low bioproductivity and high climatic risk. The question, however, is not whether credit is useful but whether it is the only route to capitalisation. Small-scale incremental investments in NRM need also to be seen in the broader context of income diversification, in which poor households adapt to change in their livelihood opportunities. This dimension is discussed in the next section.

7.5 Change in dryland livelihood systems

Policies for dryland development have often emerged from a diagnostic-prescriptive mode which focuses on a perceived need for transformation in two stages: present (which is deemed by experts to be unacceptable) and future (in which natural resource degradation has been brought under control and sustainable practices introduced). The inadequacy of such a framework is increasingly obvious as the continuity of change and its multiplicity of sources becomes apparent. Ongoing sources of change may be social or environmental or in natural resource management. Among social changes may be listed:

- *demographic transitions*, including those from higher to lower levels of human fertility (Gould and Brown, 1996). As family labour becomes scarcer, the relationships found between labour and land, which are critical in smallholder systems the world over (Netting, 1993), also change;
- *AIDS*: the fearsome truth about AIDS is that its seemingly remorseless ascendancy strikes at the very heart of the adaptive livelihood systems developed in the drylands, by removing family labour (of either sex) entirely. Furthermore, it transfers the burdens of production and of child care on to the elderly, or on to other children, plunging households below the threshold of economic viability, even under good rainfall;
- *increasing participation in education and migration*: education is often seen as a necessary preliminary to employment outside the community in eastern African communities like the Akamba, where scarce funds may be invested in it (Tiffen *et al.*, 1994). Migration takes away labour from intensive agriculture but may also bring back investment capital;
- *breakdown of family units and/or roles*: in dryland systems, where much evidence suggests that the large, integrated, extended family is best placed to manage risk, change may intensify the insecurity of small, poor or female-headed households, even while enhancing some employment opportunities;

- *hiring labour*: in some dryland systems in West Africa, the ability to hire labour has been recognised as a social discriminator (Hill, 1972). This process is linked to technical changes in farming. Hired labour, because of its cost, cannot provide a solution to labour shortages in the poorest households.

Environmental changes which may be occurring include:

- *climate change*: scenarios for the tropical drylands, under continued global warming, are still uncertain. However, the Sahel has already undergone rainfall decline since the 1960s of a magnitude greater than that predicted in climate change models (Hulme, 1996);
- *increasing land scarcity*: notwithstanding urbanisation, rising rural population densities are more common than the converse. These bring other changes in train. As a scarcity of cultivable land develops, average labour inputs per hectare increase. As with each generation more households enter the land market, the demand for land increases further;
- *the soil fertility crisis (or transition)*: in many rainfed ecosystems, the opportunity to manage low natural fertility with long fallows disappeared years ago. Can a transition to sustainable management of nutrients be accomplished by smallholders (Harris, 1998, 1999)? Is dependency on inorganic fertilizers unavoidable?
- *woodland clearance (or conversion)*: alarmist scenarios of 'deforestation' (which have been issued, and challenged, for at least 60 years in West Africa) are yielding place to more positive evaluations of farmers' potential as conservators (Cline-Cole *et al.*, 1990). Unprotected woodlands, on the other hand, are vulnerable to commercial destruction for urban markets, and forest reserves are hard to protect except on uncultivable land.

Changes are also taking place in the way natural resources are managed, in particular as a result of:

- *crop–livestock integration*: interactions between crop production and livestock management have implications for nutrient cycling, farm energy, income generation, diet and wealth accumulation. A relationship with rising population density, land scarcity and intensification has been advanced (McIntire *et al.*, 1992; Mortimore and Turner, 1991);
- *biodiversity conservation*: NGOs are promoting, with farmers, the conservation of local cultivars and wild plants in Africa. However, such conservation forms a part of the system, and cannot continue without it. While consistent with labour-intensive, low external input, multicrop farming on small holdings, conservation does not advance the interests of the large-scale or specialist market producer;
- *privatisation of natural resources*: the private appropriation of cultivation rights on common land is widely reported. Graziers cannot defend customary rights for their herds. Access to crop residues, fallows and farm trees by grazing animals is questioned or denied. Title is registered in the names of individuals to previously shared (or disputed) resources. Enclosures are erected around private territory. The dilemma is that, while often providing the necessary security for investment, such processes as these, where they occur, may also form a part of the poverty scenario in systems under transformation.

7.6 Ongoing adaptation in the Nigerian Sahel

At the centre of poor households' decisions regarding livelihood strategies and NRM is how to allocate scarce resources. The decisions they take are small (in scale), frequent (in occurrence) and sequential (in pattern). Decisions taken on one day form a part of the next day's decision-making matrix. Of course, they reflect the resource endowments of the household (of labour, livestock, arable land and so on) at the time in question. But they must also respond to the complex patterns of change which have been referred to above, and to the recurring variability of the rainfall. Many of these are perceived to be exogenous factors over which smallholders have no control. In livelihood management, therefore, three properties are advantageous: diversity, flexibility and adaptability, as shown in Figure 7.3.

Diversity in the options available to a household – whether access to natural, economic, technical or social resources – has to be painfully constructed in a context of poverty and risk. It is well known that such diversity has been a condition of economic sustainability in semi-arid African household economies during historical times. Each year, containing its own hazards or opportunities, must be negotiated, with as much *flexibility* as possible, and the 'accounting' – of food stocks, income, and capital – forms the starting point for the next year. From year to year, and in the longer term, such decisions, seen in aggregate, represent a progressive *adaptation* of the livelihood system as constraints or opportunities change. These three properties situate the household between a framework of constraints, on the one hand, and a developmental path, on the other. This path may prioritise any of the three classic alternatives that face a growing population under constrained natural productivity: intensifying NRM, diversifying away from primary production or migrating to other areas.

The Nigerian research on which this model is based suggested six interactive areas of strategic adaptation:

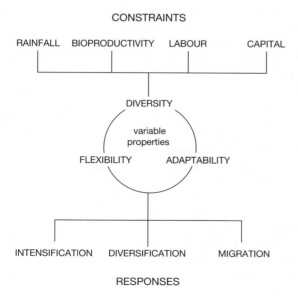

Figure 7.3 Diversity, flexibility and adaptability in dryland livelihoods
Source: Mortimore and Adams (1999)

1   *Negotiating the rain* (Washington and Downing, 1999) Development interventions – whether projects or policies – have tended to focus on the economic or technical constraints that are perceived to face smallholder agriculture, in order to increase its average output or value. Research into optimising the use of variable rainfall has focused on forecasting (Mortimore and Adams, 1999) and (in Kenya) on 'response farming' (Stewart, 1991). Much still needs to be done to transform technical findings into solid improvements on indigenous practice, which reduce risk without increasing costs.

2   *Managing biodiversity* Notwithstanding many years of converting woodland to cultivation, even in apparently degraded natural vegetation, there is greater natural biodiversity than might be expected. In famine, the routine use of gathered foods (in addition to medicine, fodder and construction materials) is extended dramatically in hungry households (Mortimore, 1989: 67–74). On farms, dozens of tree species are preserved (Cline-Cole *et al.*, 1990). Cultivated plants include as many as twelve named types of pearl millet, twenty-two of sorghum and forty-two other cultivated plants in a single village. Many lines of livestock are also maintained. Indigenous farmers and breeders are sustaining a genetic resource of great importance for formal breeding, and at the same time maintaining an insurance against new risks introduced by dependence on high-yielding or input-dependent exotics, hybrids or crosses.

3   *Integrating animals* In northern Nigeria, livestock densities increase with population densities (Bourn and Wint, 1994; Hendy, 1977). More animals are kept by farmers (whether livestock producers with farms, or farmers who also keep livestock) than by specialised (and usually nomadic) pastoralists. Everyone owns, or aspires to own, livestock. They are a depository for savings, a reserve for contingencies, an appreciating asset (growing, fattening and reproducing), a source of current income, and a source of energy for farm, well or road.

The potential of crop residues as feedstock for animals increases with rising population density (de Leeuw, 1997). In addition to this, they can support intensification on the farm (by cycling nutrients through crop residues and manure). During the last 20 years, there has been a shift from cattle to small ruminants in northern Nigeria, as they are less costly, more hardy, easier to feed and reproduce faster than cattle (de Leeuw *et al.*, 1995). This shift supports increased integration between crop and livestock production, but at the price of a higher labour requirement.

4   *Cycling nutrients* The potential of such labour intensification with low external inputs is shown in the Kano Close-settled Zone where in a farming system with a residential population density of $223/km^2$ and a mean rainfall of *c*.600 mm, bioproductivity has been sustained for several decades (Harris, 1998). The question is to what extent such methods can or should be extended to systems with lower densities (for example, $43/km^2$) and rainfall (*c*.400 mm), where nutrient stocks are maintained by transfers from rangeland and fallows (Harris, 1999). Subsidies have been withdrawn from inorganic fertilizers, which are often uneconomic except for high value crops. In Senegal, where export groundnuts used to thrive on fertilized fields (*champs de brousse*, formerly fallowed), there is insufficient organic material to extend biological fertilization from the more intensively managed *champs de case* (Badiane *et al.*, 2000). Scenarios of falling yields and 'soil mining' can oversimplify a complex reality. Yet labour, manure and livestock feed are obvious constraints affecting the possibility of a transition to sustainable management of nutrient resources (as illustrated above in Figure 7.2).

5   *Working the land harder* Rural population densities have increased almost everywhere in dryland Africa notwithstanding out-migration, as shown, for example, in the Sahel

during the 30 years 1961–90, both in urban hinterlands and in more remote areas (Snrech *et al.*, 1994). The inter-generational challenge for rural households is to make a living from a diminishing portfolio of natural resources. Converting land from natural woodland or grassland into farmland is not necessarily degradational, as plant biomass production on cropland may compare favourably with that of natural rangeland under comparable rainfall conditions (Mortimore *et al.*, 1999). 'Indigenous intensification', to stabilise or reverse the natural decline of nutrients under cultivation (Adams and Mortimore, 1997), and necessarily accomplished largely without benefit of inorganic fertilizers, reflects conscious decisions about the allocation of labour, the ownership of livestock and the acquisition of new technologies. Because of the adaptive nature of farmers' strategies, it should not be assumed that the full potential of such intensification can be predicted.

6 *Diversifying livelihoods* The critical importance of securing grain for subsistence ensures that, in the Nigerian villages, almost all family labour is available for farming tasks each year; but immediately after this objective has been achieved much labour is allocated to income diversification and (especially that of younger men) to migration. In some other drylands, such as in eastern or southern Africa, migrants may not return home for farming. This suggests a greater proportionate dependence on off-farm incomes. Such incomes may be recycled into farm investments or inputs where returns are attractive owing to the development of markets, or into obtaining educational qualifications for more competitive entry to off-farm incomes.

7.7 Differentiation among households

The particular circumstances of individual households in most dryland communities are extremely diverse. Household incomes and assets in the Nigerian villages are differentiated in terms of a matrix of variable indicators (Mortimore and Adams, 1999: 187):

- a household's demographic endowment of labour;
- its endowment of access rights to land and other natural resources;
- the level of capitalisation of the farm or other productive enterprises;
- the farm's primary production or output;
- the level of cash income secured from sales;
- the level of food sufficiency maintained, especially during times of general scarcity;
- the livestock holdings of household members; and
- the access enjoyed by household members to off-farm incomes.

This configuration of local households must be understood to sit embedded within a larger economic system, and to operate also within a specific *social* context which affects the ways in which individuals can respond to opportunities to better themselves.[4]

The above suggests that a precondition for sustainable NRM in African drylands, under present conditions of population growth, is land of sufficient value (either for crop and livestock production or for social uses) to justify private investments in farm intensification, housing or other improvements. The returns to farm or livestock labour need to be high enough to compete with off-farm alternatives,while the returns to investments must be high enough to attract savings earned in other places. Failure to invest does not necessarily mean that priorities have not been met. The Akamba of Kenya bring back a large proportion of earnings made in the cities to invest at home, even in farms too small to be viable, reflecting a high social value of land. Some evidence suggests that the Wolof of Senegal, by contrast, have exported a high proportion of their groundnut earnings and acquired the habit of investing their migrants' incomes in real estate in Dakar. The policy lessons of such contrary behaviour – if substantiated – are subtle.

7.8 Directions for development policy

Development policy for the African drylands is constrained by the same fundamentals as the strategies of dryland householders: those of low bioproductivity and rainfall variability. Concentration of development efforts on high potential areas is not an option: far too many people live outside them; population distributions cannot be altered significantly; and poverty is widespread and persistent.

Past perceptions of these constraints amongst many development practitioners and policy makers have, however, been defective. In place of scenarios of overpopulation, mismanagement of natural resources and environmental degradation, a new paradigm is now required which takes proper account of the capabilities of small crop and livestock producers to invest in real improvements to bioproductive potential, to diversify, to manage variability in a flexible way and to adapt to change. In place of top-down programmes, sometimes thinly disguised as 'participatory' and addressed to externally assessed priorities, we need ways of facilitating autonomous solutions. Local capabilities to manage, invest, innovate and diversify have been underestimated and, as the activities of many NGOs testify, these can be remobilised, though 'scaling up' is certainly problematic and may require changes in the policy environment.[5]

Here a transformational view of change, driven by new technologies, is inappropriate and inaccurate. Change must be set in a longer time frame and recognised as incremental; the time horizons of most programmes or projects are too short to influence this process effectively. Natural resources are, for the most part, managed privately by smallholders, or through various institutions, as common access resources. Decisions about their management are, and will continue to be, reached at this level. In a context of structural adjustment and liberalisation, governments have less and less effective power to direct behaviour or reorganise land use. Investment efforts of development agencies, through public investments and credit provison, are insufficient on their own and policy must take on board a broader conception of investment which takes full account of the need to promote and sustain investment created by and from indigenous resources.

Income diversification is not an indicator of failure, but a strategic choice. In fact, dependency on single crops, animals, markets or inputs undermines a fundamental principle of survival in drylands, while obstruction of free mobility between rural and urban areas merely reinforces privilege. The multisectoral livelihood system, at the community, household and individual levels, which has tended to be neglected until very recently by research and development priorities driven by disciplinary specialisms, must be confronted, without abandoning the search for new technologies. How these technologies can be inserted into varied and dynamic systems is now an issue.

Some dryland populations have achieved, during the past two decades, significant advances in a number of directions:

- extreme food scarcities have been survived, managed (at the household level) and recovered from, with lower mortalities than expected, indicating that resilience is a basic characteristic of dryland livelihood systems (Davies, 1996);

- population densities have been supported in rural and semi-arid areas at levels earlier thought to indicate 'overpopulation', a concept which has been shown to be unworkable (as much by smallholder families' achievements as by academic discourse);
- farming households (if not always livestock specialists) have demonstrated a capacity to adapt to external change and variability, a capacity which we can now understand as an inherent requirement of their livelihood and production systems, rather than representing fortuitous responses to emergencies;
- indigenous technical knowledge in many dryland systems is alive and well, diverse, adaptive, innovative and experimental, as well as preserving inherited knowledge and genetic resources;
- extensive management of abundant land resources, through converting woodland to farms, shifting cultivation or bush fallowing, was justified by labour scarcity. But population growth, rather than necessarily precipitating a Malthusian disaster, can initiate intensification and stabilise soil degradation;
- some smallholders have mobilised significant amounts of private capital, especially under conditions of rising population density, for improvements in productivity or for soil conservation;
- income diversification, if unfettered access is allowed to markets, has extended households' economic options and stabilised incomes. Through regulated or spontaneous migration, they have exploited systems of interregional dependence in which the drylands have a comparative advantage in labour-intensive activities and products.

All these advances provide, in the longer term, a more realistic foundation for designing improved development policy for the drylands than the outmoded scenarios of runaway degradation and the attendant necessity for costly, authoritarian and scientifically controversial interventions. Building this foundation has been under way for more than a decade (Nelson, 1988).

Notes

1 Research in Nigeria reported in this chapter was funded by the Economic and Social Research Council's Global Environmental Change Programme and by DfID's Natural Resources Systems Programme. Bill Adams and Mary Tiffen have made major indirect inputs to the author's thinking. Responsibility remains the author's.

2 Until recently, the sprawling, ill-regulated and 'overpopulated' cities of West Africa were considered as bottlenecks on economic growth where the costs of planning, infrastructure and services far exceeded the economic contributions made by the numberless migrants, who were thought to be victims of 'rural–urban drift' or, worse, 'drought refugees'. But the true significance of capital generation by low-income groups in cities was demonstrated by Cour (1994). The application of such a counter-intuitive thesis is even more radical in the rural drylands.

3 Two linked studies funded by the Department for International Development (Natural Resources Policy Research Programme and the Economic and Social Committee on Research), and carried out by Drylands Research in association with national researchers in Kenya, Senegal, Niger and northern Nigeria, are attempting to test the 'Machakos hypothesis' in a variety of ecological, demographic and political-economic circumstances.

4 Eastern African studies suggest that the possession of formal education is also an indicator of wealth. However, this relationship has not been demonstrated in Sahelian West Africa.

5 Emphasis on the potential of smallholders to improve their productive capacity does not, of course, deny the possibility that downward economic trajectories have been the lot of specific social or geographical groups in recent decades (see, for example, Murton, 1999 on Machakos District).

However, even the poorest families need not be without adaptive potential and can also benefit from enhanced economic opportunities.

References

Adams, W.M. and Mortimore, M.J. (1997) 'Agricultural intensification and flexibility in the Nigerian Sahel', *Geographical Journal*, 163: 150–60.

Amin, Samir (ed.) (1974) *Modern Migrations in West Africa*, Oxford: Oxford University Press, for the International African Institute.

Badiane, A.N., Khouma, M. and Sène, M. (2000) 'Politiques nationales affectant l'investissement chez les petits exploitants agricoles au Sénégal: gestion des eaux et des sols', Drylands Research Working Paper. Crewkerne, Somerset: Drylands Research.

Behnke, R. H., Scoones, I. and Kerven, C. (eds) (1993) *Range Ecology at Disequilibrium: New Models of Natural Variability and Pastoral Adaptation in African Savannas*, London: Overseas Development Institute.

Biswas, M.R. and Biswas, A.K. (eds) (1980) *Desertification: Associated Case Studies Prepared for the United Nations Conference on Desertification*, 2 vols, Oxford: Pergamon.

Bourn, D. and Wint, W. (1994) 'Livestock, land use and agricultural intensification in sub-Saharan Africa', Pastoral Development Network Paper 37a, London: Overseas Development Institute.

Buonajuti, A., (1991) 'External evaluation of the Plan of Action to Combat Desertification', *Desertification Control Bulletin*, 20: 30–3.

Cardy, F. (1993) 'Desertification: a fresh approach', *Desertification Control Bulletin*, 22: 4–8.

Chapman, M. and Mansell Prothero, R. (eds) (1984) *Circulation in Third World Countries*, London: Routledge and Kegan Paul.

Cleaver, K.M. and Schreiber, G.A. (1994) *Reversing the Spiral: The Population, Agriculture and Environment Nexus in Sub-Saharan Africa*, Washington, DC: World Bank.

Cline-Cole, R.A., Falola, J.A., Main, H.A.C., Mortimore, M.J., Nichol, J.E.and O'Reilly, F.D. (1990) *Wood Fuel in Kano*, Tokyo: United Nations University Press.

Cloudsley-Thompson, J.L. (1984) 'Human activities and desert expansion', *Geographical Journal*, 144: 416–23.

Colvin, L.G. (1982) *The Uprooted of the Western Sahel: Migrants' Quest for Cash in the Senegambia*, New York: Praeger.

Cour, J.-M. (1994) 'Performances du secteur agricole et rédistribution de la population en Afrique de l'Ouest', Document de Travail no. 12, West Africa Long Term Perspective Study, Paris: OECD/ Club du Sahel.

Davies, S. (1996) *Adaptable L;ivelihoods: Coping with Food Insecurity in the Malian Sahel*, Basingstoke, Hants.: Macmillan.

de Leeuw, P.N. (1997) 'Crop residues in tropical Africa: trends in supply, demand and use', in C. Renard (ed.), *Crop Residues in Sustainable Mixed Crop/Livestock Farming Systems*, Wallingford, Oxon: CAB International.

de Leeuw, P.N., Reynolds, L. and Rey, B. (1995) 'Nutrient transfers from livestock in West African production systems', in J.M. Powell, S. Fernández-Rivera, T.O. Williams and C. Renard (eds), *Livestock and Sustainable Nutrient Cycling in Mixed Farming Systems of Sub-Saharan Africa*, vol. 2: Technical Papers, Addis Ababa: International Livestock Centre for Africa.

Dregne, H.E. (1970) *Arid Lands in Transition*, Publication 90, Washington, DC: American Association for the Advancement of Science.

Dregne, H.E., Kassas, M. and Rosanov, B. (1991) 'A new assessment of the world status of desertification', *Desertification Control Bulletin*, 20: 6–18.

Eckholm, E. and Brown, L. (1977) *Spreading Deserts: The Hand of Man*, Worldwatch Paper 13, Washington, DC: Worldwatch Institute.

English, J., Tiffen, M. and Mortimore, M. (1993) 'Land resource management in Machakos District, Kenya, 1930–1990', World Bank Environment Paper 5, Washington, DC: World Bank.

Garcia, R.V. and Spitz, P. (eds) (1982) *Drought and Man: The 1972 Case History*, vol. 3: *The Roots of Catastrophe*, Oxford and New York: Pergamon.

Gorse, Jean Eugene and Steeds, David R. (1987) *Desertification in the Sahelian and Sudanian Zones of West Africa*, World Bank Technical Paper 61, Washington, DC: World Bank.

Gould, W.T.S. and Brown, M.S. (1996) 'A fertility transition in sub-Saharan Africa?', *International Journal of Population Geography*, 2: 1–22.

Haggblade, S., Hazell, P. and Brown, J.(1989) 'Farm-nonfarm linkages in rural sub-Saharan Africa', *World Development*, 17 (8): 1173–202.

Harris, F.M.A. (1998) 'Farm-level assessment of the nutrient balance in northern Nigeria', *Agriculture, Ecosystems and Environment*, 71: 201–14.

Harris, F.M.A. (1999) 'Nutrient management strategies of small-holder farmers in a short-fallow farming system in north-east Nigeria', *Geographical Journal*, 165 (3): 275–85.

Hendy, C.R.C. (1977) 'Animal production in Kano State and the requirements for further study in the Kano close-settled zone', Land Resource Report 21, Tolworth, Surrey: Land Resources Division, Overseas Development Administration.

Hill, P. (1972) *Rural Hausa: A Village and a Setting*, Cambridge: Cambridge University Press.

Hulme, M. (1996) 'Climate change within the period of meteorological records', in W.M. Adams, A.S. Goudie and A.R. Orme (eds), *The Physical Geography of Africa*, Oxford: Oxford University Press.

Jahnke, H.E. (1982) *Livestock Production Systems and Livestock Development in Tropical Africa*, Kiel: Wissenschaftsverlag Vauk.

Johnson, D.L. (ed.) (1977) 'The human dimensions of desertification', *Economic Geography*, 53 (special issue).

Jolly, C.L. and Torrey, B.B. (1993) *Population and Land Use in Developing Countries*, Washington, DC: National Academy Press.

Kassas, M., Ahmad, Y. and Rosanov, B. (1991) 'Desertification and drought: an ecological and economic analysis', *Desertification Control Bulletin*, 20: 19–29.

Lele, Uma and Stone, S.W. (1989) 'Population pressure, the environment and agricultural intensification in sub-Saharan Africa: variations on the Boserup hypothesis', MADIA Discussion Paper 4, Washington, DC: World Bank.

Mabogunje, A.L. (1972) *Regional Mobility and Resource Development in West Africa*, Montreal: McGill-Queens University Press.

McIntire, J., Bourzat, D. and Pingali, P. (1992) *Crop-Livestock Interactions in Sub-Saharan Africa*, Washington, DC: World Bank.

Mortimore, M. (1989) *Adapting to Drought: Farmers, Famines and Desertification in West Africa*, Cambridge: Cambridge University Press.

Mortimore, M. (1991) 'A review of mixed farming systems in the semi-arid zone of sub-Saharan Africa', Working Document 17, Addis Ababa: International Livestock Center for Africa.

Mortimore, M. (1993) 'The intensification of peri-urban agriculture: the Kano Close-settled Zone, 1964–1986', in B.L. Turner II, G. Hayden and R.W. Kates (eds), *Population Growth and Agricultural Change in Africa*, Gainesville, FL: University Press of Florida.

Mortimore, M. (1998) *Roots in the African Dust: Sustaining the Sub-Saharan Drylands*, Cambridge: Cambridge University Press.

Mortimore, M. and Adams, W. (1999) *Working the Sahel: Environment and Society in Northern Nigeria*, London: Routledge.

Mortimore, M. and Turner, B. (1991) 'Crop–livestock farming systems in the semi-arid zone of sub-Saharan Africa. Ordering diversity and understanding change', Agricultural Administration (Research and Extension) Network Paper 46, London: Overseas Development Institute.

Mortimore M., Harris, F. and Turner, B. (1999) 'Implications of land use change for the production of plant biomass in densely populated Sahelo-Sudanian shrub-grasslands in north-east Nigeria', *Global Ecology and Biogeography*, 8: 243–56.

Murton, J. (1999) 'Population growth and poverty in Machakos District, Kenya', *Geographical Journal*, 165: 37–46.

Natural Resources Systems Programme (NRSP) (DfID) (1999) *The Systems Approach in the Natural Resources Systems Programme*. London: Systems Management Office, Department for International Development.

Nelson, R. (1988) 'Dryland management: the desertification problem', Environment Department Working paper 8, Washington, DC: World Bank.

Netting, R. McC. (1993) *Smallholders, Householders: Farm Families and the Ecology of Sustainable Intensive Agriculture*, Stanford, CA: Stanford University Press.

Odingo, R.S. (1992) 'Implementation of the Plan of Action to Combat Desertification (PACD)1978–1991', *Desertification Control Bulletin*, 21: 6–14.

Raynaut, C., Grégoire, E., Janin, P., Koechlin, J. and Lavigne Delville, P. (1997) *Societies and Nature in the Sahel*, London: Routledge.

Snrech, S., with Cour, J.-M., de Lattre, A. and Naudet, J.D. (1994) *West Africa Long Term Perspective Study. Preparing for the Future: A Vision of West Africa in the Year 2020. Summary report*. Paris: OECD/Club du Sahel.

Somerville, D. (1986) *Drought and Aid in the Sahel: A Decade of Development Cooperation*, Boulder, CO: Westview Press.

Stewart, J.I. (1991) 'Principles and performance of response farming'. in R.C. Muchow and J.A. Bellamy (eds), *Climatic Risk in Crop Production: Models and Management from the Semiarid Tropics and Subtropics*, Wallingford, Oxon: CAB International.

Stiles, D. (1997) 'Linkages between dryland degradation and migration: a methodology', *Desertification Control Bulletin*, 30: 9–18.

Swift, J. (1996) 'Desertification: narratives, winners and losers', in M. Leach and R. Mearns (eds), *The Lie of the Land: Challenging Received Wisdom on the African Environment*, Oxford: James Currey.

Tiffen, M., Mortimore, M. and Gichuki, F. (1994) *More People, Less Erosion: Environmental Recovery in Kenya*, Chichester, Sussex: John Wiley.

United Nations Environment Programme (UNEP) (1990) *World Map of the Status of Human-induced Soil Degradation*, Nairobi: United Nations Environment Programme.

UNEP (1992) *World Atlas of Desertification*, Nairobi: United Nations Environment Programme.

UNEP (1993) 'Good news in the fight against desertification', *Desertification Control Bulletin*, 22: 3.

UNEP (1997) 'Summary of the tenth session of the Intergovernmental Negotiating Committee for the Convention to Combat Desertification', *Desertification Control Bulletin*, 30: 3–6.

Washington, R. and Downing, T.E. (1999) 'Seasonal forecasting of African rainfall: prediction, responses, and household food security', *Geographical Journal*, 165 (3): 255–75.

Western, D. (1982) 'The environment and ecology of pastoralism in arid savannas', *Development and Change*, 13: 183–211.

8 Paying for agricultural research and extension[1]

Stephen Akroyd and Alex Duncan[2]

8.1 Introduction

Shortcomings in the provision of agricultural research and extension services have contributed to poor agricultural performance in sub-Saharan Africa (SSA). Given that research and extension in SSA have been relatively well funded in the past, and despite some deterioration in the 1990s, the limited impact of these services is due as much to the poor management of resources as to inadequate financing levels.

In SSA there is now growing interest in alternatives to the standard model of a publicly financed and provided agricultural research and extension system. This chapter reviews the guidance provided by economic theory in defining the role of the state in the provision, and particularly the financing, of these services. It presents an economic framework in which the concepts of market and state failure are developed, and applies this framework to the provision of different forms of agricultural research and extension, from which conclusions regarding the appropriate role for the state in the financing of research and extension services are drawn.[3] The framework set out here has had wide application in Latin America and is currently being applied to the reform of research and extension services in SSA.

8.2 Economic framework

Drawing on the principles of welfare economics, the case for state intervention generally rests on the need to correct for various forms of 'market failure', and to compensate for the results of unconstrained market forces on the distribution of income, wealth and other resources that might be considered socially unacceptable. Defining the appropriate role for the state in the financing and delivery of a specific service requires an assessment of potential market failures, and consideration of the most effective and efficient actions by the state to overcome such failures and to achieve distributional objectives. This also requires an assessment of the 'state failures' associated with particular policy instruments.

Such an assessment is informed by describing markets as social institutions, that is, a framework of rules governing individual behaviour. Institutions may fail to achieve efficiency or other social objectives because these are not the objectives of the dominant classes or groups that shape them, or because the process of reforming social institutions to reflect changing circumstances is an inherently costly and difficult process.

In the rural economy of a low-income country, the institutional framework within which both markets and other institutions (including state agencies) operate is likely to be prone to multiple 'failures'. In an environment of high transactions costs, a number of important rural social institutions (such as sharecropping, other forms of land and livestock tenancy, tied credit arrangements and patronage networks) can be interpreted as substitutes for functions that would in other contexts be performed by markets. The absence of markets does not therefore necessarily point to a role for the state in remedying market failures as alternative institutional arrangements may be devised or may emerge. Indeed, many 'state failures' in rural development policy have been the result of inadequate understanding of the nature and functions of rural institutions.

8.3 Market failures

The welfare economics literature identifies different forms of market failure.

8.3.1 Public goods

Public goods have two attributes that discourage private markets because profits or benefits cannot be appropriated by the supplier:

- they are *non-excludable*: once produced, non-paying consumers cannot be excluded from using public goods, so that it is impossible to prevent 'free-riding;
- they are *non-subtractable* (or non-rival): the consumption of a public good by one individual does not diminish its supply to others.

Classic examples of public goods include street lighting, national defence, and collective political action. However, many goods and services are neither entirely public nor entirely private. For example:

- *toll goods* are excludable, but non-subtractable: the ability to exclude those who have not paid means that profits can be fully appropriated and does provide incentives for private provision (for example, a satellite TV broadcast which requires a decoder to be received);
- *common pool goods*, on the other hand, are subtractable but non-excludable: increased consumption diminishes supply for others but there are no incentives for private supply because access cannot be restricted (for example, unregulated communal grazing).

Whether an activity has the characteristics of a public good depends not just on its technical features, but also on the structure and nature of property rights. These in turn depend not just on formal legislation, but on the costs, effectiveness and reliability of enforcement mechanisms.

8.3.2 Externalities

Externalities exist when the production or consumption of a good or service has spillover effects on other individuals that are not fully reflected in the market price, so that the good may be either over- or under-provided by the market. Externalities may be positive (for example, the reduced risk of infection for one farmer's livestock resulting from the vaccination of a neighbour's animals) or negative (for example, pollution of underground water resources caused by excessive levels of fertiliser use).

8.3.3 Market power and economies of scale

Barriers to entry create market power, enabling monopoly rents to be earned and implying that production is below the level required for by efficient resource allocation. *Economies of scale* (where a high proportion of fixed costs cause unit production costs to fall as the scale of operation increases) may create barriers to entry, either because of credit market failures or because fixed costs are likely to have a significant sunk element. Similarly, one can distinguish *economies of scope* (lower unit costs from producing a combination of products compared to producing each independently) and *economies of time* (for example, from the accumulation of experience and knowledge).

8.3.4 Information and risk

'Asymmetries' of information (where parties to a transaction have different information about the nature of the exchange being made) are a critical source of market failure. Two aspects of this are *adverse selection* and *moral hazard*, often illustrated with reference to insurance markets: an insurance company unable to distinguish between high- and low-risk clients may adversely select the former, while low-risk clients may find the premiums too onerous and not insure at all. Similarly, the inability of the insurer adequately to monitor behaviour creates incentives for the insured to be less careful or even to make fraudulent claims (moral hazard). In some cases, information failures may deter transactions so completely that markets do not exist at all.

8.4.5 Costs of establishing and enforcing agreements

In the absence of low-cost mechanisms for establishing and enforcing agreements, 'time inconsistency' may increase exposure to problems of adverse selection and moral hazard, and act as a major deterrent to transactions that involve sunk costs. The problem is that one party to a transaction (for example, a farmer contracted to sell his or her crop to a processor who supplied 'free' inputs and extension, or a trader selling improved seed) may have incentives to renege on an agreement once the other party has incurred sunk costs. However, a variety of institutional arrangements (such as self-regulation and establishment of industry standards) may substitute for enforceable contracts. In particular, there will be incentives for parties to agreements not to renege on them if the relationship is an ongoing, repeated one (for example, input sales occurring each season or through repeated extension contacts).

8.4 State intervention and state failure

It is useful to distinguish two types of government intervention. Productive policies correct for market failures to improve efficiency and so bring about increases in social welfare.[4] Redistributive policies alter the distribution of income, wealth, and other resources so that some gain at the expense of others. State intervention may therefore redress a market failure which causes the level of financing in the absence of such intervention to be lower than is socially optimal, or it may redistribute income towards a target group of recipients, at the cost of some other group.

The complete separation of efficiency and distributional objectives and policies is rarely feasible or desirable, but distinguishing between them is important for assessing the common argument that the public sector should provide free or highly subsidised agricultural services to the poor. Addressing failures in related markets (especially credit) or other measures (such as targeted income transfers and public works programmes) may be more effective in relieving rural poverty than subsidising services or inputs with a high private good content, particularly as it is in the provision of such services to the poor and remote that governments have tended to have had least success in the past (Carney, 1997).

The state possesses three broad types of policy instrument for intervening in the supply of services, whether for productive or redistributive purposes:

- direct government provision (financing and delivery) through a state agency;
- the use of taxes and subsidies to influence private behaviour;
- regulation of private service providers.

The appropriate choice or combination of responses will depend on the nature of the market failure and prevailing institutional framework and capacity. The legislative framework, historical role of the private sector, administrative capacity of the public sector, and infrastructural conditions will all affect the ability of the public and private sectors to provide, and the public sector to regulate.

8.5 Finance and delivery

An analysis of these options and the role of the state is aided by the separation of the provision of any service into two components, financing and delivery. The question of who should *finance* what relates largely to the public/private characteristics of the good or service and the degree and nature of any externalities. In the case of a purely private good with no externalities, private financing would be appropriate. At the other extreme of a pure public good, for which it is not possible to introduce elements of excludability (for example, national defence), financing would need to come from general taxation. Many public services embody a range of public/private characteristics and externalities. In these cases a mix of user charges, central government financing and earmarked taxes may be appropriate subject to the prevailing administrative constraints.

The question of who should *deliver* what (and how) is a separate issue to be informed by criteria of cost-effectiveness and efficiency. For example, the public sector may continue to provide the service, or delivery may be contracted out to the private sector (including non-profit-oriented agencies such as universities and NGOs). The best institutional arrangement for the delivery of even public goods will depend on the trade-off between potential efficiency gains of private delivery arising from superior managerial incentives and greater input flexibility (for example, in terms of hiring and firing staff) on the one hand, and the transactions costs of monitoring and regulating on the other. As for *how* the service might be delivered, a range of methodological approaches to service delivery may be considered. For example, more participatory approaches to research and extension are widely held to result in more accountable, relevant and cost-effective services.

It is often argued that public provision of services 'crowds out' private provision, and that this creates a *prima facie* case for state withdrawal. However, the issues involved are more subtle than is sometimes implied. Three scenarios can be considered:

- where the state finances a service for which users would in fact be prepared to pay, state financing provides a subsidy to users and withdrawal has an entirely distributional effect;

- if, in addition, state provision is less efficient than private provision, then there will also be a net efficiency saving from privatisation of the service;
- if public provision is inadequate, then parallel private provision may emerge, implying duplication of costs. In this case the private sector may not be completely 'crowded out' of provision, but competition for resources with the public sector (or regulatory restrictions designed to protect a public monopoly) may raise the costs that private providers face.

The case for transferring the provision of services from the public to the private sector depends on an assessment of the costs and the distributional impact of different service provision options. Even for public goods, state provision may not be the best option compared to regulatory or legislative reform that overcomes barriers to private financing and delivery, or the combination of state financing with private delivery.

8.6 State failure

There are two fundamental dangers associated with the concept of the state as an impartial and efficient provider of public goods and agent of regulation and redistribution, seeking to maximise social welfare. First, the power of the state may be used for predatory purposes, in favour of a particular class or other sectional interests. This may reflect direct control of the state apparatus by a particular group or be the outcome of competition between politicians for the support of interest groups. Second, even if the objectives of policy makers are in fact developmental rather than predatory, there is a problem of monitoring and control of the agents to whom the exercise of state power is delegated, especially since competitive pressure facing state agencies is likely to be limited.

These problems imply tendencies within state organisations towards limited accountability to users or beneficiaries of services, soft budget constraints, diversion of resources to private or sectional ends, wasting of resources on unproductive activities (for example, lobbying), weak incentives for efficiency and limited flexibility.

8.7 Application to agricultural research

Three main reasons are commonly cited to explain why private investment in research is sub-optimal (Thirtle and Echeverria, 1994):

- public good characteristics of research: the profits resulting from innovation cannot in general be fully appropriated by those funding the research;
- the inherent riskiness of research (in the absence of effective insurance mechanisms);
- indivisibilities and increasing returns (and hence monopoly tendencies) in applied research.

A description of the economic characteristics of agricultural research and the application of these principles to agricultural research is facilitated by the following two classifications. The first, established by the Consultative Group on International Agricultural Research (CGIAR, 1981) and used fairly universally in the literature, adopts the following terminology:

- *Basic research* may be experimental or theoretical in character but is aimed mainly at increasing knowledge, without any particular application in view (which effectively precludes it being patented) (for example, the study of biochemical and physiological attributes of plants, and the characterisation of genotypes by molecular markers).

- *Strategic research* addresses issues which influence the efficiency with which other research further downstream can be carried out (for example, the development of new tissue culture techniques for genetic improvement, and the generation of new diagnostic methods for plant diseases).
- *Applied research* is directed at the discovery of new knowledge that has specific commercial objectives and applications (such as the genetic improvement of plants and animals, or the biological control of pests, weeds and diseases).
- *Adaptive research* is designed to adjust technology to specific environmental or socio-economic conditions (for example, the testing of agronomic practices, pesticide and fertilizer applications, and the evaluation of introduced genetic material or agricultural machinery).

These categories (which apply equally well to industrial and other forms of research) do not represent clear-cut groups, but rather point along a spectrum from the basic to the more applied or adaptive which is based on the direct commercial applicability of the research results.

Knowledge, the output of research, is clearly non-subtractable as one individual's consumption does not diminish the availability of that knowledge to others. The extent to which research may be considered a public good is largely dependent therefore on the amenability of that knowledge, or the invention or product in which it is embodied, to various exclusion mechanisms that overcome the free-rider problem and enable the appropriation of returns to research investments. This amenability to exclusion will be affected by the natural characteristics of the technology, various marketing strategies (such as guarantees of consistent premium quality and performance) that promote brand loyalty, and the existence and enforceability of intellectual property rights (IPR) legislation (Umali, 1992).[5] The boundary between public and private research, and the costs and returns to research activity, are thus influenced significantly by institutional factors that are likely to differ widely between countries.

It is generally argued in the literature that the nearer to the basic end of the spectrum one is, the less excludable and subtractable are the research outputs, and the closer research comes to being a public good. Indeed, basic research typically is described as being a public good (Umali, 1992).

A second widely used classification more specific to agricultural research uses four main categories according to the nature of the technology (Evenson and Puttnam, 1990; Pray and Echeverria, 1991):

- *Managerial (agronomic) research* involves the study of crop and livestock management techniques (such as planting dates and densities, weeding strategies and feed regimes) and other managerial practices.
- *Biological research* focuses on the generation of improved plant varieties and animal breeds, hormones and vaccines.
- *Chemical research* involves the development of new fertilizers, fungicides, insecticides and herbicides.
- *Mechanical research* includes the development of agricultural machinery and equipment such as tractors and implements.[6]

Mechanical and chemical research are generally private as patenting arrangements are comparatively straightforward. Biological technologies (the major focus of agricultural research in LDCs) are increasingly patentable, although the characteristics of self-pollinating seeds (which can be planted repeatedly by farmers without diminishing the crop's yield potential) make patent enforcement impractical even where legislation does exist, whereas hybrid seeds have natural protection and do not require patent protection. Managerial (or agronomic) research is also much less amenable to exclusion mechanisms and is usually considered to be public in nature. Available evidence on patterns of public and private research expenditures confirms that the private sector does tend to spend a smaller proportion on basic/strategic research, and on managerial and biological research, than does the public sector.

Strengthening intellectual property rights (IPR) should encourage private investment in research and development, and may therefore reduce the need for public funding of research. Most studies confirm the importance of IPR legislation in stimulating private activity, although evidence from developing countries is scarce. However, developing countries that do not have a comparative advantage in research activities may be net losers from a strengthening of IPR, if it raises the costs of access to imported technologies.

8.8 Political economy of public good provision

A variety of institutional arrangements may be developed to provide public goods for agriculture, including cooperation between farmers and government provision financed by taxpayers or producers. A key question is how the benefits of public good provision are divided between producers and consumers of the product to which the public good is an input.

Partial equilibrium models provide a useful analysis of the allocation of research benefits measured in terms of the effect upon consumer and producer surplus. These models show that the division of research benefits between consumers and producers will vary according to the elasticities of demand and supply. For example, at one extreme, if demand for a good is perfectly elastic (horizontal demand curve), then all the benefits of research which reduce its cost will accrue to producers. This will be the case, for instance, with most export crops, which are also generally amenable to levy financing by producers. If perfectly inelastic (vertical demand curve), research benefits will accrue entirely to consumers. This suggests that state finance should be focused on research in goods with inelastic demand, such as non-tradable staple food items.

8.9 Implications for the role of the state in financing research

The discussion above suggests that the gap between the private and social returns to agricultural research, hence under-investment in research, will be greatest under the following circumstances:

- where potential beneficiaries of research are poorly organised politically;
- where the ability of private financiers of research to appropriate the returns is weakest, primarily because of weak intellectual property rights or the technical nature of the innovation;
- for commodities where elasticities of demand are low, elasticities of supply are high, and the productivity of research expenditure is high, so that consumers obtain a relatively high proportion of the benefits of research;

- in relation to distributional objectives, for those agricultural products that are disproportionately produced or consumed by the poor.

In practice in Africa, this might be interpreted as implying a focus for state financing of research on non-tradable staple foods that are produced by smallholders and consumed by the poor (for which demand will be relatively inelastic); and on basic and managerial research, and those aspects of biological, chemical and mechanical research where exclusion mechanisms are weakest. The near absence of private agricultural research in many African countries, and the concentration of what does exist on a small number of commodities, and the significance of positive externalities and economies of scale, time and scope mean that a broader public sector role in research may be justified. However, the fact that the gap between social and private rates of return may be highest for such commodities does not imply that the highest social rates of return to public research finance in fact apply to these products.

The case for public financing of basic research in developing countries is weak, especially if relevant basic research results are readily available (from developed countries and from the Consultive Group on International Agricultural Research (CGIAR) network of international research) and scientific resources are limited. However, if basic research fails to push the relevant technological frontier forward, the pool of scientific knowledge to be exploited will diminish and applied research will suffer from diminishing returns (Thirtle and Echeverria, 1994).

The state also has an important role to play in creating an enabling environment and alleviating some of the constraints that inhibit private sector research. This would serve to stimulate incentives for private sector research by reducing the costs of such research (hence raising potential profitability) and making research benefits easier to appropriate by the innovator. Until this environment is in place, any curtailment in state activity is unlikely to be replaced by an emerging private sector and should be considered with extreme caution. Areas that should be addressed include:

- a consistent macroeconomic and sectoral policy framework that removes at least some of the risks inherent in much research activity;
- liberalisation of input supply markets;
- investment in human capital to ensure that there exists a pool of trained researchers that both public and private sectors can use;
- the judicious strengthening of intellectual property rights legislation and mechanisms for enforcement;
- the relaxation of restrictions on imports of technology, foreign ownership and use of expatriate staff;
- tax breaks on research activities;
- more open disclosure of public sector research findings to the private sector.

With increasing institutional pluralism in the implementation of research, there should also be a coordinating role for government to avoid unnecessary duplication of research. The maintenance of administrative and technical capacity to manage a more pluralistic research system, and to regulate where necessary, is therefore critical (Carney, 1997). In addition, regional cooperation to increase market size and reduce unit costs of research (by reaping economies of scale), and to minimise the free-rider problem arising from cross-country spillovers may be actively promoted by governments, a function that donors may also have a valuable role in assisting.

8.10 Application to extension

Numerous definitions of extension exist, with the emphasis shifting away from the simple 'transfer of technology' interpretation to those which embrace growing consultant–client relationships and participatory approaches to extension. Röling (1996) uses the following three broad categories:

- *Technology transfer* The traditional interpretation of extension as the transfer of advice, information, knowledge and skills to farmers. Here, research and extension are treated as part of a unified and linear agricultural technology system.
- *Advisory service* In this model the extension service comprises a cadre of experts who farmers use as a source of advice in relation to specific problems they face. The approach relies upon active problem solving for farmers who are able to articulate their needs and extension officers only visit farmers on request.
- *Facilitation* Rather than extension providing answers to farmers, the aim of this model is to enable farmers to define their own problems and develop their own solutions. The approach uses participatory learning techniques that encourage farmers to become experts on their own farms.

Extension is associated with information, the public good character of which depends on:

- the nature of its provision (mass media or personal contact);
- its speed of diffusion and time sensitivity;
- whether it is tied to, or embodied in, physical inputs; and
- whether the supplier of information is also buyer of the produce (Table 8.1).

Table 8.1 Economic classification of agricultural technology transfer via extension

EXCLUDABILITY

		Low	High
		Public goods	*Toll goods*
	Low	Pure agricultural information (l-term): • cultural/production techniques • farm management • market information	Pure agricultural information (s-term): • cultural/production techniques • farm management • market information
SUBTRACTABILITY		Information relayed through mass media channels	Specialised and/or client-specific information: Extension under contract farming
		Common pool goods	*Private goods*
	High	Modern technologies: • self-pollinated seeds • commonly available/used inputs	Modern technologies:[a] • e.g. new machinery, ag. chemicals, hybrid seeds, vet. supplies and pharmaceuticals

Source: Adapted from Umali and Schwartz (1994)

Note
a May involve some externalities.

Thus pure information, which is not embodied in a physical product and would include cultural and production techniques, farm management procedures, market information, and community development (for example, organisation of farmer associations), is generally regarded as being both non-subtractable and non-excludable.

In the short term, however, it may be possible to exclude non-payers (free-riders), particularly where it involves individual contact (as with the advisory model) and covers techniques which are not directly observable or cannot be copied by neighbouring farmers, or market information which can be easily concealed or withheld, or is quickly outdated (time-sensitive). In such cases, extension of information may be considered a toll good and potentially attractive to private suppliers.

In the long term, such information is likely to become a public good through diffusion via farmer-to-farmer contact. If the information is diffused very rapidly, there is an incentive for farmers not to pay for the information and to 'free-ride' by obtaining the information from other farmers. This problem may be overcome by encouraging farmer associations that are able, through membership fees, to recoup the costs of extension information.

Information provided through mass media, such as radio, is inherently public in nature and is unlikely to be provided by the private sector, although private financing through advertising is possible. Where farming practices are more commercialised and specialised, the corresponding extension services needed to support these activities also become more client and situation specific, and therefore more exclusive. Under such conditions, extension is a toll good and amenable to private provision and user charges.

Where information is embodied in or associated with physical inventions or inputs which are essentially private goods (such as new machinery or processing equipment, agricultural chemicals, hybrid seeds, veterinary supplies or farm management software), there is no presumption that the supply of the technology and the supporting technical information will be less than is socially optimal (Umali and Schwartz, 1994). However, where extension involves technology that is itself non-excludable or can be easily replicated (e.g, self-pollinating seeds), there will be limited incentives for private supply and extension can be considered a common pool good. Moreover, many inputs, once widely known and available off the shelf, will need little or no specific extension at all.

Finally, it will be easier for the supplier to appropriate the benefits of extension where the supplier is also the buyer of the produce. For example, private agro-processors will provide extension if the resulting benefits to the processor of a more reliable and higher quality crop exceed the cost of extension provision. Extension is therefore a common component of contract farming systems (Schwartz, 1994).

Hence, as with research, various types of extension activity occupy different points on a public–private good spectrum, which should help inform the focus of public resources for extension. In addition, positive externalities arising from spillover effects from one farmer to another (for example, public information encouraging the use of vaccines) may justify some public subsidy of extension activity. The merit good argument may also be used to justify public subsidy if it assumed that farmers underestimate the true value of extension advice. This underlies the ethos of much colonial extension (and continues in some services). Public extension services have also been used to deal with negative externalities associated with the management of common resources, such as soil erosion and livestock dipping. Regulation may be necessary to protect consumers from negative externalities associated with the use of some technologies, such as fertilizers and pesticides.

Problems of asymmetric information may arise with private extension services if customers are unable to assess the quality of advice (and inputs) at the time of purchase and do not have recourse to redress in the event of being defrauded. This would create a tendency towards under-consumption which may create a case for the state to certify quality by, for instance, licensing private extension practitioners. However, since extension tends to be based on repeated contacts with farmers within an area, a considerable degree of self-regulation would be expected, as suppliers will need to pursue repeat sales by establishing a reputation for quality and reliability.

Economies of scale are inherent in many forms of extension. Larger (and more specialised) farmers will in general have a greater incentive than poorer (or more diversified) farmers to purchase extension advice, because they are able to spread the costs over a larger output. The capacity of smallholders to purchase extension advice is therefore likely to depend on their ability to organise collectively so as to reduce the costs per individual of extension contacts. Likewise, collective organisation may be required to exert influence on monopoly providers of extension to improve the quality and relevance of advice. Economies of scope may also apply: extension agents are often the only permanent government presence at village level and may be used for a variety of purposes other than providing information to farmers.

8.11 Implications for the role of the state in financing extension

In summary, extension activities most suited to private provision are those which are highly specialised and client-specific (though a client may be a group with homogeneous interests) and those associated with the sale of physical inputs. This suggests that incentives for the private financing and delivery of extension services are greater the more commercialised is the farming system, as characterised by the following features:

- larger and more specialised farms;
- a higher proportion of marketed output;
- greater use of purchased inputs;
- high levels of formalised cooperation between farmers;
- close links between farmers and customers;
- more exacting quality standards from customers.

Where these features exist, government may be able to withdraw from the public financing and delivery of extension services. This is reflected in the high proportion of private extension financing and the success of recent cost-recovery programmes for public extension services in many countries with commercialised agriculture. In developing countries, it suggests that government may be able to withdraw from the public financing of extension services in some areas that may be adequately serviced by commercial agencies.

The promotion of private sector involvement in extension activities to which it is suited, and the continued provision and protection of various public interests, suggests that the state should:

- remove controls on input marketing, to which much extension activity is tied;
- ensure that policy statements and actions are consistent and instil private sector confidence in the future marketing and regulatory environment;
- provide a sound regulatory environment that ensures that negative externalities are minimised and consumers protected;
- focus state financing on public good activities such as mass media extension and the provision of general (pure) agricultural information;

- provide environmental extension and regulatory services and health and safety monitoring.

Where the state continues to provide extension services (for public or merit good reasons), widespread 'state failure' in the provision of these services also needs to be addressed. Typical problems are inadequate management incentives preventing extension delivery from being sufficiently responsive to client needs, and ill-trained and under-resourced staff with an inadequate technical message to extend.

8.12 Conclusion

In seeking alternatives to the standard model of a publicly provided and financed agricultural research and extension system, economic theory has much to offer in defining the role of the state, particularly in terms of identifying *what* the state should be financing. The analysis presented here indicates that the private good characteristics of some research and much extension may be greater than commonly perceived, and suggests that there is considerable scope for harnessing greater private contributions to the financing of research and extension. At the very least, the allocation of public funds between different forms of research and extension needs to be carefully examined to ensure that the public sector is not unnecessarily funding activities that the private sector is able and willing to finance, thereby reducing the availability of public funds for genuine public good activities. Social and distributional considerations will always be important and have often been used to justify public provision of free or heavily subsidised research and extension services to particular target groups, but this needs to be considered in the context of alternative forms of support that may be more effective.

These insights are potentially of much value in promoting a more efficient and equitable allocation of resources, particularly for public institutions under pressure from growing financial constraints.

Notes

1 This paper draws on Beynon *et al.* (1988: ch. 3).
2 The authors gratefully acknowledge the support of DfID, who funded the original research for this chapter. The findings and conclusions expressed here are entirely those of the authors.
3 Two World Bank discussion papers (Umali, 1992; Umali and Schwartz, 1994) are particularly useful and form the basis of some of the discussion here.
4 In the sense that at least some people can be made better off by the policy, while fully compensating any losers (that is, a 'Pareto improvement').
5 The appropriability of benefits will also be influenced by the competitive nature of the market: the ability to exclude rivals from benefiting from new technology becomes less important in a monopolistic or highly oligopolistic environment. However, the absence of competition may reduce the pressure to undertake research. The level of research chosen by a monopolist will be lower than is socially optimal (since the monopolist by definition faces less than perfectly elastic demand for his product).
6 Other categories are also sometimes identified separately, such as post-harvest storage and processing technologies.

References

Alston, J.M., Pardey, P.G. and Roseboom, J. (1997) 'Financing agricultural research: international investment patterns and policy perspectives', revised draft (25 July 1997) of invited paper for presentation at August 1997 conference of International Association of Agricultural Economists, Sacramento, California (forthcoming in *World Development*).

Beynon, J.G. (1995) 'The state's role in financing agricultural research', *Food Policy*, 20 (6): 545–50.

Beynon, J.G. (1996) 'Financing of agricultural research and extension for smallholder farmers in sub-Saharan Africa', ODI Natural Resource Perspectives no.15, London: Overseas Development Institute.

Beynon, J.G., Akroyd, S., Duncan, A. and Jones, S. (1998) *Financing the Future: Options for Agricultural Research and Extension in Sub-Saharan Africa*, Oxford: Oxford Policy Management.

Carney, D. (1997) 'Alternative systems for financing agricultural technology research and development in Uganda, Tanzania and Kenya', Report to DfID, 1 December, CNTR 97 3851A, London: Overseas Development Institute.

Consultative Group on International Agricultural Research (CGIAR) (1981) *Report of the Review Committee*, Washington, DC: Consultative Group on International Agricultural Research.

Evenson, R.E. and Puttnam, J. (1990) 'Intellectual property management', in G.B. Persley (ed.), *Agricultural Biotechnology: Opportunities for International Development*, Oxford: CAB International.

Pray, C.E. and Echeverria, R.G. (1991),'Private sector agricultural research in less developed countries', in Pardey, P.G., Roseboom, J. and Anderson, J.R. (eds), *Agricultural Research Policy: International Quantitative Perspectives*, Cambridge: Cambridge University Press.

Roling, N. (1996) 'What to think of extension?', paper presented at the extension workshop on alternate mechanisms for funding and delivering extension, held at the World Bank, Washington, DC, 18–19 June.

Schwartz, L.A. (1994) 'The role of the private sector in agricultural extension: economic analysis and case studies', Agricultural Administration (Research and Extension) Network Paper 48, London: Overseas Development Institute.

Thirtle, C.G. and Echeverria, R.G. (1994) 'Privatisation and the roles of public and private institutions in agricultural research in sub-Saharan Africa', *Food Policy*, 19 (1): 31–44.

Umali, D.L. (1992) *Public and Private Sector Roles in Agricultural Research: Theory and Experience*, World Bank Discussion Paper 176, Washington, DC: World Bank.

Umali, D.L. and Schwartz, L. (1994) *Public and Private Agricultural Extension: Beyond Traditional Frontiers*, World Bank Discussion Paper 236, Washington, DC: World Bank.

9 New institutional economics, agricultural parastatals and marketing policy

Jonathan Kydd, Andrew Dorward and Colin Poulton

9.1 Introduction

In a recent book (Dorward *et al.*, 1998), the authors embarked on a study of cash crop marketing in sub-Saharan Africa and South Asia, concerned and puzzled by the generally weak response of the smallholder agricultures of sub-Saharan Africa to the structural adjustment and market liberalisation which began to be implemented from the early 1980s onwards, gathering pace in the early 1990s. Structural adjustment involved reforms to macroeconomic and trade policies that were designed, among other objectives, to improve price incentives for producers of tradables. As the output mix of the agricultural sector, including many smallholder sub-sectors, has a higher share of tradables and near-tradables than most other key economic sectors, a vigorous agricultural supply response had been anticipated, via the improved terms of trade brought about by liberalisation.

Likewise, by scaling back, or entirely removing, parastatal and state-sponsored cooperative agricultural marketing structures and by lifting restrictions on private sector entry to these activities, it was hoped that there would be a strong private sector response in supplying inputs and in purchasing, storing, processing and (where appropriate) exporting produce. Additionally, it was thought that parallel financial sector reforms (encompassing monetary management at the macro-level, through banking sector reforms down to the commercialisation or privatisation of state-supported agricultural finance organisations) would catalyse the other elements of structural adjustment by channelling funds to emerging opportunities for profitable farming and trade. However, as Jones's (1994) survey shows, although African governments were slow to respond to external advice to liberalise the agricultural sector, most have now travelled a fair distance along this path but have realised, at best, only modest success. In particular, input and credit supply systems for smallholder agriculture are in a parlous state and, in general, are failing to make progress towards the World Bank's objectives (World Bank, 1997) of 'sustainable intensification'.

The authors have suggested that insights derived from New Institutional Economics (NIE) theory can help explain this state of affairs and that both its Northian and Williamsonian traditions are useful. The Northian tradition (North, 1990, 1995; Davis and North, 1971; North and Weingast, 1989) gives a historical perspective on the influence of different paths of institutional change on economic development. The influence of institutions is pervasive and, because they affect transactions costs powerfully, institutional change can take the 'anti-development' form of structuring transactions to create rents or the 'pro-development' form of reducing transactions costs, thereby providing incentives for more trade and investment. Specifically, investment is induced by: (a) the influence of changes in transactions costs on the prospective volume of trade; and (b) the creation or improvement of institutions that improve investee performance to their investors, that is, institutions for reducing transactions costs in intertemporal transactions (Bates, 1997).

The Williamsonian tradition has a 'governance cost orientation'. It examines economic organisations as institutions for the ordering of contractual relationships. For example, firms exist as a solution to the problem of minimising transactions costs in the face of: (a) limited information (because information is costly); and (b) the human propensity to opportunism. Williamson argues that where market agents face high risks, asset specificity and recurrent transactions, then bilateral contracts and vertical integration may be effective at reducing overall costs (including both transaction and production costs) as compared with standard spot market transactions and may, in this sense, be efficient (Williamson 1985, 1991, 1995). This provides a different view from that of neoclassical theory, which would argue that such non-standard contracts restrict competition and are therefore inefficient. Thus, alternative forms of economic organisation of a chosen sector or the economy can be compared in terms of their ability to solve aspects of contractual problems (screening, monitoring and enforcement). In the low-income country context, NIE analysis is open to the possibility that, under circumstances where standard competitive spot markets for inputs and credit fail due to high transactions costs, interlinked or interlocked transactions may be transaction cost-efficient modes of organisation, that is, that they are responses that allow imperfect markets to develop where the alternative is complete market failure. (An alternative explanation for the development of these institutions is that they are a means by which traders with monopolistic and/or non-market power extract surpluses from poor farmers.)

Both Williamson and North give weight to private activity (autonomous of the state) in driving institutional change and adaptation. North's work is richer on the interplay between private actors (sometimes organising collectively for an objective) and the state in changing institutions in directions which are either anti- or pro-development. Williamson's work is richer on insights into the 'nuts and bolts' of ordering contracts (that is, crafting firms and other forms of economic organisation) in order to reduce transactions costs.

There is debate (for example, Harriss *et al.*, 1995) about the extent to which NIE supplements or invalidates neoclassical economics. There is much in common in the way in which both schools see the world, including the interest in individual consumers and producers, the view that increased trade and economic specialisation is key to raising potential welfare, the assumption of optimising behaviour, and the analytical interest in and normative value attached to competition. However, NIE and neoclassical analysis of concrete policy problems can lead to different recommendations.

With regard to the effects of market liberalisation on cash crop production, it can be argued that many proponents of liberalisation, coming from a neoclassical tradition, had correctly diagnosed problems of state failure in service provision to support smallholder agriculture in sub-Saharan Africa. However, they had been naive in their assumptions about, or lack of consideration of, institutional constraints to private sector engagement. The neglect of institutional constraints by most advocates of liberalisation, also seen in the operational work of most mainstream policy advisers, reflects the blind spots deriving from a training in neoclassical economics. Neoclassical economics is 'a-institutional': it does not provide an explanation of the existence of economic organisations or of the role of institutions. Therefore neoclassical economics is only equipped to provide a very partial view of the reasons for the failure of economic organisations or of approaches to their reform. In common with most writers on NIE, the authors are in sympathy with the neoclassical emphasis on the role of the individual and the importance of processes of competition and optimisation. But the central deficiency in neoclassical economics is the absence of a distinction between transformation costs (the costs of the processes of making or growing things) and transactions costs (all the costs associated with contracting). Institutions, conceived of by North as the 'rules of the game', and modes of economic organisation (for example, firms), conceived of as 'players of the game', have massive influence on transactions costs, for good or ill. The implication is that economic organisations need to be analysed *in and with* the institutional context in which they operate, while also recognising that most economic organisations also have, to some degree, institutional characteristics: that is, they make rules within the domain of their governance as well as 'play games' in which others have determined the rules.

In its advocacy of agricultural market liberalisation, neoclassical economics correctly pointed to the baleful effects on agricultural parastatals of lack of competition and/or selective protection through government subsidies. These have included poor (too often disastrous!) operational performance and unnecessarily high costs, the consequences of which have to be borne by farmers, consumers and the government. However, neoclassical analysis has failed to identify one of the key reasons why African governments created agricultural parastatals. This was to promote increased production for the market in smallholder agricultures, which had previously been subsistence oriented and which had used little or no purchased inputs. Where intensification of production required large increases in purchased inputs, parastatals had been created to overcome the near-absence of input and credit markets through interlocking finance, input and output markets. Thus the abolition of parastatals also implied dismantling a particular 'organisational fix' to fundamental problems in the institutional context, most notably the absence of credit markets. Once this point is accepted, it follows that an institutional approach is likely to be productive in explaining: (a) the disappointing supply response to structural adjustment and liberalisation; and (b) possible solutions.

These arguments lead to a number of research and policy questions about modes of economic organisation in smallholder agriculture following liberalisation:

- What institutional forms may provide both efficient and equitable means of overcoming the high transaction costs that otherwise lead to market failure in input and credit markets for smallholder cash crop production?
- What conditions are necessary for these institutional forms to operate efficiently and equitably?
- What is the role of the state in supporting the development of these institutions in liberalised economies?

9.2 Towards a theory of 'institutional development policy'[1]

Institutional change is explained within a model of political economy in the Northian tradition. As relative prices and technological change affect transaction costs, powerful groups respond by modifying institutions in ways that they perceive to be in their interests. It is quite possible that in different countries the same sets of changes in relative prices and in transactions technology could stimulate radically different types of institutional change, as there is a strong 'path dependency' in institutional change:

- a 'fortunate' country, for example, may have had past institutional changes that have been of a 'pro-development', transactions cost reducing nature, enabling a recent history of satisfactory economic development. If political elites have benefited from the economic growth resulting from these past reductions in transactions costs, it is likely that the balance of forces within these political elites will desire to continue with institutional innovations that reduce transactions costs. In summary, those countries that, in the recent to medium-term past, have exhibited a pro-development momentum are likely to see this continued, although of course this is not predestined.

- an 'unfortunate' country, on the other hand, may have an 'anti-development' political economy with a history of political elites deriving their dominant source of income from rents extracted from the population by institutions that tax, control and restrict economic activity, in other words. In this case, elites are likely to respond to market and technological change by promoting institutional changes that attempt to preserve or even extend the possibilities for rent extraction. Again, however, this anti-development path of institutional evolution is not predestined, and there are some cases in recent history of countries having succeeded in altering the direction of their institutional development on to a more positive path.

North's writing on path dependency suggests that it may prove very difficult for countries with 'anti-development' political economies to embark on a reform path of transactions cost-reducing institutional innovation. The central challenge of development is then to achieve a 'change of path'. If in many low-income countries changes in policies and institutions were purely endogenous, determined wholly and mechanically by the self-interests of rent-extracting elites, then the outlook for these countries would be ominous. Fortunately, there may be other factors that influence policies and institutional changes in poor countries in ways that create scope for more hopeful outcomes:

1   Technological changes may fracture governing rent-extracting coalitions, by giving latitude for some elements within the coalition to anticipate more attractive opportunities through loosening controls to expand trade. Alternatively, the empowerment of previously subordinated groups may stimulate the expansion of their economic activities, thereby creating wealth, which in turn develops influence over policy. (Progress in information technology may have such effects, by increasing general access to information in the populace and by reducing transactions costs.)

2   Actors take decisions using limited information and models of the world around them that are imperfect and difficult to validate, and decisions are made on the basis of 'bounded rationality', with crude and often incorrect views of how the world works and of what policies are in the interests of their interest group or of society as a whole. Contests of ideas can therefore have powerful influences on policies and on the direction of institutional change. There is therefore a potentially powerful role for advocacy that persuades people to change their views both of how the world works and of the morally preferable course of action.

3   There is a variety of external pressures for institutional change in governments of low-income countries. International Monetary Fund and World Bank loans, for example (if they are conditioned on reforms), requirements of membership of the World Trade Organisation or regional free trade areas, and the standards required by international capital markets in order to lower risk premiums (and therefore the cost of capital) may all make demands for institutional change.

There are, therefore, strong arguments for the analysis and advocacy of 'institutional development policies', and for this reason it is worthwhile analysing different paths of institutional change. Figure 9.1 (from Dorward *et al.*, 1998, from which the following discussion is drawn) provides a framework for illustrating the ways in which the status of

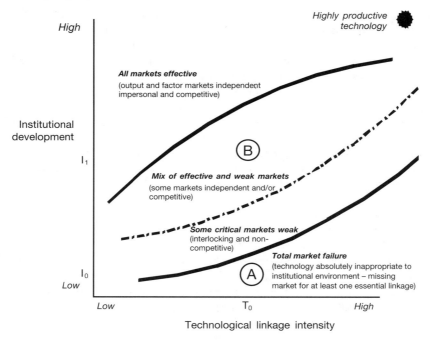

Figure 9.1 Technological linkage intensity, institutional development and market forms
Source: Dorward *et al.* (1998: 246)

'institutional development' can constrain technological possibilities. It is developed with the smallholder agriculture of low-income countries in mind, but may have broader application.

The argument developed here is based upon a fundamental proposition that more productive agricultural technologies are more dependent on the performance of a range of markets (for inputs, outputs, finance, labour, insurance and so on) because such technologies are, in general, likely to require specialisation and complex linkages between specialist operations. The horizontal axis is therefore an index of 'technological linkage intensity', which describes a spectrum from pure subsistence farming, that makes no demands on markets and involves no linkages, through to highly productive technology requiring independent, impersonal and competitive output and factor markets, with multiple linkages in a 'supply chain'. The vertical axis is an index of 'institutional development' reflecting the costs of information flows and the costs and effectiveness of contract enforcement mechanisms. This describes the extent to which formal and informal institutions are able to support markets required by different technologies across their spatial and intertemporal supply chains.

The figure shows three curves, which illustrate institutional constraints on the application of technologies in the smallholder agriculture of a particular country or region. Below the lower bold line is the zone of 'total market failure'. This occurs if at least one critical market completely fails, leading to failure along the rest of the supply chain. The movement of the zone of total market failure out of the south-east corner of the graph shows that the more 'linkage-intense' a technology, the greater the extent of institutional development required to support it. The other curves trace out bands, running from south-east to north-west, which represent different degrees of market and organisational adaptation to the status of institutional development and the demands of a particular technology. In the first two of these bands: 'some critical markets weak' and 'mix of effective and weak markets', non-standard modes of organisation (that is, markets that do not conform to the standard competitive markets) would be likely to be found, as these non-standard modes of organisation may provide mechanisms for lowering transactions costs, in the absence of strongly developed institutions. It is only in the extreme north-western band that technological choice is unconstrained by the status of institutional development, with all markets effective. It should be noted that movement in a north-easterly direction within one of the bands above the 'total market failure' curve implies the use of increasingly intensive technology: but the extent of institutional 'fit' to technology remains approximately constant within each band, as the greater institutional demands of more linkage-intensive technology are only just matched by greater institutional effectiveness. As a consequence, the optimality of standard or non-standard modes of economic organisation remains fairly constant within each band.

Two important points may be drawn from this highly stylised analysis. First, it makes explicit in the development context the insights of Williamson that the transaction cost-economising features of non-standard contracts may make them more efficient than standard neoclassical forms of perfect competition. This presents profound and important challenges to current policy thinking about market liberalisation, as will be discussed further in the last section of the chapter, and suggests that there will be situations where policy makers should promote these forms of (sometimes non-competitive) contracts. Second, it highlights the importance of institutional development rather than technological development as a major constraint to development, as the status of institutional development is seen to constrain the adoption of technology. (This is not to deny the importance of public sector support for technology research and development, but to argue for greater emphasis on understanding institutional development and for developing guidelines for appropriate institutional development.)

As will be clear from the discussion above, the figure cannot be taken to suggest that the easiest and fastest development path would be entirely within the 'all markets effective' band, where rapid institutional development enables the fast progress in the introduction and adaptation to local conditions of ever more productive technology. These circumstances are unlikely to be common, and may not necessarily be desirable, as they could require the conquering and subsequent intense assimilation of a 'low institutional development' culture by a 'high institutional development' culture. It also requires very high investment in institutions, but the 'path dependency' and the political nature of institutional change often militates against such rapid institutional development. It is more likely that countries will experience a development path in a north-easterly direction through the middle of the figure, in which the interaction of technological possibilities and the current state of institutional development give rise to a range of non-standard forms of organisation in the market. This may then lead to an iterative process of development as non-standard organisational forms enable the introduction of technology that otherwise would have failed, and this then enables economic growth that in turn can stimulate further transactions cost-reducing institutional development.

This framework can be used to develop the earlier argument about the creation and failure of marketing parastatals and the institutional problems that have become evident following liberalisation. Under circumstances where markets were absent or weakly developed, governments wishing to intensify smallholder agriculture faced two broad alternatives: (a) to leave market activities to the private sector and to try to foster market entry, investment and technological progress via interventions that promoted broader institutional development (such as appropriate legislation, improved transport and social infrastructure, and administrative and legal services); or (b) themselves to set up parastatal agencies which incorporated the institutions necessary for the provision of inputs, finance, output marketing and storage services needed for agricultural intensification.

This second option of government service provision is illustrated in the figure for the situation where a government wishes to promote a crop with a moderate level of technological linkage intensity (shown as T_o) under circumstances of low institutional development (shown as I_o). This combination of circumstances is shown on the diagram as point A. It is clear that at this point existing institutions will not support the market linkages the crop needs, with consequent market failure. The 'parastatal solution' is then to create artificially by means of the parastatal organisation a set of institutions and non-standard contracts to support the required linkages. This raises the status of institutional development to I_1, allowing the system to function at point B.

Parastatal marketing boards have generally been a problematic means of achieving this institutional fix, as they have shown poor operational performance and imposed high margins. The analysis of Figure 9.1 suggests, however, that with their collapse there will be a move by most agricultural systems back towards the south-west of the diagram, as the more intensive technologies they supported cannot, in the broader institutional environment, be supported without them, and hence these more intensive technologies may be abandoned. Not all agricultural systems will take the same route in decline, nor will they decline to the same extent. Some agricultural systems, based on technologies that are not viable with certain missing markets, will fail completely. Others will continue to be used but, under private management, will be characterised by non-standard organisational forms (and perhaps much lower volumes of activity).

Under these circumstances, if option (b) above is ruled out, government policy makers and their advisers have to focus on option (a), the intensification of agriculture via the fostering of broader institutional development. This is not the same thing as *laissez-faire* as it is a fundamental insight of the 'theory of institutional development policy' that some kinds of government activism are critical. The key questions that follow are: what should governments do (or not do) and how much should be left to the private sector?

9.3. A Malawian case study

The earlier discussion about parastatals as non-standard 'institutional fixes' can be illustrated and developed with some stylised facts and arguments in respect of Malawi. Malawi is a very poor country with an exceptionally high level of dependence on maize which, over the period 1988–92, occupied 70 per cent of arable land (Malawi Government, 1992) and provided over 70 per cent of total calories (UNICEF, 1993). Holding sizes are small and declining, due to population growth against a fixed land frontier. It is conservatively estimated that in an average year about 60 per cent of farm households run out of own-produced maize at least three to four months before the next harvest (ibid.). With declining soil fertility and government policies since independence (supported by international agencies) to introduce higher yielding varieties of maize, the use of fertilizer has become an increasingly important strategy for smallholders, in the absence of attractive organic soil fertility-enhancing technologies.

From the mid 1960s to about 1980, fertilizer use in the smallholder sector grew quite rapidly (partly illustrated in Figure 9.2). Growth levelled off in the 1980s before advancing

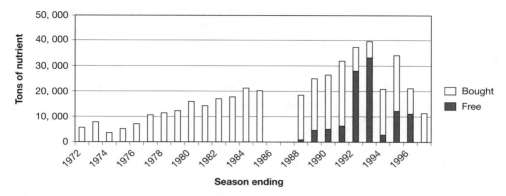

Figure 9.2 Purchased nutrients used on smallholder maize in Malawi[a]
Sources: Kydd (1989); Carr (1997)

Note
a Crop seasons ending 1972–97; no data available for 1986 and 1987.

further in the early 1990s, this being associated with expanding credit facilities made available by aid agencies and a growing euphoria in some research circles about an incipient Malawian green revolution. Unfortunately, fertilizer use peaked in 1993 and by 1997 had declined to about the level of nearly 20 years ago, when the smallholder population was only three-fifths of its present level. Furthermore, the 1990s saw widespread default on credit, initially as a result of drought, but subsequently compounded by the pressures bearing on parties competing in the 1994 multiparty elections (Carr, 1997). Free distribution of seed and fertilizer also occurred as a relief measure. In 1998–9 there was a large-scale scheme for free distribution of fertilizer and hybrid seeds, known as the 'starter pack scheme'.

The title of the 'starter pack' scheme is ambiguous, being capable of a range of interpretations. One possibility is that the scheme is designed to demonstrate to smallholders the benefits of fertilized hybrid maize, so that in subsequent years they will be motivated to buy these inputs. However, this is unrealistic, as few smallholders are able to finance input purchases due to the collapse of smallholder credit schemes and a general inability to auto-finance because of the depressed state of the urban and rural non-farm economies.

Another possibility is that the 'starter pack' scheme is seen as a cost-effective means of providing donor-financed welfare supplementation to the smallholder population. If this is the case, then free input distribution schemes, progressively expanded and administratively fine-tuned, may be a central element of support to rural Malawi for quite a number of years to come.

From the perspective of the late 1990s, it can be seen that Malawi has failed to develop institutions able to support the much-needed intensification of its smallholder agriculture. This is not only a matter of great concern but also a puzzle, because in the 1970s and 1980s Malawi achieved growth in smallholder fertilizer use financed by group-based credit schemes which, in narrow terms at least, were viable and had a very high repayment record. We discuss the puzzle by describing, in stylised terms, two institutional landscapes: before liberalisation (BL) and after liberalisation (AL).

BL was characterised by a parastatal which was an all-embracing, multicrop monopsonist and multi-input monopolist known as ADMARC, dealing exclusively with smallholders. ADMARC's operations were integrated with (mainly) group-based credit schemes administered by the Ministry of Agriculture with strong field support from agricultural extension workers. The system performed remarkably well in terms of:

- raising farm productivity;
- high repayment rates;
- delivery of a *linkage-intensive* package consisting of finance, inputs, technical advice, group formation, market coordination (for finance, inputs and outputs) and a market for the crop (linked to repayment).

The BL system worked in some ways, but not in others. Its negative features included the fact that ADMARC suffered from some of the problems of parastatals: overstaffing, poor incentives for efficiency, and an undesirable link to the personal finances of the then president of an oppressive state. Additionally ADMARC finances suffered when the government used it as an instrument for achieving control of the maize consumer price. Furthermore the government budget, largely donor financed, subsidised a number of features of the BL system, including extension workers' salaries, occasional transport of inputs and produce, fertilizer subsidies and moderate credit subsidies.

Nevertheless, despite these shortcomings, over an extended period of years the BL 'institutional fix' expanded geographically and over a decade and a half delivered to increasing numbers of farmers a stable and valuable service, with only moderate subsidy.

The transformation from the BL to AL system took place in the 1990s. To sketch in some elements of this: in the late 1980s ADMARC's monopolistic and monopsonistic role was increasingly attacked by academics and influential donor agencies, and liberalisation began at the primary level. This, together with adverse price movements and increased political interference, undermined its profitability and financial capacity for delivering services to farmers. The result was that a number of wedges were driven into the institutional structure which had delivered a productive and linkage-intensive maize system. Fundamentally, farmers no longer had incentives to repay credit, because access to future credit and/or inputs was no longer linked to repayment performance. The collapse in the integrity of the credit system was precipitated by political competition linked to the 1994 multiparty election, which caused both its bankruptcy and the breakdown of the repayment ethic and the development of the culture of strategic default. Other wedges included primary liberalisation, which distanced ADMARC from farmers and removed its ability to recover credit for the government through stop-orders; the refocusing of extension workers away from being 'system coordinators' to a more purely technological role; and ADMARC's weakened capacity to deliver inputs with credit support eroded farmer loyalty to credit groups. Presently in Malawi there is virtually no credit for smallholder maize inputs.

We are left with the question as to which is the preferable system: BL or AL? With hindsight the BL system appears to have had the benefit of delivering to Malawian smallholders a valuable service not previously or subsequently available to them, at the cost of a number of subsidies. Perceptions of the crisis of state-led development focused exclusively on the subsidy costs and concerns about the extractive capacity of monopolies and inefficiencies of public sector organisations. Our analysis balanced these costs against the benefits of a functioning system that delivers services.

Looking to the future, the challenge will be to devise an 'institutional fix' that has some of the features of the BL system, that delivers services at an acceptable subsidy cost while controlling the undesirable features associated with a public sector monopoly. Clearly the development of workable new models will be informed by recent development in public sector management theory and practice.

Note

1 This section draws heavily on Dorward *et al.* (1998: 244ff.).

References

Bates, R. (1997) 'Institutions as investments', *Journal of African Economies*, 6 (3): 272–87.

Carr, S.J. (1997) 'A green revolution frustrated: lessons from the Malawi experience', *African Crop Science Journal*, 5 (1): 93–8.

Davis, L.E. and North, D.C. (1971) *Institutional Change and American Economic Growth*, Cambridge: Cambridge University Press.

Dorward A., Kydd, J. and Poulton, C. (eds) (1998) *Smallholder Cash Crop Production under Market Liberalisation: A New Institutional Economics Perspective*, Wallingford, Oxon: CAB International.

Harriss, J., Hunter, J. and Lewis, C. (eds) (1995) *The New Institutional Economics and Third World Development*, London: Routledge.

Jones, S. (1994) 'Agricultural marketing in Africa: privatisation and policy reform', Report to ESCOR, University of Oxford, Queen Elizabeth House.

Kydd J.G. (1989) 'Maize research in Malawi: lessons from failure', *Journal of International Development*, 1 (1): 112–44.

Malawi Government (1992) *Annual Survey of Agriculture, 1988–92*, Lilongwe.

North, D.C. (1990) *Institutions, Institutional Change and Economic Performance*, Cambridge: Cambridge University Press.

North, D.C. (1995) 'The new institutional economics and third world development', in J. Harriss, J. Hunter and C. Lewis (eds), *The New Institutional Economics and Third World Development*, London: Routledge.

North, D.C. and Weingast, B.W. (1989) 'The evolution of institutions governing public choice in 17th century England', *Journal of Economic History*, 49: 803–32.

UNICEF (1993) *Situation Analysis of Poverty in Malawi*, Lilongwe: UNICEF.

Williamson, O.E. (1985) *The Economic Institutions of Capitalism*, New York: The Free Press.

Williamson, O.E. (1991) 'Comparative economic organisation: the analysis of discrete structural alternatives', *Administrative Science Quarterly*, 36 (2): 269–96.

Williamson, O.E. (1995) 'The institutions of governance of economic development and reform', in *Proceedings of the World Bank Conference on Development Economics 1994*, Washington, DC: World Bank.

World Bank (1997) *Rural Development: From Vision to Action*, ESSD Study Series 12, Washington, DC: World Bank.

10 Agricultural marketing in Africa since Berg and Bates

Results of liberalisation

Paul Mosley

10.1 Introduction[1]

This chapter seeks to understand the extent and influence of agricultural marketing reform in Africa since the onset of structural adjustment at the beginning of the 1980s. The jury is still out, of course, on what structural adjustment as a whole has achieved (Sachs and Warner, 1997; Fischer *et al.*, 1998); still less do we know what has been the payoff to particular elements within the reform package. And yet such knowledge is vital if we are to learn from the experience of the years since 1980 and achieve more effective and more humane policies in future. The chapter focuses on one of the key elements of 'bad policy' as visualised by the architects of structural adjustment in Africa – price distortions in agriculture – and asks what reform has done to correct them, what have been the determinants of progress or the lack of it, and what the impact of reform in this area has been. In particular we wish to understand what agricultural price reform has contributed to poverty reduction by lowering the cost of food for poor people.

10.2 The problem posed

The essential problem visualised by (in particular) Elliot Berg (in World Bank, 1980) and Robert Bates(1981) consisted of agricultural price control as a vehicle for exploitation of rural by urban areas. African post-independence governments, on this view, were dominated by urban interests anxious, above all, to keep the price of food down. Their tax base was weak; hence in order to secure both revenue and political support, they created (or took over from the colonial power) monopoly marketing boards that paid the farmer only a fraction of the export price. The remainder (anything from 25 to 85 per cent of the export price in different African countries, according to the appendix to Bates, 1981) could be used as a slush fund to finance any purpose, however non-developmental, and there is no better source than Killick's *Development Economics in Action* (1978) to document the kind of white elephants to which public expenditure got diverted, at the expense of pro-poor growth, by one typically rapacious government, that of Ghana from Nkrumah onward. Meanwhile, an overvalued exchange rate could be used to keep imported food cheap.

If this is the problem, then the solution is obviously decontrol of agricultural prices and exchange rates, by analogy with the decontrol of other prices such as energy prices and interest rates then being advocated by the architects of structural adjustment. If monopoly marketing boards could be overthrown in favour of private trade and exchange rates decontrolled, farmers would receive a 'fair price' for their product, the terms of trade would cease to be biased in favour of the cities and, even in the short term, the expectation would be of higher agricultural production and lower food imports. In the longer term, two even more attractive outcomes could be expected: a reduction in poverty and inequality as rural incomes began to rise towards the urban level, and a process of broad-based green revolution as price incentives provided a rationale for technical change. In all of this there was the example of Asia to inspire hope: as will be remembered, two of the most significant green revolutions in modern times, with wheat in north India in 1965–7 and with rice in China in 1976–8, began with a sharp increase in food prices, in the latter case the consequence of price liberalisation (Yao, 1994). Agricultural price reform thus provided, in Africa, the cutting edge and some of the most idealistic hopes of the structural adjustment revolution.

It is now a commonplace that many of these hopes did not come to pass. Liberalisation there certainly was, but the supply response was weak and long-delayed, the technical response was erratic and unsustained, and the poverty impact, as a consequence, muted. This chapter asks why. It does not attempt a full answer but finds part of an answer in three factors: (a) the lack of political support for the liberalisation process; (b) the continuing disincentive to local production provided by aid-led imports; and (c) the weakness of public-good services and infrastructures required to maintain the momentum of technical change in agriculture, brought about in many cases by the same forces of liberalisation and stabilisation that had brought price decontrol into being.

10.3   Explanatory hypotheses examined

At the outset we need to make a distinction between two different types of price repression practised by African post-colonial governments. In 'peasant export economies', to use Myint's terminology, where there was no significant white-settler presence and agricultural interests were generally politically weak (Ghana, Nigeria, Cameroon, Uganda, Tanzania), repression was applied to the *producer price* in order to siphon off into the public purse the difference between this and the export price, as documented by Bates (1981). In 'settler' or 'mine/plantation' economies (Kenya, Zambia, South Africa, Zimbabwe) repression (or rather subsidy) was applied to the *consumer price* in order to keep powerful consumer interests at bay: but the producer price, so far from being repressed, was often sustained at a level above the export price, a device originally introduced during the 1930s' world depression in order to protect the threatened population of white settler-farmers. And this price was offered to large farmers only by a monopolistic marketing board imposing restrictions on the operations of private traders and millers, so that not all could take advantage of it. These contrasting tendencies are illustrated by Table 10.1, which also documents the extent to which liberalisation has made inroads into the wedge between the producer and the export price.

In the 'peasant export economies', these inroads are variable, but small on average. In Uganda and Cameroon, explicit taxation on coffee and cocoa was removed (although temporarily re-imposed in Uganda during the 1994 coffee boom). Monopoly marketing arrangements were dismantled for palm oil and rubber in Nigeria, but retained in Ghana and Côte d'Ivoire for cocoa and in Madagascar for vanilla, in spite of the insistent pressure of the World Bank and other aid donors. In Ghana the wedge between producer and export price was substantially diminished by the reform process,[2] but in Madagascar it was not: the monopoly state marketing board insisted, as late as 1996, on offering a price of about US $25 per kg of vanilla extract, to producers, and then, to compound the felony, attempting to sell medium-grade vanilla on the export markets at US $60 per kilo of extract, well *above* the world market. The inevitable consequence was the growth of a huge internal black market and the collapse of Madagascar's export share from over 60 per cent to below 20 per cent of the world market between 1975 and 1995. In other words, reform of price structures for African export crops has not, except in a few countries, deprived rent-seekers of their rents, nor provided a substantially improved incentive to exporters: the 'price wedge' identified by Berg and Bates remains, apparently at an average value of about 47 per cent. Urban bias remains alive and well: the donors were unable to confront political opposition to decontrol when to confront it, and refuse refinance, would have prejudiced repayment of international debt to those donors and others. This must be one reason why the distributional consequences of adjustment have proved so disappointing.

In the 'mine/plantation economies' (formerly settler economies) of eastern and southern Africa the problem was a different one. There was no perceptible urban bias, and no 'Bates wedge' between the producer and the export price, either for cash crops such as tea and coffee or for food crops such as maize, millet and sorghum (Table 10.1), since larger rural producers – originally white settlers – have since colonial times been a sufficiently powerful interest group to obtain prices equivalent to, or often well above, export prices (Mosley, 1983: ch. 2). Rather, there were disincentives to efficiency of a different type. First, bias against small producers, barred by capital-market obstacles from driving their crop to a depot on the line of rail, and hence forced to surrender their crop to a trader acting as agent for the monopoly marketing board at, on average, around half the export price. Second, bias against consumers, for the

Table 10.1 Producer price shares (ratio of official producer price to international reference price)

Country	Commodity	Average ratio			
		1975–9	*1980–5*	*1986–93*	*1994–9(prov.)*
Peasant export economies					
Angola	Coffee	0.22	0.45	0.86	0.92
Benin	Cotton lint	0.45	0.41	0.54	0.56
Burkina Faso	Cotton	0.42	0.34	0.56	0.56
Burundi	Coffee	0.51	0.60	0.60	0.60
Cameroon	Cotton	0.42	0.37	0.40	0.40
CAR	Coffee	0.29	0.18	0.34	0.31
Côte d'Ivoire	Cocoa	0.40	0.51	0.79	0.78
Ethiopia	Coffee	0.45	0.39	0.32	0.34
Gambia	Groundnut	0.54	0.62	0.71	0.71
Ghana	Cocoa	0.30	0.87		0.89
Guinea	Palm kernels	1.08	0.86		
Madagascar	Coffee	0.40	0.29	0.38	0.43
Mali	Cotton	0.34	0.39	0.50	0.51
Nigeria	Cocoa	0.53		0.49	0.54
Rwanda	Coffee	0.58	0.89	0.81	0.82
Swaziland	Cotton	0.46	0.29	0.27	0.32
Tanzania	Coffee	0.39	0.55	0.36	0.37
Uganda	Coffee	0.13	0.22	0.34	0.65
Average		0.48	0.53	0.54	
Mine/plantation economies					
Kenya	Maize	1.05	1.15	1.17	1.14
Zambia	Maize		1.03	1.06	1.13
Zimbabwe	Maize	0.94		1.05	1.06
South Africa	Maize		1.13	1.09	1.04
Malawi	Maize		1.06		1.06
Average		0.99	1.09	1.09	

Sources: Cleaver (1993: table A-7); Mosley (1994: table A-1); individual country surveys as specified

consumer price was higher than the export producer price by a large element of monopoly profit (and averaged, according to Table 10.2, nearly three times the small-farmer producer price over the period from the 1940s to the 1980s): consumers of course wished to avoid paying this monopoly rent, but were inhibited from doing so by legal and sometimes physical barriers on inter-district movements of food crops. Hence they grew their own maize, millet and sorghum if they possibly could, rather than pay exorbitant prices, even in the driest and most inhospitable environments. The evolution of these price structures in Kenya and Zimbabwe, two typical settler economies, between the colonial and the immediately pre-reform period is shown in Table 10.2. The only changes which occurred in the basic 'Eastern/Southern African system' between the 1930s and the beginning of the 1980s were the spread of cooperatives and of statutory grain board buying depots in areas of African small-farm production. Where a farmer could get his grain to a cooperative or one of these depots, he obtained the full local price, subject to a standardised transport-cost deduction; otherwise, if still forced to sell to a trader acting as buying agent for the marketing board, he continued to receive what is described in Table 10.2 as the 'small farmer price'.

The third bias that this system contained was a bias in favour of imports, conveyed partly through this high consumer price itself, partly through the availability of imported foodgrains offered at concessionary prices through aid programmes, and partly through a gradual shift of consumer taste amongst urban populations from locally grown sorghum and millets to easily importable wheat and maize, and from coarse hammer-milled maize flour to importable fine-milled maize flour.

These biases were strongly challenged by the World Bank and other donors from the outset of structural adjustment in the early 1980s: indeed, in 1985 the World Bank withdrew for a time from lending to Kenya over a dispute concerning movement controls for maize. The process of liberalisation in food crops has indeed been contentious and often resisted for precisely the reasons mentioned in relation to cash crops. It has, however, gone a great deal further than

Table 10.2 Maize price structures in 'mine/plantation economies': evolution from colonial times to the early 1980s (prices in shillings per 200 lb bag)

	1940	1981
Kenya		
Price to African small-farm, producer, delivering to trader at village store	4.90	56.4
Middleman's deductions (railway transport, Native/African Development Fund, trader's commisssion, bags, export loss cess)	4.05	38.9
Price to large-farm producer, delivering at railhead	8.95	95.3
Consumer price	16.50	155.4
Ratio, consumer price to small-farmer producer price	3.36	2.75
Zimbabwe		
Price to African small-farm, producer, delivering to trader at village store	5.00	61.50
Middleman's deductions (railway transport, Native/African Development Fund, trader's commisssion, bags, export loss cess)	4.35	
Price to large-farm producer, delivering at railhead	9.35	120.00
Consumer price	15.45	174.00
Ratio, consumer price to small-farmer producer price	3.08	2.82

Sources: Mosley (1983: table 2.12; 1994: appendix table A1)

Note

For Kenya and for Zimbabwe in 1940, the price is given in current shillings; for Zimbabwe in 1980, the price is converted from Zimbabwe dollars to shillings at the current exchange rate.

the equivalent process in relation to cash crops. None of the Southern African countries now has any controls on the consumer or producer price of maize, and in South Africa the main grain marketing board has been abolished. The evolution of the marketing system since colonial times is illustrated in Table 10.3. These reforms sustained the on-farm price for farmers delivering direct around the export parity price but, much more importantly, increased the share of the export price received by small farmers, reduced the spread between producer and consumer prices and even, to a modest degree, increased the share of the local maize and sorghum crop processed by local hammer-mills. This process was not driven by the small farmers who were meant to be its beneficiaries, who together with cooperatives tended to be suspicious of liberalisation,[3] more particularly if the sequence of liberalisation worked out 'wrong' and the costs of liberalisation of inputs such as seeds and fertilisers were felt earlier or harder than the benefits of liberalisation of output prices. A Malawian case study of this 'perverse sequencing' is presented as a footnote.[4] The main driving force behind liberalisation of foodcrop markets was neither pressure from the beneficiaries nor from the donors, whose political weakness we have already noted; rather, it was implemented as the only visible escape route from fiscal crisis. The deficits of the statutory boards accounted for a large proportion of the government fiscal deficit in every African country by the mid-1980s, and the disposal of some or all of their assets and powers came to appear the only feasible exit route. Thus financial crisis in the age of economic reform triggered a withdrawal of subsidy, just as in the colonial period (Mosley, 1983: ch. 2, app. 1) it triggered increases in government financial support.

In all kinds of ways, therefore, the incentive to local production and productivity increase provided by liberalisation has been large, consistent and comprehensive. And yet, as documented by Jayne and Jones (1997) and Mosley (2002) the production response has been patchy, hesitant and reversible. Let us focus on the illustrations of this provided by our case-study countries – for the time being, those in the mine/plantation economies of Eastern and Southern Africa only – and recapitulated in Table 10.4.

There was a brief smallholder maize revolution in Zimbabwe (1980–6) which was no sooner documented (by Eicher, 1995) than it fizzled out amidst structural adjustment-induced cuts in government support services. A similar story, with slower rise in the early 1980s and slower decline thereafter, applies in Zambia. Malawi's moment of glory came later, in the early 1990s, and was almost at once punctured by the collapse of the agricultural credit system it currently appears to be reviving. Only in Kenya with maize and in South Africa with sorghum do we see sustained growth in production over the entire 1980–98 period; and even in Kenya population growth has latterly outstripped grain production growth. 'Getting prices right', clearly, is not sufficient, and the Indian and Chinese comparisons invoked earlier in this chapter are already beginning to look somewhat hollow. But if 'pricism' is not enough, what is the missing factor? And what are the implications of the answer for policy, in particular towards poverty reduction? These questions are reviewed in the following two sections.

10.4 Supply response and productivity response: interpretations and tests

It is a commonplace that a range of supportive factors need to be in position for the agricultural economy to respond effectively to price stimuli. The literature on this issue (for example, Mosley *et al.*, 1995) suggests one important negative influence (social and political instability) and three potential positive ones: physical capital (especially investments which fall into the public good category such as infrastructure and financial systems), human capital and complementary policies such as exchange rates. Other authors also invoke various measures of local social and economic organisation, often collectively known as 'social capital'.

Table 10.3 Six African 'mine/plantation' economies: crop marketing arrangements and their liberalisation

	Phase 1 *Colonial period, to 1960s*	*Phase 2* *Expansion of marketing services to smallholders, 1960s–mid-1980s*	*Phase 3* *Liberalisation and privatisation, 1980s onward*
Kenya	Kenya Farmers' Association statutory board established in 1923 to buy grain from European producers. Monopoly marketing board established 1942, with sole right to sell grain to registered millers. Restrictions on movements of maize and other food-crops between districts. Prices for European delivery to maize board 36% higher, on average, than official price for African maize offered by provincial boards (1941–62)	Maize and Wheat Boards combined to form national Cereals and Produce Board (1980). 500 new NCPB buying centres created (1980–2)	1988 NCPB financially restructured. Proportion of grain that millers are required to buy from NCPB declines. Interdistrict maize movements loosened (1988 and 1991), tightened (1992) and finally abandoned (1994). 1994 NCPB restricted to limited 'buyer and seller of last resort' role
Zambia	Maize control board established in 1936 serving European producers. Single price for commercial farmers close to the line of rail, 41% higher than for Africans, on average (1936–58). Restrictions on inter-district maize movements	Nationalisation of grain mills (1986). Increases in consumer subsidies. Expansion of panterritorial pricing to smallholder farmers	1986 Liberalisation of inter-district trade. 1989 NAMBOARD abolished. 1992 Deregulation of small-scale milling, restrictions on external trade relaxed. 1993–4 Consumer subsidies abolished
Zimbabwe	Maize Control Board established 1931, replaced by Grain Marketing Board (GMB), 1950. Buying stations built almost exclusively in European farming areas. African producers barred from selling outside of their 'reserve' area, except to GMB or licensed agents. GMB monopoly on sales to registered millers. Prices for European delivery to GMB 56% above prices paid to African smallholders	Development of limited GMB depot network in smallholder areas from 1970s, rapidly accelerated after 1980. Consumer subsidies increased in late 1970s, phased out 1985, reintroduced 1991–3	1991–2 Phased elimination of controls on trade between smallholder areas. 1993 Maize meal subsidies abolished, consumer prices decontrolled. 1993 GMB monopoly-supplier status restricted to large mills, then eliminated (1994). External trade still controlled by GMB
South Africa			1995 Removal of maize price controls. Removal of restrictions on private maize purchases and sales, subject to stabilisation levy. Imports liberalised (with zero tariff)

Source: Adapted and updated from Jayne and Jones (1997: tables 1–3)

Table 10.4 Trends in cereal production per capita, area, yield and net exports, selected SSA countries

		Production (000 m t)	Production per capita (kg)	Net exports (000 tons)	Fertiliser use (000 tons)
Zimbabwe	1970–74	1863	340	628	
	1975–79	1866	295	429	378.0
	1980–84	1980	267	205	471.0
	1985–89	2307	263	314	443.0
	1990–95	1340	134	–82	446.0
	1995–99	2179	135	—	—
Zimbabwe	1970–74	612	116	—	8.6
(smallholder	1975–79	731	117	—	27.1
sector)	1980–84	948	127	—	97.2
	1985–89	1562	177	—	119.0
	1990–94	1108	105	—	92.6
	1995–99	—	—	—	—
Zambia	1970–74	808	224	–78	47.9
	1975–79	753	160	–94	65.3
	1980–84	1056	188	–181	74.3
	1985–89	1618	235	–161	80.4
	1990–95	1354	183	–239	68.2
	1995–99	997	160	—	—
Malawi	1970–74	1185	328	14	14.1
	1975–79	1240	286	–5	21.8
	1980–84	1315	267	30	33.4
	1985–89	1351	228	–24	43.9
	1990–95	1346	182	–215	58.0
	1995–99	1969	228	—	—
Kenya	1970–74	4215	102	77	144.2
	1975–79	5771	133	71	130.2
	1980–84	6928	132	59	155.7
	1985–89	8533	126	120	235.1
	1990–95	7427	92	–102	241.5
	1995–99	—	81	—	—
South Africa	1970–74	7681	327	2435	—
	1975–79	9031	332	2909	—
	1980–84	8476	311	3069	—
	1985–89	7817	206	1428	—
	1990–95	7420	208	1090	—
	1995–99	10,980	195	—	—

Source: Jayne and Jones (1997: table 4), expanded and updated

Jayne and Jones (1997: 10) specifically mention as a causal influence on farm output the collapse of the 'unsustainable . . . state-led model of service provision'[5] especially in Zambia and Zimbabwe. (This of course was, however, precisely the model which successfully launched the green revolution in north India, China, Malaysia and Indonesia.) In terms of our basic model, what is being argued is that if even price liberalisation does succeed in pushing out the budget constraint (XX' in Figure 10.1) agricultural intensification may not occur if it requires complementary factors of production which are not available (such as those listed above) or would entail an increase in risk.

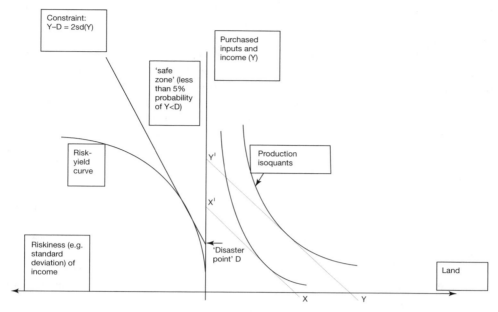

Figure 10.1 Liberalisation and technical innovation

10.5 Liberalisation and technical innovation

A decline in the availability of capital, of whatever type, occurring at the same time as liberalisation may simply push the budget constraint in by more than liberalisation is able to move it out, leaving the farm household overall in a more and not less liquidity-constrained position.

Empirical evidence from the early chapters of our study of the African green revolution (Mosley, 2002) provides evidence enabling us to nuance this general picture. Not all components of 'capital', it will be recalled, were crucial: access to credit and to physical infrastructure were universally very strong and significant influences on yields; but state agricultural extension was a very weak influence, even though the style of extension did have a role to play. We wish to understand whether these factors continue to exercise an influence in the context of agricultural price liberalisation, essentially by increasing the supply elasticity. We shall be using here a time-series analysis, and hence will be able to assess the influence of variables which cannot be assessed in cross-section analysis because they are the same across all observations, for example policy variables. Exchange rates were mentioned above, but there is another causal factor from our earlier analysis that needs to be brought in here. This is the influence of import vulnerability on the effect of import liberalisation, and behind that the influence of overseas aid.

The reasoning works through two channels, at farm and at governmental level. First, the higher the on-farm price of imported food (after the imposition of any producer subsidies, etc.) the further out the farmer's liquidity constraint (XX' and YY' in Figure 10.1) and the higher the probability that the farmer will be able to purchase indivisible inputs such as fertiliser and irrigation equipment. But second, the higher and the more unstable the price of imported food, the greater the government's incentive to take serious steps to achieve self-sufficiency in food (for example, through research and extension programmes) in order to avoid political disturbance occasioned by food shortage. This was precisely the trigger for the north Indian green revolution in 1965–7, and more broadly Timmer (1994) shows that instability and thinness in world rice markets has been a powerful influence behind the programmes of east Asian governments to bring about self-sufficiency in rice production.

However, the role of overseas aid in response to food import vulnerability is crucial. If aid supports the *production side*, as for example in the case of India above, there is almost by definition a positive incentive to agricultural production, subject to the possible presence of moral hazard problems. But if aid supports the *consumption side*, as with food aid, it depresses food prices and thus the long-term incentive to increase production and yields. And of course aid flows, both overall and in the form of food supplies, have been much more significant in Africa than elsewhere, having risen from 3 per cent of recipient GNP in the 1970s to an average of 15 per cent in the 1990s. It is not only in Eritrea, where food aid on its own was 22 per cent of GNP between 1990 and 1997, that aid of this type can be seen as a major negative incentive to food production.

In seeking to model production and productivity in foodcrops, then, we identify the potentially relevant influences as set out in Table 10.5. Before proceeding to estimate this model, certain caveats must be noted. The first is that several variables in this list are very probably multicollinear: for example research expenditure, fertiliser sales and even credit uptake are very likely to be intercorrelated, as are liberalisation and the main thing which it is supposed to influence, namely producer prices. As we have argued, the main influence of liberalisation in one large subset of African economies – the 'ex-settler' or 'mine/plantation' group – consisted not of the removal of implicit taxation on producer prices, but rather of the removal of controls on the activities of private traders and on consumer prices which had inhibited specialisation and the diffusion of incentives to commercialise. These more subtle influences are by no means easy to model, but at this stage we shall try to cope with the multicollinearity problem by separately entering into the estimating equation each of the variables within each intercorrelated group.

The results of ordinary least-squares estimation on the set of variables listed above are presented in Table 10.6. The following main tendencies emerge:

- there is a significant response of productivity (not only production) to price in all countries and periods, although the measured elasticity is quite low, generally within the 10–25 per cent range;

Table 10.5 Variables employed in the regression analysis

Influence on production	Proposed proxy variable
Crop price	On-farm price
Standard factors of production	Agricultural labour force Land area
Physical capital	Research expenditure Fertiliser sales
Institutional development	Access to credit
Human capital	Extension expenditure
Food aid	Disbursements by OECD countries in year
Other policy variables	Liberalisation measure (raw) Sequencing of liberalisation Spread between producer and consumer prices

Table 10.6 Time series analysis of foodcrop productivity, Southern Africa, 1970–98[a]

Country, crop and estimation period	Regression coefficients on independent variables (Student's t-statistics in brackets):						Liberalisation indicators	
	Constant	Price[b]	Aid[c]	Credit disbursements	Research expenditures	Extension expenditures	(1) liberalisation overall measure	(2) sequencing dummy variable
Kenya								
maize (1960–95)	1.09* (6.72)	0.042* (2.17)	−0.67** (3.45)	0.063* (2.06)	0.17* (2.12)	−0.005 (0.61)	0.047 (1.13)	0.61* (1.89)
sorghum (1960–91)	0.60* (2.33)	0.029 (0.96)		0.012 (.217)	0.45 (1.61)	−0.34 (0.23)	0.13 (0.22)	0.34 (1.65)
Zimbabwe								
maize (1960–96)	0.34 (1.32)	0.048* (2.16)	−0.33 (1.97)	0.018** (2.45)	0.34 (1.66)	0.01 (0.39)	0.17* (2.11)	0.74* (1.85)
sorghum (1960–95)	0.61** (2.80)	0.007 (0.38)		0.012 (0.60)	0.45** (2.77)	−0.0013 (0.56)	−0.54 (0.51)	0.66 (0.31)
Malawi								
maize	0.244** (2.59)	0.0042** (8.37)	−0.21* (1.65)	0.0016 (1.25)	0.36** (1.91)	0.03 (0.71)	0.34 (1.65)	0.69** (4.81)
Zambia								
maize	0.51** (6.31)	0.029* (2.98)	−0.61** (4.21)	0.34* (2.13)	0.65* (1.91)	0.67 (0.85)	0.56 (0.33)	0.45* (1.13)
South Africa								
maize	0.37* (2.13)	0.32* (1.95)	−0.51 (1.34)	0.13 (1.65)	0.77* (3.44)	0.65 (0.91)	0.61 (0.74)	0.76** (0.41)
sorghum								
pooled data	0.46* (4.15)	0.17** (3.61)	−0.57** (5.12)	0.22** (2.11)	0.71* (2.02)	0.43 (0.34)	0.34 (0.42)	0.65** (0.54)

Sources: Mosley (2002: ch. 7, tables A2 to A4)

Notes: a Dependent variable: yield per hectare of crop stated. All variables in first differences. b On-farm price to producer. c Food aid flows (in US $000).
* Significance at 5% level. ** Signifance at 1% level.

- measured price responsiveness is slightly higher in the peasant export economies where, it will be remembered, liberalisation took the price paid to farmers (a little) closer to the export price;
- credit availability and (with a lag) research expenditures retain their significance from the cross-section exercises of Mosley (2002);
- liberalisation on its own is in all countries except Zambia and Uganda not a significant influence on price, but there is a significant impact of the consumer price/producer price ratio, and the sequencing variable is also significant. In other words, what matters is not the carrying out of liberalisation actions (especially privatisation actions) but their effect on the allocation of resources and the quality of management;
- food aid is a significant negative influence.

10.6 Poverty implications

Finally, we wish to understand the influence of price liberalisation on agricultural intensification and its poverty impact. As mentioned at the beginning, one important strand in the structural adjustment literature, often unfairly lampooned, was the proposition that price reform could be distributionally progressive: could reduce poverty and inequality by redistributing income from those who had protection and privilege (often in cities) to those who had none (often in rural areas). Whether anything like this has happened, for the structural adjustment experience as a whole, has recently been the subject of acrimonious debate, with Sahn (1994) and Demery and Squire (1996) insisting that the evidence favours a positive verdict, Weeks (1997) arguing otherwise, and arbitration between those opposed points of view being difficult on account of data problems.

There is no intention here to revisit this broad debate on the rights and wrongs of structural adjustment. The focus here is on a component of that process: the liberalisation of agricultural prices. As we have discussed, this has had a mixture of effects – short-term negative ones on the input side and long-term positive ones on the output side – and the resultant of these effects has varied from country to country. However, the output-side effects are potentially important in relation to poverty. The influence of wheat and rice productivity on food prices and the ability of the marketing system to diffuse this influence have been a major channel of poverty reduction in Asia – possibly, according to Lipton (1989), the most important channel by which the green revolution reduces poverty, as food grains figure so importantly in the consumption of the very poor and most vulnerable. Those below the poverty line, to the extent that they consume the foods made cheaper by green revolution technology, will experience an increase in real income and a decline in the value of their poverty gap and other higher-order measures of poverty. We measure, in this section, whether this is an important influence in Africa also. Other influences on poverty, for example those operating through production and through the producer price, are examined in Mosley (2002).

The argument to be tested is twofold: that green revolution technology lowers the price of foods consumed by people below the poverty line, and that liberalisation increases their availability, for example by removing barriers on inter-district trade, and by improving the pattern of specialisation so that the production of foodcrops – and substitutes for them – concentrates in areas of comparative advantage. The second of these effects is difficult to test, but is examined in Mosley (2002), which uses a computable general equilibrium model. However, we can produce an approximation by using the following procedure. Across a cross-section of twenty-six African countries, we regress the real change in the consumer price on agricultural productivity change, an index of liberalisation, and other obvious independent variables such as food aid inflows and weather. We then apply the estimated 'price change due to agricultural productivity growth' to the numbers below the poverty line to estimate the extent of poverty reduction which is due to the introduction of green revolution technology (and indeed to liberalisation in the relevant crops).

The estimated cross-section equation is the following:

change in the real consumer price of maize (1980–97) =
0.14 – 0.30E – 03** (productivity change in maize 1980–97)
(1.66) (4.64)

– 0.0034* (liberalisation index) – 0.016** (food aid inflows)
(2.19) (3.79)

+ 0.07(growth in real personal disposable income)
(1.14)

+ 0.06 (weather index) r^2 = 0.59, n = 26

We note that the estimated effects of productivity change and liberalisation on consumer prices are both as expected, that is, significant and negative. The estimated impact of productivity change on crop prices is then, in Table 10.7, fed into data on the composition of poverty for individual countries, in order to compute the extent of poverty reduction due to changes in the price of foodgrains. It seems certain that in this specific context both technical change and liberalisation of output prices act as progressive influences. This is of course a very partial estimate, since productivity change impacts through other channels as well, notably the labour market. The more complete picture, which emerges in (2002),

Table 10.7 Price effect on poverty of liberalisation and technological changes in foodcrops

	(1) Estimated price effect of modern varieties of foodcrops (regression coefficient)	(2) Number of beneficiaries (thousands)	(3) % of beneficiaries poor	(4) = ((1)×(2)×(3))/ poverty line) Change in poverty gap (%) induced by estimated reduction in foodcrop prices
Cameroon	0.66E-03	2100	27	0.60
Uganda	0.31E-03	2400	53	0.61
Kenya	0.17E-03	8600	46	1.03
Malawi	0.22E-03	1900	72	0.47
Zimbabwe	0.31E-03	3650	44	0.76
Lesotho	0.12E-03	850	51	0.08
Pooled data	0.30E-03	3250	49	0.59

Sources: Price effect from regression coefficient; maize consumption data and 'number of beneficiaries poor' from national household budget surveys

Notes
Col. 1: time-series regression coefficient (pooled data: cross-section regression coefficient) of modern varieties supply on retail price of foodcrops (average of maize, sorghum and cassava), see Mosley (2002: table 7.ix).
Col. 2: estimated number of individuals buying crops mentioned for cash; data from household budget surveys as specified in Mosley (2002: table 7.iv).
Col. 3: estimated percentage of poor among food consumers; data from household budget surveys as specified in Mosley (2002: table 7.iv).

suggests that between a quarter and a third of all poverty reduction induced by technical change in African foodcrops derives from this source

10.7 Conclusions

We can now begin to summarise. Liberalisation of agricultural marketing in Africa since 1980 has taken two forms: the first of them, characteristic of 'peasant export economies', attempts to get rid of a price wedge between the producer price and the export price, and the second, characteristic of 'mine/plantation economies', seeks to reduce price differentials between large- and small-scale producers, improve specialisation and reduce real consumer prices. The second of these, on the available evidence, has been more successful than the first, in the sense of the scale of the price restructuring it has brought about; but the test of this is, of course, whether that in turn has improved welfare, and on this account liberalisation has been much criticised, typically because adjustment has been thought to wipe out the physical and institutional infrastructure improvements required to make a success of price reform.

Whether this has been the case appears to depend on:

- the sequencing of reform;
- the supply of some public infrastructural services (notably credit, research and physical infrastructure);
- the influence of liberalisation on import vulnerability and the complementary influence of aid flows. Evidence is presented to suggest that this is an important and under-reported element of the 'African problem' with liberalisation.

In its function of removing barriers to trade and producer/consumer price spreads, 'liberalisation of the second kind' has played a valuable ancillary role by increasing the impact of technical change on the consumer price of food. This has reduced one component of poverty, even though other dimensions of liberalisation, especially the mistimed and uncoordinated withdrawal of input subsidies, have indeed constituted 'a reaction too far'. The spread of the green revolution in maize, sorghum and cassava is the most promising technology yet invented for mass poverty reduction in Africa, and in the places where price spreads remain high and restrictions on the flexibility of price movements remain, further liberalisation, subject to intervention by an appropriate buffer-stock agency, may have a valuable supportive role to play in facilitating the spread of this technology.

Notes

1 Financial support from DfID (ESCOR and the Natural Resources Policy Research Programme) and from the Gatsby Charitable Foundation is gratefully acknowledged.
2 Data from Toye (1995).
3 Jayne and Jones (1997: 19) suggest that this may be because farmers had been suspicious since colonial times of village traders who served, at that time, as vehicles for exploitation by marketing boards, and refused in many cases to acknowledge that they could offer a better deal when freed from their role as agents of a state marketing organisation.
4 *The Malawi Rural Finance Company* Malawi is a very poor country with a 1995 per capita income of US \$140. As such it is heavily dependent on small-farm agriculture, which in turn is heavily dependent on maize. The development of Malawian agriculture and poverty reduction from its current level of around 80 per cent have been constrained by low crop yields (approximately 1 ton/hectare through the 1980s and early 1990s, by contrast with 2–4 tons in small-farm areas of Kenya and Zimbabwe), in spite of near-Asian population densities and the availability of suitable

hybrid seed varieties. Most farmers are too poor to buy hybrid seed or fertiliser for cash, hence their ability to invest in high-productivity inputs depends on their access to credit.

As in many countries the structure of the financial market is dualistic, with companies, estates and the few upper-income personal customers borrowing from commercial and development banks and the rest from traditional moneylenders. But there have been two recent experiments in microfinance which depart from this norm. One of them, the Farmers' Clubs (subsequently Smallholder Agricultural Credit Administration or SACA) dates back to the days of the Lilongwe Land Development Programme in the early 1970s: this was a government-sponsored group credit scheme for farmers lending at a government-controlled interest rate, with loan instalments collected out of the proceeds of the harvest. The other, the Malawi Mudzi Fund, set up in 1989, was also a group scheme, sponsored by the International Fund for Agricultural Development (IFAD); but this was aimed more at small traders and manufacturers than farmers, and modelled on the principles of the Grameen Bank of Bangladesh, with weekly repayments to a member of the bank's staff at a 'centre', containing six groups, near the borrower's workplace. Although represented as quasi-autonomous, it too had strong government representation in its management, in particular a steering committee chaired by the permanent secretary of the Ministry of Finance; hence it lacked freedom to determine its own interest-rate and personnel policies.

Both schemes got into difficulties in the early 1990s. SACA (as shown in the table), after two decades of very good loan recovery, suddenly lost all control over overdues in 1992, under the stress of a very bad drought and promises by the newly democratised Malawi Congress Party to offer a moratorium on overdues if elected. For its part the Mudzi Fund failed to achieve either the outreach or the financial discipline of its Grameen parent and remained as a consequence very subsidy-dependent: a calculation by Buckley in 1993 suggested that it would have had to raise its interest rate by a factor of 18 to break even.

The approach of the World Bank and donors to financial-sector reform was conventional. Financial repression was to be done away with, and private-sector capital to be introduced into all parts of the sector; the assets of the Mudzi Fund and SACA were handed over in 1994 to a new private entity, the Malawi Rural Finance Company MRFC), recapitalised by the Bank, which now undertook to take deposits as well as make loans to the rural poor. The MRFC decided, in pursuit of financial security, to go upmarket, and in respect of agricultural loans to lend only to those farmers who farmed a cash crop (usually tobacco) as well as other food crops; this automatically tended to disqualify the poorest farmers, and we calculate that the proportion of MRFC borrowers below the poverty line has fallen from 45 per cent in 1992 (under SACA and the Mudzi Fund) to 11 per cent in 1997. Loan volumes have shrunk, and financial stability has not returned: the MRFC had overdues of more than 30 per cent in 1996. One aggravating factor is that when, under the stress of heavy IMF budgetary conditionality following a major macrofinancial crisis in 1994, the fertiliser subsidy was finally removed, this made the application of modern inputs to maize and other crops less attractive, and depressed yields (see the bottom rows of table).

Malawi 1980–96: agricultural output and financial conditions

	Maize smallholder sector:			*Financial institutions lending to the poor:*		
	Output yield (kg/ha)	*Fertiliser use (tons)*	*Maize/ fertiliser price ratio*	*Credit volume (1996 m. kwacha)*	*Overdue ratio*	
					SACA	*Mudzi Fund*
1980s av.	1.13		0.89	SACA: 86	7	n.a
1990	1.00	99,400	0.38	SACA+Mudzi Fund: 435	13	48
1991	1.14	106,884	0.31	510	15	35
1992	0.49[a]	144,235	0.31	546	76	
1993	1.53	150,087	0.30	- - - - - - - - - - MRFC - - - - - - - - - -		
1994		95,219	0.32	36	5	
1995		139,939	0.34	150	7	
1996		90,874	0.22[b]	230	34	74

Notes
a Drought year   b Subsidy removed

6 For a particularly vivid account of the state-led drive for smallholder agricultural growth in Zimbabwe between 1980 and 1985, see the essay by Rohrbach (1989).

References

Bates, R. (1981) *Markets and States in Tropical Africa*, Berkeley: University of California Press.

Cleaver, K. (1993) *A Strategy to Develop Agriculture in Sub-Saharan Africa and a Focus for the World Bank*, Technical Paper 203, Washington, DC: World Bank, Africa Technical Department.

Demery, L. and Squire, L. (1996) 'The effects of adjustment on poverty and inequality in Africa', *World Bank Economic Review*, 11.

Eicher, C. (1995) 'Zimbabwe's emerging maize revolution', *World Development*.

Fischer, S. *et al.* (1998) *Has Africa Recovered?*, IMF Discussion Paper, Washington, DC: International Monetary Fund.

Jayne, T. and Jones, S. (1997) 'Food marketing and pricing policy in Eastern and Southern Africa', *World Development*, 25 (December).

Jones, S. *Liberalised Food Marketing in Developing Countries*, Oxford: Oxford Policy Management.

Killick, Tony (1978) *Development Economics in Action*, London: Heinemann Educational.

Killick, Tony (1988) *A Reaction Too Far*, London: Overseas Development Institute.

Lipton, M. (1989) *New Seeds and Poor People*, London: Allen and Unwin.

Mosley, P. (1983) *The Settler Economies*, Cambridge: Cambridge University Press.

Mosley, P. (1994) 'Policy and capital market constraints to the African Green Revolution: a study of maize and sorghum yields in Kenya, Malawi and Zimbabwe, 1960–91', in G.A. Cornia and G. Helleiner (eds), *From Adjustment to Development in Africa*, London: Macmillan.

Mosley, P. (2002) *A Painful Ascent: The Green Revolution in Africa*, Cambridge: Cambridge University Press.

Mosley, P., Subasat, T. and Weeks, J. (1995) 'Assessing adjustment in Africa', *World Development*, 23 (September): 1459–75.

Rohrbach, D. (1989) 'The economics of smallholder maize production in Zimbabwe: implications for food security', International Development Paper 11, Michigan State University.

Sachs, J. and Warner, A. (1995) 'Globalisation and economic reform in developing countries', *Brookings Papers on Economic Activity*, 1: 1–117.

Toye, J. (1995) 'Ghana', in P. Mosley, J. Harrigan and J. Toye (eds), *Aid and Power*, 2nd edn, London: Routledge.

Sahn, D. (1994) 'Effects of adjustment on poverty in Africa', in G. Cornia and G. Helleiner (eds), *From Adjustment to Development in Africa*, London: Macmillan.

Timmer, P. (1994) 'Policy influences on agricultural productivity', Yale University, unpublished.

Weeks, J. (1997) 'Analysis of the Demery and Squire "Adjustment and poverty" evidence', *Journal of International Development*, 9: 827–37.

World Bank (1980) *Accelerating Development in Sub-Saharan Africa*, New York: Oxford University Press.

World Bank (2000) *Adjustment in Africa: Reforms, Results and the Road Ahead*, New York and Oxford: Oxford University Press.

Yao, S. (1994) *Technical Change and Agricultural Development in China*, London: Macmillan.

11 The dimensions of food aid in sub-Saharan Africa

Policy lessons[1]

John Shaw

11.1 Introduction

Food aid is an important and undervalued resource for development, as well as to meet emergencies, in Africa. The net value of food aid to Africa in 1985–90 averaged US $1 billion a year, about the same as the net transfers of the International Bank for Reconstruction and Development (IBRD) and the International Development Association (IDA). Yet food aid remains a controversial form of development assistance. Detractors have pointed to the political and commercial motives of donors that have sustained food aid flows, to its possible disincentive to local agricultural production and disruption to trade, and to the risk of increasing dependence on imported commodities, especially wheat and rice. These dangers should not be ignored. But its potential value deserves equal prominence. Food aid has played a vital role in saving lives in emergencies, addressing malnutrition, and helping countries achieve economic growth and greater social equality.

Africa has the potential to feed itself and even export food, as it did in the past. The vast assets of land and water could be harnessed to expand agricultural production, but this transformation will take time. In the interim, external assistance, including food aid, will be required to assist the transformation process and expand food supplies to the food insecure whose numbers have been projected to increase from 215 million in 1990–2 to 264 million by 2010 (FAO, 1996). Child malnutrition is expected to fall in all major developing regions except Africa, where the number could increase by 45 per cent between 1993 and 2020 to reach 40 million. Population increase has exceeded the food-production growth rate in Africa since the early 1970s, and the gap is widening as a result of declining per capita food production (Pinstrup-Andersen and Pandya-Lorch, 1998; World Bank, 1997). Projections show that even with increased agricultural production, African countries will need to import more food in the new millennium. With few exceptions, however, they will not have the necessary foreign exchange to import all their requirements on commercial terms. Food aid, or aid for food, will be needed to bridge the import gap.

A joint World Bank/World Food Programme study on food aid in Africa in 1991 concluded that food aid was a significant development resource in Africa and should be deliberately used to attack hunger and poverty (World Bank/WFP, 1991: 1). Owing to its history, constituency and inherent nature, the study found that food aid had special advantages in sustaining a poverty focus, supporting food security programmes and attenuating the social costs of economic adjustment. The study also recognized that Africa's food import needs would grow significantly and poverty, hunger and malnutrition, already widespread, were likely to increase further. More aid, including food aid, would therefore be required, together with an expansion of food production and commercial imports.

There was no evidence that food aid had been a disincentive to national food production. Indeed, the evidence suggested that its effect on agricultural production and on economic growth had been positive. There were cases of disincentive effects at the local level, however, mostly as a result of bad timing and mismanagement of food aid shipments. They had been usually short-term and had been largely avoided by tying food aid into overall food security plans and controlling releases of food aid on to the market. There was also concern that food aid supplies of wheat and rice (as well as large direct and indirect consumer subsidies on these commodities) would shift consumer preferences, and thereby decrease the incentive to increase domestically produced foods. The study noted, however, that food habits were not immutable. They could be changed by factors such as: government import and pricing policies; changes in the relative prices of food commodities; increasing income; transport and logistical improvements; migration to urban areas; fuel costs; and changes when women participate in income-earning activities outside the home. To the extent that food aid commodities substituted for imports that would have occurred in any case, the causes of changes in demand must lie elsewhere. And a change in commodities could lead to more efficient diets.

11.2 The dimensions of food aid

11.2.1 Food aid supplies

Food aid to Africa increased progressively from the early 1970s, when it was a fraction of total food aid supplies.[2] By the mid-1980s, Africa was the major food aid recipient region. By 1992, food aid deliveries in cereals had reached almost 6 million tons, representing 39 per cent of global cereal food aid supplies.[3] Since then, it has declined precipitously, in common with all food aid, to 2.3 million tons of cereals in 1997, still accounting for about 35 per cent of global cereal food aid (Figure 11.1).

Cereal food aid represented a growing proportion of the volume of cereal imports in Africa in the 1980s, reaching over 44 per cent in 1985–6. During the first half of the 1990s, cereal food aid was between 23 per cent and 32 per cent of cereal imports before falling sharply to between 14 per cent and 19 per cent after 1995–6. On the other hand, non-cereal food aid commodities have consistently represented less than 10 per cent of the volume of non-cereal food imports (Figure 11.2).

11.2.2 Composition

Food aid has traditionally been divided into three categories: emergency relief during natural and man-made disasters; project aid in support of specific development projects; and programme aid for balance of payments and budget support. Emergency food aid has been a much higher proportion of total cereal food aid to Africa than to other developing regions, accounting for 59 per cent of the total volume of cereal food aid to Africa at its height in 1992. Since then, emergency food aid has represented more than half of declining annual cereal food aid deliveries. At the same time, programme food aid has dropped markedly, while project food aid has increased as a proportion of total cereal food aid to Africa.

11.2.3 Recipients and donors

Over the past 10 years (1988–97), forty-seven countries in Africa have received food aid. Many of these countries have received only small amounts of food aid, mainly for emer-

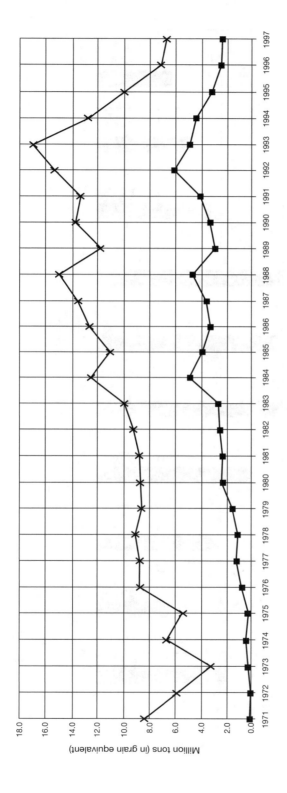

Figure 11.1 Cereal food aid to SSA compared with global cereal food aid, 1971–97

Sources: To 1988: FAO, *Food Aid in Figures*; post-1988: World Food Programme, *The Food Aid Monitor*

Million tons (in grain equivalent)

—✶— Total developing countries   —■— Sub-Saharan Africa

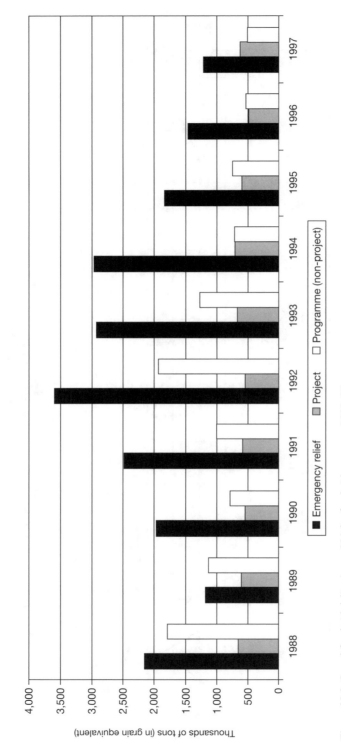

Figure 11.2 Cereal food aid deliveries to SSA by food aid category, 1988–97
Source: World Food Programme INTERFAIS database

gency relief. Only a small number of African countries have received significant amounts of food aid on a regular, annual basis. The seven main food aid recipient countries (Angola, Ethiopia, Kenya, Lesotho, Liberia, Mozambique and Sudan) have received between 43 per cent and 63 per cent of total cereal food aid to Africa annually in the past 10 years. Each of these countries has received markedly different amounts annually, mainly in response to the occurrence of drought or man-made disasters. Ethiopia, the main food aid recipient country in Africa, suffered from a devastating combination of both types of disaster simultaneously.

Similarly, while many suppliers have provided food aid to African countries, the bulk of food aid has come from a small number of donors. In 1997, for example, fifteen donors provided over 50,000 tons of food aid. Over 80 per cent was provided by seven donors (United States, the Commission of the European Communities (CEC) – on behalf of countries of the European Union – Japan, Germany, Canada, the United Kingdom and Australia). The United States alone provided 40 per cent, while the CEC supplied 20 per cent.

In addition, there has been an increasing trend for food aid to African countries to be provided more through multilateral channels, mainly the UN's World Food Programme (WFP), and by non-governmental organisations (NGOs) than by bilateral donors.

11.2.4 Food aid commodities

Cereals have made up the bulk of the food aid commodities supplied to Africa. Until the early 1990s, wheat and wheat flour were the main items in cereal food aid. Since then, coarse grains (sorghum, barley, oats) have become the main constituents of cereals food aid deliveries, while rice and blended and fortified cereal-based commodities have remained small. Non-cereal food aid commodities have also been provided, but always in small amounts. Of these, pulses have been the main commodity, followed by vegetable oil and fats, while other non-cereal commodities have been provided in very small quantities .

11.2.5 Recent trends

Africa has witnessed dramatic changes in food aid supplies and their deployment during the 1990s. There has been a precipitous decline of almost two-thirds in the total volume of food aid to Africa between 1992 and 1997. Also, major changes have occurred in the categories of food aid to Africa. Well over half of all food aid deliveries have been for emergency relief, reaching over 60 per cent of the total volume of supplies in some years as large-scale and complex man-made disasters, sometimes combined with widespread drought, have taken their toll. At the same time, the volume of programme food aid fell from about 39 per cent of total food aid supplies at the end of the 1980s to 21 per cent in 1997, while project food aid increased in volume from 14 per cent to 27 per cent over the same period.

As programme food aid has been provided exclusively on a bilateral, government-to-government basis, the proportions of food aid supplied multilaterally, mainly through WFP, and through non-governmental organisations (NGOs) have correspondingly increased. Overall trends in cereal production and food supplies have remained disturbingly low throughout the 1990s. Aggregate cereal production in Africa has remained little changed, while production per capita has steadily fallen as population has increased. Net imports of cereals have more than tripled from 3 million tons at the end of the 1980s to over 11 million tons in 1997–8, and per capita cereal supplies have declined as cereal food aid has dropped sharply in recent years.

11.3 Issues in food aid to Africa

The World/WFP study referred to above identified five key issues in food aid to Africa that are worth commenting on again. The first relates to the level and reliability of food aid supplies. The volume of non-emergency food aid to Africa has fluctuated sharply over the past two decades, and has depended in part on global surpluses and food prices rather than on the need for food aid in African countries. For most major donors, food aid has been additional in the sense that an increase in financial aid to compensate for diminished food aid is unlikely. It is important to preserve this element of additionality.

A second issue is that of absorptive capacity and disincentives. Food imports to Africa are expected to grow under all realistic scenarios. It will be increasingly difficult for most African countries to import food commercially because of balance of payments problems and high debt and debt servicing levels. Thus, even larger amounts of food aid will be required to support traditional assistance programmes. Absorptive capacity is also constrained by a country's administrative and logistical capacities.

A third issue is how to ensure that the aid contributes effectively to poverty alleviation. Food aid has seldom made a large direct contribution to alleviating poverty and hunger because deliveries have often been irregular, because funds generated from the sale of food aid commodities have not been targeted on the most needy, and because benefits have often been dissipated in general food subsidies, mainly in urban areas. Project food aid has the capacity to reach the poor and hungry directly through food-for-work, supplementary feeding and other targeted programmes, though the costs of reaching target populations through direct distribution may be prohibitively high.

Fourth, as the study acknowledged, monetisation of food aid is in principle preferable to the direct distribution of commodities because of formidable logistical problems associated with the latter and because financial resources are more flexible to use. At the same time, experience had shown that monetisation has its own problems: overvalued exchange rates, support for general subsidies, poorly functioning markets, inappropriate use of the funds involved, and difficulties in targeting the poor: problems that have inhibited donors from moving in this direction. Where these obstacles cannot be overcome or where special objectives are sought, such as giving command over the food aid resources to women, there would be a continuing need for direct distribution of food aid commodities.

A fifth issue is that of food aid allocation among African countries. As with all aid, inter-country allocations of food aid, especially programme food aid, often reflect the political and commercial interests of donors rather than the degree of need of recipients. The separation of food and financial aid also impede adjustments needed to maintain appropriate levels of total aid. The levels of usual marketing requirements (UMRs), a specific undertaking made by a recipient country to maintain at least a specified level of commercial imports of a food commodity in addition to that provided as food aid, could be further reduced.

The World Bank/WFP study made a number of recommendations, including:

1   Emergency food aid should respond more quickly.
2   Non-emergency food aid should be provided on a predictable and flexible basis.
3   Inter-country allocations of food aid should clearly reflect the developmental needs and opportunities of African countries, and stay within the limits of absorptive capacity.

4   This indicates a need to coordinate food aid and financial assistance, as part of a national development strategy. Food aid serves development in a variety of ways: easing foreign exchange constraints; providing support for development projects; facilitating cereal market restructuring; and protecting and developing human resources. Food aid should thus be brought into the mainstream of aid planning and programming, with a common set of programmes and projects.

5   Food aid should be a flexible resource. Joint or common funds generated from the sale of food aid commodities, the capacity to switch between financial and food aid, monetisation, triangular transactions, local purchases and commodity exchanges should be incorporated into policies.

6   Greater timeliness of emergency aid requires the strengthening of arrangements for disaster preparedness and prevention, with improved information bases in African countries to provide early warning of impending food shortages. Food-aided, labour-intensive public works programmes that could be activated quickly should be part of such a strategy. The International Emergency Food Reserve (IEFR) should be revised to provide adequate resources in advance of emergency food needs. Quick access to donor food supplies, as well as a financial reserve to purchase food close to where disasters occur, is essential.

11.4 Policy lessons for future food aid

With the prospect that food aid to African countries will be required on an increasing scale in future years, what are some of the main policy lessons to be learned from past experience in providing food aid to them? A number of these policy issues have been advocated in the past, some more than once (WFP, 1995; Puetz *et al.*, 1995); and they apply to other developing regions as well. These policy lessons take note of the considerable constraints on both the demand and supply side of food aid that have impaired its usefulness.

11.4.1 Country programming

The first, and basic, policy lesson from which all others flow is that food aid should be planned and executed as an integral part of individual country aid programmes, which are recipient-oriented and focused on their developmental objectives. In other words, food aid should be determined by demand in the recipient countries and not, as in the past, by supply considerations in donor countries.

While there are similarities, there are also contrasts among African countries, including their need for food aid. It makes little sense, therefore, to discuss and resolve Africa's food aid problems as though they were one, or to suggest solutions as though they would benefit all African countries equally. As the World Bank/WFP joint study pointed out (World Bank/WFP, 1991: 29–30), donors have handled food aid separately from financial and technical aid. As a result, food aid has acquired its own institutions, procedures and legislation. This has imposed a different mind-set, led to difficulties in coordinating food aid with other forms of assistance and made food aid transfers unacceptably rigid. If international assistance were planned and programmed within a common policy framework, with a common set of programmes and projects, the cost-effectiveness and impact of all aid transfers, including food aid, would be considerably increased.

Country programming of food aid in Africa should be facilitated by the fact that the bulk of food aid is received by a relatively small number of African countries and provided by a relatively small number of donors, increasingly through multilateral channels. Furthermore, aid consortia arrangements have been set up for many African countries by the World Bank and the United Nations Development Programme through which food aid can be coordinated with other forms of aid. In addition, resolutions have been passed by the UN General Assembly designed to bring about an overall improvement of the effectiveness and efficiency of the UN development system in delivering its assistance at the country level (UN, 1992), and in strengthening the coordination of UN humanitarian emergency assistance (UN, 1991).

11.4.2 Redefining food aid

The conclusion of the Uruguay Round through the signing and ratification of the Final Act, and the setting up of the World Trade Organisation (WTO) have provided a major opportunity for redefining food aid, reviewing and revising the FAO Principles of Surplus Disposal governing the use of food aid, and establishing a new food aid regime within a liberalising global economy (Shaw and Singer, 1995). The present statistical convention is that food aid is distinguished from food trade (or commercial transactions) by an element of concessionality. The FAO Principles of Surplus Disposal, first drawn up in 1954 as a 'code of conduct' to avoid 'harmful interference with normal patterns of production and international trade', refer to 'sales on concessional terms' as distinguished from 'commercial sales'. When the Principles were first adopted 45 years ago, the distinction between the two types of transactions was 'assumed' to be self-evident. However, with experience gained from applying the Principles, it became apparent that views on the meaning of 'normal commercial practice' differed among governments. Furthermore, as the objective of providing economic assistance gradually took precedence over that of surplus disposal in the use of food aid, issues arose as to whether certain kinds of transactions should be regarded as 'concessional' or 'commercial' sales.

Various attempts have been made to find a generally acceptable distinction between concessional and non-concessional transactions. None has provided a definitive answer and the issue remains unresolved. Faced with this dilemma, a list of transactions was drawn up that were regarded as constituting 'food aid' and, therefore, fell within the area of responsibility of the Consultative Subcommittee on Surplus Disposal (CSD), a subcommittee of FAO's Committee on Commodity Problems set up in Washington, DC specifically to monitor the implementation of the Principles and to provide a forum for consultations.[4] Subsequently, there has been an expanding 'grey area' between food aid, thus defined, and outright commercial transactions with no concessional element. The Development Assistance Committee (DAC) of the Organisation for Economic Cooperation and Development (OECD), which comprises the major donor countries, have agreed on a benchmark of 25 per cent below the commercial price as an arbitrary definition of the grant element of Official Development Assistance (ODA). But a large part of so-called 'trade' has not taken place as straight market transactions at free international prices. It has been conducted through a labyrinth of various forms of bilateral arrangement that have provided discounts from the 'commercial' international price (itself reduced by overhanging surpluses and domestic production subsidies) in many direct and indirect ways. However, this 'grey area' food aid, which has been significantly higher than the level of statistically recorded food aid,[5] has been provided mainly for the short-run political and commercial (market protection and penetration) objectives of donors rather than the developmental purposes of poor and food insecure recipient countries.

If this hidden food aid is now forced out into the open as a result of the Uruguay Round, and brought within the disciplines of food aid as defined in the Final Act, a major step would have been taken in dealing with the world hunger problem and, by extension, the eradication of poverty. This could be achieved by redirecting programme and 'grey area' food aid for these purposes. The Final Act provides an opportunity to use the unadjusted surplus production of exporting developed countries that can no longer be sold at subsidised prices, or otherwise disposed of, as food aid for the benefit of needy people in poor and food-importing developing countries. Advantage could be taken of the food aid provisions of the Final Act relating to public stockholding for food security purposes to facilitate this objective and for revising the provisions of the FAO Principles relating to UMRs. But whereas other developing regions would be able to absorb large increases in food aid for their development programmes relatively rapidly, assuming that it was provided in appropriate ways and on acceptable terms, African countries would require more pre-investment support to raise absorptive capacity (WFP, 1994).

The Final Act of the Uruguay Round recognizes the need:

> to establish appropriate mechanisms to ensure that the implementation of the results of the Uruguay Round on trade in agriculture does not adversely affect the availability of food aid at a level which is sufficient to continue to provide assistance in meeting the food needs of developing countries, especially least-developed and net food-importing developing countries.
>
> (GATT Secretariat, 1994: 448–9)

To this end, and specifically for the benefit of least-developed and net food-importing countries, it was agreed to:

- review the level of food aid established periodically by the Committee on Food Aid under the Food Aid Convention 1986 and to initiate negotiations in the appropriate forum to establish a level of food aid commitments sufficient to meet their legitimate needs during the reform programme;
- adopt guidelines to ensure that an increasing proportion of basic foodstuffs reaches them in fully grant form and/or on appropriate concessional terms; and
- give full consideration in the context of aid programmes to requests for the provision of technical and financial assistance to improve their agricultural productivity and infrastructure.

It was further agreed that these countries should receive differential treatment in their favour in any agreement relating to agricultural credits. Recognising that certain countries might experience short-term difficulties in financing normal levels of commercial imports as a result of the Uruguay Round, these countries may be eligible to draw on the existing facilities of international financial institutions or such facilities that may be established in the context of adjustment programmes in order to address their financing difficulties. These provisions are to be regularly reviewed by WTO's Ministerial Conference, which will normally meet every two years, and action is to be monitored by WTO's Committee on Agriculture.

11.4.3 *Assured food aid supplies*

Not only is an adequate level of food aid necessary but its reliable supply should be assured. The availability of food aid has been mainly influenced by market surpluses and food prices.

When food production and stocks have fallen in donor countries and food prices have risen, the amount of food aid available has tended to shrink at a time when it is most needed (Rosen, 1989).

At present, the Food Aid Convention (FAC), which, together with the International Wheat Agreement, forms an integral part of the International Grains Arrangement, provides the only guaranteed supply of food aid in cereals as defined under the FAC. The first FAC, which came into force in 1967, was engineered by Canada and the United States to induce other developed countries to share the burden of providing food aid to needy countries. Each signatory to the FAC has guaranteed to provide a minimum physical level of food aid in cereals annually, irrespective of changes in the levels of their production, stocks or prices. Aggregate commitments to the 1967 FAC amounted to 4.25 million tons of cereals a year, which was increased to 7.6 millions under the 1980 FAC, but reduced again to 5.35 million tons in 1995 when Canada and the United States cut back their commitments by 33 per cent and 45 per cent respectively. However, signatories have regularly surpassed their minimum annual commitments: annual aggregate shipments have ranged between 8.8 million tons and 13.6 million tons since the mid-1980s.

While there has been commitment to overall minimum cereal food aid supplies, it has been left to each signatory to determine to which countries the supplies have been delivered, on what terms and for what purposes. And while conditionality has been imposed by donors as to how the food aid they have provided should be used, the principle of 'contractuality' has not been observed, which would apply conditions as much to the giver as to the receiver of food aid, including assurance of supply.

These and related issues may be addressed in the context of a new FAC that comes into force on 1 July 1999. Echoing the Final Act of the Uruguay Round, the declaration that emerged from the first WTO Ministerial Conference in Singapore in 1996 requested that action be taken within the framework of a new FAC 'to develop recommendations with a view toward establishing a level of food aid commitments, covering as wide a range of donors and donable foodstuffs as possible, which is sufficient to meet the legitimate needs of developing countries during the reform programme'. In accordance with the Final Act of the GATT Uruguay Round, these recommendations should include guidelines to ensure that an increasing proportion of food aid is provided to the neediest countries as well as means to improve the effectiveness and positive impact of food aid.

The 1999 FAC, which has an initial three years' duration, will provide a guaranteed aggregate minimum annual commitment of 4.9 million tons (in wheat equivalent). In addition, financial aid is being provided for transport and other operational costs and for food purchases in food aid recipient countries.[6] Priority will be given to the least-developed and low-income countries in the allocation of FAC food aid. Particular importance is attached to ensuring that food aid is directed to the alleviation of poverty and hunger of the most vulnerable groups.

11.4.4 Appropriate modalities

In addition to ensuring an adequate level and assured supplies of food aid as an integral part of overall country aid programmes in Africa, careful consideration should also be given to the most appropriate mechanism to use to transfer food aid in order to maximise its cost-effectiveness and impact for the benefit of poor and food-insecure people and reduce transfer costs. Food aid commodities may be transferred through different modalities. They may be: delivered directly in kind to beneficiaries from donor supplies; sold (monetised) and the funds generated used for development purposes; purchased or exchanged in one developing country for use as food aid in another developing country through what are called 'triangular transactions'; or purchased or exchanged in the same recipient country. Generalisations about these various modalities are unhelpful in operational terms. The most appropriate modality will depend on the specific conditions of each country, and the location within a country. It must also relate to the specific objective of maximising benefits for the target population, the poor and food-insecure.

Monetisation

The World Bank/WFP study found that the complex question of monetisation had come to assume pronounced importance in the food aid debate. It has acquired even more importance since as food aid resources have sharply decreased. Some analysts believe that the most efficient use of food aid is to sell (monetise) the commodities provided and use the funds to help finance well-designed development projects. Others believe the poor can be targeted more efficiently if food aid is distributed directly. The debate has practical and operational importance.

At the beginning of the 1990s, about three-quarters of all non-emergency food aid to Africa was monetised, including most programme and a fifth of project food aid. A small proportion of emergency food aid was also monetised but most was distributed in kind free of charge. The proportion differed by donor and source: the United States monetised about 90 per cent of its food aid, for example, while WFP monetised no more than 15 per cent. During the 1990s, the proportion of food aid that was monetised would appear to have declined as programme food aid has decreased and emergency and project food aid has increased, but precise details are not available.

Programme food aid has been the main source of monetisation in Africa. When it has substituted for commercial imports, it has released foreign exchange, and if no conditions have been attached by the donor to the use of the fund, has come close to untied financial aid. The local currency generated through monetisation has provided much-needed revenue, especially for the least-developed countries of Africa. To the extent that these funds have been invested soundly, they have contributed effectively to development. Because of the volume of such aid, it is important that recipient governments include it in the wider policy dialogue with donors. This would allow officials to coordinate programmes to support policy adjustment. Within a sound macroeconomic policy framework, and taking fungibility into account, the funds generated could be targeted at economic and social programmes to support vulnerable and food-insecure groups, and attenuate the social costs of structural adjustment (Shaw and Singer, 1988; Singer, 1991).

However, effective use of monetised food aid depends on a number of conditions. Donors must be fully informed about, and responsive to, the requirements of recipient countries. To maximize the local currency generated, monetised food aid should be valued at import parity prices, and the funds used quickly and in conformity with recipient countries' priorities. This needs a capacity to coordinate and integrate food aid into agricultural, economic, social and fiscal policies, which, in turn, requires administrative and managerial skills that may not be adequately available in many African countries.

Much depends on whether local conditions are conducive to efficient monetisation and whether the funds generated can be effectively channelled to the target group. Experience has shown that the following issues require particular attention. Exchange rates can affect the value of the funds generated. If the exchange rate correctly reflects real scarcity values, food aid sold in the market at the import price will generate funds that reflect the real resource transfer. If the exchange rate is overvalued, however, the local currency funds will be less than the real value of the food aid. This has been a constraint on monetisation in Africa.

Similarly, when a general subsidy is placed on food prices, food aid may not be monetised at its full value. In addition, the subsidy may be captured by less needy groups, or may be applied to imported wheat and rice, leading to changes in diets and dependency (Pinstrup-Andersen, 1988). In such cases, the provision of food aid may make it more difficult to reform the food system. The solution may be to use food aid in support of food policy reform, which may include a targeted subsidy to needy groups in place of a general subsidy.

Stable currencies and well-functioning markets are particularly important for monetisation but these prerequisites are conspicuously absent in many African countries. Where a resource transfer to poor households is the primary objective and where the local currency is reasonably stable and local markets are functioning, cash will, as a general rule, be more valuable to the beneficiaries than food commodities. But transferring cash to poor households may raise local food prices if food supplies are not increased; in which case delivering food directly may be preferable.

Another issue concerns the problem of fungibility. Recipient governments may place monetised funds in the general budget, which may thwart donors' intention to see them used specifically for poverty and hunger alleviation. To the extent that national development policies and plans are already oriented in that direction, however, fungibility should not be a problem, and donors should support policies and programmes rather than attempt to direct their aid to specific purposes. This reinforces the need for food aid to be fully integrated into national plans and donor strategies.

Finally, the monetisation process itself, that is, the ways in which food aid commodities are sold, can have an important effect on the benefits accruing from monetisation. For example, sales through the private sector at full market prices might stimulate development and the operation of markets, while sales through public sector channels at subsidised prices might support high-cost and inefficient parastatals with a monopoly on the grain trade. Careful consideration should also be given to how the funds generated through monetisation are used. Experience suggests that the funds are best placed in a special, interest-bearing account and used promptly for priority items agreed upon at the design and approval stages of food-aided programmes and projects (WFP, 1987).

Delivering food aid in kind

Food aid in kind provided to targeted beneficiaries through specific development programmes or projects has faced some of the same problems as monetised food aid but they have generally been: more limited in scope; related to local rather than national constraints; involving efficiency at the operational rather than the policy level; and more dependent on administrative and managerial, than on legislative, solutions.

Direct distribution of food aid in kind to beneficiaries in African countries faces formidable difficulties. Ocean transport of small consignments; port and intermediate handling charges; overland transport covering long distances on poor road, rail and river routes, and to land-locked countries; and storage and re-storage; all these can make the costs of delivering food aid very high, often more than the value of the food commodities provided by donors. Some donors finance part of the internal delivery costs but this does not make direct delivery of food aid commodities more cost-effective. Some practical ways have been applied to reduce these costs, such as achieving economies of scale by shipping consolidated consignments of cereals in charter vessels with lower freight rates to a number of projects in the same developing country simultaneously, using common delivery systems, and hiring private contractors at competitive rates.

Triangular transactions, local purchases and commodity exchanges

Faced with the difficulties and costs of direct food aid distribution, donors have looked for other modalities to transfer food aid to beneficiaries in developing countries which might also have additional developmental and food aid benefits. Three main modalities have been used: triangular transactions, whereby a donor buys food in a developing country and ships it to another developing country, where it is used as food aid;[7] commodity purchases in a developing country for use as food aid in the same or another developing country; and commodity exchanges whereby a food commodity supplied as food aid to a developing country is exchanged for another commodity which is used as food aid in the same country.

These modalities are perceived as having important potential benefits for food aid. They include: reducing transaction costs; increasing the speed and timeliness of food aid deliveries; and providing food commodities more appropriate to the food tastes and habits of recipients than those directly provided by donors. In addition, they could carry wider developmental implications with them, including: stimulating increased food production in food aid recipient countries; promoting exports; fostering intra- and inter-country trade; supporting food strategies and food security programmes by helping countries manage food surpluses; assisting in the restructuring, revivication and liberalisation of markets; encouraging and assisting the improvement of transport, storage and logistics in supplying countries and with neighbouring countries; and strengthening food management and administration in association with technical assistance and training (Shaw, 1983).

Despite these potential benefits, procuring food commodities in developing countries has remained a modest, if growing, part of total food aid supplies. In 1997, local purchase, including triangular transaction, for food aid in Africa amounted to just under 420,000 tons, which was about 18 per cent of total food aid deliveries. Only in eleven African countries were purchases above 10,000 tons, the most significant being in Ethiopia, Malawi, Niger, Sudan and Uganda. In thirty-one food aid recipient countries in Africa, no purchases were made or only negligible quantities were purchased. Donors and aid agencies have carried out these transactions at different prices and exchange ratios, through different marketing channels, and at different times of the year, with insufficient consideration of their potential developmental benefits or of building up capacity in African countries to undertake them more efficiently and on a larger scale.

Ultimately, the amount of food procured in African countries for use as food aid will depend on the cash resources donors are prepared to invest in these food aid modalities. This, in turn, will be influenced by donors' assessments of the opportunity cost of using their limited financial aid for these purposes rather than for other developmental uses. But experience has shown that there are other formidable barriers that relate to the considerable diversity of African country situations (RDI, 1987, 1990; WFP, 1989a, 1990; Clay and Benson, 1990).

Two policy lessons have emerged from the experience gained in implementing these food aid modalities. First, donors and aid agencies should develop a common strategy, based on a code of conduct, for each African country in which these modalities are implemented. A common information and market intelligence system should be established and maintained for individual African countries on food availabilities and prices for traditional staples as well as cereals and other food commodities that are not so widely traded. Standard guidelines would facilitate coordinated purchasing and exchange operations, including price and exchange ratio negotiations, marketing channels, transport and logistics arrangements, and monitoring and evaluation systems that would facilitate the assessment and comparison of individual food aid transactions, and also the aggregate effect of all such operations in a country.

Second, the full developmental benefits of these food aid modalities could be achieved if a coordinated programme of financial, technical and food aid could be provided to improve the capacity of African countries to undertake their operation. Such a programme might include: technical and training services to strengthen administrative and managerial support; develop trading skills and market analysis; improve grading, handling and storage techniques to improve quality; and improve transport and logistics systems.

11.4.5 Linking relief and development

The major part of food aid to Africa since the beginning of the 1970s has been for emergency relief, first in response to widespread drought and, more recently, since the mid-1990s for refugees and displaced persons, the victims of man-made disasters (Benson and Clay, 1998b; Messer *et al.*, 1998). The division of external assistance into 'development' and 'emergency' aid, each with its separate agendas, terms, legislation, financing and operating agencies (even separate units within the same aid agency), has dichotomised what in the real experience of African countries has not been separate: the inter-relationship between disasters and the development process. This raises at least three policy lessons.

A broadened definition of emergency assistance

Much could be gained if the definition of emergency assistance were expanded from an immediate, short-term response to providing relief to encompass pre- and post-disaster action in the 'continuum' between relief and development (Singer, 1994; Shaw, 1998). Conceptually, disasters would no longer be seen in isolation but in their full setting, and their effects on development would be taken into account. Resources and assistance would be provided for disaster prevention, preparedness and mitigation measures and would not dry up when required for rehabilitation and reconstruction activities after disasters had occurred, thereby helping to maintain the development process. And the planning, design and implementation of assistance programmes for both relief and development would be executed by integrated government and aid administrations within common legislative and execution procedures and financial provisions.

Improved early warning and response systems

Experience has repeatedly shown that accurate, timely and commonly available early information of an impending disaster, coupled with sound and speedy response, are key factors in mitigating their effects. Drought-related crises in Africa and elsewhere have been largely alleviated by the food security arrangements approved at the World Food Conference in 1974, including implementation of FAO's Global Information and Early Warning System (Shaw and Clay, 1998). No such system has been developed for man-made emergencies.

The former UN Secretary-General pointed to the need for an early warning system with political indicators to assess whether a threat to peace existed in a country, and to analyse what action might be taken by the United Nations to contain it (Boutros-Ghali, 1992). In the meantime, the linkages between providing humanitarian assistance and peacekeeping and peace-making operations remains dangerously confused. There are as yet no clearly established rules, guidelines and modalities. There are no easy solutions, and much will depend on the nature of the conflict in each situation. A major problem has been the principle of state sovereignty, which is enshrined in the United Nations Charter, although the Charter also states that 'this principle shall not prejudice the application of enforcement measures under Chapter VII of the UN Charter on "Action with respect to Threats to the Peace, Breaches of the Peace, and Acts of Aggression"' (UN, 1945). This provision has assumed particular significance not only as man-made disasters have increased in incidence, scale and duration but as many of them have occurred within, rather than between, countries. This has had important practical consequences.[8]

There are now good prospects for major improvements in early warning and tracking systems for all kinds of disasters through the application of remote sensing and satellite imagery, linked to a worldwide computerised information superhighway. Such systems should have common multilateral ownership. They would produce common information that would have the confidence of all concerned and would produce a common response. However, for the full benefits of these systems to be realised, they will need to be backed up by response systems with adequate resources in order to react quickly and effectively to the onset of emergencies.

The returns from national and donor investment in these early warning and response systems would be considerable. Human and economic suffering and damage could be avoided or mitigated, and the enormous cost and diversion of resources in providing protracted relief and peacekeeping operations saved. An international programme combining financial aid and technology and skills transfer, therefore, should be given the highest priority.

A truly multilateral and fully subscribed International Emergency Food Reserve

Current multilateral arrangements for responding rapidly to emergencies, whenever and wherever they occur, would be largely adequate if donors respected the provisions they have approved. An International Emergency Food Reserve (IEFR) with a target of not less than 500,000 tons of food commodities was established in 1975 as a continuing reserve with yearly replenishments. Contributions to the IEFR were to be placed at the disposal of WFP to provide it with an initial, quick-response capability. The modalities of IEFR operations were approved by WFP's governing body in 1976 and revised in 1978. An Immediate Response Account (IRA) of at least US $30 million annually was approved in 1991 as an integral part of the IEFR to purchase and deliver emergency food aid and special provisions agreed for protracted refugee and displaced person operations lasting for one year or more.

While the IEFR and its IRA have improved and increased WFP's ability to respond quickly to emergencies, it has not fully lived up to its original expectations as a multilateral standby facility. The IEFR is not like a bank account or stock of food readily available for WFP's use; it is a voluntary facility to provide emergency relief from food stocks and funds kept in donor countries. In contradiction of its agreed modalities, contributions to the IEFR have not been announced by all donors in advance. Contributions in food commodities have fluctuated considerably. Cash contributions have fallen short of requirements. And a high proportion of contributions have been tied and designated by some donors to certain food commodities and to specific emergencies after they have occurred. These shortcomings have weakened the IEFR, eroded its multilateral nature and made it difficult for WFP to respond fully and quickly to emergencies.

While often generous, the international response to emergencies has been inconsistent, sometimes based more on political considerations than on real need. The avoidance of high malnutrition and mortality rates in certain man-made emergencies, for example, contrasts starkly with the lack of success in others (Seaman and Rivers, 1988; Toole, 1993). Also, there are real practical constraints. The logistics involved in providing adequate food rations consistently to food-insecure people in remote areas of African countries are formidable and costly. But a truly multilateral and fully subscribed IEFR and its IRA would do much to take the politics out of the provision of emergency assistance and help coordinate the international community's response.

11.4.6 Institutional coherence

Institutional complexity and duplication in the field of food aid remain a cause of wonder (Maxwell and Shaw, 1995; Shaw and Clay, 1998). Various aspects of food aid policies and their implementation are considered in parallel in different forums. WFP's Executive Board, which meets in Rome, has a wider mandate to help evolve and coordinate national and international food aid policies and programmes. The FAO Committee on World Food Security, which also meets in Rome, discusses food aid in the context of world food security issues. The Consultative Subcommittee on Surplus Disposal, a subcommittee of FAO's Committee on Commodity Problems, meets regularly in Washington, DC to monitor the trade effects of food aid on a continuing basis. This is now also a concern of the Agricultural Committee of the World Trade Organisation in Geneva. The Food Aid Committee, a non-UN body serviced by the International Grains Council secretariat, meets in London mainly to oversee implementation of the Food Aid Convention. The IMF Executive Board in Washington, DC decides on the application of the Fund's special compensatory and contingency facilities to assist developing countries in financing food imports in times of need. The Office of the Coordinator of Humanitarian Affairs within the UN Secretariat, in New York and Geneva, has an interest in food aid in the context of large-scale and complex emergencies, and convenes international meetings in relation to its wider responsibilities. UNICEF in New York and the UNHCR in Geneva have special interests in food aid in relation to their mandates for children and women and for refugees respectively. Up to 1993, the UN World Food Council (WFC), whose ministerial conference met annually in different locations, had responsibility for coordinating the work of the UN system concerning food production, food security, nutrition, food trade and food aid.[9]

There is no single forum or one recognised body through which overall food aid policies and priorities might be negotiated and reviewed or through which a major international or national crisis would automatically be considered. Many proposals have been made for such a body since the first Director-General of FAO suggested the setting up of a 'world food board' in 1946. Among the proposals for international action presented to the World Food Conference in 1974 was one for the establishment of a 'world food authority'. This was not approved. Instead, the WFC was created. The demise of WFC in 1993 has shown that the solution does not lie in the creation of a separate body without executing authority and with a mandate that cuts across that of other bodies. The need for some central multilateral authority, and for improved coordination among the various bodies involved in food aid, remains.

11.5 Future prospects

The policy issues identified above could have far-reaching effects for future food aid generally, and for African countries in particular. But what are the prospects for radical change? Some recent developments give rise to guarded optimism regarding both the environment within which food aid would have to operate and the future of food aid itself.

11.5.1 UN Special Initiative on Africa

The special and acute problems of Africa, and the need to give it priority in the allocation of multilateral development cooperation, have long been recognised (Shaw, 1999). Yet concern has been expressed over what appears to be the increasing global marginalisation of Africa. Two facts are particularly telling. Of the forty-seven least-developed countries in the world, thirty-two are in Africa; and it is the only region in the world where poverty and food insecurity are expected to increase.

Faced with its acute problems, repeated calls have been made for a 'special programme of assistance', a new 'green revolution' or a 'Marshall plan' for Africa to provide coordinated international assistance on the scale required and in the most appropriate ways, and to build local and national capacities by investing in people and improving infrastructure within sound macroeconomic policies. But the international community's generous response to meeting emergencies in Africa has been in marked contrast to that of providing development aid; while there has been an improvement in the coordination of emergency assistance, coordination of development aid still leaves much to be desired.

11.5.2 Disaster mitigation and rehabilitation

In 1992, WFP's governing body endorsed a special programme for Africa that gave more extensive and systematic application of food aid to support disaster mitigation and rehabilitation activities (WFP, 1992).[10] Conditions in many African countries, and particularly the poorest among them that suffer from recurrent emergencies, call for a programme of assistance which explicitly supports activities that contribute toward disaster prevention, preparedness, mitigation and rehabilitation.

There is a close and growing link between poverty and vulnerability to recurring emergencies. If the food security of the most vulnerable people could be improved at the household and community level through development projects that provided employment, income and assets, the need for emergency aid could be considerably reduced (WFP, 1989b; Messer *et al.*, 1998). The major focus of WFP's special programme is on supporting labour-intensive works that provide simultaneously: employment, income and food, thereby alleviating poverty and helping to strengthen self-help capacity; and the construction and improvement of the infrastructure necessary to increase agricultural production, stimulate rural development, and strengthen protective measures against drought and other disasters. Together with such programmes, targeted food, income and health interventions could help improve the wellbeing of the poor and vulnerable and enable them to withstand future food shortages. In addition, WFP emergency interventions may be adapted, if circumstances allow, to facilitate developmental initiatives. Several approaches have proved particularly successful in Africa, including: the improvement of water management; environmental protection; strengthening transportation infrastructure; food security measures, including market restructuring, pricing policy changes, and the establishment of cereal reserves and grain banks.

And development projects have been designed to be expanded rapidly when emergencies occur to provide additional food and employment when household food production or income collapse.

11.5.3 Trade and food security

The conclusion of the GATT Uruguay Round and the setting up of the WTO provide a major opportunity for improving world trade in food commodities as an essential contribution to achieving world food security (Konandreas and Greenfield, 1996). The outcomes of the Uruguay Round will be only one of a number of factors affecting world food production, trade and prices, but will it prove to be the swing factor? A major change has been brought about in the way food markets are viewed and in the rules under which countries must operate their national agricultural policies. This should prevent the return to old habits, ensure the emergence of improved opportunities in agricultural trade for developing as well as developed nations, and provide a basis for the next episode of multilateral negotiations on agriculture under the auspices of the WTO (UNCTAD Secretariat, 1994). Much will depend on how effectively WTO will actually carry out its mandate.

11.5.4 A new food aid regime

The future of food aid is clouded by a number of factors, especially relating to its supply (Clay *et al.*, 1998; IFPRI, 1998). The watered-down agreement reached on partial agricultural trade liberalisation under the Uruguay Round seems likely to result in the continued generation of food surpluses and high subsidy costs in the major food-exporting developed countries. The 1996 farm bill in the United States ended a farm management regime started by President Roosevelt under the New Deal of the 1930s by transferring responsibility for decision making on crop selection and on stock holding from the federal government to the farmer. Yet President Clinton financed a US $2 billion food aid programme in 1998 to help US farmers out of their economic difficulties caused by low commodity prices. Attempts have been made to reform the Common Agricultural Policy of the European Union, which has been largely responsible for the generation of structural food surpluses, with limited success. Both the US and the EU have been recently involved in a large food aid programme, reportedly involving 4.8 million tons of food at a cost of US $1.5 billion, to the Russian Federation.

It has been suggested that a future food aid regime should have four main features (Mellor Associates, 1992). It should be seen to serve both recipient and donor countries; it should be demand-enhancing in recipient countries; adequate provision should be made for food aid for stabilisation and emergency purposes through safety-nets in times of disaster and economic adjustment; and close coordination should be established with financial and technical assistance.

Providing needy countries with adequate levels of food aid would have three main benefits for developing and developed countries alike. It would remove a substantial quantity of food from overhanging commercial markets, relieving pressure over a period of time during which agricultural adjustment could take place in developed countries and commercial markets could grow in developing countries: in this way, food aid would help the shift towards market orientation and liberalisation, rather than being a hindrance through dumping and surplus disposal practices. It would serve to prevent market marauding and the wasteful and damaging use of food and cash resources in export-enhancement and credit programmes often directed to countries where the demand for food is highly inelastic. And it would help developing countries speed up the implementation of their development programmes, thereby increasing food consumption and the demand for commercial food imports. Developing countries, where the demand for food is highly elastic, provide the main growing markets for food exports. But they currently lack the foreign exchange to buy all the food they need on commercial terms.

The establishment of the WTO could provide an institutional framework for food aid that has hitherto been lacking. As seen above, a number of bodies are currently concerned with food aid. Their work is largely uncoordinated, sometimes conflicting. A closer relationship between food aid and food trade, which should be an essential feature of a new food aid regime, should automatically be addressed by the WTO. And the coordination of food aid with financial and technical assistance should be taken care of in the special relationship envisaged between WTO, the Bretton Woods institutions, and other UN agencies in the Marrakesh agreement establishing the WTO.

11.6 General prospects

Ultimately, much will depend on the outcome of the debate that is now developing on the future of food aid itself, for which there are contradictory signals (Shaw, 2000). Global food aid, as statistically recorded, has fallen dramatically to its lowest level for over 20 years. Such declines have occurred in the past, as in the 1960s and early 1970s, calling into question whether the current decline is a temporary phenomenon or of a more structural nature. The effects of the US 1995 'freedom to farm' legislation are far from clear (Harvey, 1998). Some see it as a clear and unambiguous signal that the United States is committed to full liberalisation of farm policy and agricultural trade, thus ending over 50 years of federal government involvement in domestic agricultural policy that resulted in massive food stocks and large-scale food aid. However, others see the new legislation as releasing the full crop productive capacity of the United States and marking an abdication of US responsibility for stabilising world markets, which might trigger their collapse in the medium term, consequently renewing pressure for greater levels of protection and subsidised production and leading back to structural surpluses and a renewal of market marauding. Attempts to reform the Common Agricultural Policy (CAP) of the European Union, which has been largely responsible for generating structural food surpluses, have been postponed and large food surpluses are beginning to re-emerge. Enlargement of the EU by the inclusion of more countries in eastern Europe could result in increasing food surpluses even further as farmers in those countries seek to enjoy CAP benefits. A new Food Aid Convention was also signed in 1999 committing signatories to another three years of food aid of at least 4.9 million tons (in wheat equivalent), plus financial aid to meet transport and other operational costs and for food purchases in developing countries, with priority given to the least-developed and low-income countries; and the WTO has to arrive at a decision for implementing the food aid provisions of the 1994 Final Act.

While food aid for meeting emergencies is not questioned, calls are now being made to end food aid for developmental purposes, to be replaced with 'aid for food' (Reutlinger, 1999). But many of the arguments put forward in support of this transformation are spurious. Genuinely untied financial aid when used for food imports is preferable to food aid. However, most financial aid is tied, explicitly or implicitly. Financial aid tied to food imports (food aid) is better than financial aid tied to dubiously required high-priced capital goods or to armaments. Some of the advantages of financial aid can be obtained by monetising food aid.

And food aid, if properly targeted, has the advantage of directly addressing the needs of poor, food-insecure people and of being more gender-friendly to women and children. Rather than increasing, financial aid has fallen in real terms and as a percentage of donor countries' national incomes, and there is little evidence that financial aid has focused increasingly on the alleviation of hunger and poverty, other than rhetorically.

A comprehensive policy review of the future of food aid concluded that 'The balance of evidence is that food aid is no longer an additional resource but must justify itself in competition with other uses of scarce aid funds' (Clay *et al.*, 1998: 63) and a call has been made to 'grasp the nettle, and reform food aid' (ODI, 2000). The review found that food aid (as statistically recorded) had quickly become a marginal and uncertain component of aid globally, making it difficult for food aid to have a significant food security impact at the international level. The Food Aid Convention had been largely ineffective in assuring stability of food aid levels. Links to agricultural surpluses were a major source of uncertainty, and the relationship between international price variability, levels of stocks and donor commitments overall made food aid 'the most unstable element in ODA'.

The review did not take into account, however, that the real level of food aid, including the 'grey area' food aid referred to above, has been much larger and a more significant part of ODA, with a much larger target potential to address problems of hunger and poverty (Shaw and Singer, 1995). There was also an apparent contradiction in the review, which stated that development food aid 'has proved relatively ineffective as an instrument for combating poverty and improving the food consumption and nutrition and health status of the very poor and vulnerable people', but that 'robust evidence for both NGO activities and WFP on impacts of humanitarian and project aid is, surprisingly, lacking', ignoring the surveys of the food aid literature commissioned by WFP in the 1970s and 1980s (WFP, 1978; Clay and Singer, 1985). In addition, the review did not recognise the considerable constraints that donors have placed on the use of their food aid resources, and the formidable difficulties of reaching and benefiting abjectly poor people in low-income developing countries. The developmental expectations of food aid, therefore, should be tempered accordingly.

The review tended to focus on the supply side of future food aid. If the demand side were also examined, a more balanced perspective would emerge. For example, if serious attempts are made to meet the targets set at the various UN international conferences during the 1990s, particularly those relating to the reduction of malnutrition among children in particular and the hungry poor in general, aid in the form of food will need to play an important role. The gender and development benefits of food aid for women and their children might also have been emphasised. And the role of food aid in 'preventative diplomacy' in addressing the problems associated with man-made emergencies, and in preventing their occurrence might also have been mentioned. While the ground rules for the deployment of food aid in such situations remain dangerously confused, food aid could play a strategic role in preventing or limiting the outbreak of war and civil strife, much of which is caused, or exacerbated, by problems of access to adequate food supplies (Messer *et al.*, 1998). Food aid alone will not provide the solution, but it has a distinctive role to play in combination with other aid resources and measures. Nothing can replace it. And the prospects of gaining food security in future years cannot compensate for inadequate nutrition today (WFP, 1999).

11.7 Conclusion

There is general agreement that African countries will require help in meeting their food import requirements in the years ahead. Whether this food is provided as food aid in kind or as aid for food has to be determined, as well as the modalities of transactions, so that transfer costs are kept to a minimum and maximum benefit is obtained for poor and food-insecure people. Given the disarray and lack of focus in the debate, an international conference should be held to reconcile the significant differences in outlook and perspective that exist between donor countries on both sides of the Atlantic with a view, *inter alia*, to arrive at a comprehensive policy on future food aid to sub-Saharan Africa as a matter of priority.

Notes

1 I am grateful to Hans Singer, Edward Clay and Simon Maxwell for their views and ideas, to George-Andre Simon for providing the food aid data. I alone am responsible for any errors or shortcomings.
2 In this chapter, 'Africa' means sub-Saharan Africa. The chapter builds on a joint study by the World Bank and WFP on food aid in sub-Saharan Africa that was undertaken between 1989 and 1991, of which the author of this chapter was co-director (World Bank/WFP, 1991). See also Shaw (1991), Thomas and Maxwell (1989), WFP/ADB (1986).
3 The food aid data referred to in this chapter do not include 'grey area' food aid, that is, credit, guarantee and export enhancement food aid programmes with different levels of subsidy between food aid as currently statistically recorded and outright commercial sales (see Shaw and Singer, 1995, 1996 and 1998).
4 The transactions that, according to the FAO Principles of Surplus Disposal, constitute food aid include: gifts or donations of food commodities by governments, international organizations (principally WFP) and private, voluntary or non-governmental organisations (NGOs); monetary grants tied to food purchases; and sales and loans of food commodities on credit terms with a repayment period of three years or more (FAO, 1992: 7–9 and Annex F).
5 Moving from partial to full market liberalisation, and defining any discount from a much higher fully liberalised commercial international free market price for food commodities as 'food aid', the true volume of food aid was estimated to be of the order of 40–50 million tons, significantly higher than the statistically recorded level of 12.6 million, in 1994 (Shaw and Singer, 1995: 6). For the United States alone, in 1993 agricultural commodity exports under credit, guarantee and export enhancement programmes totalled US $7.4 billion (three times the value of agricultural exports designated as 'food aid' at US $2.4 billion) and 18 per cent of the value of total US agricultural exports (USDA, 1995).
6 The signatories (and their annual commitments in wheat equivalent) to the 1999 FAC include: Argentina (35,000 tons); Australia (250,000); Canada (420,000); European Community and its member states (1.32 million tons); Japan (300,000 tons); Norway (30,000 tons); Switzerland (40,000 tons); United States (2.5 million tons).
7 A variant to triangular transactions used by the United States has been 'triangular operations', whereby food commodities provided by the US have been exchanged in a developing country for other commodities that have been shipped to another developing country, where they have been used as food aid.
8 WFP's General Regulations were amended in 1992 to allow it to provide humanitarian relief assistance to a country at the request of the UN Secretary-General instead of waiting for a request from a national government, which might never come (WFP, 1993).
9 Coordination has been established between some of these bodies, for example, memoranda of understanding have been signed between WFP and UNHCR and between WFP and UNICEF governing their working relationships. But overall coordination among these bodies still requires strengthening.
10 This special programme takes into account the research and policy work of the International Food Policy Research Institute (IFPRI) on famine prevention in Africa (Braun, 1991).

References

Benson, C. and Clay, E. (1998a) 'Additionality or diversion? Food aid to Eastern Europe and the former Soviet Republics and implications for developing countries', *World Development*, 26 (1): 31–44.

Benson, C. and Clay, E. (1998b) *The Impact of Drought on Sub-Saharan African Economies: A Preliminary Examination*, World Bank Technical Paper no. 401, Washington, DC: World Bank.

Boutros-Ghali, B. (1992) *An Agenda for Peace*, New York: United Nations.

Braun, J. von (1991) *A Policy Agenda for Famine Prevention in Africa*, Food Policy Report, Washington, DC: International Food Policy Research Institute.

Clay, E. and Benson, C. (1990) 'Aid for food: acquisition of commodities in developing countries for food aid in the 1980s', *Food Policy*, February: 27–43.

Clay, E. and Singer, H.W. (1985) *Food Aid and Development: Issues and Evidence. A Survey of the Literature since 1977 on the Role and Impact of Food Aid in Development Countries*, WFP Occasional Paper no. 3, Rome: World Food Programme.

Clay, E., Pillai, N. and Benson, C. (1998) *The Future of Food Aid: A Policy Review*, London: Overseas Development Institute.

FAO (1992) *Principles of Surplus Disposal and Consultative Obligations of Member Nations*, Rome: FAO.

FAO (1996) *Food, Agriculture and Food Security: Developments since the World Food Conference*, World Food Summit Technical Background Document, Rome: FAO.

GATT Secretariat (1994) *The Results of the Uruguay Round of Multilateral Trade Negotiations: The Legal Text*, Geneva: General Agreement on Tariffs and Trade.

Harvey, D.R. (1998) 'The US Farm Act: "fair" or "foul"? An evolutionary perspective from east of the Atlantic', *Food Policy*, 23 (2): 111–21.

International Food Policy Research Institute (IFPRI) (1998) 'The changing outlook for food aid', News and Views, November, Washington, DC: International Food Policy Research Institute.

International Grains Council (IGC) (1999) *Food Aid Convention, 1999*, London: International Grains Council.

Konandreas, P. and Greenfield, J. (eds) (1996) 'The implications of the Uruguay Round for developing countries', *Food Policy*, 21 (4/5), Special Issue.

Konandreas, P. and Greenfield, J. (1998) 'Policy options for developing countries to support food security in the post-Uruguay Round period', *Canadian Journal of Development Studies*, xix (Special Issue): 141–59.

Maxwell, S. and Shaw, D.J. (1995) 'Food, food security and UN reform', *IDS Bulletin*, 26 (4): 41–53.

Mellor Associates (1992) *Food for Development in a Market-oriented World*, Washington, DC: John Mellor Associates for World Food Programme.

Messer, E., Cohen, M.J. and D'Costa, J. (1998) *Food for Peace: Breaking the Links between Conflict and Hunger. Food, Agriculture and the Environment*, Discussion Paper 24, Washington, DC: International Food Policy Research Institute.

Overseas Development Institute (ODI) (2000) *Reforming Food Aid: Time to Grasp the Nettle?*, ODI Briefing Paper no. 1, London: Overseas Development Institute.

Pinstrup-Andersen, P. (ed.) (1988) *Food Subsidies in Developing Countries: Costs, Benefits and Policy Options*, Baltimore, MD: Johns Hopkins University Press for the International Food Policy Research Institute.

Pinstrup-Andersen, P. and Pandya-Lorch, R. (1998) 'Assuring a food-secure world in the 21st century: challenges and opportunities', *Canadian Journal of Development Studies* xix (Special Issue): 37–54.

Puetz, D., Broca, S. and Payongayong, E. (1995) *Making Food Aid Work for Long-term Food Security: Future Directions and Strategies in the Greater Horn of Africa*, Proceedings of a USAID/IFPRI Workshop, Addis Ababa, Ethiopia, March, Washington, DC: International Food Policy Research Institute.

Relief and Development Institute (RDI) (1987) *A Study of Triangular Transaction and Local Purchases in Food Aid*, WFP Occasional Paper no. 11, Rome: Relief and Development Institute, London for World Food Programme.

Relief and Development Institute (RDI) (1990) *A Study of Commodity Exchanges in WFP and other Food Aid Operations*, WFP Occasional Paper no. 12, Rome: Relief and Development Institute, London, for World Food Programme.

Reutlinger, S. (1999) 'From "food aid" to "aid for food": into the 21st century', *Food Policy*, 4 (1): 7–15.

Rosen, S. (1989) *Consumption Stability and the Potential of Food Aid in Africa*, Economic Research Service Staff Report no. AGES 89–29, Washington, DC: United States Department of Agriculture.

Saran, R. and Konandreas, P. (1991) 'An additional resource? A global perspective on food aid flows in relation to development assistance', in E. Clay and O. Stokke (eds), *Food Aid Reconsidered: Assessing the Impact on Third World Countries*, London: Frank Cass.

Seaman, J. and Rivers, J. (1988) 'Strategies for the distribution of relief food', *Journal of the Royal Statistical Society*, 151 (3): 464–72.

Shaw, D.J. (1983) 'Triangular transactions in food aid: concepts and practice. The example of the Zimbabwe operations', *IDS Bulletin*, 14 (2): 12–24.

Shaw, D.J. (1991) 'Food aid for Africa in the 1990s', *Food Policy*, 16 (6): 431–5.

Shaw, D.J. (1996) 'Development economics and policy: a conference to celebrate the 85th birthday of H.W. Singer', *Food Policy*, 21 (6): 561–7.

Shaw, D.J. (1998) 'The World Food Programme: linking relief and development', in D. Sapsford and J. Chen (eds), *Development Economics and Policy: The Conference Volume to Celebrate the 85th Birthday of Professor Sir Hans Singer*, London: Macmillan.

Shaw, D.J. (1999) 'Multilateral development cooperation for improved food security and nutrition', in U. Kracht and M. Schulz (eds), *Food Security and Nutrition: The Global Challenge*, Munster: Lit Verlag and New York: St Martin's Press.

Shaw, D.J. (2000) *The United Nations World Food Programme and the Development of Food Aid*, Basingstoke, Hants.: Macmillan.

Shaw, D.J. and Clay, E. (1998) 'Global hunger and food security after the World Food Summit', *Canadian Journal of Development Studies*, XI (Special Issue: 55–76.

Shaw, D.J. and Singer, H.W. (eds) (1988) 'Food policy, food aid and economic adjustment', *Food Policy*, 13 (1), Special Issue.

Shaw, D.J. and Singer, H.W. (1995, 1996, 1998) 'A future food aid regime: implications of the Final Act of the GATT Uruguay Round', Discussion Paper no. 352, Institute of Development Studies. See different versions of this paper in: *Food Policy*, 21 (4) 1996; and H. O'Neil and J. Toye (eds), *A World Without Hunger?*, London: Macmillan for UK Development Studies Association, 1998.

Singer, H.W. (1991) 'Food aid and structural adjustment in sub-Saharan Africa', in E. Clay and O. Stokke (eds), *Food Aid Reconsidered: Assessing the Impact on Third World Countries*, London: Frank Cass.

Singer, H.W. (1994) 'Two views on food aid', in R. Prendergast and F. Stewart (eds), *Market Forces and World Development*, London: Macmillan for UK Development Studies Association.

Thomas, M. and Maxwell, S. (1989) *Food Aid to Sub-Saharan Africa: A Review of the Literature*, WFP Occasional Paper no. 13, Rome: World Food Programme.

Toole, M. (1993) 'Protecting refugees' nutrition with food aid', in *Nutrition Issues in Food Aid*, ACC/SCN Symposium Report, Nutrition Policy Discussion Paper no. 12, Geneva: UN Administrative Committee on Coordination, Subcommittee on Nutrition.

United Nations (UN) (1945) *Charter of the United Nations*, New York: United Nations.

United Nations (UN) (1968) *Inter-agency Study on Multilateral Food Aid*, Document E/4538, New York: United Nations.

United Nations (UN) (1991) *Strengthening of the Coordination of Humanitarian Emergency Assistance of the United Nations*, UN General Assembly resolution 46/182, adopted on 19 December, New York: United Nations.

United Nations (UN) (1992) *Triennial Policy Review of the United Nations Development System*, UN General Assembly resolution 47/199, adopted on 22 December, New York: United Nations.

United Nations (UN) (1996) *United Nations System-wide Special Initiative on Africa*, New York: United Nations.

United Nations (UN) (1997) *Renewing the United Nations: A Programme for Reform*, Report of the UN Secretary-General to the Special Meeting of the UN General Assembly on UN Reform, New York: United Nations.

UNCTAD Secretariat (1994) *The Outcomes of the Uruguay Round: Support Papers to the Trade and Development Report 1994*, New York: United Nations.

United States Department of Agriculture (USDA) (1995) *Agricultural Export Programmes: Background for 1995 Farm Legislation*, Agricultural Economic Report no. 716, Washington, DC: United States Department of Agriculture.

World Bank (1988) *The Challenge of Hunger in Africa: A Call to Action*, Washington, DC: World Bank.

World Bank (1989) *Sub-Saharan Africa: From Crisis to Sustainable Growth*, Washington, DC: World Bank.

World Bank (1997) *Status Report on Poverty in Sub-Saharan Africa: Tracking the Incidence and Characteristics of Poverty*, Washington, DC: World Bank.

World Bank and World Food Programme (WFP) (1991) *Food Aid in Africa: An Agenda for the 1990s*, Washington, DC: World Bank and Rome: World Food Programme.

World Food Programme (WFP) (1978) *A Survey of the Studies of Food Aid*, by H.W. Singer, Document WFP/CFA: 5/5C, Rome: World Food Programme.

World Food Programme (WFP) (1987) *The Management of Funds Generated by Food-assisted Projects*, Document WFP/CFA: 23/5 Add. 2, Rome: World Food Programme.

World Food Programme (WFP) (1989a) *Food Aid Triangular Transactions and Local Purchases: A Review of Experience*, Document WFP/CFA: 27/INF/3, Rome: World Food Programme.

World Food Programme (WFP) (1989b) *Anti-hunger Strategies of Poor Households and Communities: Roles of Food Aid*, Document WFP/CFA: 27/P/INF/1 Add. 1, Rome: World Food Programme.

World Food Programme (WFP) (1990) *A Review of WFP and Bilateral Food Aid Commodity Exchange Arrangements*, Document WFP/CFA: 29/P/INF/2, Rome: World Food Programme.

World Food Programme (WFP) (1992) *Disaster Mitigation and Rehabilitation in Africa*, Document CFA:34/P/7-B, Rome: World Food Programme.

World Food Programme (WFP) (1993) *Basic Documents for the World Food Programme*, 5th. edn, *General Regulations, Part D. Procedures. Eligibility for Assistance*, Rome: World Food Programme.

World Food Programme (WFP) (1994) *Food for Development*, Beijing: China Agricultural Press.

World Food Programme (WFP) (1995) *Food Aid for Humanitarian Assistance*, Proceedings of the United Nations World Food Programme Africa Regional Seminar, Addis Ababa, Ethiopia, February, Rome: World Food Programme.

World Food Programme (WFP) (1999) *Enabling Development*, Document WFP/EB. A/99/4 – A, Rome: World Food Programme.

World Food Programme and African Development Bank (WFP/ADB) (1986) 'Food aid for development in sub-Saharan Africa', seminar held at ADB Headquarters, Abidjan, Côte d'Ivoire, Rome and Abidjan: World Food Programme and African Development Bank.

12 Rural poverty reduction in Africa

Strategic options in rural development

Deryke Belshaw

12.1 Introduction

Of the forty-eight sub-Saharan African countries which were member of the United Nations in 1997, thirty-one possessed consistent sets of national statistics and had more than half of their population classified as rural dwellers (World Bank, 1999). The combined population of these thirty-one countries accounted for nearly 500 million of the estimated 600 million people living in sub-Saharan Africa. However, although rural development had been adopted by the World Bank as a poverty-focused project-level intervention in 1973, it was as rapidly abandoned by the Bank and several other major donors in the early 1980s in the context of the second OPEC oil price shock and other associated adverse international changes. After two decades of structural adjustment, the second of which saw the Bank's 'New Poverty Agenda', the role of the rural sector in development and poverty reduction strategies remains unspecified. On the questions of poverty causality and the choice of poverty-reducing development strategies there is little agreement.

During the 1990s the dominant group of multi- and bilateral donors (which included the World Bank, the IMF, USAID and UK's ODA/DfID) perceived the open growth-oriented strategies pursued successfully by the newly industrialising economies (the NIEs) as a universally applicable model, the benefits of which could be brought to less-successful countries through the adoption of appropriate strategies of the structural adjustment programme (SAP) type. These were typified by macroeconomic stabilisation measures (elimination of macroeconomic price distortions) and structural reforms involving market liberalisation, privatisation of public enterprises and significant reductions in the developmental role of the state. Ensuing high rates of economic growth were expected to benefit the poor through 'trickle down' or 'spread effects' operating in factor and product markets. Directly poverty-focused investment should be limited to employment-based safety nets and to the development activities of non-governmental organisation (NGOs) as set out in the World Bank's 'New Poverty Agenda' (see World Bank, 1990; Wilmshurst *et al.*, 1992; Lipton and Maxwell, 1992). According to this view, efficiency need not be sacrificed in the pursuit of equity.

This position has been strongly contested, however, on three main grounds: (a) in the light of evidence that the state did play a decisive early role in the transformations of the NIE economies; (b) on the lack of evidence that the economic and institutional preconditions for successfully emulating NIE performance are yet in place in the poorest developing countries; and (c) robust evidence that in the latter countries large concentrations of poor people are bypassed by market forces and that the resulting socio-political instability counters or reduces the rate of market-led growth elsewhere.

At the centre of this debate is the following question: Can deliberately rural poverty-focused development strategies – that is, appropriate rural development strategies – be designed and implemented which will combine positive gains in equity without sacrificing economic growth, achieving a win–win outcome? Many development agencies were still committed to such an approach in the 1990s (see, for example, IFAD 1994; Commonwealth Secretariat, 1993;GTZ, 1993; Netherlands Development Cooperation, 1992, and most, if not all, development NGOs). But rural development itself as a poverty alleviation strategy has a mixed record starting with the showcase integrated projects of the 1960s (see Lele, 1975) and their widespread adoption in the 1970s, especially by the World Bank (see World Bank, 1975).

The interpretation of the record of this 'first generation' of interventions and an assessment of the currently dominant paradigm are the primary tasks of this chapter. Its purpose is to provide a succinct overview of the rural development record and the accompanying debate. In discussing this agenda it is hoped that the need for the use of a broad framework of economic, social and technical analysis will become apparent. This chapter first examines the important relationships between rural development and the macroeconomic dimension. Section 12.3 identifies the main approaches employed in the 'first generation' period (1965–90) and section 12.4 critically examines the highly decentralised approach to poverty alleviation that is in fashion at the turn of the century. The main implications for policy are summarised in the final section.

12.2 Macroeconomic aspects of rural development

Keith Griffin's (1989) seminal study of the empirical country-level experience of economic development identified six alternative strategies which had been put into effect for periods long enough to evaluate their impacts. While rural development was not amongst them, the broader category labelled 'redistributive strategies', with the key objective of increasing equity through achieving a direct impact on the poorest, can accommodate it alongside similar strategies, such as 'basic needs provision' and 'redistribution with growth'. Nevertheless, suspicion is widespread amongst economists that a rural development strategy is pure populism, that is, it panders to the political demands of a large segment of the (voting) population for the immediate satisfaction of wants, through increased consumption of scarce private and public goods. Certainly much of the 'rural' literature emanating from other cooperating disciplines strengthens that impression by neglecting to relate the design of a rural development strategy to an economy's long-term transformation path.

At least six possible ways in which rural poverty alleviation strategies may assist long-term economic growth can be identified. First, in situations where rural poverty can be alleviated in a cost-effective way by improving access by the poor to underutilised parts of the natural resource base to produce larger marketed surpluses using more productive technology and/or higher value products, the familiar neoclassical balanced growth model is relevant (Johnston and Mellor, 1961). This strategy usually implies the implementation of some combination of land reform, land tenure codification, assisted land settlement, profitable technological innovation packages, improved marketing arrangements and/or rural physical infrastructure.

Second, where 'urban bias' has prevailed in the past (Lipton, 1977), a positive redressing of the imbalance in productive investment and social infrastructure provision could contribute to overall national efficiency and equity objectives simultaneously.

Third, in a more constricted environment, a low-income economy may be inhibited from growing 'normally' through the switchback effects of domestic food supply instability (typically dominated by fluctuating surpluses from poor farm households). Diversions of foreign exchange into food imports, foreign financial aid into food aid and national budget expenditures into consumer food subsidies are typical hazards to be avoided. Improved food security would help the poorest families directly and reduce the welfare diversions of resources from economic investment and growth.

Fourth, continued neglect of the non-monetary subsistence sector under rapid population growth can lead to accelerating natural resources degradation and still greater diversion of investible resources into relief and re-employment activities (Cleaver and Shreiber, 1994; for an Ethiopian case study, see Constable and Belshaw, 1989). Reducing these contingencies in a cost-effective fashion will be beneficial to economic growth elsewhere in the economy.

Fifth, while in extreme degradation situations formerly productive areas can only be abandoned, in many apparently 'low-potential' areas, severe neglect of technological and infrastructural provision in the past may allow cheap labour-intensive investments, for example, in micro-irrigation, agro-forestry or feeder roads, to create site-specific comparative advantage in tradables production, thus contributing positively to the growth objective as well as to poverty reduction.

Sixth and finally, diversion of resources into rural poverty alleviation could be justified on growth grounds by reducing future social welfare expenditures. For example: (a) avoiding excessive rural–urban migration rates which would have increased urban infrastructure and agglomeration costs (additional commuter transport costs, air and water pollution costs and so on); (b) lower levels of fertility in the rural population, that is, reducing the pension-substitute demand for children by poor rural parents; and (c) raising the capacity of the rural economy to share public social service costs and to fund civil society institutions' welfare activities, reducing the net burden on public expenditure and the risk of excessive rates of inflation across the economy as a whole. More empirical studies and carefully monitored pilot projects are required in these areas.

In most situations, therefore, an appropriately designed rural development strategy should not only be capable of making positive contributions to both the efficiency and equity objectives, but also be able to reduce the rate of degradation of, or enhance the protection of, renewable natural resources. At the core of the strategy, there must be food security-improving and/or income-generating activities utilising the comparative advantages deriving from poor people's low subsistence costs and, in the case of the younger generation at least, their keen motivation to create improved livelihoods and life chances. The implication of this discussion is that a strategy with a welfarist or 'hand out' emphasis, quite apart from facing problems of financial non-sustainability or continued donor/government dependency, will be most unlikely to deliver efficiency benefits to the economy as a whole.

In practice, the size and structure of a rural development strategy that an economic advisory mission could recommend to the government of a given country at a particular time are matters for pragmatic judgement rather than generalised assertions. But such advice should reflect an ability to identify and design interventions that are likely to contribute to economic growth and environmental objectives as well as to rural poverty alleviation *per se*.

12.3 Types of rural development interventions: lessons from the First Generation

Table 12.1 summarises eight approaches to rural development which have been widely adopted or advocated since the 1950s. Various kinds of sectoral or land reform activity (redistribution of land, land tenure codification/modification, land resettlement schemes, irrigation development, assisted rural migration and so on) have been omitted from the table

Table 12.1 Major alternative instruments for implementing rural development strategies

Strategy emphasis or key instrumentality	Agencies particularly associated with strategy	Key planning concept(s)	Typical plan components
1 Self-help community development	Ministries and departments of community development; NGOs	Local participation and small group self-help with technical and in-kind assistance	Small-scale production and social service activities outside main sectoral programmes
2 Rural development funds	Rural Development Ministries or local governments	Self-help and assistance in kind	Community infrastructure or production activities identified by rural communities or their leaders
3 Administrative area-based public investment plans	Government departments in provinces, districts, etc. horizontally coordinated by administrative officers	Project 'shopping lists', physical plans	Government buildings, roads, water supply, power, land settlement, etc.
4 Local-scale rural 'equity institutions'	Governments with strong rural equity ideology (for example, Tanzania 1967–92)	Economies of scale; cooperation; mechanisation; modernisation; mobilisation and control	Collective farms, communes, state farms, village councils, etc. sometimes incorporating diversification (non-agricultural) activities
5 Rural employment creation	ILO, WFP, UNIDO assisted	Local-level development projects to diversify rural economy and create out-of-season wage employment or permanent self-employment	Diversification projects including crafts and small-scale industry; physical infrastructure through labour-intensive methods; grants/credit for self-employment of poorest
6 Integrated Rural Development	For example, GTZ, SNV donor agency implemented or NGO assisted	Multisectoral area-based plans; horizontal integration; participation; self-determination	Multi-sectoral development plans with strong agricultural components based on family farms
7 Poverty alleviation via small farm productivity gains	IFAD assisted	'Agriculture plus' projects; target groups below US $x per capita; small farmer rationality and efficiency	Agricultural projects for small farmers, usually cash crop based, with supporting infrastructure
8 Basic needs provision	ILO initiated; NGOs	Target groups below thresholds for nutrition, income, housing, etc.; participation; self-help	Comprehensive planning with multiple sectoral targets, including social services (health, education, water, housing)

Note
ILO = International Labour Office; WFP = World Food Programme; UNIDO = United Nations Industrial Development Organisation; GTZ = Gemeinschaft für Technische Zusammenarbeit; SNV = Netherlands Development Cooperation; NGO = Non-governmental organisation; IFAD = International Fund for Agricultural Development.

because, however necessary for achieving a significant impact on rural poverty, they are single-issue interventions and have rarely been incorporated into or classified under rural development projects or programmes. Also omitted are rural micro-finance programmes, because of the lack of reliable evaluation studies to date. But the possible impact on poverty of these activities, if present or proposed, must be taken into account when designing rural development interventions and rural poverty reduction strategies. Table 12.2 provides an evaluative framework for the eight approaches consisting of eleven criteria; the subjective scoring pattern can be replaced or modified by scores based on alternative experiences.

Tables 12.1 and 12.2 summarise in succinct form the essential features of the eight major approaches to rural development focusing on rural poverty reduction outside the traditional single-sector interventions. A few additional comments relating to their objectives and main strengths and weaknesses may be useful on each of them:

1   Self-help community development: this initial public sector approach aimed to mobilise cheap village labour on a self-help basis for additional activities which could contribute to subsistence and cash-based livelihoods. Craft work, vegetable production and beekeeping were typical activities. Women's groups tended to predominate. But poverty was not explicitly addressed and rural elites often captured most of the benefits. Geographical coverage was 'spotty' and influence on the major development ministries was minimal. However, simplicity and smallness of scale made this approach, usually with an emphasis on community participation, popular with NGOs as in the 1990s donors channelled more aid funds through them.
2   Rural development funds: some African governments set up local development funds that were allocated by local governments or district/area development committees. These 'slush funds' tended to fall under local political control rather than meeting monitored poverty reduction targets.
3   Public investment programmes have been constructed in several African countries by local development teams or committees, sometimes comprising district or area multiyear

Table 12.2 Subjective evaluation of major alternative rural development instruments, by eleven criteria

		CD 1	RDF 2	PIP 3	RE$_q$I 4	REC 5	IRD 6	SSA$_g$ 7	BN 8
1	Poverty focus: Local level/intra-institutional	?	X	X	✓	✓	✓	?	✓
2	Poverty focus: National level/ Inter-institutional	X	X	X	X	✓	✓	?	✓
3	Decentralisation of economic planning capacity	X	?	?	?	?	✓	✓	?
4	Popular participation	✓	?	X	?	?	✓	?	✓
5	Integration and coordination across sectors	?	X	X	✓	?	✓	?	?
6	Focus on production and income gains	✓	?	X	?	?	✓	✓	X
7	Area relevance and coverage	?	✓	✓	✓	✓	?	?	✓
8	Influence on national resource allocation	X	X	X	✓	?	?	✓	X
9	Reduction in 'blue print' plans	?	✓	X	X	?	✓	?	X
10	Suitable for NGO implementation	✓	?	X	X	✓	?	✓	✓
11	Simplicity of implementation	✓	✓	✓	?	✓	?	✓	?

Note
Columns 1–8 use acronyms to refer to the eight emphases or instrumentalities shown in the first columns of Table 12.1.

plans. Poverty reduction was not a central objective. In practice, they operated as 'shopping lists' for resources allocated by ministries and/or donors operating at central government level.

4   Rural equity institutions incorporated both modernisation concepts and socialist equality principles (see Table 12.1). Their replication on a nation-wide basis could achieve greater economic equality within each local institution but, as in Maoist China, mechanisms to level up poorer regions and communities with richer ones were lacking (Table 12.2: column 4 and rows 1 and 2).

5   Rural employment creation includes both additional wage employment – usually through dry-season public works – and improving the productivity of self-employment in rural production and service enterprises, usually through credit and/or grants to access assets such as improved grades of livestock, improved seeds/plants, water pumps, stores and so on. Poverty impact is to be achieved either by self-targeting, for example for 'ungenerous' food rations, or by identifying low-income families or individuals.

6   Integrated Rural Development (IRD) originally took the form, in the late 1960s, of large but isolated area-based multisectoral projects. The World Bank's Lilongwe Project in Malawi was the best known in Africa (Lele, 1975). Poverty reduction began to be a development objective after President of the World Bank McNamara's speech in Nairobi in 1973, but many ongoing growth-oriented projects were merely relabelled as rural development projects without any change to content or resources. Area coverage depended mainly on whether sufficient donors were investing in this 'large project' approach. In principle, by managing IRD within a sub-national regional level framework, comprehensive coverage of the rural poor could be achieved.

7   Small-scale agriculture 'plus': projects with a focus on the intensification of small-scale agriculture, usually plus some physical infrastructure inputs, can target the rural poor if they have access to suitable land and water resources, appropriate technology and product markets. They may take an area-based form, for example using agro-ecological or river basin criteria, or may focus on particular crop or livestock enterprises across the country as a whole. The first variant fades into the IRD approach.

8   The Basic Needs approach adds the other components of social welfare which the poor cannot command to targeted improvements in subsistence and disposable income. Priorities are not easily agreed between competing sectoral agencies and the actual budget allocations have tended to fall far short of the needs-based financial requirements. The relevant sector agencies tended to work in a centralised and top-down manner. There is a risk with this approach that the maintenance and recurrent costs of social services may exceed the incomes of local governments and contributions of poor families (via cost-sharing arrangements). Either the level of investment achieved cannot be sustained or poorer groups will be excluded from these services (or both eventualities occur).

Significant changes in the frequency of use of each approach have occurred over time, as is summarised in Table 12.3. The major changes have been:

1   The abandonment by the public sectors of both community development (approach 1) and decentralised departmental investment plans (approach 3) as the major approaches to rural development by the end of the 1960s (Holdcroft, 1984). But community development was taken up on an increasing scale by development NGOs from the mid 1970s – especially as increasing volumes of donor funds became available in the later period.[1]

Table 12.3 Approximate peak periods of implementation for rural development instrumental options, SSA, 1950–2000

	1950s	1960s	1970s	1980s	1990s
1 CD	Public sector (partial)				
				NGOS/communities	
2 RDF				Donors/local governments/communities	
3 PIP	Public sector (nationwide)				
4 RE$_q$I			Public sector (nationwide)		
5 REC			Donors/public sector in poor areas		
					Donors/NGOs (credit to poor)
6 IRD			Donors/public sector		
					Donors/NGOs
7 SSAg		Donors/public sector			
					Donors/NGOs
8 BN				Donors/Public sector	Donors/NGOs

Key:
—— Widespread adoption
- - - - Scattered initial or residual implementation

2   The provision of non-earmarked rural development funds (approach 2) in the public sector budget has often been discredited in practice due to political or bureaucratic manipulation and corruption; for the case of Kenya's District Development Funds see, for example, Alila (1988) and Rutten (1990). In a recent phase, however, UN agencies have introduced the concept of participatory 'social' or 'local' development funds on a pilot basis in seven or eight African countries

3   The most significant change has been the almost universal abandonment of collective 'equity institutions' (approach 4) favoured in socialist countries. This is compared in Table 12.4 with the rival and now dominant 'equity measurement' strategy, made up usually from several of the other approaches listed in Table 12.1, together with any single sector activities which are also pro-rural poor. This change was heralded by the post-Mao shift in China after 1978 from the rural commune/brigade/work team hierarchy of institutions to the 'household responsibility system', that is, family smallholdings (for an assessment, see Longworth, 1989). In Africa the collapse of 'eastern block' influence and disappointing production performance have been the main causes of a similar flight from collective rural institutions

4   The use of labour-intensive public works schemes (approach 5) to generate seasonal wage employment as 'safety nets' for the rural poorest (see, for example, Gaude and Watzlawick, 1992) is an accepted component in the World Bank's 'New Poverty Agenda' (World Bank, 1990). Counterpart funds generated from food aid are being used to meet the domestic currency costs of such public works programmes, as well as their extension on to private land (and into private asset formation[2]) in the cases of micro-irrigation, water-harvesting and agro-forestry investments (see Maxwell and Belshaw, 1990; Belshaw, 1992). The provision of credit and/or grant funds to secure access to assets for rural self-employment, on the other hand, has taken a variety of forms, increasingly

Table 12.4 Two approaches to rural poverty alleviation

A: The 'equity institutions' approach		B: The 'equity measurement' approach	
Stage 1	Adopt (design) equity institutions to be replicated	Stage 1	Identify activities promising to meet needs of identified target population
2	Identify outputs generally consistent with national growth	2	Check against private profitability criterion (financial appraisal)
3	Design programme scale, content, etc. to be consistent with equity institutions (and vice-versa)	3	Identify concomitant production infrastructure investment required (if any)
4	Provide government resources to institutions	4	Check 2 and 3 against growth and equity criteria (economic and equity appraisals)
5	'Persuade' etc. private individuals to join institutions	5	Extend programme resources to: (a) Production infrastructure agencies (b) individuals willing to use them
6	(a) control: from below (b) inspection of institutions from above	6	Evaluate impact of official programmes on private income distribution
7	Reward/penalise institutional officials or change institutional design	7	Change profitability structure or change programme content or change target population

Source: Belshaw (1974)

administered by NGOs replacing public sector rural credit institutions. Credit for women has frequently received high priority here. Problems remain over the accuracy of targeting the poor ('leakage' to non-poor groups) and over credit repayment.

5   Reliance in the 1970s on 'integrated rural development *projects*' (approach 6) was adversely evaluated by the World Bank and other agencies in the aftermath of the international shocks of the late 1970s and early 1980s (Blackwood, 1988; World Bank, 1988). The reliability of these evaluations is discussed in section 12.4 of this paper. An unnecessary source of confusion was provided by the word 'integrated'. This should more accurately be taken to mean 'comprehensively multisectoral' (Livingstone, 1979; European Commission, 1993). It often resulted in excessively complex projects which were difficult to implement effectively. The alternative view that 'integrated' should refer primarily to the *analysis* of poverty causation and to the *identification* and *design* of problem solutions, that is, to the processes of rural development *strategising* (Belshaw, 1977), has also found support (for example, Conyers *et al.*, 1988; Netherlands Development Cooperation, 1992). This view justifies, for example, a series of single-sector projects (including approach 7) introduced sequentially, each clearly focused so as to alleviate a set of layered constraints;

6   Finally, 'basic needs' (approach 8; ILO 1976) rapidly ran into the clash between rising demand for public expenditures of a welfarist type based on redistributive taxation on the one hand, and falling public sector revenues as a result of both international shocks and the resulting SAPs on the other. Nevertheless, in the 1990s development NGOs increasingly took up the role of provider to rural communities of missing social service infrastructure (primary education and primary health care buildings and equipment, and domestic water supply, typically).

Overall, the main choices for future rural development strategies lie amongst approaches 5, 6 and 7 in Table 12.1, with smaller contributions coming from NGOs espousing approaches

1 and 8. A few NGOs have implemented area-based projects of types 6 and 7, but this often requires closer cooperation with governments and official aid donors than is acceptable to most NGOs.

12.4 The new rural development paradigm: an assessment

The approach to assisting rural development followed by several major donors – the World Bank, USAID, the UK's ODA/DfID, and FAO, in particular – underwent significant change in the course of the 1990s. Earlier public sector project-based approaches have been rejected in favour of an emphasis on grassroots level activity in support of local or peoples' organisations and NGOs. 'Empowering the rural poor', long a slogan amongst NGOs, has become part of the official terminology of major donors.

The New Poverty Agenda (NPA) has been systematically developed by the World Bank in a series of documents elaborating and operationalising the ideas originally advanced in the *World Development Report* of 1990 which was entitled *Poverty*. This has succeeded in injecting an equity focus into the efficiency orientation of structural adjustment programmes with minimal change in the composition of the latter. In essence, the NPA adds a concern with livelihood creation on a sustainable basis to the three main goals of structural adjustment, that is, sound macroeconomic policy, expansion of tradable goods production and a switch in the private sector/public sector roles in economic development in favour of the former. In order to achieve accelerated livelihood creation, four components were added to the usual SAP configuration. These were:

- an emphasis on the use of labour-intensive techniques in the tradable production sectors;
- investment in human capital formation, especially in primary education and preventative medicine/public health measures;
- the provision of anti-poverty safety nets, as far as possible on employment-based lines (public works offering unskilled jobs at self-targeting wages);
- transfer of donor funding to those NGOs which are implementing poverty alleviating micro-projects at local community level.

As Lipton and Maxwell (1992) pointed out, this left a number of gaps and weaknesses in the poverty alleviation agenda, reproduced here in Table 12.5. Of particular concern in achieving an adequate magnitude of rural development activity is the de-emphasis on labour-intensive and anti-poverty measures at sector and project level (rows 3 and 7). De-emphasising large area-based projects, as previously pursued by major donors in the 1970s, in favour of a radical decentralisation to the level of localised community-based institutions, raises a number of important questions:[3]

- How reliable a guide to the design of future rural development strategies can be obtained from review of the experience of poor countries, especially those in sub-Saharan Africa, in the 1980s? And how sound were the evaluation methodologies?
- What is new about a highly decentralised set of development activities undertaken at grass-roots level? Even if NGOs have replaced the state, has not the wheel merely returned full circle to the community development approaches of the 1950s and 1960s, with their well-documented weaknesses?
- While 'empowerment' of the poor at local level may be a necessary condition for rural poverty alleviation to occur, will it be a sufficient condition when, so often, adverse

Table 12.5 Strengths and gaps in the new poverty agenda

Strengths	Gaps and weaknesses
1 A new emphasis on livelihoods, livelihood strategies and sustainability	1 'Good governance' and relations between communities, NGOs, state and the private sector
2 Practical procedures to measure and develop poverty profiles	2 Relieving external barriers to labour-intensive growth
3 Linkage of poverty reduction to labour-intensive growth	3 Issues of labour-intensive growth in agriculture: farm size, organisation and land reform; rural credit, 'resource-poor' areas; agricultural research, especially for stagnant areas; 'fallacy of composition'
4 Emphasis on potential of 'human capital' for poverty reduction	4 Reducing urban and off-farm poverty
5 Analysis of safety-nets against shocks	5 Content (and cost) of pro-poor health education, and population policies
6 A realistic approach to targeting	6 The thorny issue of redistribution
7 Awareness of sustainability issues	7 Anti-poverty strategies at project level
8 A balanced approach to the role of the state	8 Gender issues

Source: Lipton and Maxwell (1992)

macro-policy, missing sectoral programmes and inadequate physical infrastructure are contributory causes of the persistence of rural poverty?

- Since (a) there are many linkages and externalities between one local community and its neighbours, and between them and the nearest urban service centres, and (b) market forces which could stimulate investment in shared physical infrastructure are weakest in regions where the poorest are concentrated, is there not a need to strengthen public sector capability at this middle level: the region, the district, the province?
- Why should local institutions attempting to meet the pressing needs of poor members of the current generation place much weight on the protection of the environment on behalf of generations yet to be born? Will the state not have to facilitate a set of location-specific decisions – as well as play a general policy role – in such environmental conflict of interest situations?

It seems necessary to explore these questions to ascertain how far the dominant approach to rural development is truly learning from experience or is being blown off-course by the winds of development ideology (populist, nationalist and anti-statist currents being prevalent across the 1990s).

In relation to the first question raised above, in 1979 FAO convened the World Conference of Agrarian Reform and Rural Development (WCARRD) which summarised the public sector project approach to rural development. But that same year ushered in the negative macroeconomic, trade and indebtedness impacts of steep energy price increases and global recession which seriously affected the majority of poor developing countries throughout the 1980s. Many of the existing rural development and small-scale agriculture projects were undermined by adverse domestic terms of trade or the collapse of public sector budgets, rather than by internal design weaknesses.[4] Since careful counter-factual analyses, of the 'with and without project' type, have rarely been made, evaluations have often drawn 'wrong sign' conclusions by ignoring the negative effects of the external variables. It is a truism that in Africa the late 1970s were the least favourable time since the Second World War to introduce a strongly equity-focused programme. This was the period when increasing rates of economic growth, from which the bulk of resources and jobs for the poor had to be redistributed, went into reverse. But a closed attitude to the case for restarting projectised rural development is built upon the false belief that this approach has been a proven failure on the grounds of the evidence of the 1980s.

The second question suggests both differences and parallels between the present donor patronage of large numbers of openly competitive national and international NGOs and the fashion for public community development departments thirty years previously. Similar doubts arise about (a) the limited capacity of the NGO sector to 'scale-up' to handle sector-wide or region-wide problems; (b) the doubtful commitment of many individuals and organisations to sustain the lengthy effort – 5–15 years – required in most rural development activity, as opposed to short-term emergency/relief work; (c) innate conflicts between the more 'value-driven' NGOs over the priority of different ends and means; (d) the lack of enthusiasm by activists for investment in information systems and monitoring and evaluation procedures to ensure progress up a 'learning-by-doing' curve.

The third issue concerns situations where local community empowerment is not sufficient to resolve a poverty-perpetuating situation. This usually relates to a local dearth of sustainable income-generating opportunities. This may be due, for example, to lack of appropriate technology, to missing physical infrastructure (irrigation, transport and so on), to persistently unfavourable domestic price terms of trade or to individuals' lack of entitlement to productive assets. National and sectoral policy frameworks, and more complex strategic mixes of policies, projects, programmes and institutional capacity building, often need to be designed and implemented to secure the required improvement. A strong opportunity for rural poverty alleviation is usually present in smallholder agriculture, whether the upstream and downstream linkages are under cooperative, farmer-association or contract farming arrangements. Where agricultural sector policies and projects are missing from the poverty alleviation strategy, however, rural development often degenerates into non-sustainable welfarist or 'hand-out' approaches where the beneficiaries are 'disempowered' and are prey to the dependency syndrome. The archaic term 'social analysis', as something distinct from economic analysis, is sometimes used to protect this separation of function. Current economic analysis, however, has developed techniques to incorporate equity as well as efficiency and environmental objectives. More importantly, locking poverty alleviation into an economics-free 'social zone' is a recipe for minimising the influence of equity-focused proposals in the key economic policy-making arenas.

The fourth question highlights the risk of isolating through extreme decentralisation the rural periphery from the centre of the national economic and political processes. One FAO discussant (Meliczek, 1993) referred to 'the need for a strong government commitment to rural development, preferably in the form of a national plan'. While such commitment is important, the presence of strong working links between central ministries and local development agencies is not guaranteed by rhetoric in the form of a national plan document. Even in medium-size countries, a central ministry is unable to comprehend the key features of site-specific problems or opportunities in the rural areas; blanket policy making prevails. The case for middle-level institutional capacity which can reinforce local initiatives and relay policy reforming information upwards can only be mentioned here; space precludes detailed discussion (for a recent overview of this area, see Belshaw, 2000). Certainly, the successes of many of the state and district-level rural programmes in India, such as Operation Flood and the guaranteed employment schemes, should provoke more thought on the choice of level at which to site different poverty alleviation functions.

Last, there is a growing recognition of the often very local problems of environmental degradation, especially of the soil, water and energy resources on which the majority of the rural poor subsist. Here, the main need is to raise the productivity of subsistence agriculture *at the same time* as reducing the rate of degradation of the resource base. The market can see no profit from this, while the traditional engineering and timber monoculture practices espoused by many NGOs throw increased costs on to the poor through reducing resource productivity in the short term. Fortunately, potential is usually present for achieving both aims through agro-forestry, micro-irrigation and other intensified cropping practices (green manuring, intercropping and so on). The (often-overlooked) pay-off incentive for undertaking such investments is avoiding the future cost – to national governments and donors – of emergency relief or alternative employment creation for people whose current rural livelihoods are damaged or destroyed through environmental degradation. This non-market perspective is consistent with an ongoing state-led role in identifying preventative investment projects through 'middle-down' development initiatives that take place simultaneously with sound 'bottom-up' decision-making processes.

12.5 Conclusions

There appear to be strong grounds for viewing the current approach to poverty alleviation and rural development, reflecting a 'rolling back' of the responsibilities of state agencies, as, to use the title of a study by Tony Killick (1989), 'a reaction too far'. The fear is that the next set of evaluations of the impact achieved by the new approach will be disappointing due to (a) NGOs' partial and 'spotty' coverage of still growing African rural poverty; (b) its inability to address causes of poverty located at the national, sectoral and regional levels of economy and society; and (c) its failure to resolve situations where the short-run survival needs of the rural poor for food and energy are damaging or even destroying the natural resource bases of their livelihoods.

This is not to advocate an unthinking return to the approaches of the 1970s. Greater simplicity of design, public accountability and participatory engagement with intended beneficiaries are high priorities if rural development project performance is to achieve its potential.[5] At the same time, hitherto separate discourses on land reform, food security, rural technology, rural infrastructure, micro-finance and rural social services delivery need to be incorporated into rural development policy. This should lead to greater location-specificity in development strategy choice at national and sub-national levels, leading to more focused and simpler programmes capable of achieving a quantum improvement in the reduction of mass poverty. This requires a selectively strengthened state role and development capacity in the poorest developing countries. The New Poverty Agenda seems likely to leave unmet the needs of the majority of the rural poor and to result in a lower rate of economic growth, the risk of continuing damage to the natural environment, and continued exclusion of the poorest from education and health services where cost-sharing prevents their access.

In the task of designing development strategies for the rural sector which will achieve improved impact on growth, equity and environmental protection objectives, multi-disciplinary studies should provide appropriate analytical frameworks for diagnosis and prescription. Understanding the interrelatedness between natural resource potential and use, agricultural systems, economic variables and rural people's structures, institutions and values is a prerequisite for success. In addition, the ability to identify and prioritise the key constraints and opportunities prevailing at each pertinent level – farm household, gender, farming system, factor and product market, rural region, sector, national and international – is vital if simple and effective strategy designs are to reduce rural poverty significantly.

These strategies are quite different from centrally administered 'blueprint plans'. They allow networks of the major rural institutions from the public sector, private sector and the civil society organisations (CSOs, including NGOs, CBOs (community-based organisations) and FBOs (faith-based organisations)) to interact at appropriate decision levels within flexible guidelines and revisable impact targets.

Notes

1   NGOs implementing community development projects sometimes seem to have parallels with an episode in the cartoon strip featuring Garfield (the cat). Owner: 'Garfield, why did you destroy the sofa?' Garfield: 'It's something I happen to be good at.' Questions about the sustainability of these projects by local communities are usually answered less convincingly.
2   The use of public funds for private capital formation has been defended on grounds of externalities, in the case of natural resource protection across catchments and hillsides, and future public relief expenditures avoided in all livelihood-protecting or creating outcomes.
3   These questions were used originally to assess FAO's new rural development strategy in Belshaw (1993).
4   It is relevant to note that the World Bank's main rural development evaluation concluded that rural development project success rates had been significantly better in Asia, where international shocks were less severe, than in Africa (World Bank, 1988).
5   Despite the important improvements in conceptualising and analysing poverty causality and poverty reduction made in the World Development Report sub-titled *Attacking Poverty* (World Bank, 2001), to the present time (August 2001), rural development strategies have yet to be recognised by the major donor agencies as useful, let alone necessary, components of effective pro-poor development in sub-Saharan Africa.

References

Alila, P. (1988) 'Rural development in Kenya: a review of past experience', *Regional Development Dialogue*, 9 (2): 142–65.
Belshaw, D.G.R. (1974) 'Dynamic and operational aspects of the equity objective in rural development planning in East Africa', in V.F. Amann (ed.), *Agricultural Employment and Labour Migration in East Africa*, Kampala: Makerere Institute of Social Research.
Belshaw, D.G.R. (1977) 'Rural development planning: concepts and techniques', *Journal of Agricultural Economics*, 27 (3): 279–92.
Belshaw, D.G.R. (1992) 'The macroeconomics of counterpart funds: the case for food-for-hunger prevention in Ethiopia', *IDS Bulletin*, 23 (2): 46–9.
Belshaw, D.G.R. (1993) 'FAO's new approach to rural development: micro-activity in a strategy vacuum?', *Entwicklung und Ländlicher Raum*, 27 (2).
Belshaw, D.G.R. (2000) 'Decentralised governance and poverty reduction: comparative experience in Africa and Asia', in P. Collins (ed.), *Applying Public Administration in Development: Guideposts to the Future*, Chichester, Sussex: John Wiley.
Blackwood, J. (1988) 'World Bank experience with rural development', *Finance and Development*, December: 12–15.
Cleaver, K.M. and Schreiber, G.A. (1994) *Reversing the Spiral: The Population, Agriculture and Environment Nexus in Sub-Saharan Africa*, Washington, DC: World Bank.
Commonwealth Secretariat (1993) *Action to Reduce Poverty*, London: Commonwealth Secretariat.

Constable, M. and Belshaw, D.G.R. (1989) 'The Ethiopian Highlands reclamation study: major findings and recommendations', in D.G.R. Belshaw (ed.), *Towards a Food and Nutrition Strategy for Ethiopia*, Addis Ababa: Office of the National Committee for Central Planning.

Conyers, D. *et al.* (eds) (1988) 'Integrated rural development: the lessons of experience', *Manchester Papers on Development*, 4 (1).

European Commission (1993) *Principles of Development Design*, Brussels: European Commission.

Gaude, J. and Watzlawick, H. (1992) 'Employment creation and poverty alleviation through labour-intensive public works in least-developed countries', *International Labour Review*, 131 (1): 3–18.

Gemeinschaft für Technische Zusammenarbeit (GTZ) (1993) *Regional Rural Development: RRD Update*, Eschborn: GTZ.

Griffin, K. (1989) *Alternative Strategies for Economic Development*, London: Macmillan and Paris: OECD.

Holdcroft, L.E. (1984) 'The rise and fall of community development, 1950–65', in C.K. Eicher and J.M. Staatz (eds), *Agricultural Development in the Third World*, Baltimore, MD: Johns Hopkins University Press.

International Fund for Agricultural Development (IFAD) (1994) *The Challenge of Rural Poverty: The Role of IFAD*, Rome: IFAD.

International Labour Office (ILO) (1976) *Employment, Growth and Basic Needs: A One-world Problem*, Geneva: ILO.

Jazairy, I. *et al.* (1992) *The State of World Rural Poverty*, London: IT for International Fund for Agricultural Development.

Johnston, B.F. and Mellor, J. (1961) 'The role of agriculture in economic development', *American Economic Review*, 51: 566–93.

Killick, T. (1989) *A Reaction Too Far: Economic Theory and the Role of the State in Developing Countries*, London: Overseas Development Institute.

Lele, U. (1975) *The Design of Rural Development: Lessons from Africa*, Baltimore, MD: Johns Hopkins University Press.

Lipton, M. (1977) *Why Poor People Stay Poor: A Study of Urban Bias in World Development*, Canberra: Australian National University Press, and London: Temple Smith.

Lipton, M. and Maxwell, S. (1992) 'The New Poverty Agenda: an overview', Discussion Paper no. 306, Institute of Development Studies, University of Sussex.

Livingstone, I. (1979) 'On the concept of "integrated rural development planning"', *Journal of Agricultural Economics*, 30: 49–53.

Longworth, J.W. (ed.) (1989) *China's Rural Development Miracle, with International Comparisons*, St. Lucia: University. of Queensland Press for the International Association of Agricultural Economists.

McNamara, R.S. (1973) Speech given to the Meeting of Governors of the World Bank, Nairobi (September).

Maxwell, S. and Belshaw, D.G.R. (1990) *New Uses for Food Aid: Report of the WFP Mission to Ethiopia*, Rome: World Food Programme.

Meliczek, H. (1993) 'FAO's approaches and concepts for rural development', *Entwicklung und Ländlicher Raum*, 27 (2): 21–3.

Netherlands Development Cooperation (1992) *The Sector Programme for Rural Development*, The Hague: Operations Review Unit.

Rutten, M.M.E.M. (1990) 'The district focus policy for rural development in Kenya: the decentralization of planning and implementation, 1983–89', in D. Simon (ed.), *Third World Regional Development: A Reappraisal*, London: Paul Chapman.

Shaw, D.J. and Crawshaw, B. (1995) 'Overcoming rural poverty: thirty years of World Food Programme experience', in J. Mullen (ed.), *Rural Poverty Alleviation: International Development Perspectives*, Aldershot, Hants.: Avebury.

Simon, D. (ed.) (1990) *Third World Regional Development: A Reappraisal*, London: Paul Chapman.

Wilmshurst, J. *et al.* (1992) 'Implications for UK aid of current thinking on poverty reduction', Discussion Paper no. 307, Institute of Development Studies, University of Sussex.

World Bank (1975) *Rural Development: Sector Policy Paper*, Washington, DC: World Bank.

World Bank (1988) *Rural Development: World Bank Experience, 1965–86*, Washington, DC: World Bank.

World Bank (1990) *World Development Report 1990: Poverty*, Oxford: Oxford University Press for the World Bank.

World Bank (1993) *Implementing the World Bank's Strategy to Reduce Poverty: Progress and Challenges*, Washington, DC: World Bank.

World Bank (1999) *World Development Report 1998/99: Knowledge for Development*, Oxford: Oxford University Press for the World Bank.

World Bank (2001) *World Development Report 2000/2001: Attacking Poverty*, New York: Oxford University Press for the World Bank.

13 The role of non-governmental organisations in African development

Critical issues

Tina Wallace

13.1 Introduction

Any attempt to assess the roles and effectiveness of non-governmental organisations (NGOs) in development in Africa is difficult for a number of reasons, not least because their history has been different in each country of Africa. Some started as relief organisations and moved on to development, later including advocacy in their agendas. In others they have always worked in the field of relief and service delivery in response to wars, natural crises and a lack of effective government provision. In some contexts their role has been to meet emergency needs for water, food and shelter. In others they have worked as key service providers. In yet others they have acted as mobilisers of disadvantaged and marginalised people, assisting them to address their own development planning and make demands on the state for support. In some contexts NGOs play all these roles side by side, different NGOs meeting different needs. In addition the burgeoning of NGOs in the past 20 years compounds the enormity of the task: they have multiplied, like mushrooms, and many have grown significantly in size. The problem is exacerbated by the serious paucity of monitoring and evaluation data at every level, which inhibits the drawing of many easy and clear conclusions.

NGOs are very diverse organisations: they have played a wide range of roles in Africa and they have interacted with different political and social histories in different African countries. Their work has not been systematically researched or monitored, though thousands of individual project reports and some comprehensive programme reviews exist for several of the larger NGOs. Drawing conclusions about their relevance and impact is a high-risk strategy – whatever is said can be contradicted by particular examples. NGOs' effectiveness is subject to constant discussion within the sector as well as to scrutiny by donors.

It would be good, in an ideal world, to compare the effectiveness of NGOs with that of donors and governments. But the data needed for such an exercise does not exist. There are no overall data on spending by these different development players, nor comparative data on their ability to reach their objectives and perform their roles well. Indeed they do not necessarily agree on the nature and meaning of development. So all that can be attempted is to explore what was expected of the NGO sector in development, and how well they have been able to perform and meet at least some of those expectations. This chapter will draw on a case study from Uganda that is fairly representative of many of the issues described in the literature[1] to draw out strengths and discuss constraints affecting their performance and relevance in development.

Their roles and effectiveness in relation to relief – now called humanitarian – work would require a review of a wide range of different literature and further case studies drawn from NGO work in complex political emergencies (CPEs). These are not considered here. While there have been attempts to draw relief and emergency work closer together (Smillie, 1995, 1999, 2000), relief has involved a different set of activities, time scales and objectives and has been funded through different channels, with different requirements for reporting, monitoring and evaluation (Duffield (1994, 1999) writes about the very different imperative and purpose of relief work).

13.2 Some key differences between NGOs

There is no need to rehearse typologies of NGOs, which have been well analysed by others (Korten, 1990; Fowler, 1997; Uphoff, 1986), but it is important to highlight some of the key differentiating features of NGOs to understand the diversity of organisations being analysed. International NGOs (INGOs) are based in Europe, the US, Japan (countries of the north), registered to work in development in Africa. Some work directly on the ground, operationally – especially in humanitarian aid – others work through partners. These partners are often local NGOs (often called southern NGOs), organisations registered in their own country to work directly with local people or to do advocacy and coordination work around development issues.

SNGOs often work with more local level organisations (community-based organisations) such as women's or farmers' groups. These organisations are not part of this discussion. Most SNGOs are currently funded through external funding, from bilateral donors or from INGOs especially. This has major implications for their autonomy, the terms of their funding and the quality of their relationships with the funders affect their ways of working. INGOs have traditionally been more flexible, because many raised funds from the public in their own countries: increasingly, however, they have taken official funding, often in recent times even becoming contractors to World Bank, the UK's Department for International Development (DfID) and others. The way they raise funds clearly has a significant impact on the way they work: there are huge differences between NGOs working to tightly written and time-bound contracts and those working with local partners according to plans set and agreed between the participating NGOs.

13.3 Growth in the NGO sector

The rise in the development (and humanitarian) work carried out by the NGO sector in different countries in Africa (as elsewhere in Asia and Latin America) over the past 20 years has been very significant. While the figures are notoriously unreliable (Smillie and Helmich, 1994), in 1984–5 NGOs worldwide spent about US $3000 million, with about US $400 million of official bilateral aid being disbursed through NGOs; soon after those figures had more than doubled. Countries like the Netherlands put over 10 per cent of their annual bilateral aid budget into their NGOs, which in turn work with local NGOs on the ground. DfID increased its recorded funding for emergency work, joint funding and work through the geographical desks from £33.6 million to UK NGOs in 1987–8 to five times that amount, £161.8 million, by 1994–5. Funding has also been disbursed directly to NGOs by DfID offices based in Africa.[2]

In addition, increasing amounts of money, hard to trace because of decentralised budgeting during the 1990s, were being funnelled through NGOs in the form of contracts or accountable grants. By 1994–6 the multilaterals were also allocating funds through the NGO sector. The World Bank recently reported that 40–50 per cent of their projects had some kind of NGO involvement (cited in Van Rooy, 1998: 34). However, it needs repeating that figures on spending through NGOs, whether based on public donations or on official aid, are notoriously unreliable.

Spending money through NGOs was attractive to donors, for different reasons at different times. It is possible to see a trend whereby funding was often initially for relief funding in times of emergencies, then a shift to funding service delivery at a time when states were seen to be weak or corrupt and aid was being privatised. Subsequently NGOs were perceived to be effective at targeting the poor, and money was spent through them to increase the poverty reduction programmes of many donors. Most recently, aid has been disbursed through NGOs as part of the promotion of the good governance/democracy project, whereby NGOs are seen as organisations for promoting civic involvement and enhancing democratic processes.

The increasing number of CPEs produced a dramatic rise in bilateral and multilateral funding to international NGOs working in humanitarian and emergency aid. Public donations also rose around these emergencies, causing agencies such as Oxfam and Save the Children Fund (SCF) to expand their operations enormously following, for example, the Ethiopia famine of the mid-1980s. Second, as donor ideology moved from a position of full reliance on public sector mechanisms for promoting development to a growing belief in the weakness and obstructiveness of the state (especially in Africa) with emphasis instead on the market, donors started to privatise a lot of aid and to make use of international and, later, local NGOs. By the late 1990s, when donors started to focus on the role of civil society in development funding, NGOs became the only form of international aid from the UK to countries where the government was deemed corrupt or inefficient, including Kenya and, until the recent elections, Nigeria. Third, the more recent donor concern about the importance of democracy as a way of regulating the state, now acknowledged after all to be a key player in development (World Bank, 1998; DfID 1997), has boosted interest in funding and strengthening the local NGO sector in each country.

International NGOs thus became increasingly significant and key players in development and humanitarian work in Africa over the past 20 years, partly as a result of this massive growth in funding from official donors. In addition INGOs have become professional at fund-raising from the public through advertising, trading and selling consultancy and training skills. This has been followed by a rise in local NGOs, especially over the past 10 years, partly in response to local needs and partly because of the availability of foreign funding, as international NGOs began to work through local partners rather than working directly on the ground themselves. Most countries in Africa now have a plethora of registered local NGOs, often several hundred each compared with perhaps only a handful in the 1980s.

13.4 Rise in funding, rise in expectations

The rapid growth of the NGO sector in development and emergency work in Africa is undeniable, though the history, funding and roles they have played in each country have varied country by country. However, funding increases certainly have been coupled to a wide range of expectations from the donors and from NGOs themselves.

In relation to humanitarian relief work NGOs are expected to be swift and effective in their responses; to be impartial; to deliver goods and services in ways which will save lives immediately; and to establish a basis for longer-term work with dispossessed and displaced communities. In relation to service delivery NGOs are often expected to be able to locate in remote areas, to work closely with local people in order to ensure their ownership of the programme or project and have a commitment to contribute through the provision of labour, materials and often money. But, in addition to the necessary community development work needed to secure the involvement of the local people in the project and their contribution to start-up costs, and so to the sustainability of the project, they are also expected to work to strict timetables and technical targets set by donors. NGOs themselves want to deliver services in ways that reach the poorest and those otherwise most likely to be excluded from access. However, this is an expectation that can well contradict the need to complete projects within given, short time frames.

Expectations in relation to the achievement of poverty reduction include their ability to work closely and directly with the poor – both women as well as men – and to find genuinely participatory ways of working. Hopes at this level are that NGOs will work in partnership with local people and work in open and transparent ways. They are also expected to be (and indeed claim to be) more cost-effective than larger government and donor bureaucracies, more nimble and flexible, and with lower administrative and overhead costs. More recently, their roles in helping people to organise, building their capacities, have been seen as important for promoting democratic ways of thinking and working, providing a counterweight to the (less democratic) state. NGOs as key players in civil society are loaded with diverse and varied expectations, which differ from donor to donor as they each define the democracy project differently (Van Rooy, 1998).

Many NGOs themselves make major claims for their ability to influence key players in development, through their increasing role in advocacy. They have developed expertise in advocacy and lobbying work intended to change policies and macro-strategies that they see as negatively affecting poor people. Issues around their effectiveness in this arena (Edwards, 1993) and in who speaks for whom, who can legitimately represent the marginalised and oppressed, and who owns and controls this agenda, are contentious ones (Chapman, 1997, 1998).

The donating public has developed higher expectations and, for instance, become more wary, indeed intolerant of spending on administration and management. Yet these are essential for delivering the range of services and resources to the least advantaged and most marginalised people. The public appears to want reassurance that their money has saved lives, has transformed the day-to-day experiences of the children or communities they sponsor and has been spent quickly and honestly. There is little recognition of the causes of poverty and distress, and the complexities of tackling these and of bringing about the kind of economic and social change that would prevent the perpetuation of conflict, poverty and exclusion in the future. This lack of real understanding of development, partly due to the neglect of education by the NGOs, serves NGOs ill when they face cuts or criticisms from governments and donors. Their home constituency is often very 'underdeveloped'.

In recent years NGOs have been subject to new conditions, conditions that operate throughout the aid chain. They include demands for clear project planning with defined and measurable outputs (often presented in a logical framework or 'logframe'); rigorous monitoring and evaluation; the ability to demonstrate impact within short time periods (Wallace *et al.*, 1997; Simbi and Thom, 2000). International NGOs in their turn have imposed more and more requirements on local NGOs for accounting for the disbursement of their aid funds, and this conditionality is sometimes felt as top-down and controlling by local organisations. The debates about the reality or otherwise of partnership have become sharp, in a context where the way in which resources have been allocated have become increasingly hedged about by conditions set outside the country.[3] How these promote or inhibit local NGOs in building on their strengths, such as flexibility, working in a participatory way, pressing for greater social equality and long-term social change (for example, around gender equity or environmental sustainability), or contribute to building strong civil societies (Van Rooy, 1998), are the subject of many discussion documents (for example, de Coninck, 1992) and new research.[4]

13.5 Constraints on NGO effectiveness

13.5.1 Donor requirements

Funding plays a critical role in shaping the work of NGOs, both international and local. As this donor funding has increased, so have the requirements imposed on NGOs in the form of conditionalities. This has led to the almost universal use now of management and planning tools which have been taken from the donor community rather than developed on the basis of NGO experience. These tools, including logframes, strategic planning, impact assessment and evaluations based on logframe outputs have become a straitjacket for many NGOs and operate against their attempts to work in more participatory and open ways with local people. The impact of these tools on NGO work, and how far they inhibit or promote good capacity building, gender sensitive work, local civil society building and so on are the subject of research now on-going (Wallace *et al.*, 1997; Wallace, 2000b).

The lack of locally available funding weakens the power and autonomy of many SNGOs, especially in Africa, though a few have been able to find alternative sources of funding, or their strong links to local communities enable them to negotiate the terms of their grants from external donors. These 'differences in leverage matter as we explore the power dynamics that run through the aid industry' (Van Rooy, 1998: 2).

Reliance on donor funding also makes NGOs compete for that funding, pushing their claims above those of the next NGO. It does little or nothing to promote NGO cooperation or working together to tackle the massive problems of poverty and injustice. In addition, it forces NGOs to make exaggerated claims about their efficiency and effectiveness: often they mould their applications to changing donor fashions and make claims that they cannot realistically expect to meet. As more donor funding has become available for NGOs to fill the service provision gaps left by weak or impoverished national governments, many NGOs have become service providers. Many do provide effective services in limited geographical areas, though debates continue about issues of coverage, sustainability and whether this is the proper role for NGOs.

13.5.2 Legitimacy and representation

There are many concerns regarding whom NGOs actually represent and whence they derive their legitimacy (Edwards and Hulme, 1995). This is true for both international and southern NGOs. They are not membership organisations and few outside the church- or religious-based organisations have roots in local communities or their home countries. There are weak links in accountability to local communities, though strong links for accountability, especially financial, to donors. These issues undermine NGO claims to be working with and for the most excluded and marginalised. Although some NGOs are able to overcome this, questions remain about who is representing whom and whose development agenda is driving the work of NGOs (Crewe and Harrison, 1998).

13.5.5 North–South relations

Relationships between INGOs and southern NGOs (SNGOs) are marked by dependency (the SNGOs depend on the former for funding), lack of trust, and aid conditionality, similar to those which exist between institutional donors and INGOs. Aid conditionality has become part of these North–South NGO relations, with requirements for reporting, planning and financial controls mirroring those of the northern institution. Respect for local NGO perceptions and ways of working have often been eroded and replaced by systems of accountability and control (Simbi and Thom, 2000; CDRN, 1997).

These tensions undermine concepts of 'partnership'. Relationships remain unequal, limiting the North's learning from the South, and promoting relations of mistrust and suspicion in many cases. These limit the effectiveness of NGO work, while the fact that people are not openly working together produces internal conflicts and stresses.

13.5.4 Problems in reaching the very poor

While NGOs do try to target the very poorest and the most marginalised, obstacles exist in finding ways to work with them effectively. Short time scales, the perceived need for quick impact and the intractable nature of extreme poverty often mean that NGOs work on issues and with people that are more amenable to change (Tvedt, 1998; McClean and Ntale, 1998). An Overseas Development Institute (ODI) study found that, while NGOs work with the poor, they do not work with the poorest (Riddell and Robinson, 1992).

13.5.5 Learning: lack of good monitoring and evaluation

One of the weakest areas of NGO performance, whether northern or southern, is the lack of systematic monitoring and evaluation. This lack of learning from experience means that the same mistakes are made again and again, while listening propensity is weak because the mechanisms for listening are not developed properly. Becoming open and learning organisations still remains a distant dream for most NGOs (Surr, 1995; Davies, 1996; Edwards and Hulme, 1995; Kaplan, 1998; Fowler, 1997; Wallace *et al.* 1997; Wallace, 2000a).

13.6 NGO experience in Uganda

We can illustrate some of the strengths of NGOs and the problems and constraints they face with reference to a particular African country, Uganda, as a case study, it being a country in which NGOs have been very active. Since the end of the civil war in 1986 INGOs have poured into Uganda, while the local NGO sector has expanded very fast also. This followed the ending of armed conflict in many parts of the country and the establishment of a new government.[5]

13.6.1 Current views regarding NGOs in Uganda

Factors favourable to the expansion of NGOs locally were a government commitment to involve the citizens of Uganda in shaping their own development, the ending of conflict in many parts of the country, the availability of foreign funding for the sector, and the presence of large international NGOs, many of whom were seeking partners to work through in Uganda:

> Foreign and indigenous NGOs have flooded Uganda since the National Resistance Army stormed Kampala in 1986. The invasion of NGOs has impacted on almost every sector of Ugandan life and every region of Uganda . . . The flood of NGOs and NGO activities has produced varying degrees of both cynicism and optimism.
>
> (Dicklich, 1994: 2)

These contrasting reactions are evident in many countries of Africa. On the government side, NGOs are seen to have a clear role in service provision, in a context where services have long since broken down due to war and economic collapse, and where cuts in public spending have been required as a result of structural adjustment policies. At the same time, in cases where certain NGOs have been concerned with a political agenda around civic education, promoting civil society and human rights, they have been viewed with some suspicion:

> Years of war have created a culture of suspicion and fear, which has even placed NGO motives and activities under public scrutiny. Consequently, many NGOs, especially those that can be considered 'political' in any way, are regarded as having ulterior motives and objectives.
>
> (Ibid.: 6)

This ambivalence of the Government of Uganda (GoU) towards NGOs was evident when the NGO Registration Board was set up under the Ministry of Internal Affairs, rather than under Planning and Economic Development, in 1989 because of security concerns about NGOs. In reality the GoU lacks the resources to monitor and control NGO work, though it attempts to do so through the NGO Coordinating Office in the Office of the Prime Minister, and has refused to register the NGO's own umbrella organisation in Uganda.

Much of the funding for NGOs in Uganda, both international and national, comes from foreign aid, directly or indirectly. This funding is critical in sustaining the majority of Ugandan NGOs and, increasingly, international NGOs, now receiving lower proportions of their funding from private sources than in the past (Wallace *et al.*, 1997; Smillie and Helmich, 1994; Reality of Aid, 1997–8). The donors have a wide range of expectations of NGOs in Uganda: amongst others, that they are close to the local people and can represent their interests fairly; that they are efficient and more cost-effective than local government or private contractors; that they have the technical competence to build and deliver services; that they can mobilise women and men to participate and take ownership of projects; and that they can argue for human rights and curb the excesses of the state. Perhaps the most significant expectation is that NGOs can reach the very poor and effectively target poverty.

A recent investigation of NGOs in Uganda was carried out as part of a wider study of poverty and the impact of the Department for International Developments (DfID, previously known as ODA)[6] programme of 1987–97 on poverty alleviation. The research focus was especially on how effectively NGOs in Uganda have been able to reduce poverty. This work was inevitably hampered by the (now famous) lack of existing evaluation data on NGO projects, reflecting the known weak record of NGOs in evaluation (Surr, 1995; Davies, 1996). The paucity of data inevitably limits the conclusions that can be drawn about the impact of NGO activity on poverty, social change and gender equity.

In assessing the value and effectiveness of NGOs, it is important to gain the views of donors. Information collected from case study material and interviews with DfID and other donor staff reflecting their views is summarised in Table 13.1. While some donor (DfID) staff, especially those relating directly to NGOs, are positive about the role of NGOs in development and committed to finding ways to improve relations to support and improve the quality of their work, others in different parts of the organisation have more critical and ambivalent views about NGOs' value and role.

Table 13.1 Donor views of working with NGOs: DfID in Uganda, 1999

Strengths	Weaknesses
Can reach poor people	Often intransigent: won't relinquish their view of development
Participatory ways of working	Lack of trust in donor (DfID) views and competence
Flexible in ways of working, can change faster than government bureaucracies	Working with NGOs absorbs great amount of time because of inter-NGO differences in approach, procedures, etc.
Alternative to government	Inflexible, lack of adaptability to donor (DfID) requirements, especially when in role as contractor. 'Culture clash'
Provide good service delivery on the ground	Lack of training and experience in aspects of project planning and management
'Cost-effective' (this is debated)	Arrogance, believe their understanding and approach is better than DfID's
	Lack of transparency around other donors
	Weak monitoring and evaluation and impact assessment – poor learning systems

13.6.2 Procedural problems faced by NGOs

A strong complaint among NGOs in Uganda relates to the procedural problems they encounter in securing funding for their operations. DfID, of course, is only one donor among many supporting NGOs in Uganda, though it is quite a typical example in terms of what it funds, the complexities of its funding requirements and its expectations of NGOs. Over thirty different international agencies fund NGOs in Uganda, including bilateral agencies, northern NGOs, UN agencies, the Ford Foundation and other foundations. Many NGOs, local and international, receive funding from several donors and have to meet the many and varied demands of each donor separately. Donors have not coordinated their funding mechanisms, often using different financial years.[7] Even within one donor organisation such as DfID, there are several sources of funding with different criteria, procedures, terms and conditions. With about 20 per cent of aid funds to Uganda now passing through the NGO sector, this lack of cooperation between donors makes fund-raising by NGOs a very time-consuming process.

Table 13.2 demonstrates the complexity of the funding from one donor. To indicate the demands that this complexity places on local and international NGOs, such a table would be needed for most of the thirty or more donors active in Uganda. From discussions with a number of NGO staff in Uganda it was possible to secure a collective view of these problems, summarised here in Table 13.3. Those aspects of greatest concern to NGOs appear to be the delays in funding and donors' desires to call the tune. In addition, contracts are seen as limiting and rigid, which often prevents learning and adaptation by the implementing NGO in the course of experience.

Most local NGOs have little contact with donors and ways need to be found to develop stronger ties between donors and the local NGO sector. This may involve looking again at complex procedural demands that may be too onerous for some NGOs, while not actually delivering better projects. Most training in capacity building provided at present is logframe

Table 13.2 Procedures for NGOs to secure DfID funding, Uganda, 1999

Type of funding	Location of funding	Terms and conditions of funding
Joint Funding Scheme (JFS)	UK, East Kilbride	Proposals sent in by NGOs; have to fit within broad strategic framework of JFS; assessment by external advisers (university contractors); 50% joint funding, for between 1–3 years usually; annual reporting system; annual workshops in UK for UK-based NGOs
Direct Funding Initiative (DFI) (1994)	Kampala	Proposals submitted by NGOs in Uganda, to fall within CSP; project assessed by DFI officer – work together on proposals; sent to advisers in Kampala and Nairobi for consideration; budgeted out of sector budgets; funding 1–3 years; mid-term reviews as well as reports; training on logframes and financial systems
Emergency grants	London	Administered through programme manager in Nairobi; can disburse large sums of money fast; variable reporting systems
Contracts	Nairobi	NGOs sub-contracted to do specified work over 2–5 years; NGOs bid for the contracts; sector budgets provide funds; tight monitoring, mid-term reviews; paid retrospectively
Accountable grants	Nairobi	NGOs have contract to deliver work which DfID wants done; grant paid 'up front'; sector budgets fund; tight monitoring and evaluation; usually given through bidding process
BHC funds	Kampala	Small grants; disbursed by High Commissioner; run by British High Commission
Beijing portfolio	Kampala	Run by British Council; NGOs submit proposals under three priority areas; small grants given; BC works closely with NGOs on writing proposals; training in proposal writing

or financially based – a very narrow definition of capacity building. Community Development Resource Network (CDRN) in Uganda has made a critique of the narrow definitions of capacity building being used by donors (CDRN, 1997), finding that these ignored many essential development capacities that local NGOs and consultants already have. Alan Fowler (1998) and Allan Kaplan (1998) had similar findings in other countries.

13.6.3 *The record of NGOs in Uganda*

The development record of NGOs in Uganda is, of course, mixed. However, using a number of criteria for assessing poverty relevance, some assessment can be made.

Understanding and analysis of poverty

Some NGOs working in Uganda have made detailed analyses of poverty, which are complex and multifaceted, based on their direct involvement with communities. ACTIONAID's strategic plans (1995, 2000) and CDRN's 'Study of poverty in selected districts of Uganda' are good examples; the strategic plans of several other NGOs include analyses of poverty based on their experience of working closely with poor people. The work of local NGOs such as Action for Development (ACFODE) and Ugandan Women's Network (UWONET) builds on their experiences of working with rural and urban women in Uganda and show an

Table 13.3 Ugandan NGOs' views of donor procedures

Strengths	Weaknesses
Professional project applications, training in logframes, support provided, encourage participatory approaches in the logframe (DFI) Involve UK NGOs in sectoral debates Hands off (JFS)	Logframes too rigid, not flexible, do not allow learning from experience. Fixed indicators limit the project and can miss real impact Do not involve Ugandan NGOs Too dominating, interfering. Advisers override NGO experiences, impose their ideas on projects, insensitive, top-down (DFI, accountable grants, contracts)
Encourage participation of people in project design	Override findings of PRA and insist on their own approach
Approachable and easy to talk to	High-handed, arrogant, do not listen; patronising
Timely remittance of funds; funding often quite large and for 3 years	Funding very slow. Bureaucratic approach makes things work very slowly. Three years is still short term for the development problems being addressed
Cover administration costs	
	Decisions about projects very protracted and slow. Too many people comment and change of advisers makes this worse. Confusing and sometimes contradictory advice given
	No clear sense of direction for the development work
	Lack of clarity in contracts. Problems around who is responsible for what, who 'owns' the project, who makes which decisions. Imposition of external consultants, reviewers
	Top-down in their approach
Only a few initiatives to look at joint funding and joint procedures	Multiple donors, multiple procedures make life complex and time consuming
	No use of local consultants
	No feedback on reports submitted

Note
Muchungunzi did widespread research with NGOs in East Africa and found that all donors look the same to local NGOs, including INGOs. The latter pride themselves on being much more sensitive to local needs, but this is often not apparent to local NGOs on the ground (Muchungunzi and Milne, 1995).

understanding of gender relations and the issues that keep women in poverty. There is certainly validity in the claim that some NGOs have a better understanding of communities and their problems of poverty than the more remote agents of development–government and donors. Working through NGOs exposes donors to different perceptions and understandings about the causes and scope of poverty.

NGOs have played a key role in widening definitions of poverty to include those meaningful to poor people themselves, and in alerting the wider development community to some of the key causes of poverty. They have lobbied on a range of critical issues relating to these, including the land issue (Uganda Land Alliance, funded by DfID); debt and the impact of debt repayment on provision for the poor in Uganda (Uganda Debt Alliance,

Oxfam); and women's legal rights and their need for better laws and access to legal assistance (ACFODE, UWONET, National Alliance of Women's Organisations (NAWOU)). They have also been consistent in raising the problem of poverty and highlighting the way poverty persists, and indeed for many is worsening, in spite of Uganda's impressive economic growth performance since 1994. They have challenged trickle-down theories of development and shown that, for many communities, poverty is worsening within a surrounding context of improved economic performance (McClean and Ntale, 1997). The NGO voice in Uganda has been critical in raising issues of poverty, challenging dominant World Bank paradigms of development, and alerting others to the depth and spread of poverty in Uganda.

However, not all NGOs in Uganda take this analytical and alternative approach to defining and conceptualising poverty. As observed by de Coninck (1992: 107): 'most programmes with an economic focus are characterised by the lack of attention given to the issue of social "differentiation" resulting in some of the most needy being bypassed'.

Achievement of scale and coverage

One of the concerns of development analysts is that NGOs are fragmented, small scale, with limited coverage (Edwards and Hulme, 1995). Some have been able to 'scale up' and work on large contracts: CARE and Plan International often implement large-scale construction projects for water or clinics. Some NGOs now work across whole districts in education planning and support (International Extension College, Cambridge: IEC), while others work in all villages in sub-districts in a range of sectors (ACTIONAID, World Vision). Nevertheless, NGOs usually work alone and not in cooperation with each other, on far smaller projects than those funded through government or the private sector. Many focus on one or two venues and work more on issues of social relations and human resource development than the provision of infrastructure or large-scale and lower-cost services.

Inevitably, their impact is limited by their small geographical focus and relatively small budgets. In addition, other evidence suggests that their coverage is not always as wide as claimed. Many NGOs appear to work better with people who have some education and with men. Women are much less likely to be consulted, despite being far more involved in rural production (Kwesiga and Ratter, 1993: 43). Many NGOs are urban-based and lack the skills and organisation required to do needs assessments in participatory ways as appropriate to the rural areas (Dicklich, 1994: 13).

On the other hand, NGO reach is extended through their 'demonstration effect' and through lobbying, so that they promote approaches and attitudes way beyond their project boundaries, many of which have become parts of the wider development discourse. In Uganda these include their work on debt, HIV/AIDS, gender equity, participation in planning and implementation, and a concern with local-level involvement and activism in projects and in promoting wider sustainable social and economic change. An ODI study of many income-generating projects in Uganda in the early 1990s found that, while indeed many NGOs projects are small, they do often penetrate into areas where the government has been unable to work and have provided services and access for people previously excluded from any help (Riddell and Robinson, 1992).

A more recent study of NGOs' influence in the field of gender, and in promoting a rights-based approach to development, concluded that:

> NGOs played an important part in taking up and advancing this position [on the empowerment of women] and in examining and challenging the gender impact of policy

. . . the influence of NGOs on broader policy debates can be testified by the fact that a rights based approach is starting to filter into a number of international agencies.

(Beall and Lewis, 1999: 3)

Their influence in these areas in Uganda was apparent.

Promoting participation

Some NGOs have proven participatory skills and train their staff in the principles and techniques of Participatory Rural Appraisal (PRA). Many have staff based in the rural areas, living and working closely with the men and women in the project area. They get to know them, listen to them and interact with them. In the projects visited it was clear that staff from CARE, ACTIONAID, Busoga Trust, Vision Terudo and the Uganda Society for Disabled Children (USDC) worked closely with people on the ground and had a good understanding of the problems that they face. ACTIONAID has undertaken extensive work on participatory monitoring in Uganda (Goyder *et al.*, 1997) and Oxfam has carried out some participatory evaluations on health (Smith and Payne, 1995) and gender (Oxfam, 1998).

However, the ability to work in a participatory way with communities is not universally proven and a donor (DfID) review of the participatory approaches used by a wide range of NGOs showed a diverse range of ability in this area. While the review found that some NGOs listen to and document the voices of the poorest, in many reports the author's voice was louder than that of the community. In addition, women's voices were heard less and cycles of poverty were not well understood. Several studies were found not to be participatory, but rather extractive in their methodology. The review showed wide disparities in the competence of the NGOs and in their analytical and facilitative skills. While some NGOs have the capacity to listen and respond to rural people's voices, some do not.

NGOs certainly have the potential – and many have the commitment – to be participatory in the way they work, though constraints on staffing, skills and even donor demands (which favour an apparently well planned project over an evolving bottom-up approach) often have the effect that participation is limited. Though the skills needed to work in a facilitative and participatory way are complex and take time to learn, these are not skills usually promoted by donors.

Increasing access of the poor to resources

Many NGO projects focus on increasing access to key resources such as water or goats, new technologies, credit, health clinics or education facilities. Some projects work to find ways to overcome the barriers to access; for instance, finding ways to promote women in decision making through working with women on civic education and confidence-building; or setting up credit schemes targeted at people without collateral. NGOs may help the poor to access health provision by providing waivers of payments for health care or by developing income-generating projects to give people the means to pay for essential services.

NGO projects do work on the ground in the rural areas, often in remote or neglected areas, and they do provide increased services for poor people in different sectors, though often this provision only reaches a small percentage of the population. They also try to tackle the barriers inhibiting access by the poor, including work on issues such as attitudes to female education or the education of the disabled, confidence-building of vulnerable groups, awareness-raising around, for example, disability or HIV/AIDS. Many focus on specific groups that they define as 'in need' or vulnerable and attempt to find ways to open up resources and opportunities for them.

13.6.4 NGO sector cohesion

The sector remains fragmented, inevitably limiting its impact and its political potential to work at the level of policy in Uganda. There are coalitions around specific sectors/issues, such as the Uganda Community Based Health Care Association, the National Union of Disabled Persons in Uganda and the Uganda Debt Alliance, but overall the NGO work is not well coordinated. As observed again by Dicklich (1994: 16):

> Given the degree of coordination, competition, dependence on foreign funding for survival, the relative youth and apolitical focus of most NGOs in Uganda, the NGO sector does not presently present a strong vehicle for the development of a democratic civil society capable of pressuring the state and keeping it accountable. The GoU has prevented the formation of an official umbrella body and many NGOs still work in isolation or even in competition with each other. The patterns of donor funding, which focus on projects, do not promote cooperation or programme thinking; often they promote competition and do little to foster cohesion in the NGO sector.

Many writers have warned of the possibly damaging relationship between donors and NGOs (Fowler, 1997; Van Rooy, 1998), where lack of alternative sources of funding mean that donors can set the terms and conditions of their aid, with NGOs as pliable players in the relationship. This usually increases NGO competition and limits their ability or willingness to work together. As many international NGOs and most local NGOs in Uganda are reliant on donors within the north–south aid chain, this affects their degree of autonomy and room to manage their own business in ways they deem most appropriate. Many of these relationships have been very constraining and have exhibited more of a lack of trust between donors and NGOs and less of mutual respect and partnership

The lack of local funding weakens the power of SNGOs in Uganda. Most NGO sectors in Africa are quite weak *vis-à-vis* international players, undermining the ability of NGOs effectively to challenge and confront issues that concern them in development, including the way it is conceptualised and delivered, the continuing dominance of the neo-liberal economic agenda – the expectation that economic growth will deliver basic needs for everyone – and the lack of respect for local people and processes.

Funding has been more readily available from donors for service delivery to make up the deficiencies of the state and fill in gaps in provision. Many NGOs have become essentially sub-contractors delivering health, education, agricultural or other work which the government cannot do, because it lacks the resources and maybe even the interest and motivation. This work is necessary: for example, only 11 per cent of farmers had any extension advice in the early 1990s in Uganda; many people live far from a clinic; distances covered for water collection remain unacceptable in many areas and clean drinking water is available to less than 40 per cent of the population. But whether NGOs are the most appropriate organisations for providing large-scale service delivery is questionable. Performing this service delivery role limits the time and will that they have for work in training, civic education, confronting injustices, asserting a rights-based approach to development, challenging social exclusion and working to build a stronger civil society, roles that they may be better suited to play in the long run.

13.7 Conclusion

NGOs have risen rapidly in size and number over the past 20 years, and have played a wide variety of roles in development and humanitarian aid during this period. Donors have increasingly channelled aid through them, for a variety of reasons, including the bypassing of corrupt politicians and bureaucracies and a desire to reach the very poor.

Expectations of NGOs have been many and varied, and inevitably they have often failed to meet many of the demands made of them. However, it is clear that they do have a range of skills and commitment to qualify them as legitimate partners in development.

Different NGOs have different strengths and weaknesses and can play some roles more effectively than others. The development and political contexts in Africa vary widely too, so that generalising about their roles and relevance is problematic. What is clear, however, is that they cannot replace the state in development provision, but need to work alongside other development players, challenging, complementing and occasionally substituting for them. They do have a role to play in the provision of services, especially in ways that are made more sustainable through involving local people in planning, monitoring and evaluation. However, their relative strengths lie in raising difficult issues, in working alongside the weakest and most excluded, and in promoting social change on the ground by building strong local institutions and strengthening civil society. They can lobby and push for change and promote development debates around contentious issues. Some of their potential is weakened by their growing dependency on external donor funding, competition between each other and a significant lack of learning from past experience.

Notes

1 Writers on this issue include Riddell, Edwards, Hulme, Korten, Fowler, Smillie, Wallace on development, and Duffield, Goodhand, Hulme, Slim for humanitarian aid work. Their work is listed in the References to this chapter.

2 Direct funding was controversial, and while promoted within ODA through the Hodges report 1992, many questioned whether bilateral donors such as ODA had the local knowledge and administrative skills for direct funding to often small local organisations. Indeed in 2000 direct funding policies were under review by DfID.

3 Some argue this is inevitable given international NGO requirements to be accountable to their donors, but many local NGOs argue that there is no equivalence in accountability. Issues of how accountable INGOs are and to whom and through what mechanisms hang in the air in these debates, as INGOs have very weak accountability systems to local people or local NGOs, and seem only concerned about accounting to their donors. There is a lack of transparency in their relations to donors, which are causing increasing concern as reports hide problems and failures and inhibit the ability of INGOs or donors to learn from their real-life experiences.

4 Lewis and Bebbington are looking at these issues for the World Bank and Wallace and Chapman are starting research on these issues with research partners in South Africa and Uganda.

5 Uganda is a good example illustrating that a strong state is indeed important for the establishment of civil society; civil society cannot easily flourish in the absence of a state or in the context of conflict and disintegration.

6 This work was carried out as part of a wider study of the impact of UK aid on poverty, undertaken by the International Development Department, University of Birmingham. The Uganda study was led by Tina Wallace working with Enzo Caputo, Alicia Herbert and Philip Amis. The final report, 'The Uganda poverty study' (Wallace *et al.*, 1998) is still under discussion within DfID.

7 Donors have even failed to coordinate their funding around sector plans which were introduced to ensure coordination of plans and funding; they are notoriously poor at working together on procedures and aid management.

References

ACTIONAID (1995) *Strategic Plan for Uganda*, Kampala: ACTIONAID.

ACTIONAID (2000) *The Reality of Aid: An Independent Review of Poverty Reduction and Development Assistance*, London: Earthscan.

Anderson, M. (1996) *Do No Harm: Building Local Capacities for Peace*, New York: Cambridge University Press.

Beall, J. and Lewis, D. (1999) 'NGOs and gender equality: a short review of the issues', draft paper presented at 'NGOs in a Global Future' conference, Birmingham University.

Brown, L.D. and Kalegaonkar, A. (1998) 'Challenges to civil society and the emergence of support organizations', *Institutional Development*, 1: 20–37.

Chapman, J. (1997) *From Practice to Policy: Towards Engagement with the Corporate Sector*, London: New Economics Foundation.

Chapman, J. (1998) *Effective NGO Campaigning*, London: New Economics Foundation.

Clark, J. (1991) *Democratising Development: The Role of Voluntary Organisations*, London: Earthscan.

Collins, P. (ed.) (2000) *Applying Public Administration in Development: Guideposts to the Future*, Chichester, Sussex: John Wiley.

Community Development Resource Network (CDRN) (1997) 'Capacity building: how culturally appropriate?', Workshop report, Kampala, Uganda: CDRN.

Crewe, E. and Harrison, E. (1998) *Whose Development? An Ethnography of Aid*, London: Zed Press.

Davies, R. (1995) 'The management of diversity: challenges posed by the scaling up of NGOs', paper presented at Development Studies Association conference, Dublin.

Davies, R. (1996) 'Evaluation report on Direct Funding Initiative, Uganda', London: Overseas Development Administration.

de Coninck, J. (1992) *Evaluating the Impact of NGOs in Rural Poverty Alleviation: Uganda Case Study*, ODI Working Paper no. 51, London: Overseas Development Administration.

Department for International Development (DfID) (1997) *Eliminating World Poverty: A Challenge for the 21st Century*, London: Stationery Office.

Dicklich, S. (1994) 'Indigenous NGOs and political participation in Uganda under the NRM regime', paper presented to workshop 'Developing Uganda', Denmark.

Duffield, M. (1994) 'Complex emergencies and the crisis of developmentalism', *IDS Bulletin*, 25 (4): 37–45.

Duffield, M. (1999) 'Reading development as security: post nation-state conflict and the creation of community', paper presented at the conference 'NGOs in a Global Future', Birmingham.

Edwards, M. (1993) 'Does the doormat influence the boot? Critical thoughts on UK NGOs and international advocacy', *Development in Practice*, 3 (3): 163–75.

Edwards, M. (1997) 'Organizational learning in NGOs: what have we learned?', *Public Administration and Development*, 17 (2): 235–50.

Edwards, M. (1998a) 'NGOs as values-based organizations', in D. Lewis (ed.) *International Perspectives on Voluntary Action: Rethinking the Third Sector*, London: Earthscan.

Edwards, M. (1998b) 'International development NGOs: agents of foreign aid or vehicles for international cooperation?', *Discourse*, December.

Edwards, M. (1999) *Future Positive: International Cooperation in the 21st Century*, London: Earthscan; Sterling, VA: Stylus.

Edwards, M. and Hulme, D. (eds) (1992) *Making a Difference: NGOs and Development in a Changing World*, London: Earthscan.

Edwards, M. and Hulme, D. (1995) *Non-government Organisations: Performance and Accountability. Beyond the Magic Bullet*, London: Earthscan.

Farrington, J. and Bebbington, A. (1992) *Reluctant Partners? NGOs, the State and Sustainable Development*, London: Routledge.

Fisher, J. (1998) *Non-governments: NGOs and the Political Development of the Third World*, West Hartford, CT: Kumarian Press.

Fowler, A. (1988) 'NGOs in Africa: achieving comparative advantages in relief and micro-development', *Discussion Paper*, Institute of Development Studies, University of Sussex.

Fowler, A. (1993) 'NGOs as agents of democratization: an African perspective', *Journal of International Development*, 5 (3): 325–39.

Fowler, A. (1997) *Striking a Balance: A Guide to the Effective Management of NGOs in International Development*, London: Earthscan.

Fowler, A. (1998) 'Authentic NGDO partnerships: dead end or way ahead?', *Development and Change*, 29 (1): 137–59.

Gaventa, J. (1998) 'Crossing the great divide: building links between NGOs and CBOs in north and south', in D. Lewis (ed.), *International Perspectives on Voluntary Action: Re-thinking the Third Sector*, London: Earthscan.

Gibson, R. (1998). *Rethinking the Future*, London: Nicholas Brealey.

Goodhand, J. and Hulme, D. (1998) *The Role of NGOs in Complex Political Emergencies: Background Report to the Steering Committee*, Manchester: Institute for Development Policy and Management.

Goodhand, J. and Hulme, D. (1999) 'From wars to complex political emergencies: understanding conflict and peace-building in the New World disorder', *Third World Quarterly*, 20 (1): 13–26.

Goyder, H., Davies, R. and Williamson, W. (1997) *Participatory Impact Assessment*, London: ACTIONAID.

Hutchful, E. (1995) 'The civil society debate in Africa', *International Journal*, 51 (1): 45–77.

Hutchful, E. and Smitz, G.J. (1992) *Democratisation and Popular Participation in Africa*, Ottawa: North–South Institute.

Jackson, C. (1996) 'Rescuing gender from the poverty trap', in C. Jackson and R. Pearson (eds), *Feminist Visions of Development: Gender Analysis and Poverty*, London: Routledge.

Kaplan, A. (1998) *Crossroads: A Development Reader*, Cape Town: Community Development Research Association.

Keen, D. (1994) *The Benefits of Famine: A Political Economy of Relief in North-West Sudan*, Princeton, NJ: Princeton University Press.

Knut, R. (1997) *Globalization and Civil Society: NGO Influence in International Decision-making*, Geneva: UNRISD.

Korten, D. (1990) *Getting into the 21st Century: Voluntary Action and the Global Agenda*, West Hartford, CT: Kumarian Press.

Korten, D. (1995) *When Corporations Rule the World*, West Hartford, CT: Kumarian Press.

Kotter, J. (1998) 'Cultures and coalitions', in R. Gibson (ed.), *Rethinking the Future*, London: Nicholas Brealey.

Kwesiga, J.B. and Ratter, A.J. (1993) *Realising the Potential of NGOs and Community Groups in Uganda*, Kampala: Ministry of Finance and Economic Planning.

Lewis, D. (ed.) (1998) *International Perspectives on Voluntary Action: Re-thinking the Third Sector*, London: Earthscan.

Lewis, D. and Wallace, T. (2000) *New Roles and Relevance: Development NGOs and the Challenge of Change*, West Hartford, CT: Kumarian Press.

McClean, K. and Ntale, C.L. (1998) *Desk Review of Participatory Approaches to Assess Poverty in Uganda*, Nairobi: Department for International Development, UK.

Maina, W. (1998) 'Kenya: the state, donors and the politics of democratisation', in A. Van Rooy (ed.), *Civil Society and the Aid Industry*, West Hartford, CT: Kumarian Press.

Malhotra, K. (1996) *A Southern Perspective on Partnership for Development: Some Lessons of Experience*, Ottawa: International Development Research Centre.

Muchungunzi, D. and Milne, S. (1995) 'Perspectives from the South: a study in partnerships', AFREDA, unpublished.

Nelson, P. (1996) 'Internationalizing economic and environmental policy: transnational NGO networks and the World Bank's expanding influence', *Millennium*, 25 (3).

Overseas Development Administration (ODA) (1992) *The Hodges Report: Report of the Working Group on ODA/NGO Collaboration*, East Kilbride: ODA.

Overseas Development Institute (ODI) (1998) *The State of the International Humanitarian System*, ODI Briefing Paper no. 1, London: ODI.

Oxfam (1995) *Review of Health Projects in Uganda*, Oxford: Oxfam Publications.

Prendergast, J. (1997) *Crisis Response: Humanitarian Band-aids in Sudan and Somalia*, London: Pluto Press.

Reality of Aid (1997–8) *An Independent Review of Development Cooperation*, London: Earthscan.

Riddell, R. (1996a) *Aid in the 21st Century*, New York: United Nations Development Programme.

Riddell, R. (1996b) *Linking Costs and Benefits in NGO Projects*, London: Overseas Development Institute.

Riddell, R. and Bebbington, A. (1995) *Developing Country NGOs and Donor Governments*, London: Overseas Development Institute.

Riddell, R. and Robinson, M. (1992) *The Impact of NGO Poverty Alleviation Projects: Results of the Case Study Evaluations*, Working Paper no. 68, London: Overseas Development Institute.

Ritchey-Vance, M. (1996) 'Social capital, sustainability and working democracy: new yardsticks for grassroots development', *Grassroots Development*, 20 (1): 3–9.

Robinson, M. (1995a) 'NGOs as private service contractors', in M. Edwards and D. Hulme (eds), *Non-government Organisations: Performance and Accountability. Beyond the Magic Bullet*, London: Earthscan.

Robinson, M. (1995b) Strengthening civil society in Africa: the role of foreign political aid, *IDS Bulletin*, 26/2: 70–80.

Salamon, L. and Anheier, H. (1997) *The Nonprofit Sector in the Developing World*, Manchester: Manchester University Press.

Scoones, I. and Pretty, J. (eds) (1993) *Rural People's Knowledge: Agricultural Research and Extension Practice*. London: International Institute of Environmental Development.

Senge, P. (1998) 'Through the eye of a needle', in R. Gibson (ed.), *Rethinking the Future*, London: Nicholas Brealey.

Simbi, M. and Thom, G. (2000) 'Implementation by proxy: the next step in power relations between north and south?', in D. Lewis and T. Wallace (eds), *New Roles and Relevance: Development NGOs and the Challenge of Change*, West Hartford, CT: Kumarian Press.

Slim, H. (1997) 'Relief agencies and moral standing in war', *Development in Practice*, 7 (4): 342–52.

Smillie, I. (1995) *The Alms Bazaar: Altruism under Fire*, Rugby: ITDG Publications.

Smillie, I. (1999) *Relief and Development: The Struggle for Synergy*, Providence, RI: Brown University.

Smillie, I. (2000) 'Relief and development: disjuncture and dissonance', in D. Lewis and T. Wallace (eds), *New Roles and Relevance: Development NGOs and the Challenge of Change*, West Hartford, CT: Kumarian Press.

Smillie, S. and Helmich, H. (1994) *Non-government Organisations and Governments: Stakeholders for Development*, Paris: OECD

Smyth, I. and Payne, L. (1998) *Soft Things and Finer Details: Final Report of the Gender Review of the Oxfam Gender Programme*, Oxford: Oxfam Publications.

Sogge, D. (1996) *Compassion and Calculation: The Politics of Private Foreign Aid*, London: Pluto Press.

Surr, M. (1995) *Evaluations of Non-government Organisations' Development Projects: Synthesis Report*, London: Overseas Development Administration.

Tvedt, T. (1998) *Angels of Mercy or Development Diplomats: NGOs and Foreign Aid*, Oxford: James Currey.

United Nations Development Programme (UNDP) (1998) *Human Development Report*, New York: UNDP.

Uphoff, N. (1986) *Local Institutional Development*, West Hartford, CT: Kumarian Press.

Van Rooy, A. (ed.) (1998) *Civil Society and the Aid Industry*, West Hartford, CT: Kumarian Press.

Van Rooy, A. and Robinson, M. (1998) 'Out of the ivory tower: civil society and the aid system', in A. Van Rooy (ed.), *Civil Society and the Aid Industry*, West Hartford, CT: Kumarian Press.

Wallace, T. (1997) 'New development agendas: changes in UK NGO policies and procedures', *Review of African Political Economy*, 24 (71): 35–56.

Wallace, T. (1998) 'Institutionalising gender in UK NGOs', *Development in Practice*, 8 (2): 159–72.

Wallace, T. (2000a) 'Is the role played by donors in supporting Uganda's NGO sector enabling it to develop effectively?', in P. Collins (ed.), *Applying Public Administration in Development: Guideposts to the Future*, Chichester, Sussex: John Wiley.

Wallace, T. (2000b). 'Development management and the aid chain: the case of NGOs', *Development in Practice Readers*, Oxford: Oxfam and Buckingham: Open University.

Wallace, T. and March, C. (1995) *Changing Perceptions: Readings in Gender and Development*, Oxford: Oxfam Publications.

Wallace, T., Crowther, S. and Shepherd, A. (1997) *Standardising Development: Influences on UK NGOs' Policies and Procedures*, Oxford: Worldview.

Wallace, T., Amis, P., Caputo, E. and Herbert, A. (1998) *Report to DfID on the Official Aid Programme and its Impact on Poverty in Uganda*, London: Department for International Development (restricted).

World Bank (1998) *World Development Report*, Washington, DC: World Bank.

14 Prospects for rural labour force absorption through rural industry

Ian Livingstone

14.1 Introduction

We explore here the scope for developing rural industry in Africa and more generally for expanding productive non-farm employment outside the agriculture sector. This is related to the serious concern which exists regarding the feasibility of needed labour absorption in the rural areas, given countries exhibiting some of the highest population growth rates in the world and given the generally acknowledged failure of attempts made across the continent at import-substituting industrialisation. Our discussion ties in with research which is ongoing into 'sustainable rural livelihoods', where the focus is on rural households diversifying their sources of income and employment, particularly through non-farm activity; and with the closely related phenomenon which has been observed of 'de-agrarianisation', a process whereby people in Africa are said to be moving out of agriculture into other occupations within the rural areas or actually out of the rural areas altogether. This process has been seen as a significant secular trend that needs research (Bryceson, 1996; Bryceson and Jamal, 1997).

14.2 Population trends in Africa

Because population densities in Africa are mostly quite low, there has never been the same concern regarding high population growth rates there as in Asia, although these rates are extremely high. Annual population growth rates projected for 1994–2000 are 2.8 per cent for sub-Saharan Africa (SSA), compared with 1.7 per cent for developing countries as a whole, and in a number of countries are near to 4 per cent. While at current population growth rates population can be expected to double in developing countries as a whole in perhaps 36 years, in SSA the period is only 19 years (*Human Development Report* (HDR) tables). The need to absorb this huge additional population at acceptable, if not increasing, levels of income per head is evident.

Because population densities are not, in fact, uniformly low or labour force absorptive capacity uniformly high, there has been some concern with the problem in the past (Livingstone, 1986, 1990). The focus, however, has been on rural–urban migration, starting most famously in 1969 with Harris and Todaro's analysis of migration (Todaro, 1969; Harris and Todaro, 1970), although the strength of this migration reflects the weakness of rural sectors' capacities to retain workers as much as urban sectors' capacities to attract. The problem of effective labour absorption in the rural areas has become the focus of attention most recently in the emerging literature on sustainable rural livelihoods and rural income diversification (Ellis, 1998; Bernstein *et al.*, 1992), which considers the problem against the background of environmental and natural resource pressures, and on the phenomenon of 'de-agrarianisation' referred to above.

The process of de-agrarianisation can be observed in published statistics. Thus the share of the rural population in the total in SSA has fallen from 85 per cent in 1960 to 69 per cent in 1990 and 65 per cent in 2000. In comparison the share in developing countries as a whole has declined more slowly: from 78 to 63 and then 59 per cent in the corresponding years. Given the comparative failure of large-scale urban-based industry in Africa, this would appear to be more the result of rural failure than urban success.

A parallel trend has been the shift of labour force out of agriculture, from 81 per cent in 1960 to 66 per cent in 1990. While this latter phenomenon has occurred, however, the percentage of the labour force in industry has stagnated at 7 per cent in 1960 and 9 per cent in 1990. This has meant in turn that the proportion in services has increased dramatically from 12 per cent to 25 per cent. This increase in the share of services relative to industry in SSA is well in excess of the changes which have been occurring at the same time in other countries (Table 14.1); the ratio of service sector/industrial sector shares went from 1.71 to 2.78 over the period, while that in developing countries as a whole actually decreased, according to the figures, from 1.55 to 1.44. SSA has a very much higher share in services and at a much lower level of per capita income.[1]

14.3 Diminishing returns?

The scale of the changes outlined above raises the spectre of diminishing returns even if, certainly, other forces will work in the opposite direction. A well-known long-run trend in a country's occupational distribution is a rising share of tertiary sector employment as national incomes per head increase. In SSA increases in per capita incomes would not appear to have been such as to produce changes on the scale observed, as indicated also by the failure of the industry share to increase significantly during the whole period.

If, then, there were diminishing returns for increases in the urban labour force, this would produce what Bairoch (1975) has called 'hyper-urbanisation' and more recently Bryceson (1996) has termed 'derived urbanisation', the consequence of an urban growth process that proceeds without proportionate urban job availability. Diminishing returns to rural labour force increases outside agriculture could simultaneously affect income levels within the rural informal sector, whether in industry or trade, in the former case (in urban areas also) perpetuating the expansion of low-earning one-person 'establishments' and part-time household enterprises.

Table 14.1 Trends in sectoral labour force distribution, 1960–90

Percentage of labour force in:	*SSA*	*All developing countries*	*Industrial countries*
Agriculture			
1960	81	77	27
1990	61	61	10
Industry			
1960	7	9	35
1990	9	16	33
Services			
1960	12	14	38
1990	25	23	57
Ratio of percentage services/industry			
1960	1.71	1.55	1.09
1990	2.78	1.44	1.73

Source: UNDP, *Human Development Reports*

What form urbanisation, or over-urbanisation, will take is another question. As Pedersen (1997) states:

> Some of the rapidly growing rural population in parts of Africa still may be absorbed in the rural areas, but most of the population growth in the future will have to migrate to the urban areas. Most of these urban migrants move to the largest towns, but in many parts of Africa the small towns today are growing relatively more rapidly . . . and are likely to receive a growing part of the urban migrants in the future.

With expanding agriculture, such towns can be important growth centres as market towns providing employment in trade, services and micro/small or even larger-scale industry. Without expanding agriculture, de-centralised over-urbanisation may be the outcome.

Discussion of these trends, in occupational distribution and in urbanisation, goes back to Bairoch and, indeed, the very early literature on disguised unemployment (under-employment in Bairoch), which would now be viewed as low-paying informal sector activities. Bairoch comments that 'the most important result has been the very great deterioration of the possibilities for employment in most urban centres' (1975: 168) and to 'hypertrophy of the tertiary sector of employment' (ibid.: 161).

Two qualifications need to be made here. First, the HDR data employed to identify these general trends may not fully reflect informal sector activity, including, as already noted, part-time activity, activity which even periodic censuses have difficulty in measuring. Census/survey coverage is certainly not uniform or consistent.

Second, improvements in infrastructure, road networks in particular, can be expected to have the effect of raising rural incomes by improving access to markets and to consumer goods, increasing the relative importance of service sector employment, in trade especially, as well as increasing demand for both rural and urban informal sector manufactured goods. This trend is associated with that of progressive reduction in subsistence production and increased association with the market over time. What the balance of these forces has been is a matter for empirical investigation in particular countries, focusing on changes in the level of informal sector incomes, and will not be uniform. Empirical investigation of income levels in either urban or rural informal sectors has been quite limited.

14.4 Rural industry and basic needs

It is important to stress, first, and before less positive analysis is presented, that rural manufacturing, however primitive, often in appearance and technical level, may none the less be of critical importance to welfare. Its content needs to be seen in the context of providing the basic needs of communities at low levels of income, rather than one of dynamic enterprise development. Its value can be seen from the following description taken from rural Zambia:

> Common products of carpenters include boats, wooden doors, window and door frames, chairs, dining and side tables, baby cots, stools, cupboards, bookshelves, wardrobes and so forth . . . Products of the two crafts [knitting and sewing] include jerseys, hats, bags, shawls, rompers, socks, table-cloths and dresses of various types. These are in high demand in local communities but their production is often hampered by difficulties with input supply to the district . . . [The district has approximately 40 registered pitsawing groups spread all over the Kabalinge to Chipungu. Of late pitsawyers have been experiencing a crisis, a limited but growing crisis of insufficient productivity in timber supply industry due to receding stock of suitable species.] . . . Products [from basket/

mat-making] are equally in high demand throughout the district. Those involved normally produce baskets of various types, drying mats, side tables, bamboo/reed bowls, sun hats, baby cots, stools, water stands, chairs, fish traps and other items for home decoration. And to get the bamboo long journeys to sources are a must.

(Shula and Maleka, 1984)

This description indicates the breadth and diversity of product provision sustained by informal rural industry, even if it shares the characteristics of the urban informal sector that its products are mostly rough in nature and quality, as well as cheap.

The quotation also makes clear that the depth of rural industry provision will depend on the richness of the local natural resource base. Conservation of this resource base, wood supplies in particular, represents a critical problem in SSA, given especially its rapidly expanding population. If this resource base is cut back, key elements for the provision of basic needs, and for their maintenance even at current levels, will be affected. The resources assist the maintenance of sustainable rural livelihoods by simultaneously supporting income-earning employment and the supply of cheap essentials bought by rural households as consumers.

The other major constraint on the expansion of rural manufacturing, and rural enterprise in general, is demand and thus the level of rural incomes. These in turn depend primarily on agricultural development, generating forward demand linkages towards other sectors.[2] Differences in the level of agricultural incomes largely account for differences in turn between districts within a country as well as between countries, such activity in semi-arid areas in particular being constrained both by scanty resource bases and by weak demand linkages from agriculture, due to the low disposable incomes generated there.

14.5 Optimism and pessimism regarding the rural informal sector

It is important, however, to take a rather balanced view of the economic contribution that can be made by rural manufacturing and, more generally, by the whole rural informal sector, incorporating all non-farm activities. Although rural small-scale enterprises together constitute the rural counterpart of the urban informal sector, practically all the informal sector literature has concentrated on the latter. For this reason it is useful to start with some observations regarding this literature.

The debate here focuses on whether or not the sector could be expected to play a positive role in development. Analysts could be divided into pessimists and, starting with the original ILO (1972) report, optimists. The former see the informal sector as a 'sink' for those unable to secure formal sector jobs, as a low-income alternative to unemployment, made up of shoe shiners, car washers and the like. Expansion of the sector could be described as involutionary, rather than evolutionary, absorbing an increasing number of persons but with diminishing marginal productivity as more and more producers compete within a limited market. The optimists, in contrast, see the informal sector as a dynamic sector of entrepreneurs and small-scale capitalists, whose contribution is significantly underestimated in official statistics, and who carry important potential for employment creation.

My own interpretation (Livingstone, 1991) is that the existence of the informal sector should be seen as largely reflecting a bimodal distribution of income. Thus a mass of low-income consumers buy in the informal sector from micro- and household enterprises a range of much cheaper and also much lower quality goods and services, while a higher income group of consumers buy more expensive, higher-quality products produced by large firms in the formal sector.[3] It is not surprising that the concept was developed in Kenya where initially the contrast was between a rich European community and a low-income African community.[4] Elsewhere, in Asia particularly, the income distribution may be less polarised, generating a spectrum of different sizes of establishment rather than just the two categories. Employment in the informal sector and the low prices of goods and services provided are maintained by an elastic supply of low-wage labour, in part associated with rural–urban migration. The effect is that the urban informal sector expands through an increase in the number of micro and small enterprises (MSEs) and self-employed, rather than through increases in the size of firms.

Turning specifically to the rural economy, a similar division between optimists and pessimists may be emerging with respect to rural MSEs. Optimism is associated in particular with what might be labelled the 'Michigan School', the MSU/Gemini research group led by Carl Liedholm and Donald Mead at Michigan State University. This group has conducted a whole series of surveys in different African economies over a decade or more.

The optimists refer to rural 'microenterprises', a more attractive label than the rural informal sector, even if the majority of such enterprises employ just 1–3 persons. These are seen as making a major contribution to the manufacturing sector of the economy, as well as to employment generally. Thus Kilby (1986) refers to the 'surprising dominance of rural industry in total manufacturing employment'. Liedholm (1990) states that 'a surprising, yet important finding is that in most (African) countries the vast majority of small industries are located in rural areas'.

As in the case of the urban informal sector, the School has played a valuable role in correcting a general neglect of the sector and a failure to appreciate its quantitative significance. Its propositions need to be qualified, however, if the actual and potential contribution of the sector is to be seen in proper perspective.

14.6 Evidence from surveys

Evidence regarding the economic value of rural industry and non-farm employment provided by the numerous surveys which have been made needs to be interpreted very carefully. We can examine, first, what the data tell us about the actual amount of employment that is created; second, what the level of incomes in rural small-scale industry might be; and third, how optimistic we can be about the growth or 'graduation' of the enterprises themselves.

14.6.1 Contribution to employment

As regards contribution to employment, in some presentations a false impression is created by quoting the number of establishments rather than the persons engaged in MSEs (Table 14.2): the very large number of one-person establishments, particularly in rural areas, greatly increases the figure for number of establishments.

Again, as regards the importance specifically of rural manufacturing, the important distinction between rural manufacturing enterprise as opposed to rural non-farm enterprise as a whole, including trade and services of all kinds, is often not made clear. Thus in a table showing the composition of rural employment in eleven African countries (Table 14.3) manufacturing appears as a respectable 29 per cent of non-agricultural rural employment. However, non-agriculture's share of total rural employment as shown in the table was just 13.4 per cent, so that if a final column is added, the share of manufacturing in total rural employment can be seen to average just 3.8 per cent.

The asserted dominance of rural manufacturing in total manufacturing depends on large numbers of extremely small informal sector enterprises. One of the most comprehensive surveys produced quite recently by the Michigan School is that of Parker and Torres (1994), carried out in Kenya in 1993, a survey which in fact covered both rural *and* urban areas. The

Table 14.2 Location of medium and small enterprises (percentage of total)

Location	Botswana	Lesotho	Malawi	Swaziland	Zimbabwe
Major cities	10.2	12.0	9.0	15.5	32.0
Secondary towns	21.2	8.4	1.5	8.9	10.2
Rural areas	68.6	79.6	89.5	76.6	57.8

Source: Liedholm and Mead (1993)

Table 14.3 Magnitude and composition of rural employment[a]

Countries	% of total employment		% of rural non-agric. employment				% of total
	Agric.	Non-agric	Mining	Mfr.	Constr.	Services[b]	Mfr.
Cameroon 1976	89.6	10.4	0.2	23.8	8.6	67.4	2.5
Malawi 1977	89.0	11.0	0.8	24.9	14.9	59.3	2.7
Mali 1976	91.7	8.3	0.8	39.8	0.8	58.6	3.3
Mauritania 1977	79.2	20.8	1.9	17.3	5.8	75.0	3.6
Mozambique 1980	91.0	9.0	0.0	50.9	5.6	43.5	4.6
Rwanda 1978	95.1	4.9	9.5	23.4	13.2	53.9	1.1
Senegal 1970/1	82.3	17.7	2.4	34.3	4.3	59.0	6.1
Sierra Leone 1974	86.2	13.8	5.4	20.3	7.9	66.3	2.8
Tanzania 1978	93.9	6.1	0.0	19.6	0.0	79.4	1.2
Togo 1970	74.5	25.5	0.1	18.6	4.2	77.0	4.7
Zimbabwe 1982	80.0	20.0	0.0	46.6	0.0	53.4	9.3
Mean	86.6	13.4	1.9	29.0	5.9	63.0	3.8

Source: Haggblade *et al.* (1987: 32, table 16) reproduced in Bagachwa and Stewart (1990), final column added

Notes
a Employment is defined as economically active population primarily engaged in each activity.
b Includes transport, commerce and other activities, plus 'other'.

average size of establishment identified by this survey was 2–3 persons (Table 14.4). A later sample re-survey based on the same data set by Daniels *et al.* (1995) identified a mean establishment size of 1.9 in urban areas and 1.6 in rural areas. In the Parker and Torres survey 47 per cent were one-person establishments (Table 14.5), while in another GEMINI survey carried out in Zimbabwe by McPherson in 1991, 70 per cent of establishments were one-person establishments. In the former survey only 5 per cent of establishments employed more than five persons and in the Daniels *et al.* re-survey just 1.3 per cent. Parker and Torres (1994: 63) state that 'in Southern Africa the majority of enterprises are one worker, women-owned and home-based businesses'.

The larger enterprises that *are* located in the rural areas are generally resource-based enterprises such as coffee factories or sawmills, though even agro-processing establishments are often located in major towns. A particularly full census of manufacturing establishments was carried out in Uganda in 1984, covering all establishments employing five or more persons, with the smallest sizes excluded, and is worth referring to despite its date (Government of Uganda, 1989). This revealed that establishments located in the three main towns in Uganda, Kampala, Jinja and Mukono (even omitting a number of other significant towns), accounted for nearly 60 per cent of employment (Table 14.6) and 80 per cent of value added. Outside these towns about 65 per cent of those employed were in resource-based processing industries.

Table 14.4 Kenya: proportion of one-worker enterprises and mean
size of micro and small enterprises (MSEs) by sector

Sector	% of one-worker enterprises	Mean no. of workers per enterprise
Manufacturing	40	2.5
Commerce	52	2.0
Services	36	3.2
All sectors	47	2.3

Source: Parker and Torres (1994)

Table 14.5 Size distribution of MSEs in Kenya and Zimbabwe

Size of enterprise (no. of workers)	Kenya (% of enterprises)	Zimbabwe (% of enterprises)
1	47	70
2	28	15
3–5	20	12
6–10	4	2
11–50	1	1
Total	100	100
Mean size (no. of workers)	2.3	1.8

Sources: Kenya: Parker and Torres (1994); Zimbabwe: McPherson (1991)

Again in Uganda, a 100 per cent census carried out in one large district, Mbarara, of establishments employing four or more persons, including the large rural town of Mbarara, found that employment accounted for less than 1 per cent of the labour force. This figure is the consequence of excluding household enterprises, including part-time activities, as well as establishments with just 1–3 persons, and the fact that a large proportion of people are engaged in agriculture as a primary occupation. A few large coffee processing establishments were also not covered (Turamye, 1994).

Accepting that there are immense numbers of 1–3 person establishments, it is important to ask: How far do these represent full-time activities? How far do they generate worthwhile levels of income? And how far do they have growth potential?

The fact that many rural enterprises are carried out on a part-time basis is related to the rural livelihoods debate focusing on the diversification of household sources of income. In terms of employment provision, one would need to know the income generated and the full-time employment equivalence of the activities enumerated. Thus data for Zambia collated by Milimo and Fisseha are presented by Kilby (1986) as demonstrating the importance of employment in rural manufacturing (Table 14.7). This suggested that 263,000 individuals were employed in rural small-scale manufacturing, compared with 150,000 in large-scale urban manufacturing. However, it can be seen that 50 per cent of rural employees were engaged in beer-brewing, which is carried out by women largely as a part-time activity: its full-time employment equivalent may be just one-fifth. Similarly, the 'wood products' item is in fact forest products, including grass products such as mats. Making different reasonable assumptions regarding full-time equivalents in rural manufacturing reduces these to a range 105,000–110,000 and the share in total manufacturing employment from 64 to 41–43 per cent. It might be noted that in the Parker/Torres survey of Kenya 66 per cent of MSEs were women-owned and 32 per cent located in households, both indicators of part-time activity.

Table 14.6 Uganda: geographical distribution of persons engaged in manufacturing, 1984

	Uganda	Kampala	Jinja	Mukono	Total	Other Central	Other Eastern	Western	Northern	Total excluding Kla, Jja, Mkno
1 Meat, fish, vegetables	778	168	176	—	344	164	129	62	79	434
2 Dairy products	516	508	—	—	508	(8)	—	16	—	8
3 Coffee processing	11,158	3,531	328	1,524	5,383	3,304	740	1,694	37	5,775
4 Grain milling etc.	2,967	578	452	115	1,145	433	839	278	272	1,822
5 Tea processing	1,744	103	69	394	566	226	—	932	20	1,178
6 Sugar and jaggery	3,981	—	1,068	1,624	2,692	282	932	50	25	1,289
7 Other food products	954	549	44	13	606	187	45	107	9	348
8 Beverages	2,084	1,413	65	568	2,046	20	—	18	—	38
9 Tobacco products	719	719	—	—	719	—	—	—	—	—
10 Cotton ginning	1,151	—	—	—	0	77	366	324	384	1,151
11 Spin and weave textiles	2,108	383	671	—	1,054	10	1,020	—	24	1,054
12 Other textile goods	4,378	532	—	3,219	3,751	—	604	—	23	627
13 Knitted etc. fabrics	40	10	—	—	10	—	—	10	—	10
14 Wearing apparel	1,069	816	156	—	972	7	33	30	27	97
15 Leather and products	136	19	117	—	136	—	—	—	—	—
16 Footwear	463	289	10	—	299	60	73	31	—	164
17 Sawmilling, planing	1,991	442	410	32	884	78	87	719	223	1,107
18 Wood, straw products	89	52	—	—	52	21	—	—	16	37
19 Paper and products	570	154	236	56	446	—	119	5	—	124
20 Publishing	301	301	—	—	301	—	—	—	—	—
21 Printing	1,356	504	241	—	745	437	84	76	14	611
22 Repro/re-ed media	6	6	—	—	6	—	—	—	—	—

Table 14.6 Continued

	Uganda	Kampala	Jinja	Mukono	Total	Other Central	Other Eastern	Western	Northern	Total excluding Kla, Jja, Mkno
23 Basic chemicals	36	30	—	—	30	—	—	6	—	6
24 Other chem. products	1,225	835	242	7	1,084	30	33	67	11	141
25 Rubber products	133	64	69	—	133	—	—	—	—	—
26 Plastic products	175	175	—	—	175	—	—	—	—	—
27 Mineral products	2,749	127	22	168	317	662	664	990	116	2,432
28 Basic iron and steel	391	—	341	—	341	—	50	—	—	50
29 Casting of metals	42	42	—	—	42	—	—	—	—	—
30 Struct. steel products	1,007	764	35	—	799	70	99	39	—	208
31 Other fab. metal products	1,729	901	227	323	1,451	—	20	96	162	278
32 Gen. purpose machinery	74	74	—	—	74	—	—	—	—	—
33 Spec. purpose machinery	1,668	124	37	—	161	1,424	74	9	—	1,507
34 Elec. machinery and apps	179	117	—	62	179	—	—	—	—	—
35 TV and radio equipt.	91	91	—	—	91	—	—	—	—	—
36 MV bodies and trailers	62	62	—	—	62	—	—	—	—	—
37 Parts etc. for MV	56	41	—	—	41	—	15	—	—	15
38 Build/repair boats	14	—	—	14	14	—	—	—	—	—
39 Furniture	3,805	1,885	—	260	2,145	379	418	654	209	1,660
40 Mfg. nec	108	48	23	—	71	25	6	—	6	37
41 MV repairs, service	3,601	2,294	110	53	2,457	402	219	363	160	1,144
Total	55,684	18,751	5,149	8,432	32,332	8,290	6,669	6,576	1,187	23,352
	100.0%	33.7%	9.2%	15.1%	58.1%	14.9%	12.0%	11.0%	3.3%	41.9%

Source: Government of Uganda, *Census of Business Establishments, Uganda, 1989: Manufacturing Sector Report*, Statistics Department, Entebbe

Table 14.7 Employees in rural and urban manufacturing in Zambia

	Large-scale industry (1980) (000s)	Small-scale industry (1985) Rural (000s)	(%)	Urban (000s)
Textiles and clothing	10.2	11.7	4.5	19.8
Wood products	6.3	82.7	31.5	11.9
Metal products	6.3	10.3	3.9	7.6
Food	16.6	14.3	5.4	17.5
Beverages	3.8	132.4	50.4	8.4
Leather products	1.0	1.4	0.5	2.1
Non-metallic products	3.5	6.5	2.9	1.2
Chemicals	7.3	—	—	2.0
Machinery	1.0	1.4	0.5	4.1
Electrical products	1.7	0.9	0.3	3.1
Transport equipment	1.3	—	—	8.0
Miscellaneous	0.2	0.8	0.3	2.5
Total	58.9	262.8	100	88.2
Persons engaged per establishment (no.)	103	1.6		1.9
Share in total mfg. employment (%)	15	64		21

Source: Milimo and Fisseha (1986)

Note
Data refer to an estimated 248,388 enterprises (grossed up from sample data) representing the approximately 38 per cent of enterprises which have expanded since establishment.

14.6.2 Levels of income

More important than employment hours generated is whether rural small-scale industries generate worthwhile levels of income, information less commonly collected by surveys. However, Daniels *et al.* (1995) found in Kenya that 65 per cent of enterprises in small towns and rural areas produced profits per person of less than K.Shs 1070 per month, or US $26 p.m., equivalent to about US $1 per working day. Approximately 80 per cent earned less than US $1.7 per working day.[5] Thus a large proportion of activities, not all, represented the pursuit of survival activities, as concluded by Daniels *et al.* (ibid.: 60).

Taking all enterprises together, 67 per cent generated levels of net income per worker below the minimum wage, 75 per cent in the case of female-owned enterprises (which were 43 per cent of the total). Female-owned enterprises generated net profits per worker equal to just under 25 per cent of those in male-owned enterprises, while net profits per worker produced in rural enterprises were just 19 per cent of those secured on average by urban enterprises.

The figure for net profits per worker per annum in all rural enterprises of K.Shs 16,350 (US $397) was only just above that for the overall figure for female-owned enterprises of K.Shs 15,552, suggesting that female-owned enterprises were heavily represented among rural enterprises (ibid.: table 5–3).

It may be noted further that less than one-third of enterprises in the Daniels survey were in manufacturing: 38 per cent in rural areas and 15 per cent in urban areas (ibid.: table 3.2) compared with 55 per cent overall in trade. Most of the manufacturing activities generated lower net profits per worker relative to trade and services. Looking at specific activities, this value relative to the average (including urban enterprises that include a relatively high proportion of higher-earning trade and service enterprises) was 60 per cent in shoe-making and 23 per cent in beer-brewing, although garment-making stood out with four times the average (ibid.: table 9.1). The contribution specifically of manufacturing, alongside agriculture, is important since the development of tertiary sector activities, trading in particular, depends on the generation of output available for trade as well as forward demand linkages.

14.6.3 Enterprise development

The third question raised is whether the enterprises concerned have potential for growth as enterprises. The general conclusion of the Michigan surveys is that they do, Parker and Torres asserting that the vast number of one-person establishments enumerated constitutes a 'seed bed' for larger enterprises of 3–5 persons and 11–50 persons (Parker and Torres, 1994: 31).

Against this, first, is the exceptionally high mortality rate among MSEs in Africa; particularly useful information contained in the Parker/Torres study was of age at close of discontinued establishments (not usually covered in surveys because of the obvious difficulties), equal on average to only 4.3 years.

The question generally posed is how far enterprises can be expected to 'graduate' to larger size. Here GEMINI studies, and certainly that of Parker and Torres, are perhaps unjustifiably optimistic. Based on evidence covering 38 per cent of sample enterprises which had shown an increase in numbers engaged since establishment (Table 14.8), Parker and Torres (ibid.: 31) conclude that 'the vast proportion of one-worker enterprises is indeed the seed-bed for 3–5 worker and 11–50 worker enterprises' such that 'the best way of encouraging small enterprises (those with over ten workers) is through assistance to enterprises that start as micros, and particularly those with fewer than six workers'. This is true so far as an estimated 8000 enterprises have originated in this way, although the percentage of enterprises starting with one worker reaching the larger size was only 3 per cent *among those that have experienced growth*, and not much more than 1 per cent, probably, among all such enterprises. Among the initial 2–5 person categories, an estimated 7 per cent reached the larger size, out of those expanding, perhaps 2.5–3 per cent of all those starting at this size.

The 8000 enterprises might together generate employment of some 110,000–135,000 nationally, *including* the proportions that are in other, non-manufacturing sectors and those which are in urban locations. The evidence provided earlier from Uganda does not suggest

Table 14.8 Number of expanding enterprises by starting and current size, Kenya, 1993

| Current enterprise size (no. of workers) | Initial enterprise size (no. of workers) | | | | |
	1	2	3–5	6–10	11–50
2	100,845	—	—	—	—
	(54.1/40.6)				
3–5	70,810	35,433	6,669	—	—
	(38.0/28.5)	(86.8/14.3)	(36.7/2.7)		
6–10	8,740	5,323	8,475	1,309	—
	(4.7/3.5)	(13.0/2.1)	(46.7/3.4)	(48.4/0/5)	
11–50	5,863	69	3,016	1,398	438
	(3.1/2.4)	(0.2/..)	(16.6/1.2)	(51.6/0.6)	(100/0.2)
Total	186,258	40,825	18,160	2,707	438
	(100/75.0)	(100/16.4)	(100/7.3)	(100/1.1)	(100/0.2)

Source: Parker and Torres (1994: table III.20)

that the numbers of rural-located industrial enterprises of this size would be very large, although the number of such enterprises being formed in rural towns is clearly of importance in local development.

14.7 Conclusions

Despite problems of measurement, evidence does exist of a strong secular trend towards 'de-agrarianisation' in sub-Saharan Africa, involving progressive change in occupational distribution within the rural areas, as well as of rural–urban migration of household members or whole households. This raises the question of how effectively increases in the rural (and urban, for that matter) population and labour force are being absorbed. The research that has been carried out has concentrated on the urban rather than the rural informal sector and only now is attention beginning to be paid to the securing of 'sustainable livelihoods' in the rural areas, whether in Africa or elsewhere.

The signs are that, just as writers on the urban informal sector can be divided into 'optimists' and 'pessimists' in assessing potential, a similar division of view is appearing in respect of the rural informal sector, and in particular, of rural small-scale industry.

The contribution of small-scale industry to the provision of basic needs and employment needs to be acknowledged. Reviewing earlier and more recent survey evidence here, however, it is suggested that presentations frequently exaggerate the actual quantitative contribution in terms of full-time job equivalents; that incomes derived in the sector, though varying widely, are often low and could go lower were diminishing returns to set in or new dynamic elements not to be introduced; and that, while some graduation of enterprises goes on and is certainly of some significance, this is small in proportion to the vast numbers of one- or two-person enterprises observed to exist. Rural industry and the rural informal sector are important, just as is the urban informal sector, in a number of ways: it is necessary also that this importance be placed in perspective.

Notes

1 The published figures used here need to be treated with a pinch of salt. In particular they will relate to primary occupations only (see Chapter 19 by O'Connor in this volume). Nevertheless the magnitude of the changes indicated compensate for the rough nature of the figures.
2 The nature of the agricultural production function, and with it the technologies employed in agriculture, is also a factor, determining the extent of backward linkages with other sectors, including manufacturing. Such linkages are much weaker in Africa compared with Asia, for example. See Bagachwa and Stewart (1990).
3 The dichotomy here is not complete. Cheap plastic sandals, for example, are produced in large factories for mass consumption.
4 An element of price discrimination may also be involved, with higher-income consumers paying more for substantially similar products.
5 This figure for full days worked should be increased to the extent that a portion of activities were engaged in on a part-time basis.

References

Bagachwa, M.D. and Stewart, F. (1990) 'Rural industries and rural linkages in sub-Saharan Africa: a survey', *Ld'A–QEH Development Studies Working Papers*, Oxford: Oxford University, Queen Elizabeth House.
Bairoch, P. (1975) *The Economic Development of the Third World since 1900*, London: Methuen.
Bernstein, H., Crow, B. and Johnson, H. (eds) (1992) *Rural Livelihoods: Crises and Responses*, Oxford: Oxford University Press for Open University.

Bryceson, D.F. (1996) 'De-agrarianisaton and rural employment in sub-Saharan Africa: a sectoral perspective', *World Development*, 24 (1): 97–111.

Bryceson, D. and Jamal, V. (eds) (1997) *Farewell to Farms: De-agrarianisation and Employment in Africa*, Aldershot, Hants.: Ashgate.

Daniels, L., Mead, D.C. and Musinga, M. (1995) *Employment and Income in Micro and Small Enterprises in Kenya: Results of a 1995 Survey*, Nairobi: K-Rep.

Ellis, F. (1998) 'Household strategies and rural livelihood diversification', *Journal of Development Studies*, 35 (1): 1–38.

Government of Uganda (1989) *Census of Business Establishments, Uganda: Manufacturing Sector Report*, Entebbe: Statistics Department.

Haggblade, S., Hazell, P. and Brown, J. (1987) *Farm/Non-farm Linkages in Rural Sub-Saharan Africa: Empirical Evidence and Policy Implications*, Discussion Paper, Agriculture and Rural Development Department, Washington, DC: World Bank.

Harris, J.R. and Todaro, M.P. (1970) 'Migration, unemployment and development: a two-sector analysis', *American Economic Review*, 60: 126–42.

ILO (1972) *Employment, Incomes and Equality: A Strategy for Increasing Productive Employment in Kenya*, Geneva: ILO.

Kilby, P. (1986) *The Non-farm Rural Economy*, EDI Course Note, Washington, DC: World Bank.

Liedholm, C. (1990) 'Small-scale industry in Africa: dynamic issues and the role of policy', *Ld'A–QEH Development Studies Working Papers*, Oxford: Oxford University, Queen Elizabeth House.

Liedholm, C. and Mead, D.C. (1993) *The Structure and Growth of Microenterprises in Southern and Eastern Africa*, Gemini Working Paper no. 36, Bethesda, MD: Growth and Equity through Microenterprise Investments and Institutions, and East Lansing, MI: Development Alternatives, Inc., Michigan State University.

Livingstone, I. (1986) *Rural Development, Employment and Incomes in Kenya*, Aldershot, Hants.: Gower.

Livingstone, I. (1990) 'Population growth, rural labour absorption and household viability in eastern and southern Africa', *Population and Development Review*, special issue.

Livingstone, I. (1991). 'A re-assessment of Kenya's rural and urban informal sector', *World Development*, 19 (6): 651–70.

McPherson, M.A. (1991) *Micro and Small Scale Enterprises in Zimbabwe: Results of a Country-wide Survey*, GEMINI Technical Report no. 25, Bethesda, MD: Growth and Equity through Microenterprise Investments and Institutions, and East Lansing, MI: Development Alternatives, Inc., Michigan State University.

Milimo, J.T. and Fisseha, Y. (1986) *Rural Small-scale Enterprise in Zambia: Results of a 1985 Country-wide Survey*, MSU International Development Working Paper no. 28, East Lansing, MI: Michigan State University.

Parker, J.C. and Torres, T.R. (1994) *Micro and Small Scale Enterprise in Kenya: Results of the 1993 National Baseline Survey*, GEMINI Technical Report no.75, Bethesda, MD: Growth and Equity through Microenterprise Investments and Institutions, and East Lansing, MI: Development Alternatives, Inc., Michigan State University.

Pedersen, P.O. (1997) *Small African Towns: Between Rural Networks and Urban Hierarchies*, Aldershot, Hants.: Ashgate/Avebury.

Shula, E.C.W. and Maleka, P. (1984) *Crafts Development and Self-support Skills Base for the Out-of-School Youth in Nchelenge District: A Report to the IRDP/LP*, Mansa, Zambia.

Todaro, M.P. (1969) 'A model of labour migration and urban unemployment in less developed countries', *American Economic Review*, 59: 138–48.

Turamye, B. (1994) *The Performance of Rural Small Scale Industries in Mbarara District*, MA dissertation, Economics Department, Makerere University, Kampala, Uganda.

United Nations Development Programme (various years) *Human Development Report*, New York: Oxford University Press.

Part IV

Industry and the urban sector

15 An overview of manufacturing development in sub-Saharan Africa

Michael Tribe

15.1 Introduction

The objective of this chapter is to review some of the characteristics of the manufacturing sector in sub-Saharan Africa. For most countries the sector is still comparatively undeveloped and there are very low proportions of manufactures in total exports. The view that structural adjustment and trade liberalisation, or other economic factors, have threatened a process of 'de-industrialisation' is considered.[1]

15.2 African industrial development to the mid-1990s

Table 15.1 presents a comprehensive set of data relating to most sub-Saharan African countries from the World Bank's *World Development Report 1998/99* (World Bank, 1999), a readily accessible and comparatively reliable source. Perhaps the most obvious feature demonstrated by the data is the considerable variation in country characteristics and in growth performance. The highest population is well over 100 times the lowest, the largest land area is well over 1000 times the smallest, and the highest per capita GNP is nearly 50 times the lowest (this last ratio falls to a factor of nearly 20 if the purchasing power parity definition of GNP per capita is used). For the period 1980–90 the highest GDP growth rate experienced by any of the countries was 6.2 per cent per annum (Mauritius), and the lowest was 0.1 per cent (Niger). For the period 1990–7 the highest GDP growth rate was 7.6 per cent per annum (Lesotho) and the lowest was –6.6 per cent (Congo Democratic Republic).

The diversity is just as marked for the industrial characteristics of sub-Saharan Africa. The highest proportion of industry recorded in GDP in 1980 was 60 per cent (Gabon) and 68 per cent in 1997 (Angola), and the lowest proportion was 4 per cent in 1980 (Uganda) and 7 per cent in 1997 (Ethiopia). It is likely that these proportions are indicative of 'special cases'. It should be noted that the 'industry' statistics are not directly representative of 'manufacturing' due to the fact that some countries have high 'industrial' value added due to their dependence on minerals production (and export). In addition, the international definition of 'industry' is so wide that these industrial statistics are in no way indicative of manufacturing activity – a point that is discussed in more detail below.

The unweighted average rate of industrial growth in sub-Saharan Africa was 2.7 per cent per annum over the period 1980–90, slightly higher than the unweighted average GDP growth in the same period. However, average industrial growth fell to 1.6 per cent per annum for the period 1990–7 (GDP growth averaged 2.3 per cent per annum). Given an average

Table 15.1 The industrial sector in SSA

Country	Population (millions 1997)	Area (000 sq km)	GNP/caput		GDP growth (% per year)		Industry as % of GDP		Industrial growth (% per year)		Manuf. as % exports	
			1997 US $	1997 US $ PPP est.	1980–90	1990–97	1980	1997	1980–90	1990–97	1980	1996
Angola	11	1,247	340	940	3.7	0.7	n.a.	68	6.4	5.1	13	n.a.
Benin	6	111	380	1,260	3.2	4.5	12	14	1.3	4.1	3	n.a.
Botswana	1.5	567	3,260	8,220	n.a.	n.a.	n.a.	n.a.	1	n.a.	n.a.	n.a.
Burkina Faso	11	274	240	990	3.7	3.3	22	25	3.7	1.9	11	n.a.
Burundi	7	26	180	590	4.4	-3.7	13	18	4.5	-8.0	4	n.a.
Cameroon	14	465	650	1,980	3.3	0.1	23	20	5.9	-3.8	4	8
Central Afr. Rep.	3	623	320	1,530	1.4	1.2	20	18	1.4	0.1	26	43
Chad	7	1,259	240	1,070	3.8	1.8	9	15	8.1	0.0	15	n.a.
Congo	2	342	660	1,380	3.6	0.7	47	57	5.2	0.6	7	2
DRC	47	2,267	110	790	1.6	-6.6	33	13	0.9	-15.9	6	n.a.
Côte d'Ivoire	15	318	690	1,640	0.9	3.0	20	21	4.4	4.2	5	n.a.
Ethiopia	60	1,000	110	510	2.3	4.5	12	7	1.8	4.1	0	n.a.
Gabon	1	258	4,230	6,450	0.6	2.6	60	52	1.5	2.7	5	2
Gambia, The	1	10	350	1,340	n.a.	n.a.	n.a.	n.a.	n.a.	n.a.	n.a.	n.a.
Ghana	18	228	370	1,790	3.0	4.3	12	17	3.3	4.3	1	n.a.
Guinea	7	246	570	1,850	n.a.	4.1	n.a.	36	n.a.	3.0	1	n.a.
Guinea-Bissau	1	28	240	1,070	4.0	3.8	19	11	2.2	2.7	8	n.a.
Kenya	28	569	330	1,110	4.2	2.0	21	17	3.9	2.0	12	n.a.
Lesotho	2	30	670	2,480	4.3	7.6	29	41	7.1	11.8	n.a.	n.a.
Madagascar	14	582	250	910	1.1	0.8	16	13	0.9	1.1	6	14
Malawi	10	94	220	700	2.3	3.6	23	18	2.9	1.9	6	7
Mali	10	1,220	260	740	2.9	3.3	13	17	4.3	7.0	1	n.a.

Mauritania	2	1,025	450	1,870	1.7	4.3	26	29	4.9	3.7	0	n.a.
Mauritius	1	2	3,800	9,360	6.2	5.1	26	32	10.3	5.5	27	68
Mozambique	19	784	90	520	1.7	6.9	35	23	-5.2	2.3	18	17
Namibia	2	823	2,220	5,440	1.3	4.1	39	34	-1.1	2.9	n.a.	n.a.
Niger	10	1,267	200	920	0.1	1.5	23	18	-1.7	1.3	2	n.a.
Nigeria	118	911	260	880	1.6	2.7	46	24	-1.1	0.5	0	n.a.
Rwanda	8	25	210	630	2.5	-6.3	23	23	2.5	-11.2	0	n.a.
Senegal	9	193	550	1,670	3.1	2.4	21	18	4.1	3.7	15	50
Sierra Leone	5	72	200	510	0.6	-3.3	21	24	1.7	-6.4	40	n.a.
South Africa	38	1,221	3,400	7,490	1.2	1.5	50	39	0.0	0.8	18	49
Tanzania	31	884	210	n.a.	n.a.	n.a.	n.a.	21	n.a.	n.a.	14	n.a.
Togo	4	54	330	1,790	1.6	2.2	25	22	1.1	2.0	11	n.a.
Uganda	20	200	320	1,050	3.1	7.2	4	17	6.0	13.0	1	n.a.
Zambia	9	743	380	890	0.8	-0.5	41	41	1.0	-2.6	16	n.a.
Zimbabwe	11	387	750	2,280	3.4	2.0	29	32	3.2	-0.8	36	30
Unweighted average	14.9	516.4	748.7	2047.2	2.4	2.3	25.4	23.6	2.7	1.6	9.7	13.8
Highest	118	2267	4,230	9,360	6.2	7.6	60	68	10	13	40	68
Lowest	1	2	90	510	0.1	-6.6	4	7	-5	-11	0	2

Source: World Bank (1999: Table 1: 190–191; Table 11: 210–211; Table 12: 212–213; Table 20: 228–229)

Notes

n.a. = not available; figures in italics are for a different year to that indicated.

annual population growth rate for sub-Saharan Africa which was typically around 3.0 per cent in the 1980s and fell to about 2.5 per cent per annum in the 1990s (World Bank, 1999: 194–5), the implication is that industrial value added per capita has been falling gently throughout the two decades.

A slightly more encouraging statistic is that for the proportion of exports accounted for by manufactures, the unweighted average rising from 9.7 per cent to 13.8 per cent over the period 1980–96: the highest rising from 40 per cent (Sierra Leone) to 68 per cent (Mauritius), and the lowest rising from 0 per cent (Ethiopia, Mauritania, Nigeria and Rwanda) to 2 per cent (Congo and Gabon) over the same period. The manufacturing exports statistics are, however, also potentially among the least reliable, with twenty-six out of thirty-seven countries having no data available for 1996, and Nigeria having no recorded manufactured exports according to the World Bank source (which is simply not credible – refer to Thoburn, 2000). Of the eleven countries for which data are available seven experienced an increase in the proportion of manufactured exports in total exports and four experienced a decrease. Over the 16-year period extraordinary performances are apparent for Mauritius (increasing its manufactured exports from 27 per cent to 68 per cent of total exports) and South Africa (from 18 per cent to 49 per cent). This is a good example of the extent to which economic experience across sub-Saharan Africa as a whole defies generalisation. In the case of Mauritius, for example, the decline of the sugar industry and the determined effort to develop alternatives through export processing zones must account for much of the change, again illustrating the need to consider individual country circumstances in interpreting the data (Alter, 1991; Warr, 1990).

One of the main problems associated with the interpretation of the data in Table 15.1, which have been 'standardised' by the World Bank before publication,[2] is that of the meaning for the term 'industrial'. The United Nations System for National Accounts defines 'industry' to include mining, construction, electricity generation, water and sanitation, and gas supply, but the conventional economic definition is usually restricted to manufacturing – a distinction which is clearly maintained in the World Bank data source used for this section of the chapter (World Bank, 1999: 240). This point is also emphasised by Bennell (1998: 625). Those less-developed countries which have significant mineral extraction activity (including crude oil, gas and mineral ore production – for example Zambia and Gabon) often have a high proportion of GDP and of exports accounted for by elements which are included in 'industrial' statistics but which are not 'manufacturing'. In the context of a resource-based export promotion development strategy, which is consistent with the advice of most development economists, this leads to difficulties in interpreting export statistics. This produces a particular problem where, for example, increases in value added which are actually significantly in manufacturing (for example, sugar cane processing) are incorporated in primary product export statistics (for example, for sugar).

The changes between 1980 and the mid-1990s are obviously of considerable interest in the context of any hypothesised process of 'de-industrialisation'. First, taking the proportion of GDP accounted for by the industrial sector, inspection of the data for the thirty-seven countries in Table 15.1 shows that in 1980 industry contributed 20 per cent or more of GDP for twenty-two countries, and more than 30 per cent of GDP for eight countries. By 1997 industry contributed more than 20 per cent of GDP for twenty countries (a slight decline) and more than 30 per cent of GDP for ten countries (an increase). Without considering the individual circumstances of each country it is difficult to obtain any clear view of the reasons for these changes, but it is clear that the industrial sector is a robust element of several economies in sub-Saharan Africa.

Table 15.2 shows the extent of change in the industrial composition of manufacturing value added between 1980 and 1995 in terms of the nine ISIC (International Standard Industrial Classification) two-digit industrial groups. In the table, groups 31, 32, 33 and 34 include predominantly agro-industrial processing industries, group 35 includes chemical industries, group 36 clay and glass products, groups 37 and 38 metal and engineering industries, and group 39 comprises 'other' manufacturing.

One of the most notable features of sub-Saharan African manufacturing industry as a whole is the significance of agro-processing. This includes beer and cigarette production, food processing as well as textile production and garment manufacture. In 1995 more than half of manufacturing value added was accounted for by group 31 alone in the case of seven of the twenty-three countries for which UNIDO presents data (three of the countries in the table have no data for 1995 from this source). Two averages have been calculated from the data. The unweighted average is the arithmetic mean of the country data (not allowing for the economic size of the countries) and for this measure ISIC group 31 accounts for 46 per cent of all manufacturing value added in 1995. The weighted average has been calculated by taking the overall total manufacturing value added for all of the countries and dividing it by the total value added for each of the ISIC two-digit groups, and for this measure ISIC group 31 accounts for only 24 per cent of the total in 1995. The conclusions arising from these two statistics would, of course, be different. The difference between the unweighted and weighted averages of 46 and 24 per cent is explained by the fact that South Africa is a much larger economy than all of the other sub-Saharan African countries, having a considerably more developed manufacturing sector with a composition more related to a typical middle-income country, demonstrating that a single large country can very considerably affect continent averages.

The significance of the South African manufacturing sector in the statistics presented in Table 15.2 is made clearer by the calculation of the proportions of total sub-Saharan African manufacturing value added contributed for each of the sectors and for manufacturing as a whole over the 15-year period. For groups 31, 32 and 33 (that is, the more agro-processing sub-sectors) South Africa accounted for 40–60 per cent of the total – a considerable proportion. However, for the 'higher' ISIC groups 34 to 39 South Africa contributed about 70–90 per cent of the total. This is perhaps an argument for treating South Africa separately from the remainder of sub-Saharan Africa in this type of analysis. It is also of interest that over the period 1980–95 for group 33 (wood and wood products, furniture and fixtures) the proportion of sub-Saharan value added contributed by South Africa increased from 45 per cent to 60 per cent.

Within SSA in 1995, group 35 (chemicals) contributed more than 20 per cent of manufacturing value added in only six countries, and group 38 (metal fabrication/engineering) contributed more than 15 per cent of manufacturing value added in only four countries. For the same year, the unweighted average proportions of manufacturing value added for the twenty-three countries were 12.6 per cent for textiles, clothing, footwear and leather products (weighted average 9.4 per cent); 4.4 per cent for wood and wood products, furniture and fixtures (weighted average 2.6 per cent); and 4.9 per cent for paper and paper products, printing and publishing (weighted average 7.5 per cent). For 1995 this amounts to an unweighted average of 67.9 per cent (weighted average 43.6 per cent) of value added originating in sectors associated with processing of agricultural materials.

Overall, within the 15 years covered by the data in Table 15.2, there has been comparatively little change in the composition of manufacturing value added on average across the twenty-six countries for which data were available (this conclusion is not affected by the difference

Table 15.2 Composition of manufacturing value added in SSA (%)

Country	Year	Group 31	Group 32	Group 33	Group 34	Group 35	Group 36	Group 37	Group 38	Group 39	Total (US $ m)
Botswana	1980	43.6	18.0	0.0	0.0	0.0	0.0	0.0	7.7	30.1	39
	1995	45.3	16.0	3.8	5.7	3.8	0.0	0.0	5.7	19.8	212
Burkina Faso	1980	60.3	19.2	1.4	0.7	5.0	0.0	0.7	4.3	8.5	141
	1995	64.0	19.3	1.2	1.2	2.5	0.0	1.2	2.5	8.1	161
Burundi	1980	80.0	10.9	0.0	1.8	1.8	1.8	0.0	3.6	0.0	55
	1995	81.2	9.4	0.9	0.9	2.6	1.7	0.0	2.6	0.9	117
Cameroon	1980	57.0	9.1	6.2	5.4	5.8	3.2	6.2	5.5	1.6	691
	1995	31.7	17.5	12.4	2.2	22.0	6.8	5.4	1.3	0.7	542
Central Afr. Rep.	1980	34.3	14.3	31.4	2.9	8.6	0.0	0.0	8.6	0.0	35
	1995	62.2	0.0	16.2	5.4	8.1	0.0	0.0	5.4	2.7	37
Congo	1980	42.4	13.6	13.6	3.4	10.2	1.7	0.0	15.3	0.0	59
	1995	58.1	5.8	5.8	2.3	10.5	1.2	16.3	0.0	0.0	86
Côte d'Ivoire	1980	34.9	14.7	6.9	2.8	20.5	2.3	0.6	15.7	1.6	1271
	1995	28.0	14.0	6.7	0.8	30.0	1.9	0.2	15.3	3.2	1395
Ethiopia/Eritrea	1980	49.1	29.3	2.2	4.0	9.9	2.2	1.8	1.5	0.0	273
	1995	49.0	17.8	1.7	2.5	23.1	2.3	0.8	2.9	0.0	484
Gabon	1980	24.1	4.5	32.6	2.2	12.1	4.0	2.7	15.6	2.2	224
	1995	22.3	2.5	20.7	2.1	16.1	6.6	4.1	22.3	3.3	242
Gambia, The	1980	36.4	0.0	0.0	9.1	0.0	0.0	0.0	0.0	54.6	11
	1995	41.7	4.2	8.3	4.2	0.0	0.0	0.0	4.2	37.5	24
Ghana	1980	31.5	9.4	20.6	2.1	18.9	2.5	10.5	4.6	0.0	286
	1990	37.4	8.5	10.1	2.3	13.6	3.9	19.9	4.2	0.2	623
	1995										
Kenya	1980	33.9	12.3	3.9	7.5	15.8	3.2	1.6	20.9	0.8	743
	1995	42.8	8.7	2.5	6.8	15.6	4.1	1.7	16.1	1.7	813
Lesotho	1980	78.6	7.1	0.0	0.0	7.1	0.0	0.0	7.1	0.0	14
	1995	71.6	15.7	0.8	1.5	6.7	0.8	0.0	3.0	0.0	134
Madagascar	1980	27.0	43.7	1.8	4.5	11.7	1.8	0.0	8.6	0.9	222
	1995	27.6	40.9	1.6	5.5	15.8	2.4	0.0	6.3	0.0	127
Malawi	1980	57.3	12.1	2.4	8.1	8.1	2.4	0.0	9.7	0.0	124
	1995	42.5	7.8	1.3	7.2	21.6	1.3	0.0	12.4	5.9	153
Mauritius	1980	35.8	29.9	2.2	4.5	6.7	4.5	2.2	11.2	3.0	134
	1990	24.9	47.1	2.0	3.3	6.9	2.4	0.9	8.9	3.6	450
	1995										
Niger	1980	29.0	29.0	0.0	6.5	19.4	6.5	0.0	9.7	0.0	31
	1995	52.4	4.8	0.0	14.3	14.3	4.8	0.0	9.5	0.0	21

Country	Year	31	32	33	34	35	36	37	38	39	MVA
Nigeria	1980	17.8	11.1	6.2	4.9	21.5	4.8	1.6	31.7	0.6	2320
	1995	34.8	13.7	1.4	7.1	16.7	6.3	2.9	16.7	0.5	7881
Senegal	1980	49.6	18.8	1.5	3.8	14.7	4.5	0.0	7.1	0.0	266
	1995	48.6	5.5	0.3	2.8	30.6	6.3	0.0	6.0	0.0	399
South Africa	1980	12.3	8.7	2.4	6.4	17.0	5.2	15.1	30.6	2.3	17866
	1995	16.6	7.3	2.4	8.5	19.7	4.9	12.5	26.2	2.0	29071
Swaziland	1980	42.0	2.0	8.0	31.0	10.0	1.0	0.0	6.0	0.0	100
	1995	71.6	9.1	2.2	14.4	0.2	1.0	0.0	1.5	0.0	409
Tanzania	1980	23.3	33.2	3.6	6.1	15.2	3.1	1.7	13.3	0.6	361
	1995	27.3	19.8	2.5	6.6	24.0	5.8	4.1	9.9	0.0	121
Togo	1980	39.2	27.5	2.0	5.9	5.9	13.7	3.9	2.0	0.0	51
	1995	56.9	24.1	0.0	1.7	10.3	3.5	3.5	0.0	0.0	58
Zaire	1980	38.6	15.1	3.0	1.2	15.7	3.0	3.6	10.8	9.0	166
	1990	49.5	12.4	2.1	1.0	9.3	2.1	3.1	11.3	9.3	97
	1995										
Zambia	1980	44.0	13.3	2.6	4.1	13.9	4.7	1.5	15.6	0.3	780
	1995	43.0	11.6	4.5	3.1	22.7	3.1	1.3	10.5	0.2	449
Zimbabwe	1980	23.0	17.2	4.3	6.0	13.5	3.8	13.8	17.2	1.2	1480
	1995	39.9	13.5	3.7	5.9	10.4	2.9	9.8	13.6	0.4	1671
Unweighted av.	1980	40.2	16.3	6.5	4.8	11.1	3.1	2.6	10.9	4.5	27743
	1995	46.0	12.6	4.4	4.9	14.2	2.9	2.8	8.4	3.8	44607
Weighted av.	1980	18.7	10.7	3.4	5.8	16.2	4.6	10.9	25.7	1.9	
	1995	24.1	9.4	2.6	7.5	18.8	4.9	9.2	21.8	1.7	
S. African MVA (%)	1980	42.3	52.5	45.3	70.7	67.8	73.6	88.9	76.5	79.3	64.4
	1995	44.8	50.7	59.6	73.4	68.3	66.3	88.3	78.3	75.3	65.2

Sources: UNIDO, *Industry and Development: Global Report 1997*, Vienna: UNIDO, 1997 (Statistical appendix); for Ghana, Mauritius and Zaire the 1980 and 1990 data are sourced from UNIDO, *Industry and Development: Global Report 1993/94*, Vienna: UNIDO, 1993 (Statistical appendix)

Notes

a Group 31 = Food Products, Beverages, Tobacco Products

Group 32 = Textiles, Wearing Apparel, Leather and Fur Products, Footwear

Group 33 = Wood and Wood Products, Furniture and Fixtures

Group 34 = Paper and Paper Products, Printing and Publishing

Group 35 = Industrial Chemicals, Other Chemical Products, Petroleum Refineries, Miscellaneous Petroleum and Coal Products, Rubber Products, Plastic Products

Group 36 = Pottery, China and Earthenware, Glass and Glass Products

Group 37 = Iron and Steel, Non-Ferrous Metals

Group 38 = Metal Products, Non-Electrical Machinery, Electrical Machinery, Transport Equipment, Professional and Scientific Equipment

Group 39 = Other Manufacturing Industries

b The unweighted average is the simple arithmetic mean of individual country data. The weighted average is the total sub-Saharan African value added in each ISIC group as a proportion of total sub-Saharan manufacturing value added.

c Countries which do not appear in Table 2, such as Uganda, do not have the relevant data included in the 1997 UNIDO report.

between the unweighted and weighted averages). However, there is significant variation between individual country experiences over time. The literature on patterns of industrial development (for example, Syrquin and Chenery, 1989a and b – see below) would predict that in the long period there is a tendency for the proportion of manufacturing production arising in the 'higher' ISIC groups to increase. Focusing on ISIC group 31, out of the twenty-six countries included in the table, seven had broadly the same proportion of manufacturing value added arising in this group over the entire period, nine had a steady increase in the proportion, four showed a steady reduction in the proportion, and the other six varied markedly within the 15-year period. Focusing on ISIC groups 37 and 38 together, four countries had broadly the same proportion of manufacturing value added arising in these groups, three an increase in the proportion, nine a reduction in the proportion, and nine showed no clear change. In the case of Nigeria, one-third of manufacturing value added came from groups 37 and 38 in 1980, but the proportion had fallen to 19.6 per cent by 1995.

The textile industry perhaps justifies particular comment. It is one which tends to be regarded as a leading element of manufacturing development from a low base, perhaps because it played such a considerable role in the growth of industrial countries in the nineteenth century. However, perusal of the data for ISIC group 32 in Table 15.2 suggests that several countries have been struggling to maintain their current level of activity in textiles, or have experienced a fall in activity, rather than being able to regard this sub-sector as a strong basis for manufacturing sector growth. Further remarks on this issue will be found below in the next section.

Syrquin and Chenery's global approach is perhaps the most frequently referred to in the context of patterns of industrial development. Their analysis concludes that, over time and with increased levels of per capita income, the proportion of manufacturing value added in GDP increases and the proportion of manufacturing value added arising in the higher ISIC groups increases. This pattern is particularly modified by the factor of the economic size of countries (Syrquin and Chenery, 1989a: 29–36, 1989b) which is very relevant to sub-Saharan Africa. Syrquin and Chenery (1989b: 155) make the following remarks about patterns of industrialisation:

> During the process of industrialization, the composition of the manufacturing sector changes considerably. At a less aggregated level, country-specific features and policy become more prominent in determining the pattern of specialization. Large countries can better exploit economies of scale within their domestic markets and can more easily afford a strategy of import substitution. Variation in resource endowments is expected to generate differences in production patterns within manufacturing, particularly in small economies. Nevertheless, various studies have shown that a high degree of uniformity still remains in the pattern of change within the industrial sector.

They distinguish between *early* industries (food, textiles, clothing), *middle* industries (chemicals, non-metallic minerals) and *late* industries (machinery, paper, metal products), basically relating to the income elasticity of demand for the different groups of manufactures across various cross-sectional (and, implicitly, time series) levels of per capita income (ibid.: 156). Very broadly, the developmental patterns described by Syrquin and Chenery may be slightly detected in Tables 15.2, but the relationship is in no way strong or systematic. Other valuable discussions of the broad nature of industrial development may be found in Kuznets (1966: 127–43); Sutcliffe (1971: ch. 2); and Weiss (1988: ch. 1).

The experience of sub-Saharan Africa over the period 1980–95 does not permit easy generalisation, and there is evidence that some types of statistical analysis may indeed obscure the nature of development across the countries in the region. In quite a number of cases changes in the sub-sectoral composition of manufacturing value added reveal movements over the 15-year period which are in the opposite direction to that predicted by the global analysis cited in the above paragraph. The data in Table 15.2 make it clear that the economic size of a country is indeed a significant factor affecting the composition of manufacturing production. However, there are also other factors that appear to be very important, such as individual socio-political-economic developments (for example Nigeria, Ghana and Ethiopia) and clear development strategies (for example Mauritius). It should also be noted that data are not available from the UNIDO source that has been used in this study for quite a large number of countries in the region.

15.3 Aspects of industrial sector development in Ghana and Uganda, 1970–97

This section reviews the development of the manufacturing sectors in Ghana and Uganda over the 20-year period from the mid-1970s. These two countries have, for different reasons, experienced significant economic decline and considerable recovery over this period. They have also been regarded as prime examples of the successful application of IMF/World Bank stabilisation and adjustment programmes. They are therefore likely to give some indication of the resilience of sub-Saharan African economies (and of the manufacturing sector in particular). No countries could be regarded as 'typical' of sub-Saharan Africa, but Ghana and Uganda are representative of socio-economic structures which are quite common. Ghana pursued a vigorous import substitution programme in the period after independence, and Uganda is land-locked and has been somewhat dependent on Kenya. Both depend on primary production, principally agriculture, as the basis of their economies. As might be expected, fluctuations in industrial sector performance have exceeded those of the national economy. In the context of the possibility of African de-industrialisation these countries are of considerable interest since their experience raises the questions of the extent to which the manufacturing sector has (a) been economically resilient, and (b) moved from recovery into a period of sustainable growth and development.

Table 15.3 shows indices of manufacturing production. For Ghana the index has been published since the latter part of the 1970s, and for Uganda since the mid-1980s. It is notable that agricultural processing activities (reflecting the lower ISIC sections discussed in the context of Table 15.2) account for a high proportion of the weights: almost 54 per cent for Ghana and over 65 per cent for Uganda.

For Ghana the overall manufacturing index shows an approximate halving of manufacturing production between 1977 and 1985, and substantially more than doubling between 1985 and 1996. The level of manufacturing production at the end of the 20-year period was about 15 per cent higher than at the beginning. The most remarkable recoveries between 1985 and 1996 were experienced in textiles, wearing apparel and leather goods (nearly three-fold), chemical products other than petroleum (4.7-fold), cement and other non-metallic mineral products (four-fold), iron and steel products (12.5-fold), and non-ferrous metal basic industries (4.4-fold, which must be largely accounted for by the suspension of aluminium smelting during the prolonged drought in the mid-1980s and renegotiation of international pricing within the contractual arrangements surrounding aluminium smelting).

Table 15.3a Ghana: index of manufacturing production, 1978–96 (1977=100)

Industry	Weight	1978	1980	1985	1990	1995	1996
Food manufacturing	15.00	84.8	70.0	41.8	57.5	99.6	102.5
Beverage industries	8.11	77.0	70.2	59.3	94.0	109.0	116.2
Tobacco and tobacco products	7.75	66.1	67.0	61.3	57.1	52.0	53.1
Textile, wearing apparel and leather goods	13.71	81.5	41.4	19.2	37.7	54.8	56.1
Sawmill and wood products	7.22	92.1	52.0	75.4	74.2	100.2	105.3
Paper products and printing	1.94	103.5	80.8	65.1	53.5	45.1	49.3
Petroleum refinery	19.00	96.0	87.9	80.6	70.5	101.4	103.5
Chemical products other than petroleum	6.56	40.5	34.7	31.8	57.6	140.0	148.2
Cement and other non-metallic mineral products	2.98	87.2	52.1	63.6	117.3	258.1	258.9
Iron and steel products	3.25	39.3	73.9	46.2	5.2	581.6	584.5
Non-ferrous metal basic industries	9.62	90.3	111.8	28.4	103.8	119.8	125.6
Cutlery and other non-ferrous metal products	0.49	67.0	33.1	34.6	55.2	102.4	116.4
Electrical equipment and appliances	1.34	73.2	26.1	28.4	25.5	42.9	53.5
Transport Equipment	3.03	76.8	78.5	n.a.	—	n.a.	n.a.
All manufacturing industries	100.00	81.0	69.0	49.3	63.5	109.9	115.0

Source: Republic of Ghana, *Quarterly Digest of Statistics*, various issues, and June 1995: table 11a:13, June 1996: table 11a:13, June 1997: Table 11a:13

Table 15.3b Uganda: index of manufacturing production, 1985–97 (1987=100)

Industry	Weight	1985	1990	1995	1997
Food processing	20.7	93.9	174.9	361.8	423.6
Tobacco and beverages	26.1	84.8	155.2	308.5	398.5
Textiles and clothing	16.3	98.9	116.3	62.7	113.5
Leather and footwear	2.3	86.9	75.3	163.9	158.9
Timber, paper and printing	9.0	76.8	183.6	383.1	468.6
Chemicals, paint and soap	12.3	58.6	183.5	512.7	776.4
Bricks and cement	4.3	122.7	154.2	367.4	981.4
Steel and steel products	5.3	133.1	107.7	490.5	451.5
Miscellaneous	3.7	139.1	181.3	598.6	577.3
All items	100.0	91.3	155.5	330.9	441.3

Sources: Republic of Uganda, *Background to the Budget 1992–93*, table 61: 200; 1997–8: table 24: A26; Kampala: Ministry of Finance and Economic Planning, 1992; *Statistical Abstract 1998*, Entebbe: Ministry of Finance, Planning and Economic Development, 1998, table L1: 91

For Uganda the series for the Index of Industrial Production shown in Table 15.3b does not appear for any year before 1985, but for the period of strong economic recovery after that year the data give a clear picture, with the aggregate index of industrial production increasing by a factor of about 4.8 between 1985 and 1997. It should be noted that the base year (1987) is one in which manufacturing production was at a historically very low level. The statistics show that the overall strength of the recovery of manufacturing industry has been even stronger in Uganda than in Ghana.

Textiles and clothing (accounting for a weight of 16.3 per cent of the Ugandan index, as compared with 13.71 per cent for the Ghanaian index) have not enjoyed the same type of recovery evident in the manufacturing sector as a whole, and growth has also been less strong in leather and footwear than in other industries. The relatively poor performance of the textiles and clothing sector is perhaps attributable to two separate factors. First has been the difficulty in achieving a recovery of raw cotton production and in rehabilitating the textile industry. Second, imports of second-hand clothing have tended to eclipse the ability of domestic capacity to expand.

There has been particularly strong growth in sectors associated with construction activity, such as cement, bricks and iron and steel products. However, these sub-sectors account for only a relatively small proportion of the weights for all manufacturing activity – in Ghana just over 6 per cent of the total, and in Uganda just under 9 per cent. This is perhaps indicative of the volatility of individual sub-sectors, making the weights susceptible to variation within comparatively short periods.

Overall the data for both Ghana and Uganda are consistent with a decline in production from 1975 to 1985 followed by significant recovery through to 1996.[3] For Ghana there are a number of factors which explain the decline, including:

- government mismanagement at micro- and macroeconomic levels leading, *inter alia*, to acute distortions, significant parallel markets (including a seriously over-valued exchange rate) and general dislocation of the economy (World Bank, 1983: ch. 6);
- shortages of foreign exchange for the purchase of essential imported inputs and spare parts;
- poor performance of infrastructure (including electricity supply), of industries supplying domestic inputs and of service industries;
- decline of the economy with an associated loss of consumer purchasing power and of demand for intermediate inputs.

The subsequent recovery is largely explained by the reversal of these factors (Acheampong and Tribe, forthcoming).

For Uganda, over the period from 1970 to the mid-1980s there was a very considerable decline followed, in most industries, by strong recovery.[4] However, the reasons for the decline in Uganda were very different to those in Ghana. The dislocation of the economy in the 1970s and early 1980s was due partly to shortcomings of economic management, but principally to political disruption and an internal civil war (Collier and Pradhan, 1998). This led, *inter alia*, to the allocation of a high proportion of foreign exchange to military expenditure: the shortage of foreign exchange for the purchase of essential productive inputs was less a result of an overall shortage of foreign exchange and more of the uses to which that which was available was allocated. Only from 1986, with the ending of the major internal strife, is it possible to view the economy as being under proper control in more 'normal' conditions.

The capacity utilisation data for Ghana in Table 15.4a (only available to 1993 at the time of writing) provides further support for the view that the manufacturing sub-sector experienced a remarkable recovery from a prolonged decline through to the mid-1980s. The order of magnitude is again such that a fall of about 40 per cent to 1985 is followed by a recovery to a 1993 level which is higher than that ruling in the late 1970s. Some individual product groups have significantly better performance than others, but the unavoidable conclusion is that there has been a very strong recovery. The supposition must be that a considerable proportion of the recovery of manufacturing production since the mid-1980s was due to improvements in capacity utilisation rather than to capacity expansion and technological change.

Table 15.4a Ghana: manufacturing industries: estimated rate of capacity utilisation, large and medium scale factories (%)

Sub-sector	1978	1980	1985	1990	1993
Textiles	40.0	20.1	19.7	35.0	41.3
Garments	38.1	29.9	25.5	22.0	53.3
Metals	28.2	28.4	16.2	49.0	80.0
Electricals	32.1	17.8	33.2	13.4	23.9
Plastics	10.6	19.1	28.0	40.0	45.0
Vehicle assembly (bicycle/motor cycle)	18.4	n.a.	19.9	25.0	16.4
Tobacco and beverages	50.0	30.0	39.6	65.0	76.3
Food processing	40.8	30.0	31.2	55.0	52.3
Leather	31.3	20.9	21.5	12.0	10.0
Pharmaceuticals	25.0	16.8	16.6	30.0	40.0
Cosmetics	33.4	8.0	n.a.	25.0	16.2
Paper & printing	31.0	28.4	14.5	30.0	45.0
Non-metallic mineral manufactures	47.0	29.7	35.0	48.0	72.8
Chemicals	42.0	28.0	20.2	30.2	40.0
Rubber	21.6	16.4	16.0	48.0	54.0
Wood processing	36.0	27.3	32.5	70.0	65.0
Miscellaneous	55.9	44.9	n.a.	n.a.	n.a.
All manufacturing industries	40.4	25.5	25.0	39.8	45.7

Source: Republic of Ghana, *Quarterly Digest of Statistics*, various issues, and June 1995: table 10:12

Note from original source: Data for individual industries are obtained from Ministry of Industries. The estimate for all manufacturing industries is a weighted arithmetic average using weights proportional to the value of gross output in 1973.

Table 15.4b Uganda: manufacturing industries: estimated rate of capacity utilisation (%)

	1982	1983	1984	1985	1986	1987	1997
Sugar	2.1	2.0	1.8	0.5	—	—	64.5
Beer	19.9	28.9	30.8	17.1	14.0	34.4	182.5[a]
Soft drinks	11.4	25.3	37.1	32.2	32.5	37.4	n.a.
Cigarettes	39.2	33.9	50.8	74.5	74.7	75.5	95.3
Cement	3.6	6.1	4.9	2.3	15.0	3.1	32.0

Sources: Republic of Uganda, *Background to the Budget 1988/89*, Kampala: Ministry of Planning and Economic Development, 1988; Republic of Uganda, *Statistical Abstract 1998*, Entebbe: Ministry of Finance, Planning and Economic Development, 1998

Notes
Manufacturing capacity utilisation data are given for earlier years on a plant basis, making comparability difficult due to the problem of aggregating data consistently. Specific data are not given for years after 1987, but it is possible to derive figures from production and installed capacity data.
a This statistic is consistent with published data.

Of wider interest is the question of how these capacity utilisation figures can be interpreted. The returns to the Ghanaian Ministry of Industries request individual responding firms to estimate their own 'full capacity' as well as to report their capacity utilisation, so that the responses reflect varying perceptions of the definition of capacity. These variations are mainly accounted for by the objective circumstances of individual industries. International studies of capacity utilisation suggest that manufacturing capacity utilisation is usually considerably lower than most lay observers might expect, so that given the greater problems encountered in maintaining higher levels in countries such as Ghana, a 45–50 per cent level can be regarded as a respectable performance.[5]

The capacity utilisation statistics for Uganda in Table 15.4b cover only the period from 1982 to 1987 and then on a calculated basis for 1997. Only very limited conclusions can be drawn from the data available here, but it is clear that cigarette production has been buoyant and that capacity utilisation in soft drink production trebled to around 35 per cent over the period 1982–4 and remained at that level. The beer, cement and sugar industries had very poor capacity utilisation through the period to 1987, but the data for 1997 show that there has also been a strong recovery in these industries since then.

15.4 The issue of de-industrialisation

In recent years the literature on African industrial development has included several contributions which have dwelt on the issue of 'de-industrialisation'. Different authors have interpreted 'de-industrialisation' in different ways, and so there has not been an opportunity to reach any form of consensus in terms of either supporting or refuting a hypothesised 'de-industrialisation' process for sub-Saharan Africa. Associated with this issue is the controversy over the impact of trade liberalisation (and structural adjustment programmes in general) on the sub-Saharan African manufacturing sector. This section of the chapter will proceed by discussing a view of the protected import-substitution 'distorted growth' model of industrial development, then outlining the impact of liberalisation before moving on to consider the nature of the statistical evidence briefly.

15.4.1 De-industrialisation through distorted growth

The strategy of industrial development that was followed by most sub-Saharan African countries after independence was generally one of import substitution, with varying types of trade protection. This trade protection was initially, in some cases, an unintended by-product of tax regimes that depended on import tariffs as a major source of revenue in the absence of other viable areas of government income. As the economies have developed, systems of taxation have changed and import tariffs are, in general, no longer a significant source of government income. In other cases the trade protection was quite intentional and was, at least nominally, related to a policy of infant industry protection – with import tariffs and quotas having the instrumental policy roles. In most cases the highest levels of tariffs were placed on imports of manufactures for final consumption, medium levels on imported inputs for the productive sectors of the economy, and the lowest levels on imports of capital goods (in order to encourage investment) (Tribe, 2000b). It was this approach to the protection of industrial development, together with associated bureaucratic structures, which was so roundly criticised in Little *et al.*'s (1970) OECD study.

The intellectually based economic argument against excessive protection of import substituting industries should not necessarily be couched in terms of the transgression of GATT/WTO 'rules'. The breaking of rules may be reprehensible, but in many cases may make very good economic sense. Rather, the argument against high levels of trade protection should be in terms of the resource misallocation and the resultant 'dead-end' development that is associated with persistently high levels of protection, logically leading to stagnation. Some years ago Merhav (1969) produced a book based on this stagnation thesis, relating in the first instance to the Israeli economy, but applicable by extension to many small sub-Saharan African countries. There can be little doubt that import substitution, especially of the protected variety, has tended to lead to excessive dependence on imported inputs and to conservatism in the use of local materials.

The case against highly protected import substituting manufacturing sectors is therefore that the strategy leads to industrial stagnation and a retardation of national economic growth. Part of this argument relates to the bias of the internal terms of trade in favour of manufacturing industry and against agriculture, and other economic distortions resulting from a highly protective trade regime. Another part of the argument relates to the conservative and restricted nature of a purely import substitution approach to industrialisation for comparatively small economies. Superimposed on this argument have been ideological overtones, so that one set of arguments for the liberalisation of trade regimes (the removal of protection from imports and of taxes from exports) and for the removal of other economic controls (price and wage controls for example) has been based on the 'neo-liberal' predilection for a theoretically 'pure' market approach to economic management (without regard to market imperfections and 'failures'). A second set of arguments for some degree of liberalisation is based on the observation that the economic controls and trade protection, in general, failed to achieve their policy objectives in the long run, and were therefore counter-productive – this is a more 'structuralist' approach – but some 'structuralist' commentators have failed to recognise this second set of arguments. Liberalisation has also been directed towards the traditional export sector, of course, where it has been argued that export taxes on agricultural commodities and restricted competition within the domestic marketing of agricultural exports have lowered prices to producers and have acted as a depressing factor on export development.

Lall and Wangwe (1998: 71–2) discuss elements of this more structuralist approach in some considerable detail, explaining their position thus:

> Much of the explanation of poor industrial performance in SSA lies with exogenous shocks of various kinds: droughts, wars, internal conflict, political instability, adverse terms of trade and so on. Yet this is clearly not the w hole story. Poor policies also have to carry much of the blame. A broad range of government policies affects industrial development, from general macroeconomic management, through trade and competition policies and human resource development, to specific industrial development and technology policies.

Collier (2000) has outlined a large number of the issues which militate against the international competitiveness and development of sub-Saharan African manufacturing industry, some of which are inherent in the current structures of the economies, and others of which have been 'acquired' through the adoption of policies which could be regarded as 'perverse'. Collier therefore suggests that any relative deterioration of the international competitiveness of sub-Saharan African industry from its already disadvantaged position is based on these structural factors. These arguments imply that 'globalisation' is likely to work to the detriment of sub-Saharan Africa rather than to its advantage.

15.4.2 The impact of liberalisation

A second set of arguments relating to the potential for de-industrialisation in sub-Saharan Africa is based on the impact of liberalisation measures on the economies, and on the industrial sector in particular. These arguments regard de-industrialisation as the *effect* of liberalisation rather than the *reason* for it. Stein (1992: 86) has perhaps put this argument most succinctly:

In general, the World Bank/IMF approach is unduly exclusionary, internally inconsistent and largely underdetermined. While the [liberalisation] measures recommended will lead to a decline in import-substituting manufacturing and public ownership, these are unlikely to be replaced by the ownership or industry types desired by the World Bank/IMF (export oriented, resource processing, etc.). Overall, the policies are likely to deindustrialize, forcing countries into a problematic reliance on resource and agricultural exports.

In a review of the World Bank (1994) report *Adjustment in Africa* Lall (1995: 2022) presents the following judgement on African industry:

The growth of manufacturing value-added (MVA) during 1980–93 was only 3% per annum in real terms, and the rate of growth declined steadily over time, from 3.7% in the first half of the 1980s to 2% in 1989–93. The performance was worse in the first part of the 1990s: MVA growth fell from 3.3% in 1989–90 to 0.4% in 1991–92. Moreover, these figures conceal stagnation or drops in MVA in a large number of countries: many African countries suffered sustained 'de-industrialization' over the past decade and a half.

In a more recent article Bennell (1998: 621, 635) reviews some of the statistical evidence for the de-industrialisation hypothesis which is interpreted as follows:

Most industrial enterprises which had previously been heavily protected from both domestic and international competition as part of pervasive import substitution programmes will, it is argued, be unable to survive once this protection is removed. While it is accepted that there may be certain manufacturing activities that stand little chance of becoming internationally competitive, there are many others that could, given the right kind of support, modernize and be able to reach the 'best practice frontier'. An oft-repeated criticism of the international finance institutions (that is, the IMF and the World Bank) is that, until recently at least, they have been too narrowly preoccupied with meeting the objectives of short-term macroeconomic stabilization programmes and the elimination of market distortions which have favoured a grossly inefficient industrial sector (that is, 'getting prices right') . . . our analysis suggests that (i) MVA growth has slowed down in the large majority of economies in SSA since the start of adjustment programmes; and (ii) during the first half of the 1990s, manufacturing output has contracted in half of the countries for which data are available.

Noorbakhsh and Paloni (2000) have undertaken a very careful statistical analysis of the manufacturing sector for thirty-eight sub-Saharan African countries covering the period 1980–94. Their conclusions are very clear although, to a considerable extent, they are not consistent with the earlier discussion in this chapter based on the UNIDO data in Table 15.3, or with the country case studies for Ghana and Uganda. The significance of the great diversity within sub-Saharan Africa and the importance of individual country case studies should be emphasised. They conclude that:

The empirical results clearly highlight the depressed state of manufacturing in Africa and its poor prospects of imminent development. Over the period covered rates of growth of manufacturing and GDP dropped to negligible levels, manufacturing employment

contracted, labour productivity declined, with little evidence of technological transfer –
let alone advancement. Industrial non-diversification increased, while investment and
the productivity of investment were stagnant. These indicators show not only a
worsening of the situation in 1991–1994, but also that very little progress, if any, has
been made since 1980–1985 . . . It has been suggested that industrial development in
SSA is more difficult than elsewhere because SSA has poor infrastructure (both physical
and institutional), insufficient human capital and entrepreneurship, small and fragmented
markets. The empirical evidence in this chapter seems to support the view that in this
context, SAPs may sometimes be in contradiction with the long-run objective of building
up dynamic comparative advantage in industry.

(Noorbakhsh and Paloni, 2000: 126–9)

Jalilian and Weiss (2000) have undertaken a statistical analysis based on data for sixteen
sub-Saharan African countries over the longer period of 1970 to 1993 which reaches
conclusions contrary to those of Noorbakhsh and Paloni, based on a different definition of
de-industrialisation. They find that the analysis provides 'no support for the general
proposition that as a region Africa has been experiencing a degree of de-industrialisation not
found elsewhere'. Their relative definition of de-industrialisation 'requires that a country has
both a lower than predicted share of manufacturing in total activity, given its set of country
characteristics, and that its deviation from the predicted share grew over time' and their
absolute definition 'requires that the size of manufacturing is lower than predicted and that
this disparity grew over time'. This methodology relies on direct reference to economic
variables, rather than on the system of proxies used by Noorbakhsh and Paloni.

It is also significant that a number of studies have emphasised the role, or lack, of
technological development in the growth of the manufacturing sector. Given his particular
interest in the role of technology in economic development it is hardly surprising that Lall
gives this question a central role in discussion of the major features of manufacturing sector
performance in a recent joint publication (Lall and Wangwe, 1998: 94–6). The discussion
concludes that the evidence on imported technology, upgrading of small and medium-scale
enterprises, improving technology infrastructure and the promotion of research and
development give no grounds for an optimistic view of the technological dynamism of African
industry. Noorbakhsh and Paloni (2000) include a technology element in their statistical
analysis, using data for imported machinery and transport equipment as a ratio of total imports
as 'a proxy for the transfer of technology and the availability of imported inputs'. They find
that 'these data provide no evidence of technological advancement in SSA industry'. The
evidence that Acheampong and Tribe (forthcoming) present of negative but improving rates
of change in total factor productivity in Ghanaian manufacturing, together with substantial
increases in the level of output, is inconclusive in terms of technological change, but Teal
(1998) is much more adamant that the World Bank panel data from Ghana provides evidence
of 'zero technical progress'. Overall the consensus appears to support the rather gloomy view
that large and medium-scale manufacturing in sub-Saharan Africa has witnessed little
technological advancement in the last 10 to 15 years.

15.4.3 *The interpretation of statistical evidence*

There are several problems over the interpretation of statistics relating to African industrial
development, apart from the uncertain reliability of the statistics. If we consider again the
specific cases of Ghana and Uganda, both countries suffered a severe economic decline from

the mid-1970s to the mid-1980s, the reasons for which were significantly different in the two cases. In neither could it be said that the deterioration in the manufacturing sector was due principally to the adoption of an import-substitution approach *per se*. A considerable amount of import substitution took place as a fairly natural form of market development (Tribe, 2000b). However, trade protection and the seriously over-valued exchange rates that were associated with the strategy of economic management (in part associated with a proactive import-substitution approach) did prove to be economically counterproductive.

In the Ghanaian case it is fair to infer that the principal reason for the decline in the entire economy was mismanagement, including serious market distortions. It is likely that two of the main reasons for the manufacturing decline were a shortage of foreign exchange for the purchase of imported inputs (including spare parts and equipment), and poor maintenance due partly to cash flow problems arising from low levels of capacity utilisation associated with shortages of inputs and low levels of demand due to the economic decline (refer to Acheampong, 1996; Acheampong and Tribe, 1998 and forthcoming). In the case of Uganda the internal political and civil unrest, economic mismanagement and high levels of military expenditure associated with civil war, leading to economic decline and shortages of foreign exchange must be the main causes of manufacturing decline (see, for example, Collier and Pradhan, 1998). Further evidence on this issue is provided by the Tanzanian manufacturing sector which experienced low levels of capacity utilisation due to shortages of foreign exchange (and a number of other causes), this being a key factor affecting the economic performance of the sector (Ndulu, 1986; Wangwe, 1977).

Recovery also raises problems of interpretation. The hypothesised 'de-industrialisation' based on the liberalisation process does not appear to gain much support from the evidence presented in section 14.3 of this chapter. Both Ghana and Uganda have experienced very strong recoveries in manufacturing output over the period from the mid-1980s to the mid-1990s. Much of this recovery is based on improvements in capacity utilisation in existing industries, some of which have had significant rehabilitations, and in some cases privatisation/divestiture has contributed positively. However, more recently there has been evidence of a slowdown of Ghanaian manufacturing growth (Acheampong and Tribe, forthcoming; and Teal, 1998). There are at least three possible interpretations: first, the recovery has been based on many firms which were originally established within protected import substitution regimes, and these firms have demonstrated a long-term resilience and sustainability; second, the strong initial recovery does not necessarily provide evidence that sustained future growth based on capacity expansion and investment in new industries can be confidently expected following liberalisation; third, if the recovery has in fact been based very largely on improved capacity utilisation (as in Ghana) the implication is that total factor productivity growth has been negligible within manufacturing during the recovery and that technological development in the sector has not been a significant feature despite observed improvements in labour productivity (Acheampong and Tribe, forthcoming; Teal, 1998; Lall, 1995: 2021–6). If the recovery has been based on better capacity utilisation which arises from improved supply of foreign exchange it is not necessarily the case that this foreign exchange has been available due to better export performance following overall economic recovery – much of the higher foreign exchange inflows have been associated with aid, 'conditionality' and recovery of non-manufacturing exports. In addition, improvements to industrial performance can also be explained by higher levels of demand associated with overall economic recovery, by better operation of the economy due to improved infrastructure performance and by improved management at the level of the economy as a whole (as well as at plant level).

The overall conclusion must be that the case for the existence of sub-Saharan African de-industrialisation in the 1990s and into the twenty-first century is not strong, whether based on a long-term view of distorted growth arising from import substitution/protective trade regimes or on the detrimental effects of liberalisation. Equally, there is little evidence as yet that liberalisation has been associated with a positive transformation of the manufacturing sector – the opposite of de-industrialisation. There are three important points to be made: first, the case against the import substitution strategy *per se* (separating the basic form of the strategy conceptually from the overly protectionist trade policies which have often been adopted in the past) appears to have been weakened by the resilience of manufacturing following periods of substantial decline (for Ghana and Uganda at least, and probably for other countries); second, the recovery experienced by the manufacturing sector has been associated, *inter alia*, with trade liberalisation, and the eradication of seriously protectionist policies and other 'distortions' has contributed to the marked improvement in performance; third, the period of 10–15 years which has elapsed since the institution of the determined liberalisation of economic management and control systems is possibly far too short (given the known time-lags which exist: delays with recovery of business confidence; the time taken to implement liberalisation effectively; time lags between investment opportunities being identified and production coming on stream and so on) to permit manufacturing sector performance to exhibit the long-term gains which have been hypothesised by the proponents of the liberalisation strategy. Claims that liberalisation has 'worked' supported by reference to evidence of short-term recovery in the manufacturing sector without consideration of the more significant long-term prospects are seriously deficient.

15.5 Conclusions

We can, perhaps, extract the following conclusions, from the above:

- there is evidence that there has been fairly steady growth on average of the industrial sector (including manufacturing) in sub-Saharan Africa during the 1980s and 1990s;
- there is exceptionally great diversity within sub-Saharan Africa, so that some countries have characteristics which place them in, or close to, the category of 'newly industrialising countries', while others have very small and rudimentary manufacturing sub-sectors;
- per capita industrial/manufacturing growth has been negligible on average over the two decades reviewed in this chapter;
- there has been very substantial positive manufacturing development in a number of sub-Saharan African countries over the last 15–20 years, with several countries experiencing substantial recovery from a period of economic (and industrial) decline;
- a high proportion of manufacturing in sub-Saharan Africa is based on the processing of agricultural and other primary products;
- in general, manufactured exports are only a very small proportion of total exports, and manufactured imports tend to be a high proportion of total imports;
- interpretation of the statistics relating to the development of the manufacturing sub-sector for much of sub-Saharan Africa is sometimes difficult due to the fact that some agricultural and mineral exports contain a significant amount of manufacturing value added, and some industrial and manufacturing exports contain a significant amount of agricultural and mineral value added. In general it must be acknowledged that economic statistics vary considerably in their reliability across the continent.

A high proportion of manufacturing production in sub-Saharan African countries is targeted at local domestic markets rather than for export, and some of this production has experienced significant trade protection. Much of this production is likely to be economically viable because of natural protection (transport costs in particular), for example in the areas of food and other agricultural processing and in the production of cement and other building materials. One of the most surprising findings of this review has been the weakness of the textile and garment manufacturing industry. In general exports may be constrained by product quality and by the underdeveloped nature of international trading institutions rather than by price alone. Two of the principal reasons for the decline in manufacturing production, and for the subsequent recovery in Ghana and Uganda in particular, were the changes in the availability of foreign exchange for the purchase of essential inputs and in the strength of domestic demand during a period of general economic decline and recovery. In this context the economic debate about trade liberalisation as a basis for the development of manufactured exports is of more limited relevance to much of sub-Saharan Africa than might be thought from inspection of the recent literature on this subject.

The fact that, in general, manufactured exports are only a very small proportion of total sub-Saharan African exports may be largely attributable to the low level of production of the types of manufactured products demanded by the principal international markets – types of manufactured products which are produced in great quantities in the newly industrialised countries of Southeast Asia. Neighbouring African countries, which could be potential export markets, have similar patterns of production in their own manufacturing industry so that there is great competition between neighbours in the production of the same types of products. In addition, many of the economic institutions that would have the capacity to facilitate intra-regional trade are still in the early stages of development. There is good reason to believe that a significant amount of trading between neighbouring African countries is unrecorded, and may be regarded as 'smuggling' in many cases. Added to these factors is the question of political stability, the absence of which has made it difficult to develop long-term regional economic cooperation in some parts of the continent.

Passing on to consider the individual country data from Ghana and Uganda, it is apparent that the published statistics are somewhat patchy, particularly with respect to international trade in manufactures. For this reason it is only possible to draw very broad conclusions from the published data. Over the last two decades there has been a very considerable decline in manufacturing production in both countries, followed by an equally considerable recovery. The main long-term issue is whether this recovery and strong growth can be sustained in the longer period through the expansion of existing industries and the introduction of new industries. In all other respects the individual country data from Ghana and Uganda confirm the conclusions drawn from the broader sub-Saharan data.

Section 14.4 of the chapter considered some of the arguments and evidence which has been put forward in relation to the potential process of de-industrialisation, which some have suggested is threatening sub-Saharan Africa. The fact that a number of different versions of the de-industrialisation hypothesis exists and that statistical analysis is made difficult by methodological and statistical reliability problems makes the 'debate' inconclusive. However, most of the literature emphasises the very mixed, and often poor, performance of the sub-Saharan African manufacturing sector. It is also far from clear that the experiences of the East Asian 'miracle' countries are directly transferable to sub-Saharan African countries, so that such international comparisons are likely to be less informative than careful analysis of the development potential based on African conditions and experiences.[6]

Notes

1 A longer version of this chapter, revised from the paper presented at the SCUSA Conference and including somewhat more statistical tables, is available from the author on request. Readers are also referred to the author's chapter in Jalilian *et al.* (2000). It should be emphasised that the discussion in this chapter does not include the small-scale sector, largely on the grounds that the statistics which are presented do not incorporate data on small-scale enterprises.

2 'Considerable effort has been made to standardize the data, but full comparability cannot be assured, and care must be taken in interpreting the indicators. Many factors affect availability, comparability, and reliability: statistical systems in many developing economies are still weak; statistical methods, coverage, practices, and definitions differ widely; and cross-country and intertemporal comparisons involve complex technical and conceptual problems that cannot be unequivocally resolved' (World Bank, 1999: 233).

3 More careful analysis of the Ghanaian manufacturing sector will be found in Acheampong (1996) and in Acheampong and Tribe (1998 and 1999). A recent cautionary comment on the interpretation of the impact of structural adjustment on Ghanaian manufacturing performance may be found in Lall and Stewart (1996: 194–6).

4 For another view of Uganda's recent manufacturing sector performance, see Livingstone (1998).

5 'US manufacturing capital stock was idle over 75% of the time, even in the best of years. The other study was Robin Marris' 1964 report on British industry that found that firms planned *ex ante* and intentionally to leave capital idle most of the time as a rational part of their investment decision' (Winston, 1974: 1301). Winston (1971) represents an earlier approach to the same issues. Reference may also be made to Bautista *et al.*'s study for the World Bank (1981, esp. chs 2 and 3); Ndulu's (1986) and Wangwe's (1977) articles on Tanzanian experience, and Weiss's (1988: ch. 6) discussion of the definition of manufacturing capacity utilisation. Acheampong (1996) and Acheampong and Tribe (1998) present some recent survey data on manufacturing capacity utilisation in Ghana.

6 The article by Cline (1982) is still a major reference on this topic. More recently Chhibber and Leechor (1995) have considered broad factors of economic management and entrepreneurial development (consistent in many respects with Lall and Wangwe, 1998) in comparing East Asian and Ghanaian development. Stein's (1995) edited collection is a very comprehensive approach to the issues associated with 'lessons' from Asia for African industrialisation.

References

Acheampong, I.K. (1996) *The Impact of the Economic Recovery Programme (1983–90) on Industrial Performance in Ghana*, PhD thesis, University of Bradford.

Acheampong, I.K. and Tribe, M. (1998) 'The response of Ghana's manufacturing sector to structural adjustment', in P. Cook, C. Kirkpatrick and F. Nixson (eds), *Privatisation, Entrepreneurship and Economic Reform*, Cheltenham, Glos.: Edward Elgar.

Acheampong, I.K. and Tribe, M. (2001), 'Sources of industrial growth: the impact of policy on large and medium scale manufacturing performance in Ghana', in O. Morrissey and M. Tribe (eds), *Policy Reform and Manufacturing Performance in Developing Countries*, Cheltenham, Glos.: Edward Elgar.

Alter, R. (1991) 'Lessons from the export processing zone in Mauritius', *Finance and Development*, 28 (4): 7–9.

Bautista, R.M., Hughes, H., Lim, D., Morawetz, D. and Thoumi, F.E. (1981) *Capital Utilization in Manufacturing*, New York: Oxford University Press for the World Bank.

Bennell, P. (1998) 'Fighting for survival: manufacturing industry and adjustment in sub-Saharan Africa', *Journal of International Development*, 10 (5): 621–37.

Chhibber, A. and Leechor, C. (1995) 'From adjustment to growth in sub-Saharan Africa: the lessons of East Asian experience applied to Ghana', *Journal of African Economies*, 4 (1): 83–114.

Cline, W. (1982) 'Can the East Asian model of development be generalized?', *World Development*, 10 (2): 81–90.

Collier, P. (2000) 'Africa's comparative advantage', in H. Jalilian, M. Tribe and J. Weiss (eds), *Industrial*

Development and Policy in Africa: Issues of De-industrialisation and Development Strategy, Cheltenham, Glos.: Edward Elgar.

Collier, P. and Pradhan, S. (1998), 'Economic aspects of the transition from civil war', in H.B. Hansen and M. Twaddle (eds), *Developing Uganda*, Oxford: James Currey.

Jalilian, H. and Weiss, J. (2000) 'De-industrialisation in sub-Saharan Africa: myth or crisis?', in H. Jalilian, M. Tribe and J. Weiss (eds), *Industrial Development and Policy in Africa: Issues of De-industrialisation and Development Strategy*, Cheltenham, Glos.: Edward Elgar.

Jalilian, H., Tribe, M. and Weiss, J. (eds) (2000) *Industrial Development and Policy in Africa: Issues of De-industrialization Strategy*, Cheltenham, Glos.: Edward Elgar.

Kuznets, S. (1966) *Modern Economic Growth: Rate, Structure and Spread*, New Haven, CT: Yale University Press.

Lall, S. (1995) 'Structural adjustment and African industry', *World Development*, 23 (12):, 2019–31.

Lall, S. and Stewart, F. (1996) 'Trade and industrial policy in Africa', in B. Ndulu and N. van de Walle (eds), *Agenda for Africa's Economic Renewal*, New Brunswick, NJ: Transaction Publishers for the Overseas Development Council.

Lall, S. and Wangwe, S. (1998) 'Industrial policy and industrialisation in sub-Saharan Africa', *Journal of African Economies*, 7 (supplement 1) (AERC May 1997): 70–107.

Little, I.M.D., Scitovsky, T. and Scott, M.F. (1970) *Industry and Trade in Some Developing Countries*, Oxford: Oxford University Press for OECD.

Livingstone, I. (1998) 'Developing industry in Uganda in the 1990s', in H.B. Hansen and M. Twaddle (eds), *Developing Uganda*, Oxford: James Currey.

Merhav, M. (1969) *Technological Dependence, Monopoly and Growth*, Oxford: Pergamon Press.

Ndulu, B.J. (1986) 'Investment, output growth and capacity utilization in an African economy: the case of the manufacturing sector in Tanzania', *Eastern Africa Economic Review*, 2 (1) (New Series): 14–30.

Noorbakhsh, F. and Paloni, A. (2000) 'The de-industrialisation hypothesis, structural adjustment programmes and the sub-Saharan dimension', in H. Jalilian, M. Tribe and J. Weiss (eds), *Industrial Development and Policy in Africa: Issues of De-industrialisation and Development Strategy*, Cheltenham, Glos.: Edward Elgar.

Republic of Ghana (various issues) *Quarterly Digest of Statistics*, Accra: Central Bureau of Statistics/Ghana Statistical Service.

Republic of Uganda (various years) *Background to the Budget*, Kampala: Ministry of Planning and Economic Development/Ministry of Finance and Economic Planning.

Republic of Uganda (1997 and 1998) *Statistical Abstract*, Entebbe: Ministry of Planning and Economic Development.

Stein, H. (1992) 'Deindustrialization, adjustment, the World Bank and the IMF in Africa', *World Development* 20 (1): 83–95.

Stein, H. (ed.) (1995) *Asian Industrialization and Africa: Studies in Policy Alternatives to Structural Adjustment*, London: Macmillan.

Sutcliffe, R. (1971) *Industry and Underdevelopment*, London: Addison-Wesley.

Syrquin, M. and Chenery, H.B. (1989a) *Patterns of Development, 1950 to 1983*, Discussion Paper 41, Washington, DC: World Bank.

Syrquin, M. and Chenery, H.B. (1989b) 'Three decades of industrialization', *World Bank Economic Review*, 3 (2): 145–81.

Teal, F. (1998) 'The Ghanaian manufacturing sector 1991–1995: firm growth, productivity and convergence', Working Paper WPS/98–17, University of Oxford: Centre for the Study of African Economies, June.

Thoburn, J.T. (2000) 'Developing African exports: the case of Nigeria', in H. Jalilian, M. Tribe and J. Weiss (eds), *Industrial Development and Policy in Africa: Issues of De-industrialisation and Development Strategy*, Cheltenham, Glos.: Edward Elgar.

Tribe, M. (2000a) 'A review of recent manufacturing sector development in sub-Saharan Africa', in H. Jalilian, M. Tribe and J. Weiss (eds), *Industrial Development and Policy in Africa: Issues of De-industrialisation and Development Strategy*, Cheltenham, Glos.: Edward Elgar.

Tribe, M. (2000b) 'The concept of "infant industry" in a sub-Saharan African context', in H. Jalilian, M. Tribe and J. Weiss (eds), *Industrial Development and Policy in Africa: Issues of De-industrialisation and Development Strategy*, Cheltenham, Glos.: Edward Elgar.

UNIDO (1993) *Industry and Development: Global Report 1993/94*, Vienna: UNIDO.

UNIDO (1997) *Industrial Development: Global Report 1997*, Vienna: UNIDO.

Wangwe, S.M. (1977) 'Factors influencing capacity utilization in Tanzanian manufacturing', *International Labour Review*, 115 (1): 65–78.

Warr, P.G. (1990) 'Export processing zones', in C. Milner (ed.), *Export Promotion Strategies*, Hemel Hempstead, Herts.: Harvester Wheatsheaf.

Weiss, J. (1988) *Industry in Developing Countries*, London: Croom Helm.

Winston, G.C. (1971) 'Capital utilization in economic development', *Economic Journal*, 81 (321): 36–60.

Winston, G.C. (1974) 'The theory of capital utilization and idleness', *Journal of Economic Literature*, XII (4): 1301–20.

World Bank (1983) *World Development Report 1983*, New York: Oxford University Press.

World Bank (1994) *Adjustment in Africa: Reforms, Results, and the Road Ahead*, New York: Oxford University Press.

World Bank (1999) *World Development Report 1998–99*, New York: Oxford University Press.

16 Could import protection drive manufacturing exports in Africa?[1]

John Thoburn

16.1 Introduction

The crude liberalisation advocated for developing countries in the 1980s and 1990s under World Bank and IMF 'structural adjustment' policies increasingly has come under scrutiny by the mainstream of the economics profession. An influential push in this direction even has been provided by the World Bank's former chief economist (Stiglitz, 1998).

Trade liberalisation is at the heart of structural adjustment policies under the so-called 'Washington consensus'. Moves towards free trade are justified on grounds of static reallocative gains, on the grounds that exposure to world competition will enhance international competitiveness, and on the dynamic learning generated by export activities and also by inward foreign direct investment. The view of import protection implied by these policies dates back to the classic study by Little *et al.* (1970), which documented how import substitution policies in a wide range of developing countries had protected industries which then failed to become competitive at world prices. Protection, particularly in the form of non-tariff barriers, also provided rents that spawned domestic constituencies opposed to economic reform.

Yet, with the exception of Hong Kong,[2] no developing country that has become an exporter of manufactures has done so under a regime of free trade and *laissez-faire*. In some cases, such as Malaysia, exports were developed through arrangements such as export processing zones. EPZs encouraged a dualistic economic structure where exporting was grafted on to the domestic economy by inward direct foreign investment while protected firms served the domestic market. Such export processing arrangements merely circumvent some of the anti-export bias generated by import protection,[3] and Little–Scitovsky–Scott criticisms still apply to the protection of domestic industry. However, history suggests that protection also has been used more positively. In the case of industrial countries, few developed as manufactures exporters under free trade conditions. Britain in the nineteenth century, as the world's first industrial power, was able to do so, but latecomers such as Germany accelerated their industrial development, and subsequently their exports, under protection.

In the case of Japan too, and later South Korea and Taiwan, exports were developed while maintaining strict controls against imports. *Export-oriented import substitution* indeed may be one of East Asia's most significant contributions to development policy (Weiss, 1985). Protection was used to provide a respite from foreign competition while industrial capability was developed, with a view eventually to export.[4] In the case of Korea, the use of import protection as an export incentive is well documented (Chang, 1996; Song, 1990)

Even the negative effects of protection under import-substituting industrialisation are now being questioned.[5] In a recent and influential book, Rodrik (1999: 68–77) argues that the poor performance of many countries in the 1970s and 1980s, especially those in Latin America and Africa, can be traced primarily to the instability resulting from their being unable to handle macroeconomic shocks without distributional conflict; and not to the effects of protection. He denies that serious microeconomic distortions resulted from import substituting industrialisation policies, pointing to the evidence of strong total factor productivity growth in Latin America (and even in some African countries) prior to the 1973 oil shock.

Import protection in Africa is typically higher than in other regions,[6] and it is unlikely to be removed for some time to come, despite international pressure to liberalise. Even moves from quantitative restrictions to tariffs have led to fears that domestic manufacturing might be eliminated by import competition (Bennell, 1998). This chapter seeks to explore whether African countries could use protection in a positive way, such that particular industries could be encouraged to export on the basis of having a domestic market secured from import competition. It takes examples mainly from the textile industry, the largest manufacturing sector after food processing in many African countries. It looks at two countries – Nigeria and Zimbabwe – from which there have been textile exports, and where the textile industry has been protected against import competition. To provide a comparison with Asia, the chapter also considers the case of the Indonesian textile industry. Indonesia, with its many economic distortions and institutional shortcomings, is a better comparator than an East Asia country such as Taiwan. A development-enhancing institutional structure like that of Taiwan is unlikely to be reproduced in Africa for many years to come, if ever.[7]

It is important to stress, however, that we are here concerned with how best to achieve the *transition* to a fully open economy. There is no argument for long-term protection against imports. Once an economy is able to compete in world markets, there are great advantages to consumers of having free trade, and there are benefits to the economy more generally from the imports of technology and investment goods that growing foreign exchange earnings finance. However, premature opening to foreign competition without the establishment of a sound industrial base and market-supporting institutions is as likely to retard development as to promote it. In that sense, openness follows economic development rather than leading it, and the benefits of openness depend heavily on complementary policies and institutions (Rodrik, 1999: 137).

The chapter first considers the concept of anti-export bias, and also looks at literature on protection and exports. It then examines some econometric evidence on trade policy and exports in Africa, before presenting the case study material.

16.2   Anti-export bias?

The standard argument that protection of the domestic market against imports has the effect of discouraging exports consists of two main strands. First, protection, by reducing imports, improves the balance of trade and thereby appreciates the real exchange rate.[8] Thus the relative price of all tradables (that is, exportable and import-competing goods (ex-tariff)) is reduced relative to that of non-traded goods. Resources are thereby induced to flow out of tradable and into non-tradable activities. Second, tariffs or quantitative import restrictions tend to raise the relative price of import-competing goods in relation to exportable goods. This *intra-tradables bias* induces resources to flow away from exporting into import-competing activities. In these two senses protection against imports acts like a tax on exports.[9]

These arguments are theoretically unassailable, but their importance easily can be exaggerated. In particular, overvalued real exchange rates are not only the result of import protection. They may well, and often do, result from domestic monetary expansion. Many African countries have been able to bring about real depreciations of their currencies since the early 1990s as part of macroeconomic stabilisation programmes (African Development Bank, 1998: 212), even though they have not necessarily reformed their trade regimes. The experience of Indonesia in the 1980s suggests that exchange rates can be depreciated sufficiently to encourage exports effectively even while maintaining protection[10] (with whatever degree of RER overvaluation that protection entails) (Bevan *et al.*, 1999: 274–9). In any case, given the commonly held view that import substituting industrialisation has proved in practice to be highly import-intensive, it seems odd to argue that it also improves the balance of trade to such an extent as to cause large RER appreciations!

16.3 Import protection as a driver of exports

One case has been established in the economic literature where import protection can be shown to drive exports. Krugman (1984) shows that under conditions of oligopoly in the world economy, a national firm would increase its exports if its home market is closed to the sales of the firm of another country (in a two-country model). By closing the domestic market against foreign competition, and thereby increasing its national firm's domestic sales, a government can ensure that the national firm reaps lower marginal costs and thereby is able to increase its export competitiveness. The essential feature of the model is that all firms enjoy decreasing marginal costs. These decreasing marginal costs can be the result of static economies of scale, of increased R&D expenditure induced by prospects of a safe domestic market, or they can result from dynamic 'learning by doing' scale economies. The outcome is much the same whichever is the source of the declining marginal costs. However, Krugman does not argue that such expanded exports are necessarily welfare-increasing, and in any case domestic consumers bear the cost of higher domestic prices.

The Korean motor industry's exports were developed in a way that seems to mirror the Krugman arguments. The Korean government restricted entry to the industry in order to preserve the monopolistic profits of the protected home market, and then gave firms a strong push to use the profits on domestic sales to cross-subsidise exports. Initial losses on exports were justified by the need first to establish overseas market share, and thereby to start a process of learning more about foreign consumers' requirements with a view to raising the profitability of exports.

This chapter will show that in Indonesia a rather different mechanism drove textile exports, namely that domestic competition pushed firms into exporting. The domestic competition pushed the domestic price down to below the world price + tariff, but imports were still kept out of the domestic market because they would have to sell at world price + tariff. More on this later.

16.4 Confronting the econometric evidence on Africa

This chapter later will present evidence from firm level studies for Indonesia and Zimbabwe and material on Nigeria from sectoral reports. It will ask what drove firms into exporting despite the apparent anti-export bias generated by protection of their home market. A complementary approach would be an econometric analysis over a large range of countries. Such evidence is available in Rodrik (1999), based on a panel data set of thirty-seven sub-Saharan African countries over the period 1964–94.[11] Since this chapter has quoted Rodrik's work with approval, it is unfortunate – at least at first sight – that Rodrik's results seem quite unfavourable to the hypothesis being explored here. His evidence needs to be confronted.

Earlier in his book, Rodrik argued that openness only brings development in the presence of complementary policies and institutions. In his view, the benefits of openness accrue more on the import side than on the export (or inward DFI) side. Then, having said that import substituting policies in Latin America, South Asia and even Africa worked rather well before the macro-instability following the 1973 oil shock, he seems to take an oddly neo-liberal approach to trade policy and exports in Africa. Finally, having concluded that trade liberalisation *does* stimulate stronger export performance in Africa, he softens this conclusion by returning to the argument that export growth (and also more import competition) may well not translate into better development performance in the African context.

Two broad criticisms can be made of the Rodrik results. First, his regressions of the share of exports in GDP on import and export taxes (first both together, and then separately) do not control for real exchange rate changes in each of his countries. The parallel foreign exchange market premium does get included among his regressors and is significant (though it enters with a very small coefficient). While the premium may pick up some influence of exchange rate changes,[12] it is more an indicator of macroeconomic instability.[13] Trade liberalisation is highly likely to have been associated in practice with devaluations,[14] so improved export performance may well be the result of real exchange rate depreciations over and above any depreciation that results from the removal of import barriers. Also, trade liberalisation is often associated in Africa with drastic improvements in primary commodity marketing channels as part of a wider liberalisation package, with the result that producer prices move closer to world prices. This improves export performance independently of the effects of a reduction of import taxation (or even of the reduction of export taxation), and is potentially of great importance since African exports still are overwhelmingly of primary commodities.[15]

Second, and again since African exports are mainly of primary commodities, the mechanisms of protection-as-export-promotion would not be expected to be in operation. Few African primary commodities have large actual (or even potential[16]) domestic market sales. If protection works to encourage exports of manufactures, one might observe an increase in the share of manufactures in total exports, even if primary commodity exports (and therefore the overall export total) were being discouraged by the anti-export bias of import protection.

16.5 Textiles, exports and import protection

Most developing countries that have succeeded in becoming exporters of manufactures have done so initially by exporting goods with a mature, labour-intensive technology. Garments and footwear are the most common examples. Textile production is less labour-intensive than that of garments, but the technology of textiles is readily accessible. Basic textile products such as grey cloth do not require the sophisticated knowledge of industrial countries' markets which makes garment exporting so typically dependent on inward DFI. Also, textiles figure prominently in the manufacturing value-added of most African countries. Even when they are not export industries, they are import-competing.

In major producing countries, textiles have a complicated structure of production, which is often highly trade-dependent. This makes the industry sensitive to the structure of effective protection (that is, protection on value-added) at each stage of production, and therefore sensitive to trade policy. Textile products can be imported or exported at each stage in the value chain, starting with raw cotton, natural or synthetic fibres, and going on to unfinished fabric, finished fabric and ultimately garments. Spinning and weaving are inflexible operations needing long production runs, whereas export garment production features short production runs and frequent style changes. There are large differences in trade dependence at each stage of the value chain among major producing countries,[17] and it can be an indication of appropriate specialisation if a country is simultaneously exporting some textile fabrics while importing others for the production of export garments.

In Africa as elsewhere, imported inputs (such as synthetic fibres or raw cotton) for textile production typically have faced low or zero trade restrictions, while fabrics have faced higher duties. As a result, textile production has not needed the EPZ and other export processing arrangements required for export garment production located in a protected economy. Whereas garment exports are often made by foreign investors utilising export processing arrangements and who do not serve the domestic market, typically textiles are exported either by domestic firms or by foreign investors who originally established themselves to sell to the domestic market.

The material on Indonesia comes from firm-level interviews in 1994; the Nigeria material comes from sectoral studies in the late 1990s. The Zimbabwean material comes entirely from secondary sources.[18]

16.5.1 *Indonesia*[19]

In the 1970s Indonesia's textile industry grew behind tariff barriers and received substantial inward investment from Japan and Hong Kong. The country began to develop manufacturing exports in the 1980s as part of a policy to reorient the economy to exports following the OPEC oil shocks. There was rapid growth in textile and garment exports, and by 1992 Indonesia was the world's sixth largest net exporter. In 1993 textiles and garments were 23 per cent of Indonesia's non-oil export earnings. This impressive export performance was driven by devaluations of the rupiah. In the case of garments, a variety of export processing arrangements were important too, but we focus here on textiles.

The 25–35 per cent import duties on fabrics that made export processing arrangements crucial to the success of export garment production (undertaken mainly by foreign investors from Hong Kong and elsewhere in Asia), however, did not deter the domestic textile industry from exporting. There were roughly equal exports of textiles and garments (in volume terms). Our interest here is also whether such protection actually may have encouraged it. Since textile inputs were subject only to zero or low duties, the effective rate of protection of course was higher than 25–35 per cent. Wymenga (1991: 138–9) calculates an ERP for 'textiles, clothing and footwear' of 66.6 per cent in 1989, compared to the weighted average ERP for all non-oil manufacturing of 63.6 per cent.

A key feature of the early 1990s was that domestic competition in Indonesian textiles was pushing domestic prices to below the world price + tariff, and in some cases to below the world price. That is, there was some tariff *redundancy*. Thus exports became profitable, although in fact, it is a considerable simplification to talk of *the* world price; world prices were fluctuating according to sales from major exporters such as China. Also, the relation between domestic and world prices differed considerably between individual textile products. In the case of grey cloth, domestic prices were below world prices, whereas in finished fabrics the domestic price was above the world price. However, another factor was driving exports too, even when the apparent domestic price was above the export price. This is that Indonesian producers found it much easier to get payment quickly when they exported than when they sold to domestic customers. Overseas customers paid promptly, using letters of credit, whereas domestic customers wanted long credit periods. There was also some specialisation between export and domestic market. Most domestic sales were of cotton prints, where Indonesia could not easily compete with China and Pakistan in export markets, and most Indonesia export sales were of polyester/cotton mixes.

So what was the role of protection in these textile exports? If prices for a particular product in the domestic market fall to below the world price, why not simply remove the now totally redundant tariff? The case rests on the fluctuations in price in the world market, such that at times domestic price will be below world price and at other times above it. Since, as noted earlier, textile production tends to be inflexible and needs long, planned production runs, a secure home market allows just such planning. A 'redundant' (or partially redundant) tariff still keeps out imports: exporters to Indonesia would have to drop their supply price to the world price *minus* the Indonesian tariff, which they are unlikely to do if they can sell elsewhere at the world price.[20] Yet a (fully) redundant tariff does not give rise to (intra-tradable) anti-export bias. Also, when domestic price falls to below world price + tariff, the welfare cost to consumers is partially removed too, and completely removed if domestic price falls to world price. In other words, in these particular circumstances, domestic producers can be protected from import competition without there being strong anti-export bias and without there being heavy welfare costs to consumers!

The importance of these changing relations between world prices and domestic prices can be driven home by considering the case of vertically integrated textile producers in Indonesia. Such producers are not typical of the industry – most export garment makers are foreign investors selling only to export markets, and many textile producers do not sell textiles to be incorporated in export garment production[21] – but the case illustrates the economics of textile exporting. These vertically integrated companies do not use in their garments all the fabrics they produce, while at the same time they may buy in fabrics for garments. The decision to export fabrics depends on the relation between the world price and the domestic price. Indeed, companies may make both fabrics and garments with the intention of exporting each separately. Exported fabrics may still not be of the type or quality required for export garments.

16.5.2 Nigeria[22]

Nigeria exports textiles mainly to its neighbouring countries in ECOWAS, the Economic Community of West African States. According to Nigeria's official (and highly unreliable[23]) trade statistics, imports of textiles and clothing are greater than exports, but one source estimated that Nigeria's unofficial textile exports were many times greater than its official ones (ADCG, 1994). Unofficial trade circumvents neighbouring countries' import restrictions, and it also reflects inadequacies in data collection. Thus textile exports may be the country's second or third largest export, after cocoa and rubber.[24]

These textile exports have developed despite the apparent anti-export bias generated by an import ban on textiles which was in place up to 1997, when it was replaced by a 45 per cent import duty. In principle, the Nigerian textile industry could well be sufficiently concentrated for Krugman-style price discrimination to be practised, with higher prices being charged in the domestic market than for export. There is foreign control of 70 per cent of spinning capacity and 60 per cent of weaving capacity, dominated by two groups (one Indian, one from Hong Kong). The industry saw the privatisation of various state enterprises in the 1980s and 1990s. On the other hand, the clothing industry, which is 90 per cent Nigerian, is mainly comprised of microenterprises.

However, it is as likely that an Indonesian-style mechanism has been driving Nigerian exports. Macroeconomic restriction has reduced consumer purchasing power and pushed down prices in the domestic market. The import ban and the later tariff have probably been avoided by smuggling to some extent. Whatever protection remains would give firms some degree of price protection in the domestic market and be a base for sales to neighbouring countries.

16.5.3 *Zimbabwe*

Zimbabwe, one of Africa's most industrialised countries, has exported textiles for many decades. By the mid-1990s nearly half of the country's combined textile, clothing and footwear exports was being sold to non-African markets. While the motivation of individual Nigerian firms to export is not clear without further investigation, the existing secondary source material[25] gives useful initial pointers in the case of Zimbabwe that could be the basis for further investigation. In Zimbabwe, exports of textiles and other manufactures have been influenced strongly both by trade policy prior to structural adjustment and by the changes introduced under an 'economic structural adjustment programme' (ESAP) from 1990.

Up to the introduction of the ESAP, textile production was protected by a system of foreign exchange licensing, and in 1990 a 60 per cent tariff was introduced. This understates the sector's ERP since various imported intermediates were importable duty free.[26] Up to ESAP, the sector also had guaranteed supplies of local cotton. In addition, imports of new clothing were banned up to 1996, which helped to protect textiles since clothing companies were required to buy a yard of local cloth for every yard imported. Clothing exports grew rapidly during the late 1980s.

A full account of what drove these textile and garment exports is not yet available without further research,[27] but Riddell's (1990) firm-level studies indicate that industrial firms in Zimbabwe engaged in Krugman-style differential pricing between foreign markets and the protected home market in order to export. His interview material on differential pricing relates to agriculture machinery and to paper manufacturers, but Zimbabwean textiles also have an oligopolistic market structure, with four firms controlling 83 per cent of output; so differential pricing between home and export markets would be a strong possibility.

Trade reform under the ESAP seems to have been disastrous for Zimbabwe's textile (and clothing) production and exports. However, it is not only trade policy but also macroeconomic stabilisation that are at fault. Macroeconomic restriction reduced domestic purchasing power and sharply raised domestic real interest rates.

Carmody (1998) documents a process of trade reform under ESAP that might at first seem a logical way of reducing the anti-export bias facing the textile industry. Yet, when accompanied by macro-stabilisation, it had the effect of reducing exports as well as driving many firms out of business. In 1991, accompanying a 36 per cent nominal devaluation, the textile tariff was cut to 20 per cent. At the same time – and presumably to lower the textile industry's ERP for a given nominal tariff on the final product – tariffs were introduced on some intermediate products like chemicals. Cotton producers were given freedom to export and no longer had to sell to the domestic textile sector.

Clothing production suffered because of falling demand in the home market and from competition from secondhand imported clothing. This reduced the demand for textiles, which were being squeezed by the reduction in domestic market protection as well. Also, the higher domestic currency price of exportable goods resulting from devaluation was partly counteracted by the higher import costs resulting from devaluation and the imposition of tariffs on intermediates. In any case, inflation at times meant that devaluation was insufficient to produce real exchange rate depreciation. As some textile factories closed, clothing exporters found they had to source materials such as denim from overseas, losing the advantage of proximity to suppliers. Textile producers also found it difficult to source cotton either from domestic producers or from imports. One lesson must be that it is unwise to conduct trade policy reform until an economy has achieved macroeconomic stability.

16.6 Conclusions

This chapter has looked at the cases of three countries where exports of textiles have occurred despite the apparent anti-export bias generated by protection of the industry against imports. In all three cases it is likely (though not certain) that textiles enjoyed an effective rate of protection that was high in relation to other sectors. Textiles have been chosen because they are already an important sector in most African economies. Also, since their inputs typically do not attract import duties, they do not depend on export processing arrangements, and they are not so heavily dependent on attracting new foreign investment as are garments. Indonesia has been used as an Asian comparator because it moved from import substitution in textiles into exporting, and because good information is available from company interviews conducted by the author in the mid-1990s. The material on Nigeria and Zimbabwe is more preliminary.

We have argued that the anti-export bias resulting from import protection can be over-emphasised. Real exchange rate overvaluation arises from many factors besides import protection – and particularly from domestic monetary expansion. Also, the fact that other goods (such as garments) have been exported using export-processing arrangements casts doubt on the importance of anti-export bias. Even export processing zones generally only correct for anti-export bias in the sense that they allow exporters access to imported intermediates at world prices.[28] The anti-export bias resulting from an overvalued real exchange rate due to import protection remains,[29] and has not deterred exporters in countries such as Indonesia once RER overvaluation due to other factors has been removed by devaluation(s).

Clearly, textile exports have taken place despite import protection; the more interesting question is whether they actually may have been aided by import protection. Preliminary material from Zimbabwe suggests that, given a high degree of industrial concentration, textile (and garment) firms have been able to use the protected home market as a base from which to export at least marginal supplies. However, ill-judged trade reform sponsored by the multilateral agencies appears to have a disastrous effect on the industry in Zimbabwe, at least in the short run. The Indonesian and possibly the Nigerian cases suggest a different mechanism by which protection may have encouraged exports – that competition within the home market pushed firms into exporting. In the case of Nigeria much of that trade is unofficial and to neighbouring African countries. In the case of Indonesia, there is evidence that home market protection allowed firms to plan better and to take advantage of frequent periods when the (fluctuating) world price rose above the domestic price for particular textile products. Protection allowed them not to have to worry about losing domestic sales when world prices dipped.

In the eventual transition to free trade, the lowering of tariff protection will increase competition in the domestic economy, and thereby also increase pressure to export. The period of exporting from the secure base of the domestic market will have helped prepare firms for this transition by giving them knowledge of foreign markets and of the restructuring of production necessary to meet world standards of quality.

Consumers pay a welfare price when protection is used to stimulate exports, though the extent is difficult to measure in the context of second-best situations (Krugman, 1984). One interesting aspect of protection that arises tangentially from the present chapter and in relation to the arguments of Rodrik (1999) concerns the appropriate structure of protection. Rodrik stresses the benefits of openness on the import side, such that importing facilitates the acquisition of technology and learning through access to imported intermediates and capital goods. A consequence of this view is that the development policies of the 1960s, under which final goods were protected against imports but intermediate and capital goods were not, may not have been so foolish after all! Such policies were commonly criticised for introducing 'distortions', and the implication is that import duties should be imposed on the intermediate and capital goods as a second-best solution to equalise ERPs at different stages of production if free trade is not feasible. However, if imported intermediates and capital goods facilitate development and are therefore permitted duty free access, then a given ERP can be achieved for the final good with a lower nominal rate of protection. Since it is the magnitude of the ERP which may stimulate exports and the height of the nominal tariff which imposes the welfare cost on consumers, it seems that export promotion via import protection could be achieved at a lower welfare cost than otherwise expected. In the particular case of textiles, the products are both a final good that is purchased by consumers to make their own clothes and an intermediate input into clothing production and other textile products. However, the general point remains.

Finally, and of course, successful exporting of manufactures from an African country is not just a matter of trade policy. Achieving export competitiveness depends on a complex set of economic and institutional changes. Exporting is also inhibited by high transport costs, both to overseas destinations and within Africa itself because of the poor condition of road and rail networks (UNIDO, 1996: 113–15; Platteau and Hayami, 1998). Other aspects of infrastructural provision, especially unreliable electricity supply and inadequate telecommunications, also reduce export competitiveness. Exports to neighbouring countries are easier than to the markets of industrial countries since consumer tastes are similar.

Notes

1 *Africa* in this chapter refers to sub-Saharan Africa; South Africa is not excluded.
2 Singapore also developed its exports under a highly open trade regime. However, and unlike Hong Kong, in other respects the Singapore government intervened heavily in the economy. Another possible exception is Mauritius, whose export processing arrangements were spread to firms all over the island (Rodrik, 1999: 45–8).
3 EPZs circumvent anti-export bias by offering fast access to imported intermediates at world prices. Also anti-export bias is reduced by the subsidies provided in the form of cheap infrastructure in EPZs. However, the overvaluation of the exchange rate caused by protection remains. It would need further investigation to know whether any subsidies provided by EPZs do indeed 'neutralise' the effects of overvaluation on profitability. To the extent that they (probably?) do not, the EPZ experience shows that exports can take place despite whatever exchange rate overvaluation occurs as a result of protection. In any case, since most export industries in EPZs are highly import-intensive, the profit-reducing effects of ER overvaluation on the output side are lessened by the fact that an overvalued ER cheapens imported inputs in terms of domestic prices.
4 This is not to say that in all cases protection was used to foster exports. In Japan, for instance, some activities, such as the domestic production of rice, were highly protected and remained inefficient by world standards.
5 On the current reappraisal of import substitution policies, see also Bruton (1998).
6 Nineteen sub-Saharan African countries for which data are available had an average (unweighted) tariff of 26.8 per cent in the mid-1990s compared to 6.1 per cent in OECD countries. Also, a third

of imports into Africa were covered by non-tariff barriers compared to under 4 per cent in OECD countries. There is little reason to suppose other African countries are more open than ones for which there are data. See Ng and Yeats (1997).

7 For a discussion of the role of institutions in the development of Asia, with comparisons with Africa, see Hayami and Aoki (1998). Another useful source is Collier and Gunning (1999), which discusses the institutional framework and the 'social capital' that could enhance economic performance in Africa.

8 If the nominal exchange rate is free to float, then an improved balance of trade (*ceteris paribus*) would appreciate the real exchange rate by causing the nominal exchange rate to appreciate. If the nominal ER is fixed, then the balance of trade surplus would generate an expansion in domestic demand which would raise the price of non-traded goods while traded goods prices would be held constant by the forces of world competition; thus the real ER (price of traded goods/price of non-traded goods) would fall (appreciate).

9 The incidence of import tariffs as export tax depends on the substitutability between importables, exportables and non-traded goods. To the extent that importables are close substitutes for NTGs but not for exportables, then import protection will tend to switch consumer purchases to the NTG sector, where prices will rise relative to exportables' prices. In these circumstances, the incidence of protection ('true protection') will fall on exports as an implicit tax (see World Bank, 1987: 80).

10 According to Wymenga (1991: 138–9) the weighted average effective rate of protection for all non-oil manufacturing in Indonesia in 1989 was 63.6 per cent.

11 The 1964–94 period is actually split into three sub-periods (1964–74, 1975–84, and 1985–94), so as to give three observations for each country (Rodrik, 1997: 114).

12 This is in the sense that depreciation of the nominal exchange rate initially may reduce the parallel forex market premium.

13 This criticism is reminiscent of Rodrik's own criticisms of other econometric studies of openness and economic growth, which he accuses of confusing macroeconomic stability variables with true openness variables (Rodrik, 1999: 13, 20). For some more general comments on Rodrik's views on openness and development, see Srinivasan (1999).

14 The devaluation also may well have been associated with macroeconomic stabilisation, which allows nominal depreciations to 'stick' in real terms.

15 The World Bank (2000) cites a figure for manufacturing exports as a share of total (merchandise) exports as only 12 per cent for 1983, but unfortunately does not give an average figure for 1997, its latest year for data. However, with a few exceptions such as Zimbabwe, South Africa and Mauritius, the shares of manufactures in total exports were small for African countries in 1997.

16 Thus the domestic market of Ghana hardly could absorb a large share of the country's cocoa production. An exception, which perhaps proves the rule, is Nigeria's exports of leather, which do indeed starve the domestic footwear industry of high-quality raw materials which are in practice difficult to obtain by importing (see Thoburn, 2000).

17 For example, Korea and Taiwan are large exporters of textiles as well as garments, and both also have substantial textile imports though with a large overall export surplus. This is similar to Indonesia's trade pattern (at least up to the 1997 crisis). Both Korea and Taiwan have large synthetic fibre industries, and their exports represent the economies of scale achievable in synthetic fabric production – and also the gains from specialisation in particular types of fabrics. China, the world's largest exporter of garments, is not a *net* exporter of textiles though its gross exports are as large as Taiwan's. Domestic cotton is important to China's textile industry, as is the case also in Nigeria and Zimbabwe. Note too that, while restrictions on exports continue to exist under the Multifibre Arrangement up to end of 2004, exporting fabrics as well as garments gives access to a wider range of MFA quotas.

18 The data on Indonesia were collected as part of a consultancy on export development. The Nigerian material was also informed by a consultancy on export development, but most of the material is from other sources.

19 No attempt is made here to update the Indonesian material to take account of developments since the 1997–8 Asian crisis. What is interesting are Indonesia's lessons for Africa in the early stages of its development of textile exports.

20 This is no more than saying that an individual overseas exporter faces a perfectly elastic demand curve, the normal 'small country' assumption which justifies us in speaking of a 'world' price.

21 Purchases of export quality domestic fabrics by garment exporters, though not insignificant, were inhibited by various policy distortions. It was more difficult to get rebates of value-added tax on domestic purchases than to get rebates of duty on imported inputs. However, in other cases local

sourcing was not undertaken because suitable quality fabric could not be procured, or the overseas buyers could secure special discounts on imported fabrics for their garment suppliers, or fabric manufacturers were unwilling to supply quickly in the small quantities required for garment orders.

22 See ADCG (1994) and Thoburn (2000).

23 For instance, Nigeria's trade statistics show no exports of agricultural machinery, yet company surveys reveal that Nigeria exports such machinery to surrounding countries.

24 There are truly astonishing discrepancies between export (and production) statistics for Nigeria from different sources, and even for a particular year in different issues of the same source. It is often not possible to know either the absolute size of exports of a particular commodity or whether those exports have risen or fallen. There are also great discrepancies between exports as reported by Nigeria and imports from Nigeria as reported by importing countries (see Thoburn, 2000). Thus surveys of enterprises may produce more reliable information than can be found in trade statistics.

25 Material for this section comes mainly from Riddell (1990) and Carmody (1998).

26 Data on ERPs in other sectors are not to hand at the time of writing. Of course, protection is a *relative* concept, and if ERPs are higher in other sectors then resources would flow away from textiles into those other sectors. In that sense, a positive ERP is not a sufficient condition for (intra-tradable) anti-export bias against textiles. However, an absolute level of ERP higher than 60 per cent would suggest textiles are *probably* protected in relation to other sectors. In the case of Nigeria, good data on ERPs is hard to find, but an import ban on textiles would clearly generate intra-tradable anti-export bias, although in practice the *relative* protection would depend on how far that ban was circumvented by smuggling.

27 Two additional factors that will have driven clothing exports were the system of foreign exchange retention for exporters, and access (later temporarily withdrawn) to the South African market.

28 If exporters have to pay import duty on imported intermediates, an anti-export bias is created in the form of negative ERP on the export activity. This is in addition to the standard forms of anti-export bias already discussed. This highlights the fact that intra-tradable anti-export bias is best assessed by considering the effective rate of protection on importables relative to exportables.

29 As noted earlier, and particularly in the case of EPZs, it may be possible to counteract that overvaluation to some extent by providing subsidies in the form of (say) cheap infrastructure. Such subsidies, of course, help to move the trade regime faced by exporters towards trade 'neutrality'. In the case of Indonesia in the mid-1990s, however, exporters were complaining about the *high* cost of using EPZs: they preferred other arrangements such as licensed manufacturing warehouses.

References

African Development Bank (1998) *African Development Report 1998*, London: Oxford University Press.

African Development Consulting Group (ADCG) (1994) *The Nigerian Textile Industry*, Lagos: African Development Consulting Group.

Bennell, Paul (1998) 'Fighting for survival: manufacturing industry and adjustment in sub-Saharan Africa', *Journal of International Development*, 10: 621–37.

Bevan, D.L., Collier, P. and Gunning, J.W. (1999) *Nigeria and Indonesia: The Political Economy of Poverty, Growth and Equity*, New York: Oxford University Press for the World Bank.

Bruton, H.J. (1998) 'A reconsideration of import substitution', *Journal of Economic Literature*, 36 (June): 903–36.

Carmody, P. (1998) 'Neoclassical practice and the collapse of industry in Zimbabwe: the cases of textiles, clothing and footwear', *Economic Geography*, 74 (4): 319–43.

Chang, H.J. (1996) *The Political Economy of Industrial Policy*, rev. edn, London: Macmillan.

Collier, P. and Gunning, J.W. (1999) 'Explaining African economic performance', *Journal of Economic Literature*, 37 (June): 64–111.

Hayami, Y. and Aoki, M. (eds) (1998) *The Institutional Foundations of East Asian Economic Development*, London: Macmillan.

Krugman, P. (1984) 'Import protection as export promotion: international competition in the presence of oligopoly and economies of scale', in H. Kierzkowski (ed.), *Monopolistic Competition and International Trade*, Oxford: Clarendon Press.

Little, I.M.D, Scitovsky, T. and Scott, M. (1970) *Trade and Industry in Some Developing Countries*, New York: Oxford University Press.

Ng, F. and Yeats, A. (1997) 'Open economies work better! Did Africa's protectionist policies cause it to be marginalised in world trade?', *World Development*, 25 (6): 899–904.

Platteau, J.-P. and Hayami, Y. (1998) 'Resource endowments and agricultural development: Africa vs Asia', in Y. Hayami and M. Aoki (eds), *The Institutional Foundations of East Asian Economic Development*, London: Macmillan.

Riddell, R.C. (1990) *Manufacturing Africa: Performance and Prospects of Seven Countries in Sub-Saharan Africa*, London: James Currey.

Rodrik, D. (1999) *The New Global Economy and Developing Countries: Making Openness Work*, Washington, DC: Overseas Development Council.

Srinivasan, T.N. (1999) 'Trade orientation, trade liberalization, and growth', in T.N. Srinivasan and G.R. Saxonhouse (eds), *Development, Duality, and the International Economic Regime: Essays in Honor of Gustav Ranis*, Ann Arbor, MI: University of Michigan Press.

Song, B.L. (1990) *The Rise of the Korean Economy*, Hong Kong: Oxford University Press.

Stiglitz, J.E. (1998) 'More instruments and broader goals: moving towards the post-Washington consensus', Helsinki: United Nations University World Institute for Development Economics Research, <http://www.wider.unu.edu/publications.htm> (accessed April 2001).

Thoburn, J.T. (2000) 'Developing African exports: the case of Nigeria', in H. Jalilian, M. Tribe and J. Weiss (eds), *Industrial Development and Policy in Africa: Issues of De-industrialisation and Development Strategy*, Cheltenham, Glos.: Edward Elgar.

UNIDO (1996) *The Globalisation of Industry: Implications for Developing Countries beyond 2000*, Vienna: United Nations Industrial Development Organisation.

Weiss, J. (1985) 'Japan's postwar development policy: some implications for LDCs', *Journal of Development Studies*, 22 (2): 385–406.

World Bank (1987) *World Development Report 1987*, Washington, DC: World Bank.

World Bank (2000) *World Development Report 1999–2000*, Washington, DC: World Bank.

Wymenga, P.S.J. (1991) 'The structure of protection in Indonesia in 1989', *Bulletin of Indonesian Economic Studies*, 27 (1): 127–53.

17 Stimulating economic recovery through private sector development

T.W. Oshikoya and K. Mlambo

17.1 Introduction

After a prolonged decline in the 1980s and early 1990s, Africa's economic growth prospects have improved in recent years, real GDP averaging 3.8 per cent per annum between 1995 and 1998. The recovery has occurred on the back of economic reforms introduced since the 1980s, evidenced by a progressive reduction in budget deficits and inflation rates and the achievement of more realistic exchange rates and real interest rates. However, despite this improvement, the performance so far achieved still falls short of the continent's potential and is not sufficient to deal with the growing scourge of poverty. Current per capita growth rates are still below those achieved in the 1960s and early 1970s, when real per capita GDP was increasing at more than 3 per cent per annum. We argue here that higher and sustained levels of economic growth in Africa cannot be achieved except through the promotion of the still nascent private sector as a dynamic engine for development.

After reviewing, first, the changing perceptions of private sector development in Africa, we provide a brief profile of the private sector in Africa, considering the question why, despite the progress made in implementing economic reforms, the response of the private sector, especially in the form of new investment, has remained generally weak A number of constraints are identified including: an unstable and uncertain macroeconomic environment, raising doubts about future policy direction; a weak legal and regulatory framework; a poor administrative structure; and weak financial systems, characterised by thin and under-developed financial markets, all presenting serious barriers to private sector development in Africa. We then highlight a number of issues that influence private sector development in Africa, including the need to revive private investment; enhancing entrepreneurial capability; accelerating the privatisation of state enterprises; and strengthening the financial sector.

17.2 The changing perception of private sector development

In recent years, the role played by the private sector in the economies of African countries has expanded, reflecting the changing perceptions of appropriate economic policy and development strategy. At the time when most African countries gained their independence, there was general distrust of markets, together with scepticism about comparative advantage and international trade as foundations for growth. It was believed that the private sector, if left to itself, would neither generate enough investible resources nor allocate them in a socially optimal manner; and that government would have to take the lead in guiding the structural transformation of these economies. Moreover, the African entrepreneurial class was generally considered not up to the task of spearheading economic development.

Governments interpreted economic modernisation to mean import-substituting industrial-isation, with consequent extensive government interventions in resource allocation. This favouring of industry meant discriminating against agriculture, even though the agricultural sector was the main employer as well as the dominant exporter in most African countries. Though protected from external competition, the industrial sector was also subject to a wide array of state regulations, including price and foreign exchange controls, minimum wage and licensing requirements. Controls also led to rationing in several markets, which generated inefficient administrative methods of allocating resources.

A distrust of the private sector was combined with a strong belief in the need for state planning agencies to direct investment activity. Governments thus invested directly in a range of productive activities that subsequently became a drain on state resources. While extensive investment in infrastructure was obviously needed, most of the large investments in parastatal firms turned out to be generally less efficient than their private counterparts. Public enterprises were favoured with restrictive concessions, licences, subsidies and credit allocations and trade protection. Attempts were made also to steer private investments via licensing requirements, again with uncertain consequences for the effectiveness of resource allocation. The extension of public ownership, whether through nationalisation or the establishment of new state-owned ventures, often used agricultural surpluses generated through agricultural marketing boards with monopoly powers over agricultural produce.

The decades following the 1960s witnessed an increasing disillusionment with the performance of public enterprises. In a study of the public enterprise sectors in Ghana, Senegal, Tanzania and Zambia undertaken before privatisation measures became wide-spread, for example, Killick (1983) evaluated performance with respect to five criteria: financial performance, productivity, balance-of-payments effects, employment and income distribution. Poor management within the parastatals led to losses of revenue and to the underemployment of resources. Instead of contributing to the budget as intended, the public enterprise sector became a source of fiscal imbalance in the economy. Apart from the distortions of the incentive structures and the negative impacts on resource distribution, the financial deficits of state-owned enterprises had serious impacts on fiscal and monetary policy – with upward pressure on interest rates and prices. Exchange rates, which were usually controlled, became seriously overvalued. These factors added to the cost of doing business in Africa.

Over time, it became increasingly clear that greater efficiency and economic dynamism could only be restored through competition and the fostering of private entrepreneurship. Now, as more African countries have adopted comprehensive economic reform programmes, the private sector in many of them is beginning to emerge from the historical legacies which in the formal sector have hindered the promotion of indigenous entrepreneurial skills. Many countries in the region have implemented broad policy reforms – including liberalisation of prices, removal of exchange controls and trade restrictions, and reduction in government tariffs – changes that constitute necessary conditions for the promotion of private sector activities.

17.3 The current profile of the private sector in Africa

The private sector plays a significant role in the economies of African countries, contributing to gross domestic products, employment creation and gross investment. It promotes economic growth and development, and thus contributes to the overall strategy of poverty reduction. The development of the private sector, especially of small- to medium-scale enterprises, not only creates jobs, but can also be an instrument of 'participatory development' in the sense that it enables a wider section of the population, particularly the poor, to participate in the process and benefits of development.

The sizes of enterprises vary from micro and small-scale enterprises – which are predominantly informal – to medium and large-scale enterprises. The common pattern is of a preponderance of micro to small-scale enterprises, together with a few medium- to large-scale establishments involved mainly in trading and commerce as well as in the processing of agricultural and mineral products for the domestic market and for export. The large and growing informal sector revolves around activities such as food processing, cosmetics, building, footwear and garments, along with tertiary activities such as transport and mechanical repair, retail, food preparation and distribution. Informal sector enterprises provide important inputs and services to the rural and urban communities, and serve as major sources of employment.

The private sector in Africa is still heavily natural-resource-based, with agriculture predominating, composed mainly of traditional smallholdings and limited numbers of large- and medium-sized commercial farms. At the country level, however, there are considerable differences in the sector's characteristics. At one extreme are the dynamic, well-established private sectors of countries such as Kenya, Zimbabwe, Côte d'Ivoire, South Africa and Nigeria, which have a diversified structure of enterprises of all sizes and growing capital markets. Elsewhere there is a nascent private sector consisting mainly of small-scale businesses concentrated in agriculture and the informal sector, with a few principally foreign-owned larger enterprises, in some cases showing evidence of movement into large-scale modern activities.

17.4 Constraints on private sector development

Private sector development in Africa has been hindered by a number of constraints, some of them policy-related and others structural. In a study by Brunnetti *et al.* (1997), covering more than 3600 entrepreneurs in sixty-nine countries, African respondents listed regulatory and institutional constraints as major problems. The most commonly mentioned was corruption, followed by tax regulations or higher taxes. Poor infrastructure came third, followed by inflation, usually used as a proxy for macroeconomic uncertainty (Figure 17.1). Below we discuss some of these obstacles to private sector development in Africa.

17.4.1 Risky macroeconomic environment and private investment

It is widely recognised that the establishment of a sound macroeconomic framework is a prerequisite to the rapid and sustainable growth of Africa's private sector. In the short run, countries may realise some gains by in effect living beyond their means. But in time, the consequences of this will be manifested in high inflation, low investment and low growth, with negative repercussions on private sector confidence. A number of African countries have, by and large, improved their macroeconomic environment in recent years. Most countries have removed foreign exchange controls, and there has been some progress in liberalising credit allocation and interest rates, and in strengthening monetary management in many countries. An integral part of macroeconomic reforms in a significant number of countries has been the elimination of import licensing, of price controls and of direct allocation of inputs and foreign exchange.

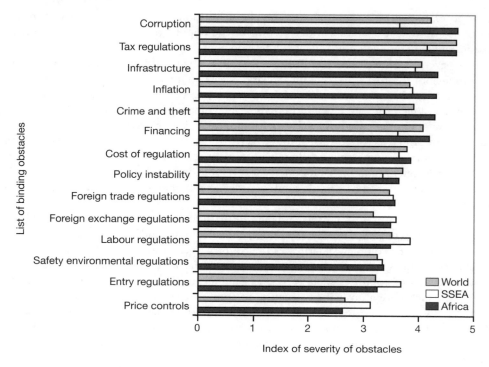

Figure 17.1 Barriers to private sector development, world and selected macro-regions
Source: Brunetti *et al.* (1997)
Note: SSEA: South and Southeast Asia.

However, despite the progress made with implementing these economic reforms, Africa has remained a high-risk, low investment region. Economic reforms have generally failed to generate sustained private sector investment and supply response, despite the progress made in the macroeconomic sphere, due to factors which still include overvalued and unstable exchange rates, loose fiscal and monetary policies, macroeconomic uncertainty and risk of policy reversals.

The imperfect credibility of policy reforms is an important source of risk or uncertainty. In a 1994 survey of actual and potential investors in East Africa by Economisti Associati, the single most important impediment to investment was non-commercial risk, mainly in the form of policy reversal fears, currency inconvertibility and the threat of civil war. In the presence of macroeconomic instability and uncertainty, investors postpone investment, in both the traded and non-traded sectors, in order to wait for additional information, while potential investors will require a large current return to compensate for the possibility of a costly mistake should policy be reversed.

One consequence has been a reduced level of investment, which has still not recovered. Gross domestic investment as a proportion of GDP fell from 26.5 per cent in 1980 to 22 per cent in 1990, and was estimated at 20 per cent in 1998. Private investment rates also show a downward trend since the mid-1970s (see Figure 17.2). Regional Programme on Enterprise Development (RPED) survey data collected for manufacturing firms in Cameroon, Kenya, Ghana and Zimbabwe similarly indicated low investment performance over the period 1991–4, confirming the macroeconomic picture (Table 17.1). Data for investment in plant and equipment for the four countries in each of the three survey years showed that almost

Figure 17.2 Private and public investment rates in Africa, 1975–97
Source: African Development Bank, Statistics Division

Table 17.1 Investment in the manufacturing sector, four African countries, 1992–4/5

Investment by country	*Proportion of firms investing*	*Investment/ value-added if firms invest*	*Investment/ capital added if firms invest*	*Investment/ value-added*	*Investment/ capital*
Cameroon					
1992–3	0.18	0.12	0.19	0.02	0.03
1993–4	0.24	0.27	0.15	0.06	0.03
1994–5	0.38	0.33	0.13	0.12	0.05
Average	0.28	0.28	0.14	0.08	0.04
Kenya					
1992	0.44	0.25	0.12	0.11	0.5
1993	0.45	0.28	0.9	0.13	0.04
1994	0.47	0.26	0.12	0.12	0.05
Average	0.45	0.26	0.11	0.12	0.05
Ghana					
1991	0.38	0.23	0.15	0.09	0.06
1992	0.51	0.27	0.16	0.14	0.08
1993	0.53	0.28	0.14	0.15	0.08
Average	0.48	0.27	0.15	0.13	0.07
Zimbabwe					
1992	0.69	0.19	0.14	0.13	0.10
1993	0.74	0.14	0.10	0.10	0.07
1994	0.70	0.14	0.12	0.10	0.08
Average	0.71	0.16	0.12	0.11	0.08
All countries	0.49	0.22	0.13	0.11	0.06

Source: Teal (1997)

half of the firms did not invest at all. Of those that did carry out some investment, the investment to value-added ratio was, on average, 0.22 and the investment to capital ratio was 0.13. The overall investment to value-added ratio for all firms – those making some investment and those making none – was 0.11, which is approximately one-third of the rates found in East Asia.

17.4.2 Weak regulatory and institutional framework

The legal and regulatory framework, the structure and administration of taxes and poor private sector dialogue feature prominently in the constraints raising the cost of investing for private businesses. These constraints include ineffective legal, judicial and financial systems, and bureaucratic administrations that impede competition in product, labour and capital markets. The lack of an appropriate legal and judicial framework hinders secure and flexible transactions, without which private sector activity cannot be effectively executed. In contrast, there is greater incentive to use resources efficiently when property rights and collateral are well defined and given appropriate legal recognition. In many African countries, where land is a critical household asset, land distribution is skewed, particularly against women, tenure is insecure, and the assignment and transfer of titles is cumbersome. With reformed land tenure, farmers would have more incentive to invest in their farms for the long term.

Moreover, in many countries, there exist special conventions allowing firms to maintain production and distribution monopolies, import protection and price guarantees, cumbersome licensing requirements and rigidities in employment practices, all of which distort the competitive structure and impose considerable costs on the rest of the economy. The performance and responsiveness of public sector institutions that are critical to business activity – such as tax administration, customs and enterprise registration and investment approval agencies – also remain weak. Entrepreneurs often complain about requirements for excessive documentation, as well as the need to bribe government officials while pursuing licensing and registration approvals from various government ministries, again serving to increase the costs of doing business. More positively, in countries such as Uganda, Ghana, Zambia and Zimbabwe, licensing of new industries has recently been simplified or removed from ministerial discretion, and investment application has been streamlined, with investment centres being established as promotional rather than regulatory organisations.

17.4.3 Governance and overt rent-seeking activities

Related to the problem caused by weak but excessive regulation, African entrepreneurs have listed corruption as the most serious obstacle to business operations. Mauro (1997) and Kaufman and Wei (1998) have demonstrated that corruption is greater in countries where there are problems of regulatory and institutional inefficiency.

The costs of corruption in terms of private sector development are enormous. By lowering private sector incentives, increasing the cost of making transactions and lowering the quality of public infrastructure provision, corruption leads to inefficient economic outcomes, impacting negatively on economic growth. Corruption has been likened to a tax on private investment, though its distortionary effects on the economy can be more serious than taxes. Officials asking for bribes from entrepreneurs before enterprises can be started can discourage new investments altogether, given that, by the nature of their secrecy, overt rent-seeking contracts are not enforceable.

Mauro (1997) has shown that corruption is harmful to private investment and that a reduction in the levels of corruption in developing countries would induce a significant positive investment response in private investment. Using a scale of 0 (most corrupt) to 10 (least corrupt), the author calculates that, were countries with a corruption index of 6 to reduce this to 8, investment would increase by 4 percentage points. In another study, covering fourteen major source and forty-one host countries, Wei (1997b) found that corruption in host countries acts as a major disincentive to foreign investment. If corruption levels were to increase from that calculated for Singapore to that for Mexico, this would be equivalent to increasing the tax rate by 20 percentage points.

17.4.4 *Inadequate and expensive infrastructure*

Costly and unreliable infrastructure is usually identified as a major obstacle both to private sector development and to attracting foreign investment specifically. The high cost of such provision in Africa has been linked to basic structural factors such as low population economic density and the fact that many countries are landlocked. Other critics, pointing to poor maintenance of infrastructure already in existence, suggest that the problem has more to do with institutions, incentives and policies.

Poor and inadequate infrastructure results in high transaction costs. For example, it has been shown that water transport facilities and services in Africa are underdeveloped in comparison with other areas of the world in terms of quality, efficiency and reliability. Thus ocean freight rates for Africa are 3–4 per cent higher than those of Asian countries competing in the same export markets and nearly double the rates of European countries. These high freight charges result from the inefficiency and the poor quality of port operations, lack of transport facilities at many ports and high port charges imposed on ship operators.

The poor state of the region's public infrastructure, therefore, a state stemming partly from a lack of autonomy, accountability and competition, impedes the development of the private sector by raising transaction costs and restricting access to both domestic and international markets. Greater efficiency in the delivery of infrastructure services can and needs to be achieved by applying commercial principles, such as cost recovery programmes, user charges and removal of price controls that set prices below cost. Transferring the responsibility for providing infrastructure services through management contracts and leases can reduce demands on limited government resources and capacity. Other options for private participation in infrastructure include build–operate–transfer (BOT), build–own–operate (BOO) and full privatisation arrangements.

17.4.5 *Weak financial systems*

African financial systems are characterised by low levels of financial intermediation and sophistication as well as by policies that 'crowd out' the private sector. A major reason for their lack of development is the existence of various elements of financial repression that thwart growth of the financial system. The role and dominance of government in the sector is still pervasive in a number of African countries, resulting in low or even negative real interest rates, necessitating credit rationing by the banking system.

A major problem faced by African financial institutions is that of enforcing loan repayments. In Africa, loan repayment has been particularly problematic in the case of subsidised credit, arising because of the breakdown of 'credit morality'. Borrowers came to regard loans made as grants and believed they did not need to repay. This problem has been prevalent in institutions where loans have been disbursed partly on non-economic grounds, or where governments had announced debt-forgiveness programmes in the past. It is also the case that credit to the private sector is mostly garnered by large and well-connected firms and traders. Most farmers and small and medium-size enterprises have little access to credit, so that their source of funds is limited to their own savings. For example, in the RPED study of Zimbabwean manufacturing firms, among micro- and small-scale enterprises over 80 per cent of the firms used own savings as start-up finance, over 50 per cent in the case of medium- to large-scale. Only 3 per cent of micro-enterprises used bank loans, 14 per cent of small-scale enterprises and 9 per cent of medium- and large-scale enterprises. On average, it has been found that in Africa about 65 per cent of start-up capital in African businesses come from own savings, with friends and relatives accounting for much of the remainder. Few firms can draw regularly from external sources, including loans from development finance institutions, equity issues and advances from parent companies and, while some firms use bank loans for working capital, the amounts involved are often far less than those desired by the firms.

17.5 Enhancing private sector development in Africa

To foster improvements in the business environment, attention needs to be focused on a number of areas. These include strengthening the legal and judicial system and improving the regulatory framework; supporting entrepreneurial development; promoting trade and investment; accelerating the process of privatisation; and strengthening the financial sector.

17.5.1 Improving the regulatory and legal framework

Efforts to develop competitive markets by eliminating monopolies, deregulating business activities, and removing barriers to entry and exit in various sectors are essential for lowering costs. Regulatory reforms should be more clearly focused to promote competition, something which is fundamental for efficient private sector development. Reform must also focus on rationalising tax and customs administration, as well as business licensing and registration requirements. Reform of labour markets is also vital to allow firms more flexibility in responding to changing competition.

A credible legal and judicial framework that supports private economic rights and is enforced equitably and transparently is essential for private sector activity. Laws relating to business contracts, property rights, collateral and debt recovery, commercial dispute resolution and arbitration will need to be enacted and effectively enforced. In many countries, inappropriate legal institutions or procedures and lack of qualified personnel have inhibited enactment and enforcement of these laws. Effective enforcement of laws will require revision of court procedures, streamlining of court registry and administrative systems, as well as greater investment in judicial personnel.

17.5.2 Promoting trade and investment

Many countries have made notable progress in liberalising both internal and external trade by reducing price and marketing controls and non-tariff barriers for both imports and exports. But in some countries tariffs and protection remain high, constraining access to capital, technology and raw materials. The emphasis should be on lowering the cost of protection, reducing tariffs, eliminating quantitative restrictions and removing impediments to internal trade. These will also be important for increasing both domestic and foreign investment.

Although the need for foreign direct investment (FDI) is greatest in low-income countries, little has been forthcoming in this respect outside the extractive sectors of mining and oil. Indeed, there is concern regarding the considerable foreign disinvestment from the region. FDI can be critical in introducing widespread technological change, complementing domestic investment, improving the agility and competitiveness of firms, and providing access to skills and global markets. Investment promotion units that are responsive to private sector needs should also be supported. Some degree of successful local business participation in a host country economy is necessary to attract risk-averse foreign private investors.

17.5.3 Supporting entrepreneurship

African entrepreneurs abound in the agriculture, industry and services sectors, although firm size is small and enterprises operate mainly in the informal sector. Many operators are women. The important role that micro, small and medium-sized enterprises play in creating employment opportunities, providing technological change to local industries and contributing to growth is now well recognised. Most small firms acquire their capabilities and expertise in the normal course of operating their businesses: through suppliers, customers, traders and larger companies.

Promoting the development of small- and medium-scale enterprises (SMEs) presents a more daunting challenge for African countries. Evidence suggests that the potential benefits of this sector for employment creation have not been fully exploited. SME development in many African countries has been hindered by excessive government control and regulations or by too much protection of large firms. The evolving policy environment and competitive market conditions are critical to the development of SMEs in Africa. Government support programmes to improve the technological, regulatory and institutional infrastructure should be designed so that they minimise the risk and uncertainty faced by SME operators, while allowing them to explore and develop their entrepreneurial capability. For SMEs to respond effectively to changes in incentives induced by recent policy reforms and to grow over time, entrepreneurs and workers must have the requisite technological capabilities. These include the skills and information required to establish and operate modern machinery, and the learning ability to upgrade these skills when needed.

Government programmes can also help to develop SMEs in a variety of ways. First, the shortage of capital faced by many small enterprises can be alleviated by making available loans at low interest, channelled through specialised public or private institutions. Second, a central small-industry institute should be established in each country to provide specialised services such as technical training, business advice, dissemination of information, management consultancy, legal counselling, procurement, product design, quality control, marketing and display, financial control and assistance in obtaining loans from financial institutions, these services to be supplied free or at a nominal fee. Third, selective controls to protect small firms from competition may be introduced, in particular industrial licensing and pricing regulations and controls against imports, as are already applied in some countries.

Besides direct government assistance, there are also increasing numbers of specialised non-governmental financial institutions which rely on local savings and lend at market rates to large numbers of small-scale clients, particularly women. These budding institutions need temporary help with seed or equity capital and technical assistance. The better ones with longer track records are already starting to tap into capital markets and attract private investors.

It is, however, important to note that the ultimate success of SMEs depends largely on the strategic decisions taken by private entrepreneurs themselves. Specifically, SMEs will have to build networks and alliances to compete effectively in domestic and foreign markets and take advantage of rapid technical changes.

17.5.4 Accelerating privatisation

An important component of private sector development is the privatisation process, which can help to reduce public sector fiscal deficits, spur additional investment and improve efficiency. Despite these benefits, progress in the implementation of privatisation programmes has been slow. Where state control and inefficient operation of key sectors – electricity, telecommunications, water – is thought to hamper the development of businesses, privatisation programmes need to be accelerated.

Four elements of a successful programme for privatising public enterprises in the region may be identified. First, there is a need to focus on large and strategic public enterprises that absorb large amounts of resources. Second, a case-by-case approach to improving efficiency and ownership change needs to be adopted. Third, fundamental socio-political constraints relating to equity and distribution issues should be tackled through public information campaigns and schemes to broaden ownership in privatised enterprises. Possible schemes to broaden public support include partial voucher schemes, trusteeship arrangements, trust funds, share give-aways and employee ownership options. Fourth, the private sector itself must be involved in the implementation process.

17.5.5 Strengthening the financial sector

The financial sector is an integral component of private sector development. If no functioning financial system exists, a strong private sector is impossible. The primary role of the financial sector is to mobilise savings and efficiently channel those savings to the most productive investments. Many countries have undertaken financial reform programmes to reduce financial repression by limiting the monetisation of fiscal deficits, liberalising interest rates, eliminating credit controls and reducing directed credit programmes. These measures constitute the first steps towards making the financial systems responsive to the needs of the private sector. While the elements necessary to improve upon this process will vary, depending on country-specific circumstances, four areas deserve priority attention: improving banking infrastructure; restructuring the banking system; developing non-bank financial instruments; and, as already mentioned, supporting microfinance.

Building an efficient financial sector for the mobilisation of domestic resources for private investment will require a sound payments and settlement system, reliable accounting and auditing procedures, more qualified and experienced personnel, and an effective supervisory system. Financial claims cannot be fully secured without proper monitoring and information. Developing effective accounting and auditing practices to ensure timely and accurate financial reporting is essential for efficient financial intermediation. While improvements have been made with respect to monetary policy, prudential supervision of the banking system is still ineffective in many countries. Also required is a system of banking supervision with enforcement power that ensures adequate capitalisation, allows for asset provisioning, prevents portfolio concentration and ensures sound management.

In addition to reforming the banking sector, non-bank financial instruments such as venture capital schemes, bond and equity, pension funds, insurance underwriting, leasing, and mortgage finance can help in promoting competition and diversification of financial markets. Foreign portfolio investment can be attracted with emerging country and regional market funds.

17.6 Conclusion

In many African countries, the progress of private sector development has been hindered by distorted and over-regulated markets or by ineffective government support programmes. Experience shows in contrast that the private sector has flourished in instances where governments have allowed the markets to operate freely and firms have been allowed to compete with each other. In particular, the continuing economic reform programmes being implemented by many African countries have led to a rapid growth of small- to medium-scale enterprises as private entrepreneurs take advantage of the relaxed economic environment. The reforms include the removal of price controls, relaxation of foreign exchange regulations, financial sector and labour market reforms, and trade liberalisation. Where adopted, these measures have improved the entry conditions for new firms and the competitiveness of existing ones

References and bibliography

Adam, C. (1997) 'Privatisation in sub-Saharan Africa: issues in regulation and the macroeconomics of transition', in J.A. Paulson (ed.), *Changing the Role of the State in Key Markets*, London: Macmillan and New York: St Martins Press.

African Development Bank (1997) *African Development Report 1997: Fostering Private Sector Development in Africa*, Oxford: Oxford University Press.

Berg, E. (1994) 'Privatisation in sub-Saharan Africa: results, prospects, and new approaches', paper prepared for the Study of African Economies in Transition, Washington, DC: World Bank.

Biggs, T., White, E.D., Moody, G. and van Leeuwen, J.-H. (1994), 'Africa can compete! Export opportunities and challenges for garments and home products in the US market', World Bank Discussion Paper 242, Washington, DC: Africa Technical Department.

Biggs, T., White, E.D., Moody, G. and van Leeuwen, J.-H. (1996), 'Africa can compete! Export opportunities and challenges for garments and home products in the European market', World Bank Discussion Paper 300, Washington, DC: Africa Technical Department.

Bigsten, A. and Kayizzi-Mugerwa, S. (1997) *Adaptation and Distress in the African Economy: A Study of Uganda's Adjustment Experience*, London: Macmillan.

Brunetti, A., Kisunko, G. and Weder, B. (1997) 'Institutional obstacles for doing business: data description and methodology of a worldwide private sector survey', background paper for the *World Development Report*, 1997.

Easterly, W. and Levin, R. (1997) 'Africa's growth tragedy', *Quarterly Journal of Economics*, November: 1203–50.

Kaufman, D. and Wei, S. (1998) 'Does "grease payment" speed up the wheels of commerce?', Washington, DC: World Bank.

Killick, T. (1983) 'The role of the public sector in the industrialisation of African developing economies', *Industry and Development*, 7.

Mauro, P. (1997) 'The effects of corruption on growth, investment and government expenditure: a cross-country analysis', in K. Elliot (ed.), *Corruption and the Global Economy*, Washington, DC: Institute for International Economics.

Mead, D. (1994) 'The contribution of small enterprises to employment growth in Southern and Eastern Africa', *World Development*, 22 (12): 1881–94.

Meade, J.E. (1952) 'External economies and diseconomies in a competitive situation', *Economic Journal*, 62: 54–67.

Sachs, J.D. and Warner, A.M. (1997) 'Sources of slow growth in African economies', *Journal of African Economies*, 6: 335–76.

Wei, Shang-Jin (1997a) 'How taxing is corruption on international investors?', NBER Working Paper no. W6030, Cambridge, MA: National Bureau of Economic Research.

Wei, Shang-Jin (1997b) 'Why is corruption so much more taxing than tax? Arbitrariness kills', NBER Working Paper no. W6255, Cambridge, MA: National Bureau of Economic Research.

World Bank (1994) *Adjustment in Africa: Reforms, Results, and the Road Ahead*, New York: Oxford University Press.

18 Economic growth, wellbeing and governance in Africa's urban sector[1]

Carole Rakodi

18.1 Introduction: changing urban opportunities

Cities and towns are spatial constructs: the sectoral structure of their economies and their demographic characteristics imply economic roles, socio-political organisational arrangements and management challenges that are different from those in 'rural' areas. This is not to say that they can be analysed in isolation. External economic and development assistance policies and processes, political changes and domestic economic and social policies impact on urban and rural areas alike, but their implications and the scope for local response differs. Spatial analysis focuses on the interrelationships between sectoral characteristics and policies as they play themselves out in areas (regions, cities or localities) with particular spatial, geographical and natural resource characteristics.

By the beginning of the 1990s about a third of SSA's population was urban, double the proportion in 1960, and some analysts expect this to pass 50 per cent in the early decades of the twenty-first century (UN, 1995).[2] Although the fastest increase occurred in most countries in the decade or two immediately after independence, high rates of natural increase have resulted in continued rapid urban growth despite economic crisis. Urban economies in SSA shared in the subcontinent's economic stagnation and decline and in its failure, on the whole, to recover from recession, despite attempts at economic and institutional reform. This is a critical development issue, given (a) the important role played by cities in national economies; (b) the impact of urban economic decline on poverty; and (c) the resulting lack of revenue for investment in the services on which enterprises and households depend.

It is contended here that the main challenge facing African towns and cities is the achievement of economic growth and its equitable distribution, so that urban economies can contribute appropriately to national economic development, provide sufficient labour market opportunities for their growing populations, and generate resources for service provision and urban management. Policy issues include, therefore, the political and organisational arrangements for improved management of urban development, as well as appropriate interventions with respect to economic development, land management, service delivery and environmental management. The chapter will, first, present arguments to support this identification of priority issues and to elaborate on the main causes of the problems before going on to examine experience with a selection of management arrangements and policies for Africa's urban areas. In section 18.2 we provide an overview of the urban situation, commenting on the economic role of urban areas, the degree of urban poverty and resource constraints on urban management. In section 18.3, in four sub-sections we review the policies needed for urban development and welfare in Africa.

18.2 Overview of the urban sector

18.2.1 *Economic trends and the economic role of urban areas*

Cities enjoy productivity advantages deriving from (a) economies of scale; (b) economies of agglomeration (i.e. efficiency advantages produced by the clustering of firms in a specific or in related sub-sectors); and (c) location-specific factors such as good communications. It can be to these factors, as much as or more than to urban-biased policies, that the disproportionate share of GDP produced in urban areas (compared to their share of population) is attributable (Peterson *et al.*, 1991). Using alternative simplifying assumptions, Becker *et al.* (1994: 25–6) estimated that between 40 and 60 per cent of Africa's GDP was produced in urban areas, though these accommodated just under a third of the population in 1990. Although they also conclude that these are (a) overestimates, due to price distortions, and (b) declining, due to implementation of structural adjustment policies affecting wages, subsidies, exchange rates and agricultural prices; they conclude that urban areas will continue to generate half or more of GDP.

In the years after independence in many African countries (the 1960s and early 1970s in most), indigenisation of the public sector, import substituting industrialisation (ISI) and increased direct intervention in production by the state led to expansion of formal wage employment and rising real wages for formal sector employees. Even in those countries where African involvement in trade, small-scale manufacturing and services had been discouraged, and where city authorities attempted to restrict and harass informal sector activities, these expanded and diversified in response to the demand generated by formal sector incomes. Rural–urban migration was fuelled by release of the pent-up desire (especially of women and children) to migrate which had been constrained by restrictive colonial policies, and also by expanded urban economic opportunities, poor performance in the agricultural sector and pro-urban policies (for example, food price controls). Despite increased in-migration, the majority of migrants and other new entrants to the urban labour force were able to find formal wage employment, especially in the capital and larger cities and mining settlements.

Successive shocks to African economies brought the economic boom to an end for most during the 1970s, culminating in the debt crisis and recession of the early 1980s (Rakodi, 1997c). Already, in the 1970s, the real wage increases of the 1960s had been eroded by inflation and government budget deficits, and growth in formal sector wage employment was increasingly falling behind labour force growth. In particular, ISI was constrained by the limited size of domestic markets, continued dependence on (falling) foreign exchange earnings for inputs and spare parts, and inefficiencies arising from political objectives and public sector mismanagement.

Falling real wages and job availability led to rising unemployment. In five of six Francophone countries studied by the ILO in the late 1980s and early 1990s, between a fifth and a quarter of the active population aged 15 and over were unemployed.[3] Unemployment is highest amongst the educated and young people. Lachaud (1994) found that 25–40 per cent of those who had received secondary education were unemployed and between a quarter and half of young people aged 15–29, the latter said to constitute 60–75 per cent of the unemployed, although by the early 1990s they accounted for only a third of the labour force (ILO-JASPA, quoted in Rogerson, 1997: 345; see also Zeleza, 1994). Increasingly, both men responsible for households and women who had not previously been involved in income-earning activities engaged in informal sector activities, which now account for between a half and two-thirds of the labour force and most new jobs in urban areas (Rogerson, 1997: 346).

Urban–rural disparities in average incomes (which had always concealed inequality, especially in urban areas) declined and disappeared (Jamal and Weeks, 1993; Becker *et al.*, 1994).

As the 'formal' part of urban economies declined and urban households became more dependent on 'informal' economic activities, the earlier view of the informal sector as driven and sustained by demand generated largely in the formal sector became increasingly untenable.[4] Evidence mounted that in many countries most parts of the informal sector were becoming overcrowded, that many activities were only marginally profitable, and that only in a few (some sectors of trade, public transport, rental housing and so on, shading into criminal activities) was it possible to accumulate sufficient profits for enterprise growth (or, more frequently, diversification) (Rogerson, 1997). However, understanding of the dynamics of urban economies struggled to keep pace with economic change, due to the dearth of relevant statistics and research.

It is frequently argued that the export of primary products has limited potential for long-term economic growth in a globalised economy. Development of higher productivity economic activities (manufacturing and services), which are located in towns and cities, is considered vital, both as an independent contribution to the national economy and because of their positive linkages with rural and agricultural development (Peterson *et al.*, 1991). Urban economic success reflects the roles of a city both in the evolving global economy (either as a site of operations important to the command and control system of global economic organisation or as a channel and receptor for international flows of investment, goods and services) and within its national economy. Further, success depends on a city's economic structure and its ability to restructure to maintain competitiveness, profit levels and employment. This, in turn, depends on policies, politics and its administrative and organisational structure and capacity. Amongst the conditions necessary for stable and prosperous urban economies to emerge are an appropriate policy environment, security for people and property, and efficient and effective infrastructure services (especially water, power, transport and communications and waste management) (Vidler, 1999).

Many of the policies relevant to the health of urban economies are national policies (subject to greater or lesser degrees of external influence). These include policies on economic liberalisation, civil service reform, labour force protection, and education and training. Rapid removal of industrial protection, privatisation of state-owned manufacturing enterprises and public sector retrenchment have had adverse impacts on urban wage employment opportunities. Yet the ability of enterprises to adjust quickly to increased international competition, the knock-on effects of job losses on demand in the rest of the urban economy, and the ability of the small-scale sector to respond rapidly to increasing numbers of job seekers are rarely adequately considered by economic policy makers. In the course of efforts to adjust formal sector employment and wages to market levels, other aspects of industrial protection have also been abandoned, leading to increased insecurity and casualisation in remaining formal sector employment, evasion of labour regulations and increased sub-contracting to small-scale (sometimes informal sector) enterprises (Rogerson, 1997; see also Lachaud, 1994). Education and training systems evolved after independence to meet the demands of the formal sector for academic and vocational credentials. Efforts to increase cost recovery have hindered access by poor people, and public sector vocational training programmes, never very large or efficient, have declined (Mulenga, 2000; see also Lachaud, 1994). Neither, in any case, had syllabuses which imparted the practical and business skills useful for participation in informal sector activities.

Security incorporates the rule of law as applied to property, an issue in Africa because of earlier nationalisation policies and problems related to land titles, as well as conflict, crime and violence. Insecurity has inhibited investment, especially the ability of cities to attract international investment (Vidler, 1999; Beavon, 1997).

Many infrastructure and service provision policies have a major impact on the efficiency of economic activity: some are prerequisites for land development (access, drainage), others affect daily business operation and production (public transport, telecommunications, solid waste removal), others may be direct inputs into the production process (power, water). Enterprises differ in their dependence on particular services. Many service provision policies are also formulated at national level and the scope for adjustment to pursue local economic development or other goals may be limited. For example, power policies and prices are generally set nationally. The scope for local policy making with respect to water, waste management and land administration varies. Land administration systems which only recognise some of the land supply mechanisms operating in African cities and which operate inefficiently hinder the access of investors to land and condemn many small-scale enterprises to insecure tenure. All cities are characterised by significant social and environmental externalities associated with urban population and economic growth, as well as by market failures with respect to the supply of land and infrastructure. Public sector investment and regulatory systems, therefore, are required. However, not only has investment in infrastructure failed to keep pace with the demands of economic activities and population growth since the 1970s, but also the infrastructure which was in place has deteriorated markedly, especially since the 1980s. These problems have been attributed to:

- inappropriate organisational arrangements, technical decisions and pricing policies;
- political interference; and
- poor cost recovery mechanisms and performance.

The regulatory systems dealing with development control, licensing and environmental health also constrain economic activity, especially in the informal sector, when they are enforced (Rogerson, 1997). The political and attitudinal obstacles to change are significant.

18.2.2 Urban poverty

In general, the incidence of poverty has been demonstrated by the World Bank-funded poverty assessments to be at least a third higher in rural than in urban areas in Africa.[5] Nevertheless, in most countries where more than half the population lives in poverty overall (for example, Zambia, Sierra Leone, Madagascar), more than 40 per cent of the urban population are poor (World Bank, 1996). Although limited at present in many African countries, the contribution of urban to overall poverty is generally growing, as the proportion of the population living in urban areas increases and urban–rural income gaps are eliminated.

Economic reforms deliberately set out to eliminate policy and price distortions which benefited urban areas. The resulting decreases in real wages, declining numbers of formal sector jobs and pressure on the labour absorptive capacity of the informal sector had adverse effects on many urban households. The ILO's studies in Francophone Africa demonstrate the association between household poverty and labour market vulnerability: the incidence and depth of poverty are greatest where 'the breadwinner is an irregular worker, a marginal self-employed person or even a non-protected employee, and in households where the head is unemployed or inactive' (Lachaud, 1994: 59) – or a woman. The result of deteriorating labour market opportunities was that the position of poor people, already relatively deprived with respect to living conditions and access to social services, generally worsened (Beall and

Kanji, 1999). Urban–rural links, always significant in migration strategies and social networks, were in some instances intensified as impoverished urban households drew on their claims on agricultural land and rural households took advantage of economic liberalisation to diversify their livelihoods (Potts, 1997; Bryceson and Jamal, 1997). In other respects, however, they were threatened; for example, urban impoverishment reduced the remittance flows on which many rural households depend (such as in Zimbabwe). Basic infrastructure installed in the 1960s and 1970s, as noted above, was not maintained and when poor residents resorted to alternatives to public sector provision, these were either poorer quality (for example, the use of water from shallow wells), more expensive (for example, water vendors, private healthcare), or both. When infrastructure was later rehabilitated, new capacity installed or services extended, cost recovery policies (often instituted without workable exemptions or cross subsidy arrangements) either excluded the poor or enabled them to gain access only at the cost of other basic expenditure, particularly on food.

18.2.3 Resource constraints on urban management

Urban economic decline and increased incidence of poverty reduce the public sector's capacity to raise revenue (via personal, property or corporate taxes and user charges) for investment in service provision, regulation and other activities. In Anglophone Africa, a model for urban local government based on the British system was developed, rather hurriedly, in the run-up to independence. This system had shallow roots, except perhaps in Zimbabwe and South Africa, and was emasculated in the years after independence. Politicians of ability were attracted to national, not local, politics. Preoccupied with forging a sense of national unity, fulfilling their political promises, and establishing control over national resources, they eroded the autonomy of local authorities. The budgets of the latter were subject to central government control, they were not permitted to exercise the powers many had to borrow or issue bonds, and responsibility for service functions was removed to central ministries, as in education, or to agencies in the case of electricity and sometimes water. Qualified local staff were in short supply and were attracted to central rather than local government because of the better conditions of service and greater prestige of employment in the former. Governments were keen to deliver national development goals, using the most buoyant of the limited available sources of revenue to do so.

Centralisation, always strong in Francophone Africa, increased in newly independent Anglophone Africa. Despite periodic recognition of its failures (uniform policies inappropriate to local conditions, implementation delays and shortfalls, and waning local support for national politics and development goals), attempts at decentralisation were long on rhetoric and short on real devolution. As politics based on tribe and ethnicity threatened the stability of many states and concern mounted over the inability of the central state to deliver on its development promises, arguments for strengthening urban local authorities' democratic structures, revenue raising powers and planning and management capacity made little headway. The project orientation of lending by the aid agencies, especially the World Bank, did not help to build local capacity to sustain and replicate what were sometimes appropriate approaches.

In many countries, local democratic politics was suspended for long periods during periods of non-democratic national government or in response to perceived mismanagement and corruption. Re-establishing political literacy amongst potential councillors and the electorate when local democracy was restored was slow and made more difficult by the paucity of resources and skills available to local authorities. Local urban politics during the one-party state era of the 1970s varied between countries: in some, especially in the early years, it made possible the consolidation of a hierarchical structure of political organisation which was top-down and directive, but also offered varying degrees of opportunity for choice of representatives and participation in decision making. Once party competition was restored, low levels of awareness of the responsibilities and roles of local government amongst those standing for election and the electorate, and poorly resourced local authorities with responsibilities which far exceeded their capacity to deliver, made it difficult to increase the legitimacy of local government.

The central view of local politics (backward, merely a stepping stone in a national political career, or a potential threat to the ruling party) and reluctance to devolve powers to match their responsibilities to local authorities they perceived as incompetent led to long-term neglect of local government. Outdated legislation was not revised (or was revised to achieve over-ambitious rural decentralisation programmes). Cities were regarded as both drains on national resources and already favoured in resource allocation, and little attention was given to exploring ways of raising more revenue from the economic activities taking place in them, to implementing appropriate systems of central–local financial transfers, or to replacing the general subsidy systems which partly accounted for inadequate service delivery with more appropriate pricing arrangements. Efforts at civil service reform, intended to improve salaries and incentive structures once surplus labour had been shed, rarely progressed to the second stage and even more rarely got around to tackling the staffing issues of local government, where typically lower levels were fully or over-staffed (often as a source of political patronage) and managerial, supervisory and professional posts unfilled. There were few attempts to review vital aspects of local authority operations including:

- organisational structures, which were sectoral, with underdeveloped coordinating arrangements;
- management approaches, which prioritised regulatory over developmental functions;
- local authorities' conceptualisation of their role as direct rather than indirect providers, regulators rather than enablers, and sole actors rather than partners;
- approaches to decision making, which were non-participatory, top-down and technocratic;
- the knowledge base for policy formulation, which was limited, especially with respect to how urban economies and societies were evolving;
- priorities, which reflected vested interests in maintaining traditional functions, even if these were failing or irrelevant;
- financial management, which was inefficient and corrupt.

The chronic weakness of local government is part of a vicious circle in which the public sector's inability to deliver appropriate interventions with respect to economic development, land management and service delivery has resulted in prolonged economic stagnation. The result is an inadequate base for raising finance to fulfil government responsibilities, leading to a lack of political legitimacy because of the failure to deliver services and, in turn, inability to achieve good revenue collection performance. Although well entrenched, such weakness is not inherent, and it is possible to envisage what form improved urban management could take, based on recent research and experience. Research into the political, economic and administrative processes by which badly managed cities break out of the vicious circle would be helpful.

18.3 Policies for urban development and welfare

Strengthening political and organisational capacity for management of urban development requires:

- reconceptualisation of roles and organisational structures, based on greater realism and an understanding of issues of political leadership and legitimacy;
- financial reworking, to provide both access to resources commensurate with responsibilities and improved methods of financial management;
- institutional capacity strengthening, especially with respect to planning and coordination;
- rethinking of ways to handle land, service delivery and regulation.

Much past research has concentrated on assessing formal planning and management structures and processes, typically documenting and explaining their shortcomings. Starting in the early 1970s, with recognition of the role played by the informal sector in economic activity, land development and housing production, increased attention has been given to understanding these informal processes. In recent years there have been attempts to improve urban management, sometimes based on accommodating/adapting to the reality of cities produced largely through informal processes, and sometimes not. There is a need to examine these experiments, and to assess the reasons for their outcomes, including not just policy and project design, but also the political and economic circumstances which were necessary to their success or help to explain their failure. Independent, systematic and preferably comparative evaluations are rare, and those which have been done have generally been commissioned by the international agencies and are not in the public domain.

18.3.1 Urban economic development

The prospects for economic growth are determined, primarily, by the state of the national economy and national economic policies, themselves linked into global processes and the policy conditionality of the international agencies. Nevertheless, there does exist, to a varying extent, scope for policy intervention at the city level to further local economic development goals aimed at attracting international or national investment or increasing the productivity of existing enterprise, by ensuring that the supply of several key factors of production (land, utilities, a healthy and appropriately skilled labour force) is adequate.

African cities are very poorly placed at present to join the competition between cities for investors in the most buoyant sectors of the global economy: hi-tech manufacturing, the command and control functions of transnational corporation headquarters, and business, producer and financial services.[6] Nor are they well placed to compete for investors seeking cheaper land and labour than are available in Europe, North America and Japan for (mainly) manufacturing. The human resources, stable political environment, and reliable and efficient services which are required by such investors are absent at present. The policy challenge is, therefore, more realistically, to provide minimum conditions for the efficient operation of manufacturing and services and to stop actions which are harmful to small-scale economic activities, while retaining sufficient regulatory powers to protect environmental resources and health (Amis, 1999).

Attempts to foster economic growth in the past have taken the (often costly and ineffective) form of providing industrial estates, offering incentives (subsidies, tax holidays, exemptions from regulation), exercising political influence over government decisions on the location of public/parastatal investment, or investing municipal resources in ventures which, it is hoped, will yield revenue/profits and employment, such as abattoirs, breweries or markets. Such strategies are characterised by:

- a focus on the physical development of sites, based on limited understanding and analysis of the economics of business or the economic position and trajectory of the city or town concerned;
- a preference for large-scale formal sector investment, despite the costs of attracting it and the resulting vulnerability of the local economy to fluctuations in the fortunes of a single firm or sector; and
- a neglect, or even harassment, of the small-scale and informal sector.

More economically well-informed policy formulation is required, which gives priority to creating conditions conducive to investment, the key tools of which are land administration, infrastructure investment and service delivery, not incentives and subsidies (Peterson *et al.*, 1991).

Support to the informal sector should, first, take the form of revising policies and regulations which constrain its operations, restricting regulation to that essential for public health, simplifying licensing requirements, legitimising mixed use in residential areas and allocating resources for the development of infrastructure considered important by small-scale entrepreneurs (typically access, electricity, water, and also waste management and telecommunications) in sites they consider appropriate (generally not industrial estates and workshop complexes). Attention may also be given to identifying informal sector enterprises with growth potential, with a view to relieving constraints on or providing support to such growth. Rogerson (1997) suggests that there is potential for technology upgrading and fostering linkages between the large and small-scale sectors (for example, subcontracting and encouraging cooperative networks of small firms), but neither the potential for these nor the desirability and appropriateness of a public sector role has yet been substantiated by research or experience.

In addition to supplies of land and services (see below), economic development requires suitable human resources: a healthy, literate and skilled workforce. Literacy and numeracy via access by all to primary education is the most basic requirement but needs to be topped up by skills and knowledge of use in assessing economic opportunities, including vocational and business skills. Traditional public sector vocational skills training provision is (a) in disarray, and (b) unsuitable, and NGO schemes are often tiny, not financially sustainable or provide training in skills areas that do not match market demand. In West Africa, especially, many gain vocational skills through the informal apprenticeship system (Lachaud, 1994), but this also has shortcomings (King, 1996). Improved arrangements for developing knowledge and skills require a good understanding of economic trends and labour market demands, as well as public–private sector collaboration. Health is addressed in sub-section 18.3.4 below.

18.3.2 Political and organisational arrangements for urban management

Despite periodic efforts to establish district-level administration in many African countries, the structures of urban local government have remained essentially the same as those inherited at independence, with some exceptions (for example, South Africa, Ghana, Uganda). Even where new arrangements have been instituted, the sectoral structure oriented to service delivery and regulation bears strong similarities to older local government systems (Davey *et al.*, 1996).[7] The traditional model has been criticised for its lack of policy making capacity, poor coordination between departments and a rigid rule-based approach to managing urban development. Most of the economic activity, land development, production of housing and social support in African towns and cities, and much service provision, is done without public sector support, in contravention of legal rules and despite public sector attempts at regulation and control. In these circumstances, it cannot be said that public sector agencies manage cities (Devas, 1999; Korboe *et al.*, 1999; see also Swilling, 1997). They continue, against all the odds and in the face of continuing lack of progress and declining legitimacy, to operate in their habitual ways, rather than reconsidering how to improve decision making and make the best use of their limited resources in response to urban management priorities.

As noted above, decision making has often been in the hands of central government appointees. Where it is representative, systems vary. Typically, in Anglophone Africa elected councillors represent wards, and the council and its committees, advised by officials, are the decision making bodies. These representative arrangements have both strengths (better prospects for accountability based on local accountability of councillors to constituents and joint decision making) and weaknesses (weak leadership and a tendency for party or ethnic rivalries to hinder decision making). Much of the political ferment of the 1980s was urban based, although it was 'concerned with broader struggles against monolithic state structures, dictatorships and economic hardship' (Aina, 1997: 431), as well as specific urban issues. Reinforced by external pressure, the outcomes included significant changes in systems of political representation, including:

- the replacement of dictatorships or single party systems with multiple parties;
- decentralisation of functions, especially in Francophone Africa; and
- the introduction of executive mayors, to provide leadership and improve accountability, in some Anglophone countries.

In addition to the broader political changes, there have been a number of attempts, usually, but not always, driven or facilitated by external donors, to develop more consultative modes of decision making. The most obvious examples are the city consultations that form the core of the planning process in World Bank and UNCHS-supported urban management, poverty and sustainable city programmes. Similar approaches have been used in planning processes elsewhere, for example, city visioning in Nairobi and Kisumu in Kenya supported by German foundations; a DfID-supported collaborative approach to design of an urban poverty project in Mombasa, Kenya. Locally initiated arrangements also exist, for example, the Informal Settlements Coordinating Committee in Nairobi. Wekwete (1997) and Farvacque-Vitkovic and Godin (1998) suggest that the outcomes of such decision making processes need to be encapsulated in 'urban contracts' which agree the allocation of responsibilities and financial commitments between collaborating partners. Such approaches to policy making and planning, which are more inclusive than traditional plan making and budgeting, are relatively recent. They will only demonstrate potential for more effective urban management if they (a) result in improved design, implementation and maintenance, and (b) are institutionalised. Positive responses have led to production of guidelines and replication of practices, with donor support. However, more independent evaluations are needed.[8]

Attempts to move beyond earlier top-down and government-centred decision making (which only occasionally allowed for consultation and participation by other actors) to more inclusive approaches are a response not just to local government weakness but also to the growth in scale and importance of NGOs and civil society organisations in the 1980s and 1990s (Aina, 1997; Mitlin, 1999).[9] Research has documented the increased scale, differentiation and evolving approaches of NGOs[10] and their relationships with governments, but little has a specifically urban focus. The overall impact of proliferating NGO initiatives is limited by poor coordination, in part because NGOs often see each other as potential competitors for funds. In addition, they and governments may regard each other with mutual suspicion. However, even where donors, sceptical of the will or capacity of a local authority to provide services to meet the needs of the poor, have decided to support an NGO instead (for example, British support for a CARE water supply project in Lusaka), it is quickly recognised that NGOs have neither the capacity nor the desire to provide services city-wide and on a permanent basis, and that the development of appropriate relationships with the local authority is essential.

Research on civil society associations in African towns and cities has been limited and selective (see, for example, Tripp, 1992). Repressed under dictatorships and one-party states and restricted to organisations seen as non-threatening by the state, many of these are longstanding organisations (trade unions, ethnic associations, women's organisations, professional bodies, trade associations), although their freedom to operate and express their views has often been curtailed (Aina, 1997). Others emerged in response to new needs and opportunities. Examples are vigilante groups or local militia, formed to compensate for the absence of an effective police presence; and associations of Community-based Organisations (CBOs), such as the National Homeless People's Federation in South Africa. At the same time, community organisation based on single-party political structures, which had been used for organisational, service delivery and even participatory purposes in the 1970s, was modified or disbanded (for example, in Zambia, Tanzania and Mozambique) (Kombe and Kreibich, 2000; Lugalla, 1995; Chichava, 1997). National political change in South Africa has had equally far-reaching but different implications for civil society organisations and their relationship to the state (Beall *et al.*, 1999). In some cases, new or modified forms of community organisation have emerged, often with NGO support (for example, in Mozambique); in others there seems to be something of a vacuum between representative democracy and associational life (social and kinship networks, religious associations, rotating savings and credit associations (ROSCAs), burial societies), with little area-based community organisation or resident involvement in local planning and management. Research is needed to examine, on a comparative basis, the extent to which broader political changes have resulted in changes in the scope and organisation of CBOs and in their relationship with the state, as well as subsequent strategies adopted to press community demands, become engaged in decision making or service delivery, and regulate community affairs.[11]

Critical to workable arrangements for urban management are financial resources. Decentralisation of responsibilities unaccompanied by financial and human resources to fulfil those responsibilities is, as has been amply demonstrated, meaningless and even counter-productive in terms of political legitimacy. The recent moves towards decentralisation, driven in part by the political liberalisation agenda of increasing accountability and in part by the economic liberalisation agenda of increasing efficiency and reducing budget deficits, run a similar risk (Wekwete, 1997). The reform agenda is being taken more seriously in some countries than others, but everywhere the scope for improvement is limited by continued recession, the failure of many economic and institutional reforms to deliver the anticipated results, political tensions and inadequate data. Donors have been urging a reconsideration of local finances, including central–local fiscal relations, sources of capital for developmental purposes, local revenue generation and financial management practices (Devas, 1999).

In Africa, improvements to the financial resource base for urban areas have focused on:

- political and technical negotiations over the basis for central–local government resource transfers and revenue sharing (for example, recent developments in Kenya, Côte d'Ivoire, Cameroon, Senegal). In Francophone countries the *unicité de caisse* principle needs to be changed (Farvacque-Vitkovic and Godin, 1998);[12]
- experiments with the establishment of capital development funds under the control of national ministries of local government, which local authorities can access if they satisfy certain conditions designed to improve their credit-worthiness (for example, in Zimbabwe, Nigeria, Côte d'Ivoire);
- broadening the scope for and increasing the efficiency of local revenue generation, especially by increasing the use of user charges (see below); improved design and collection of other buoyant and easily collectable taxes, such as market fees, while eliminating minor taxes and fees which cost more to collect than they yield in revenue and which do not perform other (non-financial) functions; and improving property tax systems (see below);
- improving local authority financial management, by introducing new approaches to budgeting, checks on corrupt practices, and improved accounting and auditing procedures.

None of these is a purely technical matter. Evaluation of their outcomes and assessment of their potential must take into account the institutional culture and conditions within local authorities, as well as the wider economic and political context.

18.3.3 Urban land management

The basic physical, economic and political resource in cities is land. It is, therefore, environmentally sensitive, subject to speculation and hoarding as well as productive investment and vulnerable to political contestation. The urban land administration system needs to deal with tenure arrangements, procedures for registration and transfer, patterns of land use, investment in and development of property, and provision of infrastructure. The instruments used are embodied in national legislation, typically based on some amalgam of imported legislation and indigenous systems, added to by land 'reforms', such as nationalisation in the 1970s. It has long been recognised that implementation of formal systems of tenure, registration, transfer, land use planning and development control exceeds the administrative and financial capacity of African local authorities and, moreover, that such systems are misconceived in the context of this limited capacity, rapid low income urban growth, and indigenous systems of property rights and practices.

In practice, most land becomes available for urban use and is developed and transferred through informal processes which operate in parallel to formal systems that are effective only in limited parts of the urban areas in most countries (Rakodi, 1997d). In recent years, research has focused either on repeated documentation of the failure of formal systems of land administration or on trying to understand the informal processes of subdivision, transfer, development and regulation (see, for example, Kombe and Kreibich, 2000, and the summary of earlier research studies in Rakodi, 1997d). It is clear that the informal processes have enabled large numbers of (even quite low-income) households to gain access to land, the development of land and property markets, and incremental investment in housing (for occupation or renting) and business premises. However, it is also acknowledged that problems result, including:

- insecure and contested tenure;
- imperfect markets characterised by limited information;
- speculation in and hoarding of land;
- increasing difficulties facing low-income people in accessing land, because of increasing commercialisation of land markets and the importance of political patronage in laying claims to land;
- unplanned urban sprawl;
- development in unsuitable areas;
- poor environmental health due to the lack of infrastructure;
- poor quality construction due to insecurity, lack of access to credit and non-enforcement of building regulations;
- failure to yield revenue to pay for extensions of roads and other urban services, and lack of land for public uses, for example, schools.

Recent policy recommendations concentrate on ways of addressing some of these problems, while recognising the advantages of informal processes of land and property development over ineffective formal systems. Some are listed below:

1   There is an ongoing debate over the best method for improving security of tenure. Some advocate that formal individual title should be universalised, to improve eligibility for credit, increase investment and simplify administration. However:

> individualisation of title discriminates in favour of those who succeed, via claiming group rights or exploiting their wealth or connections, in achieving land and house ownership, while it excludes those who previously benefited from family claims or non-commercialised access via squatting . . . The majority of urban residents are already tenants and do not stand to benefit from individualisation of title [which may also lead to increased rents].
>
> (Rakodi, 1997d: 402–3)

Others, therefore, argue for arrangements to improve security of tenure without the costs and other disadvantages of individual freehold title, including protection from eviction, alternative forms of title (leasehold, collective title),[13] simplified surveying and registration systems, and working with indigenous arrangements for transfer, registration, and conflict resolution wherever possible (Mabogunje, 1993; Kombe and Kreibich, 2000).

2   It is argued that, because bureaucratic processes for subdividing and allocating nationalised land have created bottlenecks in supply and opportunities for corruption, land should be denationalised. However, the issue of individual freehold titles to subdivided public land is generally regressive. The priority where substantial public land banks still exist, therefore, is to use them for revenue generation by issuing leases. This would enable local authorities to undertake a land supply programme on a sufficient scale to exert downward pressure on prices and reduce the main incentive for corruption arising from scarcity, without removing the land entirely from public control.

3   Priority should be given to complete registration of occupancy as a basis for taxation, rather than the preparation of a formal cadastre (Durand-Lasserve, 1993; Farvacque-Vitkovic and Godin, 1998). Generally, a formal cadastre is needed as a basis for property taxation as well as protection of property rights through individual title. However,

attempts to create cadastres failed or were abandoned half-way in the 1980s in several Francophone countries, for example, Cameroon, Mali, Senegal. It is recommended, instead, that a register of occupants be used as the basis for a residential tax. This can evolve into a multipurpose cadastre, if appropriate. Benin's experience with the establishment of land registers and Burkina Faso's with new techniques for plot subdivision and allocation, as well as the introduction of a residence tax, are considered to be promising.

4   Collaborative decision making methods (see above) should be used in the preparation of strategies and plans at the city, neighbourhood and site levels, with a greater spatial content than those promoted in the 1980s and 1990s, as guides to future land use which are more realistic and implementable than the allocations in traditional desk-based land use plans (Attahi, 1999; Farvacque-Vitkovic and Godin, 1998).

5   Major infrastructure investment (roads, drainage and water mains) should be used to guide broad patterns of development. Farvacque-Vitkovic and Godin (1998) suggest that a broad super-grid of roads and drainage should be planned and the rights of way secured, followed by incremental infrastructure provision. They suggest that zones of 25–100 ha can then be developed by public, private or customary operators.

6   Infrastructure should be installed with minimum disruption in already settled areas, by adopting appropriate standards, participatory approaches and incremental progression from *de facto* occupation to legal regularisation.

7   Methods for improving the planning of new development areas, installation of infrastructure, and reservation of sites for public uses, without government having to acquire the land, should be piloted, for example, by the use of readjustment techniques.

8   More cost-effective methods should be developed for infrastructure installation, operation and maintenance, such as AGETIPS (Farvacque-Vitkovic and Godin, 1998); community partnerships in construction contracts, operation and maintenance; or private sector participation.

9   The quality of construction can be improved by working with informal and small-scale contractors and their clients, through education and demonstration programmes, so that the limited resources for operating formal development control systems can be concentrated on the main industrial and commercial areas and buildings open to public access.

Few of these recommendations have been implemented on more than a pilot basis. The ideas need to be assessed for their local applicability, in the light of an improved understanding of how the processes which produce most of the urban built environment actually work.

18.3.4 Urban service delivery

The direct subsidised delivery by the public sector of services such as water, sewerage, electricity, solid waste management, telecommunications and health is diagnosed as being not only unsustainable and deficient, but also anti-poor, since poor people end up paying higher unit costs for inferior services (for example, water), with inadequate access to poor quality services (for example, education, health), or without publicly provided services at all (for example, solid waste management, telecommunications).

African governments and local authorities appear to have little bargaining power with respect to service delivery reforms. The models adopted tend to be those favoured by the technical assistance agencies providing support, and these in turn are heavily influenced by current developments in economic theory and northern experience: they include private sector, NGO and community participation; reduced subsidies; and increased managerial autonomy (Batley, 1997; Nickson, 1997). Resistance (of national and local government professionals such as water engineers, and politicians) to increased private sector participation has generally been substantial: because of the concern of vested interests to protect themselves; fear of a loss of public sector control; lack of public sector capacity to perform indirect provider functions, such as the design and enforcement of contracts; perceived lack of (especially domestic) private sector capacity to raise capital and manage services efficiently; fear of private sector monopoly; and concern that efficiency (read profit maximisation) may be pursued at the expense of equity (basic needs satisfaction and public health) objectives. As a result, much service provision remains in the public sector, subject to greater or lesser attempts to improve provision, with varying performance outcomes. Participation by communities, NGOs and private companies has, in practice, been uneven.

Public transport has almost universally been wholly or partly privatised. Supply has improved and consumers are offered greater choice of services, but regulation failures mean that they also suffer from the downside of competition: unsafe vehicles, overcrowding, racing and 'taxi wars'. In addition, services may be concentrated on heavily used routes with inadequate provision elsewhere, especially in peripheral low-income residential areas. This pattern of service provision and the lack of personal safety while using some services adversely affects women in particular.

Private sector participation in *solid waste collection* has increased in most cities, but has been far from the panacea hoped for.[14] It has provided more regular services, but generally to commercial, industrial and higher income areas/consumers. Services in low-income areas remain absent or irregular because of access difficulties and limited ability to pay. Community-based alternatives have only been tried on a small scale. Regulation of disposal and landfill management tend to remain with the public sector (with the exception of informal sector participation in recycling and re-use) and are often inadequate, resulting in adverse environmental impacts. For example, private operators may dump refuse in unauthorised locations or manage landfills badly.

Commercialisation of public sector *water* provision has been more common than private sector participation, especially in Anglophone Africa. Increased cost recovery, revised pricing structures, greater managerial autonomy, increased consumer responsiveness, improvements to productive efficiency by shedding labour and so on have, in some cases, resulted in improvements to supply. However, contracting arrangements, whether for limited operations or general construction and/or operation of the water treatment and delivery system, are spreading.

Only small parts of cities are served by *sewerage* systems, responsibility for which may or may not rest with the water utility. Most systems are badly maintained, have long since exceeded their capacity and service only commercial and high-income users, but do not pay for themselves. The majority of urban residents rely on individual solutions (septic tanks or pit latrines), backed up by limited enforcement of public health regulations and inadequate systems for the evacuation of facilities when they are full.

Attempts to improve delivery of utilities have become increasingly widespread in recent years, following the massive deterioration in urban infrastructure and services which occurred in most cities in the 1980s. However, evaluations of their outcomes in terms of efficiency, and especially equity, are still lacking. Such evaluations need to consider not only outcomes, disaggregated by urban settlement size and income group/residential area within cities, but also impact, especially on health (see below). Important explanatory factors beyond those often considered by the technicians concerned with delivery of a particular service are likely to include: political dynamics; capacity in the public sector to perform indirect as well as direct provision functions; the strengths and weaknesses of inter-organisational networks compared to monolithic central or local government provision or wholesale privatisation; and the extent to which sustainable financing can be achieved in a situation where domestic capital finance and private sector capacity are limited. The cost of rehabilitating, operating and extending infrastructure may, Becker *et al.* (1994) warn, exceed the potential revenue generated from user charges, given general levels of poverty and the need to satisfy political and public goals.

Poverty is closely linked to conditions in the residential areas and work environments occupied by low-income residents: poor environments have direct impacts on health, to which some groups are more vulnerable than others. Low-income groups are both more exposed to environmental hazards and more vulnerable to their effects, because of physical weakness and their lack of buffers to cope with illness or injury (UNCHS, 1996; Nunan and Satterthwaite, 1999). Improvements to environmental conditions directly contribute to improved wellbeing and may indirectly contribute to increased incomes by improving adults' ability to work and the effectiveness of children's learning. A healthy workforce is also, as noted above, important to economic development. Environmental improvements in areas occupied by the poor can be achieved by a combination of the measures discussed above (security of tenure, installation of basic services), improvements to housing quality and provision of health care (including preventative and promotive services designed, *inter alia*, to maximise the benefits of improved utilities, and basic curative care) (Harpham and Tanner, 1995; Nunan and Satterthwaite, 1999).

As with other social services, especially education, recent policies designed to extend and improve healthcare, introduced at a time when real incomes were declining, often worsened the already disadvantaged position of the poor, who suffer both from locational disadvantage due to their residence in illegal settlements and from financial discrimination. However, there is insufficient space here to discuss recent initiatives or potential alternative arrangements to counter the adverse outcomes of reforms. In the short term, marked reductions in the health burden which exacerbates poverty, increases demands on health services and holds back economic growth can be achieved by tackling the infections and parasitic diseases associated with inadequate basic services and poor living conditions. For improved health status to be achieved, an integrated approach is needed to physical infrastructure provision and preventative and curative healthcare. In practice, divisions of responsibility between departments, levels of government and providers; professional resistance; and inadequate coordinating arrangements hinder progress.

18.4　Conclusion

What can be done at the urban level to reverse economic decline, create jobs and foster enterprise is constrained, first, by the impact of external economic and development assistance processes and policies and, second, because many of the relevant policies are dealt with at national level. Recession and structural adjustment policies have resulted in declining large-scale formal sector economic activity and increased informalisation. Our understanding of the recent dynamics of urban economies is limited, but it seems unlikely that buoyant demand for goods and services produced by both the formal and informal sectors can be generated by informal sector economic activity alone. Therefore, renewed policy attention to fostering formal sector growth is required, at both national and city levels. The latter should concentrate on improving land administration and service delivery, reforming regulatory regimes so that they do not constrain either formal or informal sector activities unnecessarily and supporting the development of relevant labour force and entrepreneurial skills.

Apart from income-earning opportunities, poor people identify a range of other aspects of deprivation which have important effects on their well-being, especially access to physical infrastructure and social services; security (of income and tenure, as well as personal safety); and access to social networks and political influence. Improvement in their well-being is, therefore, dependent on increased incomes, but can also be achieved by improved security and access to services, which may be assured wholly or partly independently of income levels. Although overall resource constraints are severe, these aspects of urban life are amenable, independently, to policy intervention, even without economic growth (Amis, 1999).

So far, decentralisation and many of the technical solutions to service delivery problems have not been notably successful, despite considerable donor pressure and support. Often, approaches have been based on inadequate understanding of labour, land and housing markets, political processes and social organisation. Often, they have been imported and local resistance has been substantial, both because they are considered inappropriate for local conditions and because of vested interests. It has been obvious for years that decentralisation will not improve efficiency and accountability unless the financial base for urban management agencies, especially local government, is commensurate with their responsibilities. However, it is also clear that urban local government lacks the capacity and legitimacy to tackle problems independently of other actors. Recent political changes, the growth and differentiation of NGOs and civil society organisations, and the reallocation of responsibilities between the public and private sectors have changed the institutional context. However, the newly constituted or reconstituted relationships between the wide range of actors now seen as having a role to play in urban management (NGOs, grass-roots organisations and private sector enterprises, in addition to the public sector) are in the early stages of evolution. Whether they contain the seeds of the improved governance needed to secure renewed economic growth and improved well-being in Africa's towns and cities is, as yet, unclear.

Notes

1 This chapter draws on a recent United Nations University funded project which assembled and reviewed available material on globalisation and large cities in Africa (Rakodi, 1997a) and work underway on an an ESCOR funded research programme on urban governance, partnership and poverty, for which three of the nine first-stage city case studies were drawn from SSA (Rakodi, 1999). The former project included case studies of Lagos, Johannesburg, Nairobi, Kinshasa and Abidjan, the latter of Johannesburg, Kumasi and Mombasa.

2 Large data gaps mean that the current total and urban populations of most African countries are unknown. Regular collection of reliable census, labour market and environmental data was one of the victims of economic crisis, war and civic conflict in the 1980s and 1990s (UNCHS, 1996; Rakodi, 1997b; Simon, 1997).

3 This is likely to be an underestimate, because recession and retrenchment were thought to have increased the 'discouraged worker' phenomenon (Lachaud, 1994).

4 Except in, for example, Zimbabwe up to 1990.

5 Methodological and conceptual difficulties in defining the poverty line may result in underestimation of the incidence of poverty in urban areas. Not all poverty assessments allow for price differences between urban and rural areas or acknowledge the need for poor urban residents to pay for housing and collective services (water, refuse collection, sanitation, public transport) which are free, less costly and/or less necessary for rural residents.

6   With the possible exception of one or two in South Africa (Beavon, 1997; Rogerson, 1997; Beall *et al.*, 1999).

7   In Anglophone Africa, councils are typically organised sectorally, with a Town Clerk who is head of the legal department (responsible for licensing etc.) and also the senior chief officer (Davey, 1996).

8   For example, Doe and Tetteh (1999) discuss the progress achieved and difficulties faced by attempts at environmental planning and management instituted under the Accra Sustainable Cities Project.

9   Past research has focused on the operation of urban government, 'the nexus of public sector agencies involved in supervision, financing, planning and execution of governmental interventions in urban areas' (Davey *et al.*, 1996: 47), rather than adopting a wider conception of governance (but see Swilling, 1997). Davey *et al.*'s international comparative study of urban government included Zimbabwe and Uganda.

10 The second stage of the ESCOR funded research on Urban Governance, Partnership and Poverty examined these issues on an international comparative basis, including case studies of Kumasi and Johannesburg.

11 Religious organisations cannot be clearly categorised as NGOs or civil society organisations. Membership organisations (unlike many NGOs), they may also provide services to the wider community and catalyse community organisation (like NGOs) and may take on an advocacy role for human rights and social justice.

12 According to the *unicité de caisse* principle, the central state, in theory, collects taxes destined for municipalities, advises them how much they will receive for the year and remits these funds regularly each month. In practice, central government lacks local knowledge to estimate expected revenue accurately and generally overestimates yields, so that municipalities set budgets unrelated to funds received and accumulate deficits, while funds are invariably remitted late. As a result, municipalities have devised a range of ways to circumvent the system (Farvaque-Vitkovic and Godin, 1998).

13 The changes needed are different in Francophone Africa (see Durand-Lasserve, 1993; Farvaque-Vitkovic and Godin, 1998).

14 See Kironde's (1999) evaluation of initial attempts to privatise Dar es Salaam's solid waste collection.

References

Aina, T.A. (1997) 'The state and civil society: politics, government, and social organization in African cities,' in C. Rakodi (ed.), *The Urban Challenge in Africa: Growth and Management of its Large Cities*, Tokyo: United Nations University Press.

Amis, P. (1999) *Urban Economic Growth and Poverty Reduction*, Urban Governance, Partnership and Poverty Theme Paper 2, Birmingham: University of Birmingham, School of Public Policy, International Development Department.

Attahi, K. (1999) 'Les outils pour une meilleure gestion urbaine', in C. Stein (ed.), *Development and Urban Africa*, Barcelona: Centre d'Estudis Africans.

Batley, R. (1997) *A Research Framework for Analysing Capacity to Undertake the 'New Roles of Government'*, Role of Government in Adjusting Economies Paper 23, Birmingham: University of Birmingham, School of Public Policy, International Development Department.

Beall, J. and Kanji, N. (1999) *Households, Livelihoods and Urban Poverty*, Urban Governance, Partnership and Poverty Theme Paper 3, Birmingham: University of Birmingham, School of Public Policy, International Development Department.

Beall, J., Crankshaw, O. and Parnell, S. (1999) *Urban Governance, Partnership and Poverty in Johannesburg*, Birmingham: University of Birmingham, School of Public Policy, International Development Department.

Beavon, K. (1997) 'Johannesburg: a city and metropolitan area in transformation', in C. Rakodi (ed.), *The Urban Challenge in Africa: Growth and Management of its Large Cities*, Tokyo: United Nations University Press.

Becker, C.M., Hamer, A.M. and Morrison, A.R. (1994) *Beyond Urban Bias in Africa: Urbanization in an Era of Structural Adjustment*, Portsmouth, NH: Heinemann, and London: James.

Bryceson, D. and Jamal, V. (eds) (1997) *Farewell to Farms: De-agrarianisation in Africa*, Aldershot, Hants.: Ashgate.

Chichava, J. (1997) 'Urban management in Mozambique, with particular reference to the capital city, Maputo', unpublished PhD dissertation, University of Wales, Cardiff.

Davey, K., with Batley, R., Devas, N., Norris, M. and Pasteur, D. (1996) *Urban Management: the Challenge of Growth*, Aldershot, Hants.: Avebury.

Devas, N. (1999) *Who Runs Cities? The Relationship between Urban Governance, Service Delivery and Poverty*, Urban Governance, Partnership and Poverty Theme Paper 4, Birmingham: University of Birmingham, School of Public Policy, International Development Department.

Doe, B.K. and Tetteh, D. (1999) 'The working group approach to environmental management under the Accra Sustainable Programme', in A. Atkinson, J.D. Dávila, E. Fernandes and M. Mattingly (eds), *The Challenge of Urban Management in Urban Areas*, Aldershot, Hants.: Ashgate.

Durand-Lasserve, A. (1993) *Conditions de mise en place des systèmes d'information foncière dans les villes d'Afrique sud-Saharienne francophone*, UMP 8. Washington, DC: World Bank.

Farvacque-Vitkovic, C. and Godin, L. (1998) *Future of African Cities: Challenges and Priorities for Urban Development*, Washington, DC: World Bank.

Harpham, T. and Tanner, M. (eds) (1995) *Urban Health in Developing Countries: Progress and Prospects*, London: Earthscan.

Jamal, V. and Weeks, J. (1993) *Africa Misunderstood: or, Whatever Happened to the Rural–Urban Gap?*, London: Macmillan for the ILO.

King, K. (1996) *Jua Kali Kenya: Change and Development in an Informal Economy 1970–95*, London: James Currey, Nairobi: East African Educational Publishers, and Athens, OH: Ohio University Press.

Kironde, J.M.L. (1999) 'The governance of waste management in African cities', in A. Atkinson, J.D. Dávila, E. Fernandes and M. Mattingly (eds), *The Challenge of Urban Management in Urban Areas*, Aldershot, Hants.: Ashgate.

Kombe, W. and Kreibich, V. (2000) *Informal Land Management in Tanzania*, SPRING Research Series 29, Dortmund: University of Dortmund, Faculty of Spatial Planning.

Korboe, D., Diaw, K. and Devas, N. (1999) *Urban Governance, Partnership and Poverty in Kumasi*, Birmingham: University of Birmingham, School of Public Policy, International Development Department.

Lachaud, J.-P. (1994) *The Labour Market in Africa*, Geneva: International Institute for Labour Studies.

Lugalla, J. (1995) *Crisis, Urbanisation and Urban Poverty in Tanzania*, Lanham, MD: University Press of America.

Mabogunje, A. (1993) *Perspective on Urban Land and Urban Management Policies in Sub-Saharan Africa*, Technical Paper 196, Washington, DC: World Bank.

Mitlin, D. (1999) *Civil Society and Urban Poverty*, Urban Governance, Partnership and Poverty Theme Paper 5. Birmingham: University of Birmingham, School of Public Policy, International Development Department.

Mulenga, L. (2000) 'Livelihoods of young people in Zambia's Copperbelt and local responses', unpublished PhD dissertation, University of Wales, Cardiff.

Nickson, A. (1997) 'The public–private mix in urban water supply', *International Review of Administrative Sciences*, 63 (2): 165–86.

Nunan, F. and Satterthwaite, D. (1999) *The Urban Environment*, Urban Governance, Partnership and Poverty Theme Paper 6, Birmingham: University of Birmingham, School of Public Policy, International Development Department.

Peterson, G.E., Kingsley, G.T. and Telgarsky, J.P. (1991) *Urban Economies and National Development*, Washington, DC: US Agency for International Development, Office of Housing and Urban Programs.

Potts, D. (1997) 'Urban lives: adopting new strategies and adapting rural links', in C. Rakodi (ed.), *The Urban Challenge in Africa: Growth and Management of its Large Cities*, Tokyo: United Nations University Press.

Rakodi, C. (ed.) (1997a) *The Urban Challenge in Africa: Growth and Management of its Large Cities*, Tokyo: United Nations University Press.

Rakodi, C. (1997b) 'Introduction', in C. Rakodi (ed.), *The Urban Challenge in Africa: Growth and Management of its Large Cities*, Tokyo: United Nations University Press.

Rakodi, C. (1997c) 'Global forces, urban change, and urban management in Africa', in C. Rakodi (ed.), *The Urban Challenge in Africa: Growth and Management of its Large Cities*, Tokyo: United Nations University Press.

Rakodi, C. (1997d) 'Residential property markets in African cities', in C. Rakodi (ed.), *The Urban Challenge in Africa: Growth and Management of its Large Cities*, Tokyo: United Nations University Press.

Rakodi, C. (1999) *Urban Governance, Partnership and Poverty: A Preliminary Exploration of the Research Issues*, Working Paper 8, Birmingham: School of Public Policy, International Development Department, Urban Governance, Partnership and Poverty.

Rogerson, C.M. (1997) 'Globalization or informalization? African urban economies in the 1990s', in C. Rakodi (ed.), *The Urban Challenge in Africa: Growth and Management of its Large Cities*, Tokyo: United Nations University Press.

Simon, D. (1997) 'Urbanization, globalization, and economic crisis in Africa', in C. Rakodi (ed.), *The Urban Challenge in Africa: Growth and Management of its Large Cities*, Tokyo: United Nations University Press.

Swilling, M. (ed.) (1997) *Governing Africa's Cities*, Johannesburg: Witwatersrand University Press.

Tripp, A.M. (1992) 'Local organisations, participation and the state in urban Tanzania', in G. Hyden and M. Bratton (eds), *Governance and Politics in Africa*, Boulder, CO: Lynne Rienner.

United Nations (UN) (1995) *World Urbanization Prospects: The 1994 Revision*, New York: United Nations, Population Division.

United Nations Centre for Human Settlements (UNCHS) (1996) *An Urbanizing World: Global Report on Human Settlements*, Oxford: Oxford University Press.

Vidler, E. (1999) *City Economic Growth*, Urban Governance, Partnership and Poverty Theme Paper 1. Birmingham: University of Birmingham, School of Public Policy, International Development Department.

Wekwete, K.H. (1997) 'Urban management: the recent experience', in C. Rakodi (ed.), *The Urban Challenge in Africa: Growth and Management of its Large Cities*, Tokyo: United Nations University Press.

World Bank (1996) *Taking Action for Poverty Reduction in Sub-Saharan Africa: Report of an Africa Region Task Force*, Washington, DC: World Bank.

Zeleza, T. (1994) 'The unemployment crisis in Africa in the 1970s and 1980s', in E. Osaghae (ed.), *Between State and Society in Africa: Perspectives on Africa*, Dakar: Council for the Development of Economic and Social Research in Africa.

19 Information needs for urban policy making in Africa

Anthony O'Connor

19.1 Introduction: policy making by whom?

In the early independence years this might have seemed a silly question, the answer being so obviously national and municipal governments. Both are of course still responsible for much urban policy formulation in sub-Saharan Africa, as elsewhere; but with the widespread 'rolling back', and in many cases the weakening, of the state in the 1980s and 1990s, much decision making has been taken over by various elements of civil society (Aina, 1997; Halfani, 1997). These include private enterprise, NGOs and local communities, all of which are left increasingly to their own very diverse devices; and both national and municipal governments may themselves lack vital information on these policy makers. Even the survival strategies of extended families and individual households are also now important in shaping Africa's urban future (Potts, 1997; Tripp, 1997).

Meanwhile there has been a partial recolonisation of much of sub-Saharan Africa by overseas and international institutions, including the IMF, World Bank and national aid donors, which have taken upon themselves much policy making for Africa, including its cities. Hence a recent book (Farvacque-Vitkovic and Godin, 1998) is boldly titled *The Future of African Cities*, even though it is not just a World Bank publication but is almost entirely devoted to World Bank projects.

The range of policy makers is thus extremely wide; and while in every case the information base from which they work is very important, the types of information that are most relevant will differ greatly. Some policy makers may be in greater need of understanding and awareness of the complexities of African cities, and contrasts among them, than of precise data on the few aspects that are readily quantifiable. The discussion in this chapter relates much more to macro-scale than to micro-scale policy making, but it aims to relate equally to 'insiders' and to 'outsiders'.

19.2 Policies about what?

We might consider whether 'urban' means the same in Africa as elsewhere. Is there here a clear urban/rural dichotomy in respect of people as well as places? Or is it more realistic to say that sub-Saharan Africa today has about 50 per cent rural-dwellers, 20 per cent urban, and 30 per cent undefinable either because they move so much between the two or because they occupy the vastly expanding peri-urban areas? In almost every country the information base for this possible 30 per cent is much weaker than for the more unequivocally urban 20 per cent.

In some countries, Kenya for instance (Obudho, 1983), policies have been formulated by government on the basic issues of urban growth and its spatial distribution. But although natural increase now accounts for about half of urban population growth in most African countries, policies on this have almost everywhere been national (and often half-hearted) rather than specifically urban; while few countries have had sustained policies relating to the extent, nature or spatial pattern of rural–urban migration. More elaborate policies such as growth-centre promotion in Tanzania (Darkoh, 1994) did not prove very successful and are now largely a thing of the past.

The most ambitious forms of policy making directed towards the spatial pattern of urban growth have been decisions to establish new capital cities. This went largely according to plan in Malawi; it also materialised in Nigeria, though in few respects according to plan; but was followed by little real implementation in Tanzania or Côte d'Ivoire.

Most current policy making by both governments and other bodies, including that most discussed in major academic works such as *Urban Research in the Developing World: Africa* (Stren, 1994) and *The Urban Challenge in Africa* (Rakodi, 1997) relates to more specific aspects of urban life. These aspects include official responses to spontaneous or squatter settlement; the small public-sector contribution to housing; the provision of physical infrastructure; and reactions to the proliferation of unlicensed trading. In many cases urban policy is essentially only the urban dimension of national policies, for instance with respect to education and medical provision.

Policy making by or within local communities is also mostly directed to diverse specific issues, including harassment by the authorities, rather than to urban growth or urbanisation in general.

19.3 Why is the situation especially problematic in Africa?

One reason for a weak information base for policy makers is that in most African countries urban growth is still probably taking place more rapidly than anywhere else in the world, despite a near-certain, if poorly documented, deceleration almost everywhere following the peak growth rates of the 1960s and 1970s. Sub-Saharan Africa's urban population, however defined, certainly more than doubled, and possibly trebled, between 1980 and 1999. This in turn brought expansion in total economic activity and income, though generally not in line with population growth, resulting in increased poverty, just as the expansion of urban housing has not prevented increased overcrowding.

Shortage of information is related, second, to the material poverty of most African countries. Their governments cannot afford costly data-gathering exercises, especially those such as censuses that pose particular challenges in countries and cities with poor infrastructure and partly non-literate populations. Precisely the forms of activity which are very hard to measure and document are far more extensive and important in most African cities than in cities elsewhere. More journeys to work or market are made on foot, water is more often drawn from wells or polluted streams, cooking more commonly depends on firewood, and much more trade and manufacturing takes place within the so-called informal sector (Grieco *et al.*, 1996; Macharia, 1997).

Another issue altogether is corruption, which may be smaller in absolute magnitude than in some European or American cities of similar size, but which (however defined) affects a higher proportion of all financial dealings in Kinshasa or Lagos than in London or New York (Hope and Chikulo, 2000). The larger the extent of corruption, the less is known by anyone about the total size and nature of the urban economy, and the ways in which it is changing.

The scarcity of reliable published information is certainly not only a matter of lack of resources to collect data and the problems of measuring small-scale or illicit activity. There is also the wish of many of the interested parties to keep knowledge to themselves or their closest confidants, rather than 'disclosing' or 'revealing' (two favourite words in African English-language newspapers) what might elsewhere be readily available to all. Knowledge is power in so many ways in African cities that making it freely available is often regarded as a strange alien notion (with implications for education at all levels). Hence census data which have been officially approved may still be available only by personal application to the Director (who never seems to be 'on seat'). A basic map is often now regarded as a classified document.

19.4 What types of information?

19.4.1 *Information on the urban population*

The basic information here is simply the numbers of people in each city and town, along with the rate at which numbers are growing. Most countries have conducted at least two censuses since independence which have yielded some data on this; but there are great contrasts between those such as Senegal, Côte d'Ivoire, Uganda and Zambia, where sufficient is known for most basic policy making, and those such as Nigeria, D. R. Congo and Angola, where the population of the largest cities is not known by anyone even to the nearest million. Total numbers in Lagos, Kinshasa and Luanda may have been growing through the 1990s by 4 per cent annually – or 5, or 6, or even 7. All censuses also provide data on sex ratios and age structures, so that we know that some cities no longer have such a clear male majority or disproportionate number of young adults as was the case 40 years ago; but does Maputo conform to this pattern and what is the demographic structure in Conakry? For some demographic characteristics such as infant, child and total mortality, estimates of greater or lesser reliability are published and widely available for almost every country (UNDP, annual; UNICEF, annual), but only at a national level. My attempts to find such data for individual cities have all failed – as have my attempts to discover whether unpublished figures do exist, the real problem being only one of access to them.

Most recent African censuses have yielded little information on spatial and temporal patterns of migration, and none on the key issue of the ages at which most rural–urban migration takes place. I believe that across tropical Africa this has increasingly become a school-leaver phenomenon for both boys and girls, but cannot produce statistical evidence of this. Similarly, increasing numbers of city-dwellers may now be 'going home' (Potts, 1995), but if so, at what age?

Data on the ethnic composition of the urban population are also now very scarce in most countries, reflecting not lack of significance but conversely the political sensitivity of the subject. To what extent is Lagos a predominantly Yoruba city? How many Khartoum residents are of southern Sudan origin? Widely divergent answers can be given because the facts are not known – by anyone.

Such issues are highlighted by Nigeria's expensively planned new capital. Who was migrated into Abuja from where? What is the resulting ethnic make-up of a city explicitly intended (by some) to be free of domination by any of the three largest ethnic groups? To what extent are members of these and other groups becoming concentrated in certain parts of the new city? Little is known about these things despite their centrality for Nigerian political life.

19.4.2 Information on the urban economy

GNP estimates are made each year for most African countries, but no equivalent exercise is undertaken for individual cities. Africa is no different in this respect to other continents, but it does mean that statements such as 'the capital city accounts for over half the national GNP' are usually no more than a perception or an assumption. The aspect that most urgently requires policy response, extreme poverty (Jones and Nelson, 1999), has proved extremely hard to measure everywhere in the world, both in conceptual and in practical terms, and African cities are certainly no exception. Most estimates of the numbers of urban dwellers below an arbitrarily defined 'poverty line' are of very little value, especially when used to make comparisons either with rural dwellers or with other countries. It is extremely difficult to establish to what extent cities remain islands of privilege within African countries or to what extent rural–urban income disparities have narrowed in recent years – surely matters of potential policy relevance.

When the focus is on employment, the problems are again conceptual as well as practical. Claims that unemployment rates exceed 50 per cent in some African cities are clearly utter nonsense unless this refers to all adults not in documented large-scale employment. Figures of the proportions working in informal rather than formal activities in specific cities depend on the definitions used, as well as the coverage of the surveys undertaken. The census figures available for a few cities, such as Dar es Salaam, which differentiate all employees from the self-employed and which record people's declared main occupation, including in many cases farming, are far less arbitrary. However this leaves untouched those who have a formal, even government, job but who devote most of their working hours to informal activity and rely on this for most of their income.

One sector of African urban economies about which official policy is often ambivalent is cultivation and livestock-keeping within the municipal areas. Up to the 1980s information on this was very scarce, but the situation has changed remarkably with the publication of academic research studies on the subject for at least a dozen African cities in the 1990s, for example those in Egziabher *et al.* (1994).

There is a more urgent need now for studies of remittances both into and out of the cities. It has long been recognised that remittances sent by migrant city-dwellers to their rural home areas may represent a substantial proportion of their income: but how substantial, and to what extent are they now offset by intra-family inward flows, notably of food? Much larger remittances from overseas are a vital, but undocumented, element of the economy in cities such as Asmara, and even Accra and Kumasi. The professional brain-drain out of countries such as Ghana and Uganda is, however, a predominantly negative force and something about which much more needs to be known.

Readily available information can give a partial picture of the economy of some African cities, but this does not go far to demonstrate how it really 'works' (Chabal and Daloz, 1999). It will tell us nothing, for instance, about the smuggling which helps to fund many other activities in Lomé or Brazzaville, or how the economy of Mogadishu, Monrovia or even Freetown continues to function in a situation of extreme civil disorder. Just occasionally an academic is brave enough both to investigate and to publish (Green, 1981; MacGaffey, 1991), and hence to assist more realistic policy making.

19.4.3 *Information on social welfare*

With respect to urban housing, the strength of the information base differs greatly from one country to another. In both Zambia and Tanzania each census has devoted much attention to housing quality, tenure and occupancy rates, whereas for Nigeria there are only a few surveys in selected cities undertaken many years ago. In Nigeria, as elsewhere, the published literature is grossly distorted, the majority of it relating to the minute contribution to the urban housing stock made by government: but perhaps little public policy making is required, given that 90 per cent of urban dwellers house themselves with little or no state involvement. However, there is a dire need for an improved information base as a foundation for government policy in cities across Africa in respect of matters such as sanitation and refuse disposal. The need for an emphasis on the public provision of such social infrastructure, rather than actual housing, is a recurrent theme in the important volume *African Cities in Crisis*, edited by Stren and White (1989).

Violent crime is a phenomenon which clearly requires a policy response, but are most African cities becoming more violent? At a major conference in Nigeria (Albert *et al.*, 1994) most participants assumed ever-increasing rates of crime and criminal violence without offering any evidence to demonstrate that they are expanding any faster than the total city population.

Healthcare can be planned using an information base which is in part relatively strong. The facilities, staffing, number of patients and diagnosed illnesses in both government and private hospitals are generally well documented. However, in every city in tropical Africa there is a second, indigenous healthcare system (Good, 1987), on which very little documentation is available, however well each part of it may be known to local communities. In cities such as Kano and Ibadan it predates the 'western' system; in others such as Lusaka and Harare its growth has mainly been very recent. Any overall policy making for healthcare must acknowledge both systems, yet in no city is there any sound basis for estimating their relative importance. A third and also poorly documented element, which expanded rapidly in the 1990s, consists of poorly paid medical professionals moving into private practice, in some cases legal and licensed, in other cases diverting equipment, drugs and their own services from government hospitals: this is a newly developed private health sector.

One issue that is far more significant in many African cities than in most cities elsewhere, and also more prominent than in most of rural Africa, is the incidence of HIV infection and AIDS. This plague is now appropriately receiving much national and international attention in medical and social welfare circles, but its impacts are such that it is relevant to policy making much more broadly. It does not always have the attention it merits, partly because for no city are there remotely comprehensive or reliable figures for HIV incidence, AIDS cases or AIDS deaths. HIV testing is undertaken only in particular segments of city populations, while AIDS cases are not all diagnosed as such. To complicate matters further, many of those who are correctly diagnosed then return to a rural home area for what remains of their lives. Another force at work is a culture of denial in some government circles, so that even some of the information that has been gathered cannot be published or widely circulated.

Because there are hospitals and health centres, there are some records of the incidence of most serious diseases: but since many ill people receive no treatment at all, the recorded figures are always far from comprehensive and may be very misleading. Some very odd data on malaria incidence are explicable only in terms of spatial and temporal difference in coverage; and in reality no one knows for any city in tropical Africa what proportion of its inhabitants, and especially its infants, are suffering from malaria.

19.5 When information is misinformation

Even where reliable facts, and especially precise figures, do exist, they may mislead the policy maker, particularly one unfamiliar with the city in question. There may be problems in the definition of either the topic or the urban area. Data for Khartoum may be presented without clearly indicating whether they are for Khartoum 'proper' or for the three times larger metropolitan area including Khartoum North and Omdurman. The population for Kampala did not really grow from 47,000 to 330,000 between 1959 and 1969, as crude census figures suggest: boundary change accounts for at least half this apparent growth. Figures indicating massive growth of some Kenya provincial towns between 1969 and 1979 also may be quite accurate but equally misleading, since these too reflect boundary changes: yet they have been put forward as evidence of the success of the urban dispersal policy.

A case of misleading information that is important, because it has been widely quoted by others, is that relating to access to social infrastructure, published by UNCHS (1996). In the appendix tables in *An Urbanizing World*, we learn that 99 per cent of the urban population of Burundi and 98 per cent in Niger had access to safe drinking water in 1990, compared with only 43 per cent in Tanzania; and that 100 per cent had access to sanitation in both Cameroon and Malawi, compared with only 14 per cent in Lesotho and 12 per cent in Madagascar. These are not misprints, and while it is acknowledged in the small print (ibid.: 515) that 'definitions differ among countries', these figures have been quoted without this proviso, and so are worse than worthless.

In some other cases it is not a matter of definitions, but of gross error, deliberate or otherwise. Tanzania's 1978 census is widely regarded as remarkably accurate (a success only achieved at high cost), but conclusions reached by Barke and Sowden (1992) on a decade of change, using the 1988 census, were soon challenged by Briggs (1993) in respect of Dar es Salaam, where there was certainly a substantial undercount. However, errors in population figures for Tanzania are minor compared to those in successive Nigerian censuses. In that country gross distortion of figures for political purposes has plagued every attempt at a count, although only some have been officially declared null and void: so nobody knows whether the population of Kano, say, is close to 2 million, 3 million or 4 million.

Uncertainty would matter less if it were always acknowledged. Unfortunately, statisticians may be paid according to the precision of the figures that they produce (and the more precise the figure, the more probable that it is inaccurate). Data on urban Africa are often presented, by academics, governments and international agencies, with a confidence that is totally unjustified. Two academics writing for the ILO inform us that Mwanza's population 'will double to 644,496 in the year 2000' (Kaijage and Tibaijuka, 1996: 83). Farvacque-Vitkovic and Godin (1998: 16) of the World Bank inform us that in West Africa '63 per cent of the population will be city dwellers in 2020' ('city' undefined). They also state twice that 'the urban population is now in the majority': one would like to know how they feel able to assert this.

Many examples could be provided of published figures offering absurd precision, notably populations to the final digit, as above, or ratios to two decimal places. These are often extrapolations or other calculations from a very shaky original enumeration – including Nigerian censuses in the case of some offerings from Onibokun (1989: 78). The latter goes one stage further in providing a sequence of figures (again not misprints) showing Ibadan to be almost as populous as Lagos. It may not be wholly irrelevant that he writes from Ibadan.

A wonderful example of pure fiction from Nigeria is provided by Osaghae (1994: 34), who informs us without query that 'Igbo leaders put the number of the Igbo in Kano today at 2.5 million' and that 'the number of the Yoruba in Kano is currently estimated at 1 million'. Since these are acknowledged to be minority groups, one wonders what the estimate for the total population would have been.

19.6 Conclusions

This chapter of course has been only illustrative rather than comprehensive. While it relates to policy, it does not in general aspire to make policy recommendations. It does plead for caution in the use of available information, for instance with regard to checking definitions. It also pleads for less spurious precision when even broad orders of magnitude are in doubt. It would be inappropriate to advocate costly data-gathering exercises in desperately poor countries, especially where the capacity for macro-scale policy making is in any case very limited. However, a case could be made for more frequent censuses in urban and peri-urban areas than in rural areas, in view of the generally faster rate of change in the former. Even more clearly, efforts should be made to encourage more sharing and circulation of such sound information as does exist but is currently kept secret because this is in one person's or one group's interests.

The main recommendation is that the extent of our ignorance should be clearly recognised, whether 'we' are external advisers, national governments, municipal authorities, or even community groups (very familiar with their local situation but not the broader picture). Policy makers must distinguish what is known precisely and accurately, what is known with some reliability but not precision, what is known only partially, and what is largely a matter for speculation – especially in terms of current trends. Often what is really more important than information in a narrow sense is an *understanding* of the total multifaceted situation, including the non-quantified and sometimes non-quantifiable parts of it, rather than undue concern with precise data on the more easily quantified parts. Gross distortion of reality is not acceptable, but for broad policy making in and for urban Africa broad orders of magnitude may sometimes be sufficient – particularly since circumstances may well be very different next year, or even next month.

Perhaps the most fundamental of all policy making at state and municipal level concerns, first, how far to try to formulate policies at all, rather than abandoning this task or delegating it to others; and, second, having formulated some policies, how far to make any serious attempt to implement them. The appropriate answers to both questions are themselves affected by the probability that the information base, and especially the depth of real understanding of how the cities work, available to such urban policy makers is at present weaker than in any other major world region. It must be strengthened.

References

Aina, T.A. (1997) 'The state and civil society: politics, government and social organization in African cities', in C. Rakodi (ed.), *The Urban Challenge in Africa*, Tokyo: United Nations University Press.

Albert, I.O., Adisa, J., Agbola, T. and Hérault, G. (ed.) (1994) *Urban Management and Urban Violence in Africa*, 2 vols Ibadan: Institut Français de Recherche en Afrique.

Barke, M. and Sowden, C. (1992) 'Population change in Tanzania 1978–88', *Scottish Geographical Magazine*, 108 (1): 9–16.

Briggs, J. (1993) 'Population change in Tanzania: a cautionary note for the city of Dar es Salaam', *Scottish Geographical Magazine*, 109 (2): 117–18.

Chabal, P. and Daloz, J.-P. (1999) *Africa Works: Disorder as Political Instrument*, Oxford: James Currey.

Darkoh, M.B.K. (1994) *Tanzania's Growth-centre Policy and Industrial Development*, Frankfurt: Peter Lang.

Egziabher, A., Lee-Smith, D., Maxwell, D.G., Memon, P.A., Mougeot, L.J.A. and Sawio, C.J. (1994) *Cities Feeding People*, Ottawa: International Development Research Centre.

Farvacque-Vitkovic, C. and Godin, L. (1998) *The Future of African Cities: Challenges and Priorities for Urban Development*, Washington, DC: World Bank.

Good, C.M. (1987) *Ethnomedical Systems in Africa: Rural and Urban Kenya*, New York: Guilford Press.

Green, R.H. (1981) 'Magendo in the political economy of Uganda', Discussion Paper 164, Institute of Development Studies, University of Sussex.

Grieco, M., Apt, N. and Turner, J. (1996) *At Christmas and on Rainy Days: Gender, Travel and Transport in Urban Accra*, Aldershot, Hants.: Avebury.

Halfani, M. (1997) 'The governance of urban development in East Africa', in M. Swilling (ed.), *Governing Africa's Cities*, Johannesburg: Witwatersrand University Press.

Hope, K.R. and Chikulo, B.C. (eds) (2000) *Corruption and Development in Africa*, London: Macmillan.

Jones, S. and Nelson, N. (eds) (1999) *Urban Poverty in Africa: From Understanding to Alleviation*, London: Intermediate Technology.

Kaijage, F. and Tibaijuka, A. (1996) *Poverty and Social Exclusion in Tanzania*, Geneva: ILO.

MacGaffey, J. (1991) *The Real Economy of Zaire: The Contribution of Smuggling to National Wealth*, London: James Currey.

Macharia, K. (1997) *Social and Political Dynamics of the Informal Economy in African Cities: Nairobi and Harare*, Lanham, MD: University Press of America.

Obudho, R.A. (1983) *Urbanization in Kenya: A Bottom-up Approach to Development Planning*, Lanham, MD: University Press of America.

Onibukun, A.G. (1989) 'Urban growth and urban management in Nigeria', in R.E. Stren and R.R. White (eds), *African Cities in Crisis*, Boulder, CO: Westview Press.

Osaghae, E.E. (1994) *Trends in Migrant Political Organizations in Nigeria: The Igbo in Kano*, Ibadan: Institut Français de Recherche en Afrique.

Potts, D. (1995) 'Shall we go home? Increasing urban poverty in African cities and migration processes', *Geographical Journal*, 161 (3): 245–64.

Potts, D. (1997) 'Urban lives: adopting new strategies and adapting rural links', in C. Rakodi (ed.), *The Urban Challenge in Africa*, Tokyo: United Nations University Press.

Rakodi, C. (ed.) (1997a) *The Urban Challenge in Africa: Growth and Management of its Large Cities*, Tokyo: United Nations University Press.

Stren, R.E. (ed.) (1994) *Urban Research in the Developing World*, vol. 2: Africa, University of Toronto, Centre for Urban and Community Studies.

Stren, R.E. and White, R.R. (ed.) (1989) *African Cities in Crisis: Managing Rapid Urban Growth*, Boulder, CO: Westview.

Tripp, A.M. (1997) *Changing the Rules: The Politics of Liberalization and the Urban Informal Economy in Tanzania*, Berkeley, CA: University of California Press.

United Nations Centre for Human Settlement (UNCHS) (1996) *An Urbanizing World: Global Report on Human Settlements 1996*, Oxford: Oxford University Press.

United Nations Development Programme (UNDP) (annual) *Human Development Report*, New York: Oxford University Press.

UNICEF (annual) *The State of the World's Children*, New York: UNICEF.

Part V

International trade and transport constraints

20 Trade policy reforms in sub-Saharan Africa

Implementation and outcomes in the 1990s

Oliver Morrissey[1]

20.1 Introduction

Trade liberalisation, the removal of restrictions on imports and reduction of discrimination against exports, has been an important element of World Bank structural adjustment programmes (SAPs). Almost all countries in sub-Saharan Africa (SSA) have implemented some degree of trade liberalisation as part of an SAP since 1980. The success of the many liberalisation episodes has been mixed. Furthermore, evaluating the impact of trade reforms has been made more difficult by the fact that other economic reforms were normally attempted at the same time (see Corbo *et al.*, 1992; Greenaway and Morrissey, 1993, 1994; McGillivray and Morrissey, 1999; Mosley *et al.*, 1991; Papageorgiou *et al.*, 1991).

Although the discussion in this chapter concentrates on trade policy reforms, it is appropriate to mention also exchange rate liberalisation as this affects the relative price of importables and exportables and facilitates trade (see Greenaway and Morrissey, 1994). Import liberalisation reduces the price of imports and importables (domestic goods that compete with imports), and essentially refers to the removal of quantitative restrictions (QRs) and the reduction of tariffs. Export promotion measures can directly facilitate and increase the return to exporting. Devaluation tends to increase returns to exporters (as they receive more domestic currency for a given world price), and allowing exporters to retain export earning is a simple (and potentially effective) export promotion measure. Thus measures relating to imports, exports and the exchange rate all have implications for trade and it is relevant to consider all three in discussing trade policy reforms.

In aiming to discuss the trade policy reforms in Africa since about 1980 this chapter has a broad remit. In effect, our attention is somewhat narrower. First, we restrict attention to sub-Saharan Africa (SSA) and exclude South Africa. Second, useful data on most countries were unavailable so our actual focus is limited to twelve SSA countries. Even for these it was often difficult to obtain up-to-date information. Nevertheless, attempting to review trade reforms in ten countries in any detail would have resulted in a long, and rather tedious, chapter. Consequently, the country reviews are intentionally brief.

As the aim is to present a flavour of what has been achieved, we concentrate discussion on some summary tables. As data availability varies (considerably), the countries included vary from table to table (and often include countries not covered in any detail in the brief case studies). As the intention is to give a broad overview of trade reforms in SSA, this approach seems justified. In successive sections the chapter reviews the types of reforms implemented in a range of countries, for convenience arranged in alphabetical order; it examines summary evidence of the degree of reform implemented, and then evidence on the impact of trade reform, before concluding.

20.2   Trade policy reforms in selected SSA countries

Various influences on trade policy reform in SSA have operated in parallel since the early 1980s. Many countries, under the auspices of IMF/World Bank programmes, were under pressure to devalue and liberalise their import regimes. Implementation of, and commitment to, such unilateral trade liberalisation varied significantly. Ghana and Uganda, for example, achieved quite a lot whereas Kenya and Nigeria, for instance, achieved much less. Multilateral liberalisation under the Uruguay Round added some momentum to reforms in the 1990s. However, it is the many, often interlocking, regional integration negotiations that have provided the greatest impetus for trade liberalisation, at least within the specific preferential trading arrangements (PTAs).

One of the most important PTAs, which overlaps with the Common Market for Eastern and Southern Africa (COMESA), is the Cross-Border Initiative (CBI) involving fourteen countries to encourage 'a market-driven concept of integration in Eastern and Southern Africa and the Indian Ocean countries' (CBI, 1998a). This initiative is co-sponsored by the African Development Bank, the European Union, the International Monetary Fund and the World Bank. As its name suggests it is concerned to promote cross-border trade, investment and payments, as well as to assist in establishing effective regional integration. The policy measures promoted by the CBI include trade liberalisation and facilitation, as well as liberalisation of exchange and payments systems, deregulation of investment and measures to facilitate labour movement.

20.2.1   Ghana[2]

The reform process in Ghana began after Flight Lieutenant Rawlings took power in 1981, initiated by the Economic Recovery Program (ERP) launched in 1983. The first phase consisted of a major devaluation in April 1983 followed by more frequent modest adjustments. The second phase of exchange rate liberalisation started with an auction system in 1986. Finally, the foreign exchange auction was abolished in March 1992 and replaced with an inter-bank market. The parallel market virtually disappeared while the only remaining controls were on outward capital flows. Import reforms were realised in three phases, the first following the devaluation in April 1983. The government reduced and unified tariffs into three bands: 10 per cent on raw materials and capital goods, 20 per cent on consumer goods and 30 per cent on luxury goods, while a few quotas were replaced by equivalent tariffs. In October 1986 the government streamlined the import licence programme by moving from a positive to a short negative list. By 1989 the import licensing requirement and most restrictions were abolished. By 1991 sales and excise taxes across comparable imported and domestically produced goods were unified.

Export reforms started with the introduction of the EPR, where the large devaluation of the cedi restored the competitiveness of the industrial sector. Exporters were automatically granted 'specific' licences for imported inputs from 1986. In August 1989, the Export Finance Company Limited was created to assist commercial financing of non-traditional exports and in addition, depending on the share of exports in production, 20–50 per cent of exporting firms' corporate income taxes were refunded. The government also increased the duty drawback rate to 95 per cent. The export reform had some effect in promoting exports and export diversification. Non-traditional exports increased from 3.2 per cent of total exports in 1986 to 4.6 per cent in 1991.

20.2.2 Kenya

Kenya has engaged in structural adjustment and trade liberalisation since 1980, but its reforms have typically been only partially implemented, frequently rescinded and reversed, and always subject to political whim. Kenya's relationship with the World Bank has been 'on–off' in line with its reform programmes, and it has been presented as a country where trade policy is a by-product of macroeconomic policy (Foroutan, 1994). Import restrictions have been introduced and relaxed, not in accordance with any trade strategy but in response to internal and external imbalances.

From 1980, gradual liberalisation was introduced by reducing the range of products subject to quotas and converting the quotas on less-restricted imports into (typically high) tariffs. This was effectively reversed in 1982 when virtually all imports required a licence. No effective export promotion schemes were implemented and despite devaluations, the real effective exchange rate did not depreciate (ibid.). Liberalisation recommenced in 1988, in conjunction with a series of sectoral adjustment loans.

The import licensing system was rationalised in 1988 and most restrictions had been eliminated by 1992. The removal of QRs was almost complete, although it was often more difficult in practice to obtain an import licence and access to foreign exchange (fully controlled by the government until August 1992) than was implied by the unrestricted status of imports. At the same time, tariff rates were rationalised. The maximum tariff, 170 per cent in 1988, was reduced to 70 per cent by 1992. The number of tariff rates was reduced from twenty-four in 1988 to twelve in 1992, and the unweighted average tariff was reduced from 40 to 34 per cent by 1992, although the average tariff on manufactures remained high, at 45 per cent in 1992 (ibid.). Effective rates of protection remained high (above 40 per cent for most manufacturing sectors and 100 per cent for garments in 1992), and *ad hoc* exemptions and exceptions were widespread.

The ineffective Export Compensation Scheme (which refunded duty paid on a limited range of imports by direct exporters, and did so with considerable delays) was replaced by an import duty/VAT exemption scheme in 1990. An export processing zone was established in the early 1990s, but by 1993 had attracted only a few firms (ibid.). In sum, direct export promotion measures have been limited. Of potentially greater benefit to exporters is liberalisation of the exchange rate regime, effectively since late 1992. In 1991, non-traditional exporters were permitted to retain all of their foreign exchange earnings, and in 1992 this was extended to allow traditional exporters 50 per cent retention. The retention scheme was revoked in March 1993, but reintroduced almost immediately at 50 per cent for all importers.

20.2.3 Madagascar

The initial phase of adjustment was between 1983 and 1986. Steps towards the promotion of exports were undertaken from 1983, such as retention schemes and guaranteed access to foreign exchange. Export taxes on all manufactured goods were eliminated and exporters of manufactured goods were exonerated from all indirect taxes on raw materials and spare parts used in the production of exports. On the import side, the removal of prohibitions was achieved in stages. Further import liberalisation was not implemented (Foroutan, 1993a).

In the period 1986 to 1989, the exchange rate was devalued, the tariff structure was rationalised and export restrictions were substantially reduced. Most agricultural monopolies and export taxes were abolished. From 1992 to 1998, imports were fully liberalised, traditional non-tariff barriers (NTBs) were removed and tariffs on intra-regional trade were eliminated. Export licensing was abolished except on minerals and certain flora and fauna items. External tariffs have been classified in four rates with minimum rate at 10 per cent and maximum at 30 per cent.

20.2.4 Malawi

Most reforms in Malawi have been implemented in the 1990s (see Imani Development International, 1998). Import and export licensing and quotas were abolished. Tariffs were reduced and classified under a new structure containing a full rate (the base), an MFN rate, a COMESA rate and a zero Zimbabwe rate (based on bilateral agreement). Capital goods and raw materials are zero-rated and the maximum MFN rate is 40 per cent. For intra-COMESA trade, the country has reduced the original tariffs by 70 per cent. By 1995, the average nominal tariff (trade-weighted) was 15.5 per cent, although this varied widely. Zero, or almost zero, tariffs applied to goods as varied as fertilizers, cereals and some other agricultural products (such as live animals and meat), wood pulp, wood and raw leather. The highest rates, above 35 per cent, applied to coffee, tea and spices, fur skins, wickerwork, clothing, silk and some other products. Most of the Southern African Development Community (SADC) imports are concentrated at the lower end of the duty spectrum. Few export promotion measures were implemented although the exchange rate regime was liberalised.

20.2.5 Mauritius

Trade has been extensively liberalised in Mauritius over the past decade (see Imani Development International, 1998). Imports have almost been fully liberalised and almost all QRs have been removed. The tariff structure, which used to consist of sixty different rates, was reduced to eight rates ranging from 0 per cent to 80 per cent since 1994. The maximum tariff was reduced from 600 per cent to 80 per cent. About 60 per cent of imported goods attract duty between 0 per cent and 20 per cent, comprising mostly foodstuffs, raw materials and capital goods. However, the average nominal weighted tariff in 1995 was 21 per cent with high rates on goods such as articles of iron and steel or base metals (over 30 per cent), clothing (80 per cent), footwear (60 per cent), articles of leather (68 per cent), beverages (56 per cent) and cocoa and cocoa preparations (53 per ccent). Trade liberalisation measures at an intra-regional level have also been undertaken, notably implementing the COMESA timetable for tariff reduction. Negotiation of tariff reductions within SADC is also under consideration. Mauritius has also implemented a number of trade facilitation measures. Customs export and import procedures have been simplified. Both the export tax on sugar and export licensing requirements have been abolished.

20.2.6 Namibia

Namibia has not had any SAP or a specific reform programme and received financial assistance from the CBI for trade and investment facilitation measures. Namibia's commitments to tariff reduction and foreign exchange liberalisation are guided by its membership of the Southern African Customs Union (SACU) and Common Monetary Area (CMA). In the 1990s it significantly reduced the number of tariffs lines and abolished import and export licences, although tariffication of agricultural QRs has been slow (see Imani Development

International, 1998). Tariffs on intra-regional trade were removed by 1996, following its admission to SADC, and it is in the process of harmonising external tariffs.

20.2.7 Nigeria

Early stabilisation efforts in 1984 proposed only ad hoc measures, such as a more uniform tariff schedule, but few reforms were implemented. The liberalisation of the exchange rate regime that began in 1986 was reversed in the late 1980s but real devaluations continued (Castillo, 1993b). Nigeria displayed a poor implementation record and commitment to import liberalisation. In September 1986, the government reluctantly abolished the import licensing system and the 30 per cent surcharge on imports while at the same time reducing the negative list from seventy-four to seventeen products. In October of the same year, an interim tariff regime was implemented pending a more thorough review of the tariff and excise tax structure. Most of the new duties fell within the range of 10–30 per cent and the average rate fell to 23 per cent compared to 33 per cent in 1984. However, most of these reforms were effectively reversed in the late 1980s. In 1994, effective rates of protection were estimated at 43 per cent on capital goods, 69 per cent on intermediate goods and 89 per cent on consumer goods; the latter should be reduced by proposals for tariff reform by 2001, but both the former may increase (Thoburn, 2000).

Nigeria implemented some measures for export promotion and diversification, but these were not very effective and few firms availed themselves of them (ibid.). The removal of export duties, prohibitions and most export licensing requirements provided an incentive for the development of exports. Exporters enjoyed tax holidays and preferential tariff rates on imported raw materials and could retain 100 per cent of foreign exchange receipts. The Export Decree of 1986 also revamped the duty drawback and exemption regime to encourage non-oil exporters who were entitled to a 50 per cent refund of custom duties. Most export taxes, prohibitions and licensing requirements were removed by 1991 and exporters became eligible for generous tax holidays, accelerated depreciation and duty rebates on imported inputs (Castillo, 1993b). There is little evidence that these were effective in promoting export diversification.

20.2.8 Senegal

Senegal's trade regime was shaped by an import policy based on widespread use of non-tariff barriers, the most common being import licences, quotas, import monopolies and administrative customs valuation (Foroutan, 1993b). In 1982, 32 per cent of all imports needed a licence and administrative customs were applied to all imports. Tariffs were high and dispersed, varying between zero per cent on necessities and 175 per cent on luxuries, with an average nominal rate of 86 per cent but with many exceptions and exemptions. Overall effective protection was estimated at 165 per cent in 1985. The exchange rate regime was determined by its membership of the West African Monetary Union (UMOA), implying a fixed exchange rate *vis-à-vis* the French franc. The inability to devalue was a constraint on all franc-zone countries. Exports were subject to administrative impediments and were also discouraged by widespread government intervention in domestic markets in the form of price controls and labour regulations.

Gradual elimination of QRs and tariff rationalisation were implemented between July 1986 and July 1989. By February 1988 most of the QRs and prohibitions were removed and administrative valuation of imports was eliminated. Tariff reform also proceeded on schedule with the lowering of the highest rates and the imposition of a minimum duty on all inputs (ibid.). Faced with a shortfall in customs revenues, the government reintroduced some of the QRs or resorted to new forms of restrictions. In August 1989, the government increased the flat rate of custom duties from 10 to 15 per cent and shifted products from low to high VAT. Administrative valuation of imports was reintroduced. These policy reversals caused the average nominal rate of protection to rise from 35 per cent in mid-1988 to 43 per cent in 1989. No effective export promotion measures were implemented.

20.2.9 Tanzania

A series of adjustment programmes were adopted throughout the 1980s and the principal reforms implemented rationalised the tax regime, especially tariffs, liberalised the exchange rate regime and included successive devaluations (Basu and Morrissey, 1997). The principal reform episodes were the structural adjustment programme (SAP) of 1982–3 to 1984–5, the Economic Recovery Programme (ERP) launched in June 1986, and the Economic and Social Action Programme (ESAP, sometimes called ERP II) launched in 1989.

Export taxes were actually abolished in the 1981 budget. An Open General Licensing (OGL) facility was introduced in April 1980, but was very restrictive. Most importers continued to require a specific import licence. The 1984 budget was a turning point: a variety of import controls were lifted and 'own-funds' imports were instituted. A programme of tax reform commenced in 1985, when import duties and sales tax rates were reduced. Major devaluations began in 1986; the move towards an equilibrium exchange rate was allied to the gradual removal of restrictions on trade and access to foreign exchange. Both the own-funds and export retention schemes were simplified and widened in scope. Exporters also benefited from a duty drawback scheme (although reimbursement was slow) and simplification of export procedures. A new OGL system, easing access to import licences, was introduced in 1988. Further reductions and rationalisations of both import duties and domestic sales taxes were announced in 1990.

Major reductions and rationalisations of both import duties and domestic sales taxes were announced in 1988 and 1989. The range and levels of tariffs were reduced, and most specific sales taxes were converted to *ad valorem* taxes. The average implicit tariff (revenue relative to value of imports, data from Lyakurwa, 1992) rose from 2.9 per cent in 1986 to 4.5 per cent in 1988, and fell slightly to 4.4 per cent by 1990. Customs duty was further simplified in the 1992 budget, which reduced the number of rates to five. Emphasis was placed on the need to limit the scope of exemptions, as too many importers were exempted from tariffs and sales tax. The number of categories of importer exempted from tariffs was severely restricted in the 1992 budget, but problems on monitoring were apparent by early 1993. Exemptions for imported inputs were enacted in the 1992 budget, but effectively rescinded in early 1993 because of severe difficulties with evasion and avoidance.

Tanzanian reforms stalled somewhat in the early 1990s, but have been revived in recent years, motivated partly at least by commitments to COMESA and the EAC. Licences for virtually all imports and exports were abolished in 1993, and by the end of that year the foreign exchange market was almost fully liberalised. The number of tariff rates and the maximum tariff have been reduced a number of times such that by 1997 there were only four non-PTA rates, the maximum being 30 per cent.

20.2.10 *Uganda*

Since 1987 there has been progress in policy reforms towards market and price deregulation and trade liberalisation (Morrissey and Rudaheranwa, 1998). Significant steps have been taken in liberalising the foreign exchange market and in attaining macroeconomic stabilisation. In an attempt to encourage more production for export markets a drawback scheme was introduced in 1991 under which the custom duties initially paid are refunded once the export has taken place. Tax on coffee exports was abolished in 1992 but reintroduced in 1994 to limit the appreciation of the exchange rate as a result of the coffee price boom. This coffee tax (the stabilisation tax) was abolished in 1996.

The first bout of tariff rationalisation in 1992 established a range of 10–60 per cent, reduced to 10–50 per cent in 1994; by 1996 the highest tariff rate was 30 per cent, and in 1997 the range of rates was further reduced to 0, 5, 10 and 20 per cent. The range of rates on imports from COMESA countries, notably Kenya and Tanzania, was 0, 2, 4 and 8 per cent. This understates actual protection. When tariffs were reduced in 1996 an excise duty of 12 per cent (of the tariff inclusive price) was imposed on certain imports. The duty was reduced to 10 per cent in 1997, but its coverage extended to apply to almost all finished and consumption goods (including, especially, those from COMESA). A number of products attract very high tax rates (notably cars, petroleum, beers and tobacco). Under the third Structural Adjustment Credit agreed with the World Bank (SAC3), Uganda is to further rationalise tariffs to two rates, 7 and 15 per cent, and is expected to abolish the special excise duties on imports (the latter is a principal sticking point).

20.2.11 *Zambia*

The reform programme began in 1983 and included stepwise devaluations. From 1985, import licensing and prohibitions were to be eliminated, the maximum tariff level reduced from 150 per cent to 100 per cent, extending the 10 per cent minimum tariff to include some 300 items previously subject to zero tariff and unifying sales tax for imports and domestic production (Nash, 1993). The average nominal weighted tariff was 16 per cent by 1995, although above 20 per cent for many agricultural products, wood products and clothing (Imani Development International, 1998). The duty drawback system and the export licensing system was simplified and an autonomous Export Promotion Board with the participation of the private sector replaced the Zambian Export Promotion Council. The government also established export credit insurance and export credit guarantee schemes allowing exporters to retain 50 per cent of foreign exchange earnings for own imports. Other trade-related reforms include the linking of domestic agricultural prices to border prices and the adjustment of maize and fertilizer prices according to an agreed schedule.

Although most of the trade policy reforms cited above were implemented as planned the reform programme was abandoned after only 16 months (Nash, 1993). Import licences and foreign exchange allocation mechanisms were resurrected and requirements for the retention scheme were tightened. In February 1990, the exchange allocation system was radically altered and the export retention scheme was made less restrictive. By 1992 the retention rate had been increased to 100 per cent. Export bans were also removed (with a few exceptions). Export promotion policies included an improvement of the export board, the duty drawback and bonded warehouse schemes and streamlining of export licensing.

In recent years, Zambia has been rationalising its tariff structure. In the 1996 budget, the Zambian authorities adopted an integral package of customs duty reductions and exemptions to address the issue of regional economic cooperation. This resulted in a moderate cascading tariff structure ranging from 0 per cent to 5 per cent for most capital goods and raw materials; 15 per cent for intermediate goods; and 25 per cent for finished products. Zambia also offers 60 per cent reduction on COMESA imports. Zambia's average tariff is 16 per cent with a tariff ceiling of 25 per cent. The latter applies to clothing and textiles and certain food products only. The average applied rate for agricultural imports is 18.2 per cent and imports of food products bear an average of just above 20 per cent. The average rate in manufacturing is 13.5 per cent. Excise duties on certain products range from 10 per cent to 125 per cent. The import declaration fee was abolished in July 1988.

20.2.12 *Zimbabwe*

Since the inception of the Economic Structural Adjustment Programme, Zimbabwe has done much in an endeavour to open up the economy. Imports have been liberalised, except for a very short list of products. Further removal of non-tariff barriers is being contemplated. Zimbabwe has four tariff structures: the Most Favoured Nation (MFN), the South African structure (RSA), the COMESA structure and the Trade Agreement structure (TAG). By default the MFN applies unless there is an agreement. The average nominal tariff in 1995 was 24 per cent, but most agricultural products attracted higher rates and tariffs on clothing were above 60 per cent (Imani Development International, 1998). The tariff regime announced in February 1997 was of a cascading nature. There were major reductions in duty on raw materials, educational and medical goods (from 0–40 per cent to a flat rate of 5 per cent). Tariffs on capital goods were abolished. However, there are punitively high tariff rates on clothing, textiles and tobacco. Zimbabwe is also close to following the CBI timetable, offering effective nominal tariff reductions on intra-regional trade. The customs tariff was substantially revised in February 1997 in a bid to harmonise external tariffs. There are generally no controls on exports. Export permits are required for a few items related to food and minerals. The easing of price controls and exchange controls has improved the business environment for both importers and exporters.

20.3 Overview of trade reforms in SSA

Table 20.1 presents a broad summary of the reforms implemented by the countries reviewed above. In broad terms, quite a lot has been achieved and it is true to say that trade regimes in SSA countries are more open now than they were even 10 years ago. Almost all countries (included in the sample) have largely eliminated import licences and QRs, and all have significantly rationalised their tariff structure. However, the latter may tend to exaggerate the true degree of liberalisation. As indicated in section 20.2, most of the countries have reduced their basic tariff structure to four or five rates, usually in the range 0–30 per cent. The basic rates hide the true picture, which is a complex mixture of exemptions, exceptions, special tariffs (often not called tariffs) and variable rates for different trading partners. In all of the countries, effective tariffs on sensitive sectors (especially textiles, clothing, petrol, beer and tobacco) are much higher than implied by the basic rates. Nevertheless, significant progress has been made.

Of greater importance, perhaps, is the evidence that most countries have removed taxes and restrictions on exports, introduced direct export promotion measures, devalued and liberalised their exchange rates (even if many have not achieved full liberalisation). This

Table 20.1 Trade reforms in SSA, selected countries

Country	Import liberalisation		Export promotion	Exchange rate	
	QRs	Tariffs		Dev.	Lib.
Ghana	Y	Y	Y	Y	1992
Kenya	Y	Y(p)	Y	Y	1992
Madagascar	Y(p)	Y(p)	Y	Y	part
Malawi	Y	Y(p)	few	?	part
Mauritius	Y	Y	Y	Y	Y
Namibia	Y	Y	?	?	?
Nigeria	p	p	few	Y	part
Senegal	Y(p)	Y(p)	?	CFA	no
Tanzania	Y	Y(p)	few	Y	1993
Uganda	Y	Y	Y	Y	1993
Zambia	p	p	Y	Y	?
Zimbabwe	p	Y	?	?	Y

Source: Derived from materials discussed in text, especially CBI (1998a, b)

Notes

'QRs' refers to abolition of QRs/licences; 'Tariffs' refers to reduction and rationalisation of tariff rates; 'Export promotion' refers to implementation of measures; 'Dev.' refers to devaluation and 'Lib.' to achievement of fully liberalised exchange rate (date indicates year attained).

Y   Yes, the measures were implemented.

p   Only partial implementation (or reversal of some measures).

few Measures were few and/or ineffective.

part Partial liberalisation of the exchange rate regime only.

?   Not known (or mixed experience, questionable implementation).

CFA Franc zone country, no unilateral devaluation and not liberalised.

suggests that significant progress has been made in removing the bias against exports. We will consider in the next section if there is any evidence that this has encouraged export growth, the hoped-for result of trade liberalisation.

Table 20.2 provides some specific evidence on the extent of tariff reductions for a sample of SSA countries. Of the eight SSA countries covered, only five recorded a reduction in average nominal tariffs during reforms in the late 1980s. Furthermore, with the exception of Ghana, the recorded reductions were low. In fact, tariff reductions on average for SSA were much lower than for the other regions reported, although it is true that average rates of nominal protection are lower in SSA than in South Asia. We should note that use of the average nominal tariff may be misleading even regarding the direction of change in trade policy orientation (Milner and Morrissey, 1999). Morrissey (1995) argues that because the nominal tariff omits the effect of changes in QRs it does not pick up the true degree of liberalisation that was implemented in Tanzania. Nevertheless, changes in average nominal tariff are indicative and it is the only measure easily available (none the less, the data are somewhat dated and of limited coverage).

Table 20.2 also includes some estimates of the change in tariff equivalent in the SSA countries. It is clear that when one attempts to account for the removal of QRs and import licences, the reductions in protection seem much greater, The picture is not consistent, however. The degree of import liberalisation was greatest in Ghana, Tanzania, Madagascar and Nigeria (in this case these reforms of the late 1980s were partially reversed in the 1990s), and the major contribution came from abolition of QRs. In the case of Kenya, although average tariffs were reduced the erratic policies towards QRs and import licences imply that overall protection was increased.

Table 20.2 Tariff reforms in SSA, selected countries

Country	Average nominal tariff[a]			Tariff equivalent[b]
	Initial	*Reform*	*% change*	
Côte d'Ivoire (1985, 1989)	26	33	27	−15.9
Ghana (1983, 1991)	30	17	−43	−298.2
Kenya (1987, 1992)	40	34	−15	67.6
Madagascar (1988, 1990)	46	36	−22	−73.5
Nigeria (1984, 1990)	35	33	−6	−107.1
Senegal (1986, 1991)	98	90	−8	1.0
Tanzania (1986, 1992)	30	33	10	−114.9
Zaire (1984, 1990)	24	25	4	
Regional averages:				
SSA (8)	41	38	−6	
South Asia (4)	80	53	−29	
East Asia (5)	29	25	−18	
Latin America (8)	44	15	−65	

Source: Derived from various tables in Dean *et al.* (1994); tariff equivalents from Milner and Morrissey (1999: 69)

Notes

Years given in parenthesis are pre-reform (initial) and post-reform (reform) years used. Numbers in parentheses for regional averages are number of countries included.

a Unweighted average nominal tariff (tends to be biased upwards), figures in 'Average' rows are simple averages for each region.

b Percentage change in tariff equivalent of all restrictions on non-fuel imports, generally refers to early 1990s.

What emerges is a rather mixed picture. A number of SSA countries have achieved a significant and sustained degree of trade liberalisation in the past two decades, notably Ghana, Mauritius and Uganda, and Madagascar and Tanzania to a lesser extent. These countries have also liberalised the exchange rate regime and implemented export promotion measures. However, in terms of export growth, Mauritius is the only clear success story. Other countries have had a more erratic experience, implementing and then reversing reforms or often counteracting reforms with new measures. Kenya and Nigeria exhibit such a pattern (which would be expected to undermine business confidence). Other countries have achieved much less – Côte d'Ivoire and Senegal are good examples, both constrained by membership of the franc zone.

20.4 Evidence on the impact of trade reform

There is a shortage of convincing evidence that trade liberalisation increases growth, although there is an absence of evidence that trade liberalisation retards growth. When the evidence is examined in greater (econometric) detail, the link between liberalisation and growth is found to be weak (Greenaway *et al.*, 1998). Such econometric evidence must be interpreted with extreme caution, however, as cross-country studies are limited to using simple measures of liberalisation (such as whether or not a reform episode was in place, or changes in average tariffs); these measures are often inaccurate and misleading (Milner and Morrissey, 1999).

Onafowora and Owoye (1998), in a time series analysis of twelve African countries over 1963–93, find evidence that exports are positively related to growth in about half of the countries. There is some evidence that growth is higher in more outward-oriented economies, suggesting that trade liberalisation (represented as an index of outward orientation) offers potential for SSA countries to increase growth rates. The potential is very limited for some countries; the commodity composition of exports, in particular, can limit the contribution of exports to growth.

The basic objective of trade reform is to remove the bias against exports and the anticipated beneficial effect is that exports will increase and, in turn, fuel economic growth. However, trade policy alone is not the only constraint on exports. Most SSA countries are dependent on agricultural exports but policies have been biased against exports in general and agriculture in particular (Bautista, 1990). In addition, farmers face many constraints that can limit any export supply response (McKay *et al.*, 1997). Transport costs can be quite high for many SSA countries and this can act as an important constraint on exports. Milner *et al.* (1998) demonstrate that transport costs represent a high implicit tax on Ugandan exporters. Delays in implementing institutional reforms have partly been responsible for the lack of adequate export supply response in Uganda (Belshaw *et al.*, 1999). Thus, because of other important constraints one may not observe a quick export response to trade liberalisation. This does not mean that trade reforms should not be undertaken; it does mean that one should exercise care in interpreting the evidence.

A more general point can be made regarding the link between trade liberalisation and openness. While the latter may give rise to concerns regarding the competitiveness of domestic producers of importables, access to imported investment goods and the technology embodied in imports may be very beneficial (Rodrik, 1999). Furthermore, openness and being seen to implement trade reforms may attract foreign investment. While this is no panacea, foreign investment may be essential for exports, whether in agricultural marketing (such as coffee in Uganda) or for the few manufacturing exporters (see Grenier *et al.*, 1999; Teal, 1999).

Because trade policy reforms are often adopted as part of a wider package of structural adjustment programmes, it is difficult to isolate changes in economic variables that are attributable solely to trade policy reforms. Mosley and Weeks (1993) attempt to explain the contribution of various components of adjustment programmes in sub-Saharan Africa and argue that the impact of a full programme was more important than the individual components. Trade reforms alone, without real exchange rate depreciation, had a negligible effect on investment and exports. Liberalisation of agricultural markets had a greater impact when accompanied by an increase in public expenditure. This highlights the importance of accounting for the interrelationships between reforms.

Trade liberalisation, through increased competition from imports and removal of tariff subsidies, can have adverse effects on manufacturing industries. Manufacturing value added (MVA) contracted in more than a half of the SSA countries between 1990 and 1994 and there is no clear link with trade reforms (Bennell, 1998). However, the impact of trade reforms is likely to take some time and, with different reform periods, it is difficult to make a consistent comparison. Rudaheranwa (1999) shows that the impact of trade liberalisation (initiated in 1987) on the performance of the manufacturing sector in Uganda was gradual in the early stages of the reform period but both production and capacity utilisation (efficiency) increased significantly after 1993.

Generally, the most immediate threat of trade liberalisation to many manufacturing industries tends to be the sharp increase in the level of import competition, and therefore trade liberalisation is likely to have undesirable but inevitable side effects in the short run. Greenaway (1998) argues that these depend on initial conditions and sequencing of reforms. The short- to medium-term responses to trade reforms are on factor allocation and therefore on the composition of output, as resources switch from inefficient import substitute

Table 20.3 Impact of trade liberalisation: summary of selected studies

Study	Food crops	Cash crops	Manuf.	Exports	Imports	Total output
1		–ve		–ve		
2	+	+		+		
3	+	+	+	mix	+	+
4				+		mix
5			mix			
6	mix	mix				
7				+		mix
8	–ve	+				
9			+	+		mix

Notes

Coding is '+' where effect was positive, '–ve' where negative, and 'mix' indicates mixed evidence – either the impact of trade policy reforms impacted differently across sectors/countries or over time.

The numbers in the table correspond to the following studies:
1 Colman and Okorie (1998); Nigeria (1970–92)
2 Belshaw, Lawrence and Hubbard (1999); Uganda (1986–97)
3 Rudaheranwa (1999); Uganda (1987–97),
4 Mosley and Weeks (1993); cross-country (sub-Saharan African countries) (1980–90)
5 Bennell (1998); cross-country (mainly sub-Saharan Africa) (1981–91)
6 Guillaumont (1994); cross-country analysis (African countries) (1970–88)
7 Greenaway (1993); coss-country analysis (reform periods vary)
8 Duncan and Jones (1993); cross-country analysis (1980–91)
9 Teal (1999), six African countries (average 1985–95)

production to export oriented activities. The medium- to long-term impact of trade reforms should be reflected in capital formation and growth of real output and the growth of trade.

Table 20.3 presents an overview of the findings on the impact of trade liberalisation for a sample of studies, some for specific SSA countries and others for a sample of countries that included SSA. Two studies (both of Uganda) found positive effects on agricultural production, although only one identified generally positive effects for other indicators, and one found a negative effect (for Nigeria). Two studies had mixed results (6 and 8). Manufactures prices usually decline after (unilateral) liberalisation, with mixed impacts in Bennell (1998), reflecting the large cross-section coverage. However, Teal (1999) reported increased per capita real manufacturing exports in Ghana, Kenya, Mauritius and Zambia (negligible growth in Zimbabwe and a slight decline in Cameroon). Again Uganda seems to have experienced a positive response. What the range of studies shows is that overall impacts will depend on how relative prices change and the relative responsiveness of different sectors. Thus, while Bennell (1998) expresses concern about the impact of economic liberalisation in sub-Saharan Africa, he finds that in some countries it appears to be associated with manufacturing decline but in others the sector responds well. Grenier *et al.* (1999) for Tanzania and Rudaheranwa (1999) for Uganda find no evidence that manufacturing output declined following liberalisation.

The impact on exports tends to be positive (Nigeria being an exception) although imports also tend to increase following liberalisation, so the net impact on the trade balance is indeterminate. As the impacts on and responsiveness of sectors vary, it is perhaps to be expected that evidence on the overall effect on output is mixed. Taking the limited evidence overall, there is an encouraging tendency for more favourable impacts to be found in those countries that have sustained the most trade reforms. While the effects have not been great, they appear to have been positive for Ghana, Mauritius and Uganda (and not negative for

Tanzania). The country with a consistently poor performance is Nigeria, and that cannot be attributed to trade reform.

20.5 Conclusion

The overall story is mixed, within and across countries. This is true both in terms of the range and degree of implementation of reforms and the evidence on effects. However, many SSA countries have implemented sustained trade and trade-related reform measures in the last 10–15 years, notably import and exchange rate liberalisation. Furthermore, the impact, in terms of supply response in affected sectors and export growth, has tended to be more positive in those countries that implemented more reforms. Greenaway (1998) concludes that while outward orientation and exports contribute to growth, trade liberalisation has not consistently led to increased exports. Partly this is because the extent of liberalisation actually implemented (in the study periods) may have been quite limited, and partly it is because economies are subject to external shocks that make it difficult to evaluate the impact of trade liberalisation. We can add that the effects of trade liberalisation will only materialise over time, and it may well be too early to draw firm conclusions. Trade reform is no panacea but, given the highly distorted starting point, it appears to be a beneficial policy option in SSA countries. In particular, it is worth emphasising that reducing trade distortions and simplifying (and reducing) the tariff schedules can have significant effects in reducing the incentives for rent-seeking behaviour. Given the problems of corruption in many SSA countries, the principal benefits of trade liberalisation may be in improving the efficiency with which the economy functions. These will only be realised over time, and may not be reflected in export supply response.

Notes

1 Helpful research assistance for this chapter was provided by Nick Rudaheranwa, Verena Tandrayen and Satiumsingh Ragoobur.
2 Information on Ghana is taken from Castillo (1993a) and CERDI (1997).

References

Basu, P. and Morrissey, O. (1997) 'The fiscal impact of adjustment in Tanzania', in C. Patel (ed.), *Fiscal Reforms in the Least Developed Countries*, Cheltenham: Edward Elgar for UNCTAD.

Bautista, R. (1990) 'Price and trade policies for agricultural development', *World Economy*, 13 (1): 89–109.

Belshaw, D., Lawrence, P. and Hubbard, M. (1999) 'Agricultural tradables and economic recovery in Uganda: the limitations of structural adjustment in practice', *World Development*, 27 (4): 673–90.

Bennell, P. (1998) 'Fighting for survival: manufacturing industry and adjustment in sub-Saharan Africa', *Journal of International Development*, 10 (5): 621–37.

Castillo, G. (1993a) 'TEP study on trade reform in sub-Saharan Africa: the case of Ghana', Washington, DC: World Bank, mimeo.

Castillo, G. (1993b) 'TEP Study on trade reform in sub-Saharan Africa: the case of Nigeria', Washington, DC: World Bank, mimeo.

Castillo, G. (1994) 'TEP study on trade reform in sub-Saharan Africa: the case of Tanzania', Washington, DC: World Bank, mimeo.

Centre d'Études et de Recherches sur le Développement International (CERDI) (1997), 'Étude de l'impact de l'introduction de la reciprocité dans les relations commerciales entre l'Union Européenne et les pays ACP: les cas des pays de l'UEMOA et du Ghana', mimeo, Clermont-Ferrand: CERDI.

Colman, D. and Okorie, A. (1998) 'The effect of structural adjustment on the Nigerian agricultural export sector', *Journal of International Development*, 10 (3): 341–55.

Corbo, V., Fischer, S. and Webb, S. (eds) (1992) *Adjustment Lending Revisited: Policies to Restore Growth*, Washington, DC: World Bank.

Cross-Border Institute (CBI) (1998a) 'Cross-Border Initiative implementation and evolving agenda', synthesis paper for consideration at the Third Ministerial Meeting, Harare, 19–20 February, Harare: Cross-Border Initiative.

Cross-Border Institute (CBI) (1998b) 'Cross-Border Initiative', vol. 3: 'Country Papers and Progress Reports for the Initiative to Facilitate Cross-Border Trade, Investment and Payments in Eastern and Southern Africa and the Indian Ocean', January, Harare: Cross-Border Initiative.

Dean, J., Desai, S. and Riedel, J. (1994) 'Trade policy reform in developing countries since 1985: a review of the evidence', World Bank Discussion Paper 267, Washington, DC: World Bank.

Duncan, A. and Jones, S. (1993) 'Agricultural marketing and pricing reform: a review of experience', *World Development*, 21 (9): 1495–514.

Foroutan, F. (1993a) 'TEP study on trade reform in sub-Saharan Africa: the case of Madagascar', mimeo; later published in F. Foroutan and J. Nash (eds), *Trade Policy and Exchange Rate Reform in Sub-Saharan Africa*, Washington, DC: World Bank.

Foroutan, F. (1993b) 'TEP study on trade reform in sub-Saharan Africa: the case of Senegal', mimeo; later published in F. Foroutan and J. Nash (eds), *Trade Policy and Exchange Rate Reform in Sub-Saharan Africa*, Washington, DC: World Bank.

Foroutan, F. (1994) 'TEP study on trade reform in sub-Saharan Africa: the case of Kenya', mimeo; later published in F. Foroutan and J. Nash (eds), *Trade Policy and Exchange Rate Reform in Sub-Saharan Africa*, Washington, DC: World Bank.

Greenaway, D. (1993) 'Liberalizing foreign trade through rose-tinted glasses', *Economic Journal*, 103: 208–22.

Greenaway, D. (1998) 'Does trade liberalisation promote economic development?', *Scottish Journal of Political Economy*, 45 (5): 491–511.

Greenaway, D. and Morrissey, O. (1993) 'Structural adjustment and liberalisation in developing countries: what lessons have we learned?', *Kyklos*, 46: 241–61.

Greenaway, D. and Morrissey, O. (1994) 'Trade liberalisation and economic growth in developing countries', in S.M. Murshed and K. Raffer (eds), *Trade Transfers and Development*, Cheltenham, Glos.: Edward Elgar.

Greenaway, D., Morgan, C.W. and Wright, P. (1997) 'Trade liberalisation and growth in developing countries: some new evidence', *World Development*, 25 (11): 1885–92.

Greenaway, D., Morgan, C.W. and Wright, P. (1998) 'Trade reform, adjustment and growth: what does the evidence tell us?', *Economic Journal*, 108: 1547–61.

Grenier, L., McKay, A. and Morrissey, O. (1999) 'Exporting, ownership and confidence in Tanzanian enterprises', *World Economy*, 22 (7): 995–1012.

Guillaumont, P. (1994) 'Adjustment policy and agricultural development', *Journal of International Development*, 6 (2): 141–55.

Imani Development International (1998) 'Study on the impact of introducing reciprocity into trade relations between the EU and the SADC region', report to the European Commission, Harare: Imani Development (International) Ltd.

Lyakurwa, W. (1992) 'Fiscal implications of trade policy reforms in Tanzania', paper presented at the CREDIT–CSAE workshop on 'Trade and Fiscal Reforms in Sub-Saharan Africa', St Anthony's College, Oxford, 6–8 January.

McGillivray, M. and Morrissey, O. (eds) (1999) *Evaluating Economic Liberalisation*, London: Macmillan.

McKay, A., Morrissey, O. and Vaillant, C. (1997) 'Trade liberalisation and agricultural supply response: issues and some lessons', *European Journal of Development Research*, 9 (2): 129–47.

Milner, C. and Morrissey, O. (1999) 'Measuring trade liberalisation', in M. McGillivray and O. Morrissey (eds), *Evaluating Economic Liberalisation*, London: Macmillan.

Milner, C., Morrissey, O. and Rudaheranwa, N. (1998) 'Protection, trade policy and transport costs: effective taxation of Ugandan exporters', CREDIT Research Paper 98/13, University of Nottingham.

Morrissey, O. (1995) 'Politics and economic policy reform: trade liberalisation in sub-Saharan Africa', *Journal of International Development*, 7 (4): 599–618.

Morrissey, O. and Rudaheranwa, N. (1998) 'Ugandan trade policy and export performance in the 1990s', CREDIT Research Paper 98/12 (CPD006), University of Nottingham.

Mosley, P. and Weeks, J. (1993) 'Has recovery begun? "Africa's adjustment in the 1980s" revisited', *World Development*, 21 (10): 1583–1606.

Mosley, P., Harrigan, J. and Toye, J. (1991) *Aid and Power: The World Bank and Policy-based Lending*, 2 vols, London: Routledge.

Nash, J. (1993) 'Zambia: country study', Washington, DC: World Bank, mimeo.

Noorbakhsh, F. and Paloni, A. (1998) 'Structural adjustment programmes and export supply response', *Journal of International Development*, 10 (4): 555–73.

Onafowora, O. and Owoye, O. (1998) 'Can trade liberalization stimulate economic growth in Africa?', *World Development*, 26 (3): 497–506.

Papageorgiou, D., Michaely, M. and Choksi, A. (1991) *Liberalizing Foreign Trade*, 7 vols, Oxford: Basil Blackwell.

Rodrik, D. (1999) *The New Global Economy and Developing Countries: Making Openness Work*, ODC Policy Essay no. 24, Baltimore, MD: Johns Hopkins University Press..

Rudaheranwa, N. (1999) 'Transport costs and export trade of landlocked countries: evidence from Uganda', PhD thesis, School of Economics, University of Nottingham.

Teal, F. (1999) 'Why can Mauritius export manufactures and Ghana not?', *World Economy*, 22 (7): 981–94.

Thoburn, J. (2000) 'Developing African exports: the case of Nigeria', in H. Jalilian, M. Tribe and J. Weiss (eds), *Industrial Development and Policy in Africa: Issues of De-industrialisation and Development Strategy*, Cheltenham, Glos.: Edward Elgar.

21 The non-recovery of agricultural tradables and its consequences for rural poverty

Peter Lawrence and Deryke Belshaw

21.1 Introduction

The debate surrounding the effectiveness of World Bank/IMF structural adjustment programmes in sub-Saharan Africa is usually constructed around econometric analyses which attempt to measure the effect of these programmes on GDP growth (see, for example, World Bank, 1995; Mosley, 1996). However, a large part of most of these countries' GDP is derived from the agricultural sector. The original World Bank (1981) report (the 'Berg report'), which led to the emergence of economic liberalisation programmes, laid considerable emphasis on the importance of improving the performance of the agricultural sector, and especially of its export cash crop component. The key factors that were expected to contribute to this improvement were an increase in producer prices and devaluation of an overvalued exchange rate. The former would give appropriate incentives to producers to increase marketed output of these tradables, while the latter would allow marketing agents to increase the share of the export parity price going to producers while at the same time covering their marketing costs. Increased tradables output would lead to an increase in export revenue (and in some case a reduction in the food import bill) and so help to alleviate balance of payments deficits. Recent work on supply response has confirmed that agricultural producers respond positively to increases in producer prices and that they react positively to changes in relative prices between tradables and non-tradables (see McKay *et al.*, 1999). Similarly, recent work on price transmission has confirmed the importance of passing on to farmers a higher proportion of the export parity price (Lloyd *et al.*, 1999).

It might have been expected that, after nearly two decades of widespread implementation of structural adjustment programmes, major attention would have been given to the performance of agricultural tradables. They are still the dominant source of most SSA countries' export earnings. Increased agricultural tradables output and producer revenue have important rural employment and income multiplier effects, together with inter-sectoral linkages with manufacturing and services. However, as we have argued elsewhere in the case of Uganda (Belshaw *et al.*, 1999), that country's apparent macroeconomic success has masked the failure of the tradables sector to mount a sustained recovery. There followed from this a failure to alleviate serious imbalances on the trade account, which are currently being covered by remittances and aid, and a failure to alleviate rural poverty. The idea behind the present analysis is to see how far this 'non-recovery' situation was repeated across other SSA countries which entered into adjustment programmes with the World Bank and the IMF.

This investigation is pursued in the following way. We first describe the performance of a selected group of SSA countries over the period from 1971 to 1995, showing that recovery has been poor and the conventionally recommended policies have largely not been implemented. We then set up, test and report the results of a regression model of the determinants of exportable output. Here we show that most of the variables which one would expect to influence output do not appear to do so and that it is arguable that output recovery has not taken place because the proposed policies have generally not been properly implemented. The reasons for this are then discussed, drawing attention to the institutional and attitudinal barriers to policy change and the failure to take these into account. The concluding section emphasises the important and non-substitutable nature of the contribution tradables recovery make to the alleviation of rural poverty in sub-Saharan Africa.

21.2 The performance of agricultural tradables

Table 21.1 details the output performance of the leading agricultural tradables for a selection of country–crop combinations for five selected 3-year averages over the 28-year period 1971–98. These countries and crops constitute a sub-set of thirteen countries from the twenty-nine listed in World Bank (1994), with the addition of Ethiopia. Their selection is dependent solely on the availability of data from 1971 on all the causal variables being considered. The crops chosen are those listed in World Bank (1994) as SSA's principal exportables. The table takes the output story to the late 1990s where possible, although the complete analysis only covers the period from 1971 to 1994 because of data availability constraints across all variables. The starting date is chosen as representing the period of peak production before most countries and crops underwent a period of decline, stagnation or lower growth. The subsequent years selected for presentation in this and subsequent tables cover the time periods demarcated by World Bank (1994) as significant for comparison of the effect of the adoption of structural adjustment programmes. Of the four main cocoa producers in the sample, only Côte d'Ivoire shows a systematic increase in cocoa output (fourfold over the 25-year period). The other three – Ghana, Cameroon and Nigeria – show output levels lower than at the beginning of the period. Coffee output stagnates or falls across the whole sample of producers (except in two minor cases). Cotton output in the Francophone West African countries and Malawi has shown strong growth, but elsewhere has stagnated or fallen over the period under study. Only in Kenya has groundnut output experienced growth over most of the period. Cameroon, Kenya and Malawi have enjoyed good growth rates of sugar output; elsewhere, rates of growth have been small or negative. Of the featured tea producers, only Kenya and Malawi enjoyed relatively high output growth. Tobacco output has increased significantly in Malawi and to a lesser extent in Tanzania. It is noticeable that a number of crops have enjoyed substantial increases in output in the latter half of the 1990s.

Stagnation in exportables output has been blamed on low and declining producer prices. Table 21.2 presents producer price indices for selected 3-year averages. Two decades after the Berg report, reported producer prices are lower in real terms (in one-third of cases by more than 50 per cent) in thirty-two out of the forty-three country/commodity cases than at the beginning of the period under review (which began 10 years before the report's publication). This pattern applies to cocoa, coffee (but spectacularly not Uganda), cotton (except Chad, CAR and, spectacularly in the latest period, Uganda), groundnuts, sugar (except Cameroon and Malawi), tea, except Uganda, tobacco except CAR and, spectacularly again, Uganda). In the Uganda case, really large producer price increases came towards the end of the period rather than shortly after the implementation of the reform package began in 1987. Earlier price peaks had occurred in either the second period (seventeen cases) or in the third period (twenty-two cases). Overall, real prices were lower in 79 of the 129 inter-period changes.

Table 21.1 Tradables output for selected SSA countries, crops and years, three-year averages (metric tonnes)

Country	Product	1971–3	1979–81	1985–7	1991–3	1996–8
Benin	Cotton lint	17,667	6,667	36,333	86,353	172,000
Cameroon	Sugar	235,766	1,072,003	1,653,333	1,333,333	1,350,000
Cameroon	Coffee	94,956	107,553	104,860	77,380	62,883
Cameroon	Cocoa beans	132,600	119,525	124,736	99,735	125,242
Cameroon	Groundnuts	234,766	136,786	97,229	95,000	143,746
CAR	Coffee	8,142	16,627	18,248	12,616	15,500
CAR	Cotton lint	17,433	8,800	10,067	8,463	20,952
CAR	Tobacco	1,190	1,329	933	533	549
Chad	Cotton lint	40,947	30,158	40,181	51,628	85,750
Côte d'Ivoire	Coffee	270,115	297,833	270,803	154,388	259,020
Côte d'Ivoire	Cocoa beans	206,594	426,577	609,942	793,839	1,164,616
Ethiopia	Sugar cane	1,249,100	1,385,800	1,600,000	1,565,000	1,616,667
Ethiopia	Coffee	175,260	192,400	175,733	213,000	220,660
Ghana	Cocoa beans	409,567	268,167	202,990	269,523	394,333
Ghana	Coffee	6,261	1,567	607	2,900	4,110
Kenya	Sugar cane	1,641,786	4,197,267	4,127,533	4,383,333	4,600,000
Kenya	Groundnuts	4,005	6,973	8,833	12,943	12,144
Kenya	Coffee	64,246	88,711	104,068	82,267	80,214
Kenya	Tea	48,730	93,370	148,741	200,939	252,627
Madagascar	Groundnuts	42,655	37,408	32,205	28,000	35,333
Madagascar	Coffee	66,725	81,635	80,510	80,582	61,000
Madagascar	Cocoa beans	1,160	1,745	2,600	3,875	4,300
Madagascar	Cotton lint	9,115	10,376	14,786	9,442	10,733
Madagascar	Tobacco	5,727	3,344	4,594	1,811	3,741
Madagascar	Sugar cane	1,141,382	1,419,897	1,891,583	1,977,000	2,163,333
Malawi	Sugar cane	377,368	1,426,667	1,583,333	1,633,333	1,786,667
Malawi	Groundnuts	206,232	175,667	149,300	46,013	59,682
Malawi	Coffee	175	967	4,046	6,480	4,647
Malawi	Tea	20,960	31,496	36,945	36,044	42,054
Malawi	Cotton lint	6,652	7,812	8,953	7,533	15,400
Malawi	Tobacco	29,194	53,149	69,808	123,662	153,258
Mali	Groundnuts	138,933	140,945	91,741	153,337	141,790
Mali	Cotton lint	23,179	48,172	67,042	115,548	187,961
Nigeria	Cocoa beans	237,567	168,667	124,333	130,000	158,333
Nigeria	Coffee	3,357	3,233	2,900	2,533	4,033
Tanzania	Sugar cane	1,150,870	1,536,667	1,226,667	1,433,333	1,460,000
Tanzania	Coffee	52,274	54,039	52,739	53,204	42,340
Tanzania	Cotton lint	69,973	53,223	61,332	82,133	74,333
Tanzania	Tobacco	13,416	17,055	13,980	22,233	28,827
Uganda	Coffee	190,600	111,900	156,981	134,084	229,183
Uganda	Tea	21,133	1,667	4,201	10,161	20,299
Uganda	Cotton lint	77,267	5,100	10,426	8,714	16,433
Uganda	Tobacco	4,433	433	1,259	5,670	6916

Source: FAOSTAT

Taking the possible disincentive caused by the inefficiency in the marketing system, one partial measure of this is the share of the export parity price going to the producers. Part of this share sometimes goes in the form of an export tax, while the rest goes in state or private agents' margins along the marketing chain. As Table 21.3 shows, in around half of the cases, producer shares of the export parity price are below 50 per cent in each of the four periods, while in over 80 per

Table 21.2 Real producer price indices, selected SSA countries, crops and years (1971=100)

Country	Product	1971–3	1979–81	1985–7	1992–4
Benin	Cotton lint	98.17	82.27	93.55	69.29
Cameroon	Cocoa beans	97.00	125.07	98.83	38.86
Cameroon	Coffee	92.14	82.48	63.91	24.43
Cameroon	Groundnuts	102.15	139.10	125.59	96.53
Cameroon	Sugar cane	104.02	100.99	139.48	156.66
CAR	Coffee	93.57	42.05	54.01	33.89
CAR	Cotton lint	94.48	101.63	223.91	359.70
CAR	Tobacco	98.37	53.40	123.37	167.18
Chad	Cotton lint	97.66	159.67	233.60	153.39
Côte d'Ivoire	Cocoa beans	102.12	112.11	104.70	45.49
Côte d'Ivoire	Coffee	96.62	88.38	81.22	48.26
Ethiopia	Coffee	116.21	96.14	98.06	50.37
Ethiopia	Sugar cane	113.99	101.42	125.56	54.45
Ghana	Cocoa beans	115.31	48.15	77.90	63.85
Ghana	Coffee	100.00	34.44	81.37	55.13
Kenya	Coffee	113.05	136.21	120.66	62.64
Kenya	Groundnuts	96.74	85.11	185.54	116.95
Kenya	Sugar cane	101.19	103.90	108.30	85.75
Kenya	Tea	88.36	81.10	80.55	57.69
Madagascar	Cocoa beans	97.32	95.66	72.60	62.78
Madagascar	Coffee	95.05	67.12	65.97	45.95
Madagascar	Cotton lint	95.91	62.82	71.76	56.12
Madagascar	Groundnuts	104.30	95.39	51.20	42.34
Madagascar	Sugar cane	103.11	131.45	104.26	67.77
Madagascar	Tobacco	101.91	73.47	29.60	26.85
Malawi	Coffee	111.20	110.80	65.28	95.00
Malawi	Cotton lint	114.70	161.44	135.00	81.12
Malawi	Groundnuts	111.83	211.40	156.31	109.88
Malawi	Sugar cane	108.69	122.52	181.24	208.74
Malawi	Tea	112.86	117.11	84.37	71.39
Malawi	Tobacco	90.51	92.57	106.02	95.37
Mali	Cotton lint	94.88	68.21	62.17	65.14
Mali	Groundnuts	94.88	81.86	83.82	84.10
Nigeria	Cocoa beans	96.93	122.32	152.39	81.03
Nigeria	Coffee	96.93	95.43	116.20	63.55
Tanzania	Coffee	111.40	40.87	43.25	39.87
Tanzania	Cotton lint	109.28	141.46	92.90	47.61
Tanzania	Sugar cane	95.25	87.17	71.17	53.53
Tanzania	Tobacco	93.67	55.49	52.04	35.74
Uganda	Coffee	93.13	158.72	338.19	222.55
Uganda	Cotton lint	93.13	121.27	291.13	702.36
Uganda	Tea	93.13	56.33	202.31	128.43
Uganda	Tobacco	94.00	86.58	160.13	329.80

Source: FAOSTAT

cent of the cases, producer price shares were less than 70 per cent. While there has been some small improvement in the mid-1980s, the picture in 1992–4 was much the same as in 1971–3.

A further disincentive to tradable output has been laid at the door of the relative price of non-tradables. An overvalued exchange rate increases the relative price of food non-tradables and will discourage tradables output, *ceteris paribus*. Table 21.4 presents the data on selected food non-tradables prices for the period under study, while Table 21.5 presents composite

Table 21.3 Producer price share of export parity price, selected SSA countries, crops and years

Country	Product	1971–3	1979–81	1985–7	1992–4
Benin	Cotton	0.41	0.53	0.68	0.47
Cameroon	Cocoa	0.52	0.50	0.74	0.88
Cameroon	Coffee	0.61	0.49	0.62	0.60
Cameroon	Groundnuts	0.64	0.93	2.03	1.50
Cameroon	Sugar	0.08	0.09	0.37	0.24
CAR	Coffee	0.64	0.22	0.19	0.35
CAR	Cotton	0.60	0.50	0.89	1.08
CAR	Tobacco	0.77	0.35	0.17	0.16
Chad	Cotton	0.81	1.38	1.11	0.80
Côte d'Ivoire	Cocoa	0.53	0.51	0.60	0.64
Côte d'Ivoire	Coffee	0.50	0.47	0.47	0.93
Ethiopia	Coffee	0.31	0.10	0.03	0.20
Ethiopia	Sugar	0.23	0.33	0.88	0.27
Ghana	Cocoa	0.40	0.57	0.35	0.36
Ghana	Coffee	0.55	0.69	0.45	0.58
Kenya	Coffee	0.97	0.91	0.94	0.72
Kenya	Groundnuts	0.36	0.22	0.56	0.40
Kenya	Sugar	0.03	0.04	0.03	0.06
Kenya	Tea	0.94	1.02	1.13	0.92
Madagascar	Cocoa	0.61	0.45	0.31	0.43
Madagascar	Coffee	0.65	0.36	0.28	0.55
Madagascar	Cotton	1.04	0.66	1.07	0.66
Madagascar	Groundnuts	0.38	0.33	0.17	0.13
Madagascar	Tobacco	0.43	0.28	0.10	0.06
Madagascar	Sugar	0.06	0.06	0.07	0.02
Malawi	Coffee	0.44	0.26	0.16	0.44
Malawi	Cotton	0.44	0.48	0.97	0.31
Malawi	Groundnuts	0.46	0.46	0.70	0.20
Malawi	Sugar	0.07	0.05	0.10	0.05
Malawi	Tea	0.55	0.66	0.49	0.47
Malawi	Tobacco	0.21	0.22	0.25	0.18
Mali	Cotton	0.35	0.43	0.50	0.54
Mali	Groundnuts	0.24	1.10	0.65	0.55
Nigeria	Cocoa	0.61	1.10	0.79	0.64
Nigeria	Coffee	1.50	1.43	1.90	1.38
Tanzania	Coffee	1.10	0.36	0.41	0.34
Tanzania	Cotton	0.69	1.01	1.30	0.21
Tanzania	Sugar	0.03	0.04	0.04	0.01
Tanzania	Tobacco	0.64	0.69	0.69	0.24
Uganda	Coffee	0.21	0.28	0.18	0.19
Uganda	Cotton	0.67	0.93	0.80	2.28
Uganda	Tea	0.15	0.34	0.21	0.09
Uganda	Tobacco	0.41	0.80	0.41	0.56

Source: FAOSTAT

indices of food output. In half the countries real food prices were 10–15 per cent lower on average than at the beginning of the period, while in the others prices had slightly increased or remained at the same level in real terms. Thus, comparing with Table 21.2, relative prices for food have broadly increased over the period. As expected, food output has also increased for the countries surveyed, reflecting a mix of imported food substitution, export commodity substitution and an increased secular trend in imported food due to (especially urban) population growth.

Table 21.4 Real food producer price index, selected SSA countries
and years (1971=100)

Country	1971–3	1979–81	1985–7	1992–4
Benin	100.04	105.67	87.68	100.80
Cameroon	99.63	104.05	95.16	103.27
CAR	100.86	117.24	115.86	100.26
Chad	101.53	98.92	117.91	86.02
Côte d'Ivoire	104.87	95.40	94.95	91.05
Ethiopia	102.89	96.30	98.28	102.53
Ghana	112.49	116.79	98.56	98.27
Kenya	98.70	97.43	99.50	88.84
Madagascar	105.76	98.56	97.96	83.37
Malawi	104.54	93.09	106.64	95.09
Mali	102.17	110.67	107.60	103.87
Nigeria	111.84	118.17	93.87	68.09
Tanzania	96.25	104.02	112.76	106.13
Uganda	104.48	79.94	43.00	101.42

Source: FAOSTAT

Table 21.5 Food output index, selected SSA countries, three-year moving averages (1990=100)

Country	1972	1978	1981	1987	1990	1996
Cameroon	75.05	81.63	85.08	94.94	100	119.52
CAR	61.26	74.32	82.60	94.41	100	123.43
Chad	79.27	87.35	90.72	90.28	100	126.83
Benin	51.65	62.69	61.67	84.04	100	134.19
Ethiopia	82.77	83.66	93.75	93.45	100	120.50
Ghana	84.69	70.37	72.20	93.45	100	148.43
Côte d'Ivoire	44.68	64.33	71.55	91.42	100	125.42
Kenya	55.09	70.85	69.44	93.05	100	105.05
Madagascar	74.32	81.75	85.74	95.12	100	107.57
Malawi	78.28	90.5	94.09	99.75	100	105.93
Mali	57.95	73.03	84.01	87.84	100	114.70
Nigeria	55.95	55.7	59.19	78.45	100	136.07
Tanzania	55.44	75.54	77.44	93.06	100	99.05
Uganda	80.60	83.33	73.14	83.64	100	106.49

Source: FAOSTAT

The liberalisation of foreign exchange markets has become an important and high-profile characteristic of liberalising regimes. Table 21.6 presents selected data on real exchange rates. As these data are sourced from the IMF's *International Financial Statistics*, it follows the IFS convention that an increase in the index signifies real exchange rate appreciation. As the table shows, the well-known real exchange rate appreciation of the 1970s was followed by the real depreciations of the 1980s in the wake of the early adjustment programmes. Since the second half of the 1980s, three early depreciators have shown a return to appreciation, while half have continued to show real depreciation. Kenya shows continuing appreciation across the whole period examined here. While there has been substantial nominal exchange rate depreciation, in many cases this has been insufficient to compensate for changes in relative domestic and foreign prices.

Table 21.6 Real exchange rates index, selected SSA countries and years (1990=100)

Country	1971–3	1979–81	1985–7	1992–4
Benin	71.97	98.28	79.68	85.20
Cameroon	64.70	87.36	81.56	78.77
CAR	82.08	108.07	99.73	77.43
Chad	79.21	92.37	71.29	82.68
Côte d'Ivoire	67.80	112.92	79.16	84.83
Ethiopia	72.31	104.19	102.97	72.37
Ghana	7.71	71.36	100.29	75.02
Kenya	8.69	11.55	47.88	295.32
Madagascar	102.06	138.21	107.20	90.36
Malawi	124.28	129.50	85.53	85.09
Mali	46.80	67.80	76.03	78.83
Nigeria	155.76	313.58	240.26	96.49
Tanzania	71.95	100.76	134.93	89.78
Uganda	514.36	702.83	221.89	78.18

Source: IMF, *International Financial Statistics*

Table 21.7 Terms of trade index, selected SSA countries and years

Country	1971–3	1979–81	1985–7	1992–4	% change
Benin	165.33	104.33	91.67	100.33	–39.3
Cameroon	84.33	99.00	77.67	53.00	–37.2
CAR	102.33	97.67	91.67	76.67	–25.1
Chad	94.33	102.00	112.00	115.00	+21.9
Côte d'Ivoire	92.33	104.00	83.67	61.67	–33.2
Ethiopia	146.33	102.33	91.33	48.67	–66.7
Ghana	78.33	102.00	63.00	40.33	–48.5
Kenya	101.67	98.33	85.00	61.67	–39.3
Madagascar	110.67	94.33	98.67	63.00	–43.1
Malawi	143.33	106.00	102.67	89.33	–37.7
Mali	128.33	106.33	108.33	111.00	–13.5
Nigeria	24.33	93.33	67.00	48.00	+97.3
Tanzania	107.33	98.00	86.33	58.00	–46.0
Uganda	84.67	92.67	88.33	34.33	–59.5

Source: UNCTAD and World Bank African Development Indicators

Note: Percentage changes are over the whole period.

Much has been made by SSA governments of evidence of the declining terms of trade for the major exportables, and Table 21.7 presents terms of trade data for the countries covered. There is a great deal of *prima facie* support from these data for a declining terms of trade explanation. Twelve of the fourteen countries covered suffered a decline over the period covered here. Of those twelve, only four enjoyed better terms of trade at some time during the period, as compared with the beginning.

Finally, Table 21.8 presents yield data for the crops and years covered in this chapter. Yield per hectare is one possible way of capturing output changes due to technical change. The data show some evidence of greatly increased yields during the period under study (some spectacular), as well as greatly declining yields. On the negative side, this may reflect the effects of devaluation and/or subsidy removal on the cost of modern inputs, usually imported, such

Table 21.8 Yields per hectare, selected SSA countries, crops and periods, three-year averages (hectogrammes)

Country	Product	1971–3	1979–81	1985–7	1991–3	1996–8	% change
Benin	Cotton lint	3,739	3,559	3,756	4,321	4,265	14.1
Cameroon	Sugar cane	69,118	147,654	100,000	100,000	100,000	44.7
Cameroon	Groundnuts	8,329	4,071	3,023	2,969	4,070	−51.9
Cameroon	Coffee	3,331	2,942	3,095	2,703	2,161	−35.1
Cameroon	Cocoa beans	3,709	2,762	2,932	2,904	3,478	−6.2
CAR	Coffee	2,552	3,557	7,122	5,739	6,200	142.9
CAR	Cotton lint	3,758	3,111	2,702	4,126	4,559	21.3
CAR	Tobacco	7,933	8,996	11,707	7,958	7,760	−2.2
Chad	Cotton lint	3,744	3,644	3,812	3,999	3,897	4.1
Côte d'Ivoire	Coffee	3,832	2,868	2,471	1,266	1,539	−40.2
Côte d'Ivoire	Cocoa beans	4,582	5,071	5,211	5,522	5,677	23.9
Ethiopia	Sugar cane	1,532,005	1,399,540	1,066,667	1,049,722	910,416	−40.6
Ethiopia	Coffee	2,951	4,202	5,491	6,888	7,480	153.5
Ghana	Cocoa beans	2,817	2,235	2,396	3,770	4,209	49.4
Ghana	Coffee	2,422	1,958	1,215	2,900	3,406	40.6
Kenya	Sugar cane	758,464	1,112,548	804,388	893,256	811,664	7.0
Kenya	Groundnuts	5,875	6,269	6,465	6,513	6,523	11.0
Kenya	Coffee	7,020	7,900	6,741	5,284	4,834	−31.1
Kenya	Tea	9,799	12,211	17,519	19,567	22,116	125.7
Madagascar	Sugar cane	429,632	361,912	318,741	390,005	325,156	−24.3
Madagascar	Groundnuts	9,987	9,532	9,727	5,593	7,423	−25.7
Madagascar	Coffee	3,667	3,768	3,597	3,737	3,171	−13.5
Madagascar	Cocoa beans	3,735	3,006	3,364	7,433	9,280	148.5
Madagascar	Cotton lint	3,653	3,810	3,701	3,828	4,002	9.6
Madagascar	Tobacco	9,822	6,826	10,001	6,913	8,917	−9.2
Malawi	Sugar cane	1,109,750	1,138,580	1,117,299	965,609	1,042,423	−6.1
Malawi	Groundnuts	8,433	7,027	8,646	7,064	6,475	−23.2
Malawi	Coffee	3,653	6,588	11,851	14,333	12,109	231.5
Malawi	Tea	14,109	17,998	20,397	19,539	20,741	47.0
Malawi	Cotton lint	3,300	2,600	2,600	2,597	2,598	−21.3
Malawi	Tobacco	5,618	7,719	7,121	9,815	13,378	138.0
Mali	Groundnuts	5,959	8,525	8,997	8,751	9,675	62.4
Mali	Cotton Lint	3,137	3,574	3,354	4,131	4,157	32.5
Nigeria	Coffee	4,998	5,000	5,000	5,000	5,000	0.0
Nigeria	Cocoa beans	5,939	4,217	3,108	3,250	3,682	−38.0
Nigeria	Cotton lint	3,522	3,395	3,422	3,135	3,074	12.7
Tanzania	Sugar cane	324,961	769,023	1,095,567	1,102,564	1,042,857	220.9
Tanzania	Coffee	5,108	4,997	4,794	4,084	3,660	−28.3
Tanzania	Cotton lint	3,400	3,281	3,333	3,303	3,301	−2.9
Tanzania	Tobacco	6,788	6,270	6,444	6,932	8,189	20.6
Uganda	Coffee	7,194	4,996	6,988	5,052	8,383	16.5
Uganda	Tea	13,814	4,583	5,346	6,208	9,881	−28.5
Uganda	Cotton lint	3,062	3,063	3,121	3,086	3,081	0.6
Uganda	Tobacco	6,832	1,680	6,041	7,538	9,179	34.4

Source: FAOSTAT

Note: Percentage changes are over the whole period

as fertilizers and agricultural chemicals. On the positive side this may be a reflection of good dissemination through effective extension practices for specifically targeted crops. The descriptive statistics appear to confirm that the lack of a sustained recovery of tradables is related to a failure to sustain a consistent policy of improving incentives to producers. Next, we attempt a more systematic analysis of the contribution of the 'Berg variables' to this non-recovery.

21.3 Analysing the determinants of tradables production

The conventional economic theory that lies behind the thesis put forward in the Berg report, as indicated in our introduction, suggests that output of a tradable in country C could be explained by the following relationship:

$$X_i = f(P_i/P_f, P_i/XP_i, Y, RER, ToT, FA, WBP)$$

where:

X_i = output of tradable i;

P_i/P_f = producer price of tradable i, relative to the price of non-tradable food crops;

XP_i = export parity price of tradable, i;

Y = output per hectare (yield);

RER = real exchange rate (capturing the effects of overvalued exchange rates);

ToT = international net barter terms of trade;

FA = food aid;

WBP = dummy = 1 if there is an IFI programme operating and = 0 if there is not.

Three variables are added to those discussed in the previous section. First, we add shipments of food aid as recorded by FAO. These shipments capture previous supply shocks, but they also capture any resource reallocation effects in favour of exportable production. Second, we add dummy variables capturing any country-specific (c1–c13) and product-specific effects (p1–p6); and third, a binary dummy, capturing the existence of a World Bank/IMF structural adjustment programme. If properly implemented, such programmes should result in the kind of economic liberalisation policies that would be expected to increase exportables production. Even if they are not implemented fully, the presence of such a programme might still lead economic agents to expect policies favourable to increased output in the near future.

We would expect the first five variables to be positively related to tradables output. In the case of yields, it is possible that the relationship could be negative, where there is contraction of production in low yield areas, giving higher yields but lower output. For this reason we discount the existence of collinearity between output and yields, although we present the regression results with and without yields.[1] Food aid would be positively related in the case of a resource reallocation effect, but negatively if it reflected a supply shock which had adversely affected exportables as well as domestic food output. Tradables output should be positively associated with the existence of a structural adjustment programme.

If our explanatory variables are found to be significant and of the right sign, this would imply that the countries whose traditional exportables have not recovered have not done so because they have not implemented the reforms. In this case, given that at one time or another they have adopted liberalisation programmes, the question of compliance would arise and explanations sought for possible slow (or non-)compliance. If the explanatory variables turn out to be insignificant, then this would imply that there must be other explanations for tradable output non-recovery which lie outside of the conventional framework.

The data were structured in matrix form with consecutive country crop combinations for the 25-year period on the vertical axis and the independent and dependent variables across the top, thus giving 1032 observations.[2] Table 21.9 shows the summary statistics for the variables used. As the earlier tables suggested, there is a large variance around the mean for all the continuous price and quantity variables. Variances are accentuated because of the differences between crops and countries surveyed – within country and crop variances are likely to be much smaller. Table 21.10 presents the regression results. OLS regressions were

Table 21.9 Summary regression statistics

Variable	Observations	Mean	Std deviation	Minimum value	Maximum value
X	1032	306,805	681,658	100	4,825,000
$P_i P_f$	1032	1.125582	0.9398441	0.0142291	9.528458
P_i/XP_i	1032	0.4781642	0.3625282	0.0034671	2.770462
RER	1032	124.1084	133.4819	6.881093	867.9147
ToT	1032	94.19671	39.40522	17	994
FA	1032	78,144.44	155,450.6	0	1,248,832
Y	1032	110,355.1	310,949.9	377	2,088,710
WBP	1032	0.3972868	0.4895735	0	1

performed with and without the World Bank dummy variable and the results are presented in the first two columns. Relative producer prices are not significant, while producer prices as a proportion of export parity prices are significant, but with the 'wrong' sign. Decomposing the producer price/export price ratio showed that this result is generated by a significantly negative relationship to the producer price. Terms of trade also have the 'wrong' sign and are highly significant. Yield is positively related and highly significant as expected. It could be that the producer price variables are picking up a demand-side effect, but of the crops surveyed, only sugar is both exported and consumed internally to a significant extent.

Another possibility is that the effects of prices are lagged. Tree crops in particular have long gestation periods (3 to 10 years), but periods of price stagnation have often resulted in under-harvesting, so that price improvements should result in rapid output increases due to more intensive harvesting without the need to expand planting. Another possible explanation is that of hysteresis, with long periods of tradables price stagnation leading to a lack of response when prices rise because of a lack of credibility as to whether these price increases will be sustained. Downward movements in the terms of trade are strongly positively associated with tradables output. This again is a surprising result but consistent with a lagged response, a delay in transmission of these price effects to producers, or the income maintenance hypothesis. Further work is being done to investigate these possibilities. The World Bank variable is not significant and adds nothing to the explanatory power of the model.

The third and fourth columns of Table 21.10 repeat the analysis excluding yields. Relative producer prices are now significant and positive and the real exchange rate is now much more significantly negative as is the producer price share of the export parity price. Food aid is now significantly positively related to exportables output. This latter result could be explained by such aid encouraging resource allocation away from food output or by an increase in harvesting of perennial exportables when food harvests fail. Leaving yields out reduces the R^2 substantially and the insertion of the World Bank dummy variable does not increase it. However, yield changes have been quite marked, as Table 21.8 shows, and may be over-reducing the significance of other variables, especially relative producer prices.

The final two columns show the results of running the model controlling for country and crop fixed effects. Countries and crops are placed in alphabetical order and Uganda is the country and tobacco the crop left out. The results in column 5 show that none of the price variables (except for the producer price share of export parity price, which has the 'wrong' sign) is any longer significant, yield remains significant, while food aid and the existence of World Bank/IMF programmes become significant. The positive association with World Bank/IMF programmes being in place may be an expectations effect or the result of planned

Table 21.10 Regression results: dependent variable: exportables output

	Regression 1	Regression 2	Regression 3	Regression 4	Regression 5	Regression 6
Constant	400,356.6	386,480.2	609,549	616,453.1	58,092.49	89,978.79
	(8.267)	(7.212)	(8.951)	(8.178)	(0.633)	(0.962)
P_i/XP_i	−213,265.7	−212,402.8	−627,267.3	−627,590.1	−160,741.4	−140,583.9
	(−5.197)	(−5.172)	(−11.343)	(−11.339)	(−4.344)	(−3.737)
FA	−0.0401286	−0.0436653	0.6683305	0.6699161	0.329257	0.2566447
	(−0.431)	(−0.468)	(5.210)	(5.211)	(3.594)	(2.767)
ToT	−1,476.502	−1,409.47	−585.7428	−619.599	−412.3657	−392.2333
	(−4.152)	(−3.784)	(−1.164)	(−1.175)	(−1.315)	(−1.226)
RER	−212.2876	−205.2922	−388.883	−392.3446	177.4554	146.792
	(−2.034)	(−1.954)	(−2.629)	(−2.636)	(1.446)	(1.174)
P_i/P_f	3,604.611	2,925.536	43,004.53	43,334.65	6,362.197	−3,643.898
	(0.239)	(0.193)	(2.014)	(2.024)	(0.381)	(−0.215)
Yield	1.567044	1.567874			0.4606724	
	(32.260)	(32.254)			(6.473)	
WBP		18,199.79		−9,128.045	52,527.37	59,648.08
		(0.606)		(−0.214)	(2.231)	(2.486)
c1					15,240.69	−16,396.57
					(0.147)	(−0.155)
c2					25,104.01	−95,722.63
					(0.302)	(−1.159)
c3					12,113.69	−17,016.77
					(0.158)	(−0.218)
c4					118,566	84,854.17
					(1.167)	(0.820)
c5					316,632.7	264,916.7
					(3.579)	(2.948)
c6					−290,302.4	−160,466
					(−2.935)	(−1.624)
c7					62,563.85	11,940.52
					(0.688)	(0.129)
c8					560,038.8	551,274.5
					(7.181)	(6.930)
c9					9,519.936	−57,168.75
					(0.127)	(−0.752)
c10					−46,467.47	−40,321.98
					(−0.673)	(−0.573)
c11					110,672.1	72,572.95
					(1.196)	(0.770)
c12					45,098.16	−1,925.171
					(0.567)	(−0.024)
c13					−75,518.58	−93,086.6
					(−0.962)	(−1.164)
p1					128,622.5	165,208
					(2.458)	(3.114)
p2					−1,381.227	−1,180.442
					(−0.035)	(−0.029)
p3					47,384.38	46,266.26
					(1.063)	(1.017)
p4					−52,538.63	−30,502.32
					(−1.091)	(−0.622)
p5					1,237,153	1,583,881
					(17.342)	(32.951)
p6					−139,421.8	−158,645.5
					(−2.591)	(−2.895)
R^2	0.5771	0.5772	0.1477	0.1477	0.7691	0.7595

Note
The t-statistics are in parentheses.

resource allocation towards tradables. Among the countries, it is not surprising to find Côte d'Ivoire and Kenya showing significantly positive country effects, given their strong institutional support to export agriculture. Among the crops, sugar shows up very strongly positive, again probably because of strong institutional and commercial support for the product across SSA. The R^2 is now 0.77. Column 6 of Table 21.10 shows the results of running the fixed effects model without yields, and this has little effect on its explanatory power, slightly reducing the R^2. The model was run with and without interactive dummies to test possible interactions between the presence of World Bank programmes and the price variables, but no interaction or significant changes from earlier results were found.

The statistical analysis carried out so far suggests that the four price variables do not play the role predicted by the Berg report and that the effect of specific country characteristics and policies may be more decisive in determining exportables output in the case of the products examined here. In the next section, we turn to an analysis of non-price explanations of exportables output.

21.4 Explanations for the poor performance of tradables: proximate institutions and policy non-commitment

There are several possible non-price explanations of agricultural tradables performance. These suggest reasons why changes in the relevant price variables may have little effect or may not take place at all. First, there is the resistance of entrenched agricultural service monopolies to the reduction or termination of their roles. These institutions include statutory marketing boards, cooperatives, banks and rural credit institutions, farm input supply agencies and parastatal agro-industries. This resistance will affect producer incentives through producer prices and costs and producer price shares of the export parity price. Second, and as part of this institutional resistance, there is pressure to prioritise payments to underpaid and unpaid workers in such organisations rather than to small-scale producers. Again, this will affect the price variables used in the previous analysis. Third, there is the tendency to allocate public resources in the agricultural sector to non-traditional tradables and non-tradables in the interests of 'economic diversification'. Fourth, and related to the previous point, there are policies influenced by general export pessimism with respect to primary products, encouraged by beliefs among influential economists and among policy makers in general in the significance of declining (net barter) international terms of trade and the so-called 'fallacy of composition'. Fifth, farmers may have become increasingly risk-averse and dependent on government initiatives (which are not forthcoming), and less inclined to take on risky entrepreneurial activity. Finally, there are the effects of continuing 'urban bias', including failure to maintain rural infrastructure.

There appears to be an intellectual failure to confront the above factors because of a widely held belief in the imminent feasibility of NIC-type export-oriented industrialisation that will eliminate the need to re-engage in primary product-led export strategies for recovery. As a result, restoring the primary production economic base is regarded as inherently purposeless. These attitudes and beliefs, emanating as much from 'western' as from domestic sources of influence, are reinforced by perceptions of the degrading character of such agricultural activity (especially its associations with colonial rule) and/or the demeaning position for members of 'modern educated society' to have to rely largely on the productivity of 'lowly rural workers' to kickstart economic recovery and sustainable development.

Detailed examination of particular 'success stories' may suggest some common institutional or policy features. Table 21.11 abstracts from Table 21.1 the foremost country/crop cases

Table 21.11 Physical output of selected high-growth agricultural tradables, by country and crop
1971–3 (A) to 1996–8 (B) (000 tonnes)

	Cotton		Cocoa		Tobacco		Tea		Sugar	
	A	B	A	B	A	B	A	B	A	B
Benin	18	172								
Cameroon									236	1350
Côte d'Ivoire			207	1165						
Kenya							49	253	1642	4600
Malawi					29	153			377	1787
Mali	23	188								

Source: FAOSTAT

of significant and sustained expansion of tradable goods production. These eight cases show overall output increases in the range 280–970 per cent over the 25-year period 1971–3 to 1996–8. In half of these cases, output was steadily advancing over the whole period. Of the others, the three sugar cases recorded major growth 'spurts' over shorter periods, mainly in the 1970s; this doubtless reflects the 'lumpy' nature of investment in integrated sugar mill capacity and the associated irrigated cane estates, the dominance of transnational corporation activity in this sub-sector and the deterioration in the investment climate in sub-Saharan Africa from the early 1980s. In the case of cotton expansion in Benin, much the greater part surprisingly occurred in the 12-year period 1979–81 to 1991–3, when currency over-valuation was severe throughout the thirteen-country CFA franc zone.

Were world market conditions particularly favourable for the five (out of seven) commodities featuring in Table 21.11, or for the six (out of fourteen) countries? The major 'zero success' commodity is coffee, produced in eleven of the sample countries. Coffee experienced marked price swings across the period, from world booms in 1975–9 and 1994–8 to severely depressed international prices from 1989 to 1994. Of the seven major producer countries (Cameroon, Côte d'Ivoire, Ethiopia, Kenya, Madagascar, Tanzania and Uganda) the most significant increase occurred in Kenya, where a 63 per cent rise over base had been achieved by 1985–7. But this had fallen back to a 25 per cent gain only by 1996–8 (see Table 21.1). The difference is largely attributable to the smuggling of coffee into Kenya from neighbouring Uganda and Tanzania, where currency over-valuation, high tax and marketing margins and consumer goods scarcities had been detrimental between the late 1970s and the late 1980s. After IFI programmes were initiated in Uganda in 1987 and in Tanzania in 1989, smuggling on any scale was soon eliminated: there is even evidence of some reverse smuggling from Kenya into Uganda after the Ugandan reforms (Belshaw *et al.*, 1999). Low coffee output in Côte d'Ivoire also seems attributable in part to non-recorded illegal exports from 1989 to 1994. It seems that the combination of high-risk premiums in illegal markets coupled with domestic tax and monopoly inefficiencies and high levels of world price instability together have totally suppressed the expansion of the African share of world coffee production and trade which was taking place in the 1950s and 1960s.

Turning to the other commodities providing successful case studies, while international markets (in the cases of cocoa, cotton, tea and tobacco, or, in the case of sugar, the combination of domestic and world demand) apparently encouraged supply expansion, such favourable opportunities were not taken up in other African countries producing the same commodities, as Table 21.12 indicates. This suggests the possibility of a common variable with strong causality occurring in the major expansion group but absent, or working less strongly, in the two other groups.

Table 21.12 Tradable production performance, five selected crops in fourteen SSA countries, 1971–3 to 1996–8

	Cocoa	Cotton	Sugar	Tea	Tobacco
Major expansion	Côte d'Ivoire	Benin Mali	Cameroon Kenya Malawi	Kenya	Malawi
Minor expansion	Madagascar	Chad Malawi Tanzania CAR	Madagascar Tanzania Ethiopia	Malawi	Tanzania Uganda
Contraction or stagnation	Cameroon Ghana Nigeria	Madagascar Uganda		Uganda	Madagascar CAR

A shared institutional factor that is suggested by common observation in the 'non-successful' group of countries is the retention of horizontally integrated institutional configurations, that is, multicommodity extension, farm input delivery and farm credit agencies, and marketing cooperatives culminating in single or multiple commodity, statutory monopoly marketing boards. These institutions intervene between the farm producer (usually small-scale peasant households) and the international or national commodity buyers. When macroeconomic crises have left employees in these service and marketing agencies under-remunerated or even unpaid for long periods, the gains from devaluation and market liberalisation may not be passed on directly to the producer. Payment is deferred or the producer price is increased at below the prevailing rate of inflation. In such systems the link between the recovery of output and each agency's financial viability is obscure. Conversely, vertical integration between exporting agencies or even agro-processing or manufacturing plants and the producer is more conducive to an immediate and more transparent identity of interests. For example, the Uganda government has been implementing a realignment of the previously horizontal configuration of institutions to a vertical set (Belshaw *et al.*, 1999). Coffee institutions were restructured relatively quickly; for cotton and tea the reforms have taken considerably longer to implement.

Applying this 'vertical integration hypothesis' to the eight success cases, three institutional variations can be identified. First, in the cases of rapid expansion of sugar, international TNCs (Asian and European) were investing directly in private or joint venture large-scale integrated sugar projects. The less rapidly expanding countries have required greater levels of state control and ownership, with TNCs only given an operating contract along 'turnkey project' lines, producing lower rates of return relative to perceived risks. Second, in the cases of the cotton and cocoa successes in Francophone West Africa, production is exclusively in the hands of family farms but production inputs and credit are provided by French TNCs which have global interests in agro-processing. They also have representation on national policy making and marketing bodies. They would also have been able to use their direct access to hard currency and black market exchange rates to buffer the adverse effects of over-valued currency in order to maintain producer incentives (the converse of the frequently criticised transfer pricing facility of TNCs). In the period up to 1994, when the CFA franc was devalued by 50 per cent, significant tradable expansion by peasant producers seems likely to have required such a hidden producer price lift, with the connivance or passivity of state authorities.

Finally, the steady expansion of smallholder tea in Kenya and tobacco in Malawi has taken place under parastatal commodity authorities that operate alongside strong private production and marketing agencies and utilise their standard procedures (especially strict quality grading and open auction floor systems). Input delivery, credit, applied research, marketing and processing are organised largely on vertically integrated lines. It should also be noted, however, that Kenya in particular, and Malawi to a lesser extent, avoided the worst effects of unsound macroeconomic policy compared with many other former state-dominated countries in the sample.

Clearly, the question remains whether still other factors were at work demotivating or hindering the other eight countries from emulating the success stories in the group of six. Was rapid tradable commodity expansion not identified as an option by the eight, or downgraded or spurned as useless *ab initio*? Were horizontal structures too entrenched or TNCs too suspect for recovery opportunities to be perceived and appropriate decisions taken? Were development agencies or African governments learning from each other?

Even in the pre-crisis literature, primary commodity expansion was not advocated as a complete long-run development strategy for all developing countries. But the contributions of agriculture, especially smallholder farming, were correctly seen by Myint (1980 and earlier editions) as utilising unused land and labour resources for enhancing import capacity through expanded exports, and further expanding production by adopting new technology which would reduce unit costs, that is, shifting to a new set of production functions (Myint, 1988). Enhancing farming system productivity by further enterprise intensification or diversification is a third route that seems to have been largely overlooked in the general policy debate.

However, there is some evidence in the literature of a policy bias against using the tradable commodity expansion option in rapid recovery and short-run import capacity strengthening strategies. The possibility of taking these options seriously appears to be nullified in many cases by four strongly held beliefs. The first is that the Singer–Prebisch thesis (strictly, the tendency for the net barter international terms of trade facing primary producers to decline *in the long run*) closes the case for adopting the recovery and/or expansion of primary products as a recovery strategy. As Ndegwa and Green (1994: 4–5) put it:

> Basically, African exports have been stagnant while both African populations and world trade have grown at the rate of 3 per cent a year. The causes are not simply policy related – the collapse in the terms of trade of most African primary product exports means 'more of the same' neither was nor can it now be a viable export recovery strategy. The greatest failure is a lack of sustained transformation of the export base.

More recently, Mkandawire and Soludo (1999: 39) have argued similarly and more explicitly in favour of the Prebisch–Singer hypothesis. They press the case for diversification into 'non-traditional' exports. It is true that, as Table 21.7 shows, most countries in our sample have experienced terms of trade deterioration over the 25-year period under examination. However, it is important not to confuse net barter terms of trade with income terms of trade. These incorporate both technological and commodity diversification gains, which, as Table 21.8 shows, have been significant in many cases. Nor should mono-export commodity situations be confused with already diversified patterns in which not all prices move consistently in the same direction.

The belief in the 'fallacy of composition' (that a decline in the international price of a commodity will follow from multicountry increases in supply due to inelastic world demand, and the price decline will offset any gain from expanded output) is widely prevalent. This argument is countered on empirical and conceptual grounds by Berg (1993) and Husain (1993). They argue that hard evidence suggests this might be the case only for cocoa and, perhaps, for coffee; and that in any case long-run gains are ignored. In addition, it should be said that, viewed in a game theoretic context, as most approximating the real world situation, the more countries that strongly believe the fallacy of composition prediction, and so fail to expand exports, the less true it will be for any country which does decide to do so (consider for example, the cases of Côte d'Ivoire *vis-à-vis* Ghana and Nigeria in Table 21.1). The more influential the fallacy of composition is, the higher the probability that it is itself fallacious regarding real outcomes; it is a self-negating proposition. In any case, not all competing countries producing the same commodity can or need to expand production at the same rate – or even at all. Even if the fallacy of composition is granted, its unit price-reducing effects can be countered by productivity increases assisted by more effective farmer-support services.

In any case, traditional primary product exportables are still the main present and potential source of foreign exchange earnings and savings for most sub-Saharan African countries, and look like being so for the medium-term future. In the longer run, the alternative to traditional commodity exports, however, is seen to be export-oriented industrialisation. As Helleiner (1988: 226) puts it:

> Strategy for the non-agricultural sector must be rethought. If import-substitution industrialisation has proven costly and efforts at major 'structural transformation' have been premature, what are the sensible alternatives? Is it really sufficient to return to a primary-producing mode, wait and hope for the best?

The answer to this question is probably that, at least in the short run, returning to 'primary-producing mode' may not be sufficient, but it is necessary to buy the time needed to switch to a more diversified strategy involving industrialisation. The success of countries such as Mauritius in finding industrial export niches is often referred to in support of such a strategy and is seen as being replicable in sub-Saharan Africa as a whole (IMF, 1999: 216). Even if that were surely the case, it will not happen in the immediate future. In the meantime, making the best of what is available appears to be the most sensible strategy.

21.5 Conclusions

We have examined possible determinants of agricultural exportables output in Africa, using a sample of seven primary commodities across fourteen countries. The descriptive statistics presented in section 21.2 suggested that one explanation for the slow growth and or recovery of traditional agricultural exportables can be found in the failure to sustain improvements in price incentives to producers, in spite of the widespread adoption of adjustment programmes. The findings of our econometric analysis then suggested that, controlling for country and product fixed effects during the period, price variables did not behave in the way theory would predict. This suggests that institutional arrangements affecting producer revenue and costs were probably a more significant factor. Further analysis needs at least to incorporate lags, more countries and more crops, and to test more sophisticated measures of World Bank/IMF involvement and of institutional arrangements. So far, these results do suggest that while 'getting prices right' may be necessary to increase tradable output, it is not a sufficient policy. The absence of other factors not included in price variables appears to account for the patchy and sporadic pattern of recovery of tradables production, the enduring pattern of constrained capacity to import at the national level and continuing concentrations of deep-seated poverty in rural areas. Exploiting the primary product potential in most SSA countries seems to be the only viable option for recovery and development in the medium term.

In section 21.4 a *prima facie* case was made, it is contended here, for investigating more thoroughly the effect of some specific non-price factors in accounting for differential country tradable goods performance. These include, first, the adoption of a vertical integration pattern of production and marketing services delivery; second, the possibility that some TNCs have seen it in their interest to 'bypass' overvalued official exchange rates in order to maintain price incentives for small-scale producers; third, the possibility that donor agency and government economists have been more affected by export pessimism arguments (especially in Anglophone countries), compared to the weight placed on commercial actualities and the profit calculus in the private sector; fourth, the possibility that the understandable unpopularity of the effects of stabilisation measures amongst the African intellectual elites has led to the non-acceptability to them of the associated primary production-led strategies; and finally, and conversely, preference of elites to embrace any proposals which appear to offer a direct route to accelerated industrial development, such as the Lagos Plan, free trade areas or common markets, and Mauritius-style export zones.

The irony is that, whatever else the various structural adjustment programmes in SSA have delivered, they have failed to deliver the sustained recovery of tradables exports which was at the centre of the 1981 Berg Report and on which structural adjustment programmes were originally based. While certainly not a solution to the development problems of SSA, a recovery and expansion of the 'traditional' tradables sector, in conjunction with other policies for non-traditional agriculture and for manufacturing, is one important factor in reducing rural poverty and assisting the process of accumulation necessary to the achievement of longer-run strategies. Understanding the determinants of traditional tradables output will assist the development of appropriate policies for recovery and expansion in this crucial sector.

Notes

1   There is also the possibility of multicollinearity between the price variables. However, tests showed this not to be present.
2   The authors wish to acknowledge the assistance of Robin Bladen-Hovell with the processing of the data and his helpful comments on the model and its interpretation.

References

Belshaw, D., Lawrence, P. and Hubbard, M. (1999) 'Agricultural tradables and economic recovery in Uganda: the limitations of structural adjustment in practice', *World Development* 27 (4): 673–90.

Berg, E. (1993) 'Reappraising export prospects and regional trade arrangements', in D. Rimmer (ed.), *Action in Africa: The Experience of People Actively Involved in Government, Business, Aid*, London: Royal African Society/James Currey.

Haggblade, S., Hazell, P. and Brown, J. (1989) 'Farm–nonfarm linkages in rural sub-Saharan Africa', *World Development*, 17 (8): 1173–1201.

Helleiner, G.K. (1988) 'Comments on [Uma Lele's] "Comparative advantage and structural transformation: a review of Africa's economic development experience"', in G. Ranis and T.P. Schultz (eds), *The State of Development Economics: Progress and Perspectives*, New York: Basil Blackwell.

Himmelstrand, U., Kinyanjui, K. and Mburugu, E. (eds) (1994) *African Perspectives on Development: Controversies, Dilemmas and Openings*, New York: St Martin's.

Husain, I. (1993) 'Trade, aid and investment in sub-Saharan Africa', in D. Rimmer (ed.), *Action in Africa: The Experience of People Actively Involved in Government, Business, Aid*, London: Royal African Society/James Currey.

International Monetary Fund (1999) 'Structural adjustment in sub-Saharan Africa is focus of high-level regional gathering: Bank of Mauritius–IMF Institute Seminar', *IMF Survey*, 28 (13): 216–17.

Lloyd, T.A., Morgan, C.W., Rayner, A.J. and Vaillant, C. (1999) 'The transmission of world agricultural prices in Côte D'Ivoire', *Journal of International Trade and Economic Development*, 8 (1): 127–43.

McKay, A., Morrisey, O. and Vaillant, C. (1999) 'Aggregate supply response in Tanzanian agriculture', *Journal of International Trade and Economic Development*, 8 (1): 107–23.

Mkandawire, T. and Soludo, C.C. (1999) *Our Continent, Our Future: African Perspectives on Structural Adjustment*, Trenton, NJ/Asmara: Africa World Press for the Council for the Development of Social Science Research in Africa and the International Development Research Centre.

Mosley, P. (1996) 'The failure of aid and adjustment policies in sub-Saharan Africa: counter-examples and policy proposals', *Journal of African Economies*, 5 (3): 406–13.

Myint, H. (1980) *The Economics of the Developing Countries*, 5th revised edn, London: Hutchinson.

Ndegwa, P. and Green, R.H. (1994) *Africa to 2000 and Beyond: Imperative Political and Economic Agenda*, Nairobi: East African Educational Publishers.

World Bank (1981) *Accelerated Development in Sub-Saharan Africa* [the Berg Report], Washington, DC: World Bank.

World Bank (1989) *Africa's Adjustment and Growth in the 1980s*, Washington, DC: World Bank.

World Bank (1994) *Adjustment in Africa: Reforms, Results and the Road Ahead*, New York: Oxford University Press, for the World Bank.

22 Disadvantaged economies

Africa's landlocked countries

Ian Livingstone

22.1 Introduction

In appraising the problems and prospects of different African countries, it is important to acknowledge that their geographical circumstances differ: first, in respect of their natural resource bases and climate, and second, in respect of their location. It is necessary in particular to consider the special case of the landlocked countries: out of twenty-nine landlocked developing countries (LLDCs) in the world, fifteen are in sub-Saharan Africa (SSA).

In contrast, there has been a general tendency among economic analysts to assume that all economies should be able to pursue a similar development path, following standard prescriptions, and to appraise their performance using the same criteria – for example, the rate of GDP growth achieved – without regard to their initial and local circumstances. As Bloom and Sachs (1998: 5) comment: 'They treat economies as blank slates upon which another region's technologies and economic history may be grafted.' A number of economists are now beginning to re-emphasise the importance of geography in constraining and shaping development (see also Sachs, 1999; Krugman, 1991; and Landes, 1998).

We focus here primarily on one aspect of geography, namely location, and its effect on international transport costs. The transport cost handicap, raising the price of imported goods and lowering the price received domestically for exports, and their competitiveness, is related to the total distances involved, particularly on land, and to the fact that the countries concerned are landlocked, creating a dependence for transit on another, usually coastal, country.

22.2 The nature of international transport costs

The distances involved in most cases are extremely long (Table 22.1), for eight out of the fifteen countries in SSA between 1000 and 1700 miles. Apart from the costs associated with distance, however, a critical aspect is that inland countries lack jurisdiction: over investment in transport infrastructure between themselves and the sea; over the management and organisation of transportation facilities and industry in that sector; and over tariffs charged for services provided. Deficiencies in any of these areas can significantly increase international transit transport costs. There are thus *two* sources of such costs: the natural geographic cost burden and additional costs which are the outcome of these other deficiencies and are associated in part with lack of sovereignty over the relevant transportation route. This lack of sovereignty leaves the inland country potentially subject to monopoly or other power exercised by transit route owners and operators, as well as to charges, rules, regulations and controls imposed by transit country governments. The lack of LLDC power extends to the acquisition of the full information required for efficient transport operations along the transport network.

Table 22.1 Shortest distance to a seaport from capital or main city, SSA landlocked countries

Country	Distance (km)
Eastern Africa	
Burundi	1455
Ethiopia	700
Rwanda	1530
Uganda	1150
Western Africa	
Burkina Faso	990
Central African Republic	1710
Chad	1735
Mali	1250
Niger	1060
Southern Africa	
Botswana	1100
Lesotho	740
Malawi	815
Swaziland	220
Zambia	950
Zimbabwe	590

The costs associated with landlockedness may be either direct, comprising the actual increased costs of shipping goods to the point of destination, in either direction, or indirect: including interest charges on inventories and goods in transit; disruption of domestic production due to non-arrival of inputs; and loss of orders due to inability to effect reliable delivery of goods to market (Table 22.2). Such costs may stem from either management deficiencies within the transit transport sector or in some cases from specifically discriminating or unhelpful policies. The nature of the costs listed is self-evident except perhaps for the informal payments to facilitate the movement of goods through transport corridors, which are another aspect of lack of development. In Central Africa freight forwarders estimated that as much as 40–50 per cent of their fees were due to these and to costs of police escorts for sensitive goods. These 'transactions costs associated with transit transport' constituted between 6–30 per cent of total costs from the sea to Chad and the CAR (World Bank, 1992). Long distances from the port will specifically result in higher costs of imported petrol and diesel fuels, compared with other countries, with further effects on costs and prices.

A distinction can be made between potential regular costs as above and potential exceptional costs associated, for instance, with civil wars or sometimes hostile or disruptive policies, including border closures, deliberately pursued. Chaotic disruptions due to civil wars or political upheavals over the last two decades have affected international trade and transport in Zimbabwe, Zambia, Malawi (war in Mozambique, political relations with the Republic of South Africa), Burundi, Rwanda (war in Uganda), Uganda (political relations with Kenya), Ethiopia (becoming landlocked because of civil war in, and international war with, Eritrea), and all five West African LLDCs.

The extended lines of international communication faced by the LLDCs also render them more vulnerable to natural disaster, compared with coastal countries. In a number of developing countries major damage to infrastructure along transit transport corridors has been suffered as a result of El Niño. Heavy rains have caused major damage to rail and road communications

Table 22.2 Costs associated with landlockedness

Natural geographic cost burden:
- high cost of overland transport over long distances;
- payment for transit transport services, in hard currency;
- increased cost of imports;
- worsened terms of trade;
- reduced export competitiveness and export range/volume and reduced payments to factors for actual exports.

Additional cost elements
Potential regular costs:
- costs resulting from under investment in transit country infrastructure, rolling stock, etc., hence wear and tear on transport equipment, delays;
- high or discriminatory charges for transport or port services in transit country;
- costs of additional transhipment, loading/unloading;
- restrictions placed on routing of goods in transit;
- pilferage or damage to goods en route, raising insurance costs;
- delays associated with inadequate management of goods traffic in transit;
- inadequate communication between ports and LLDCs, absence of information on movement of goods and proper tracking systems;
- imposition of bureaucratic procedures and unnecessary or complicated documentation;
- preferential treatment of coastal country goods at port or elsewhere resulting in:
  — loss of potential orders requiring reliable delivery to market;
  — disruption of production due to unreliable delivery of inputs, requiring raised level of inventories;
  — tying up of capital in stocks and goods in transit.

Potential exceptional costs associated with civil wars, etc. in transit country or hostile or disruptive policies deliberately pursued, including border closures:
- major and extended supply and export disruptions;
- political and social costs of food insecurity;
- costs of maintaining additional 'insurance' transport corridors and of use of higher-cost routes.

---

along the Northern Corridor in East Africa, for instance. As a result of this and other factors, the transport cost of moving a container by road from Mombasa to Kampala, which used to cost US $1500, had increased in 1998 to as much as US $3500–4000.

Extended lines of communication also mean that in very many cases the transport chain has to be 'multimodal', combining successively sea, rail, road and perhaps river or lake transport along the route, requiring several unloading and reloading procedures, sometimes with repackaging of contents. This process of cargo transfer can be surprisingly expensive as a result of labour costs particularly. A multimodal transport chain will also involve a greater number of intermediaries, which also adds to costs.

Landlocked countries may also suffer from 'remoteness'. This should be distinguished from 'landlockedness'. It refers to the disadvantage of being away from main lines of communication such as shipping lanes and may thus apply as much to remote islands which are not regular ports of call as to inland countries which suffer from additional transport distance compared with coastal countries, particularly overland distance. This remoteness will also generate delays and poor communications with overseas markets and input suppliers. Inland countries, also because of being small and underdeveloped, will have problems of this type, which are not necessarily measured by distance from the nearest port. The reduced national level of total economic and in particular business activity may itself generate less-effective channels of external communication.

22.3 The size of the international transit transport cost handicap

A rough estimate of the direct costs can be made using balance of payments data (Table 22.3) which allow us to compare, on an aggregate country basis, payments for transportation and insurance with the value of exports of goods and services. Compared with a ratio of freight payments to exports of only 4.7 per cent for all developing countries in 1994, Rwanda had a ratio of 67 per cent and Malawi and Chad ratios above 50 per cent. One estimate (Amjadi *et al.*, 1996) for SSA put these payments as absorbing a substantial 15 per cent of total SSA earnings in 1990. The ratio for the landlocked SSA countries here, taken in aggregate, is 21 per cent. Thus a relatively large proportion of the foreign exchange earnings of African LLDCs is absorbed by transport payments to foreign carriers for transport services, constituting a huge drain on these earnings.

Since the landlocked countries are often 'price-takers' in world markets for their exports, these international transport costs usually cannot be passed on to overseas buyers and have to be paid for by reduced ex-works or 'farmgate' prices. Since most of the transport, warehousing, port and insurance services provided outside the country, within the transit country as well as beyond, are usually supplied by foreign enterprises, they have also to be paid for in foreign exchange, reducing the landlocked country's net foreign exchange earnings from exports. The most serious effect, however, is on the international competitiveness of African exports. The freight costs above have been compared with an average post-Uruguay Round tariff on all imports into the US of under 4 per cent (UNIDO, 1996: 113–15). Freight costs therefore act as a colossal tariff against exports.

Table 22.3 Transportation and insurance payments as a proportion of total exports of goods and services for African landlocked developing countries (LLDCs), 1994

Country	Transportation and insurance payments (US $ m)	Exports of goods and services (US $ m)	Ratio (%)
Rwanda	68.6	102.0	67.2
Malawi	213.7	384.8	55.5
Chad	98.6	190.0	51.9
Mali	183.6	386.5	47.5
Burundi	39.8	95.3	41.8
Uganda	173.3	500.6	34.6
CAR	58.7	179.0	32.8
Ethiopia	212.7	666.6	31.9
Niger	79.6	255.8	31.1
Burkina Faso	70.4	249.9	28.2
Lesotho	35.8	181.3	19.8
Zambia[a]	215.0	1,255.0	17.1
Zimbabwe	379.3	2,344.2	16.2
Botswana	192.7	2,064.5	9.3
Swaziland	24.2	842.8	2.9
Least developed countries[b]	3,344.0	21,547.0	15.5
All developed countries[b]	59,435.0	1,254,534.0	4.7

Source: UNCTAD (1998) based on *Handbook of International Trade and Development Statistics, 1994* and IMF, *Balance of Payments Yearbook, 1996*

Notes
a Ibid., 1991
b Ibid., 1993

The ratios above varied very widely among LLDCs, however, from as little as 2.9 per cent for Swaziland to the 67 per cent mentioned for Rwanda. It should be realised, therefore, that the international transit handicap is highly variable between countries. Comparison of freight costs with payment for imports is similarly revealing of the transport handicap (Table 22.4). For industrial countries freight costs were only 3.2 per cent of the value of imports and even among developing countries only 5.7 per cent. This compares with 21.0 per cent overall for the eleven landlocked countries in Africa which are also categorised as least developed. This in turn compares with 13.1 per cent for nineteen other countries in Africa which are least developed, but not landlocked, and is almost three times the figure for 'All other Africa'.

The transport cost handicap of the landlocked countries in relation to coastal neighbours is exacerbated by the fact that the extra distance over which their goods travel, whether exports or imports, is over land rather than ocean, overland travel being very much more expensive than by sea. Moreover, industrial development in developing countries, if they are not landlocked, is often concentrated around ports, minimising internal transport costs for imported materials used in manufacturing. Estimates made of the costs of shipping a container of export cargo from different African cities within landlocked countries to North West

Table 22.4 Freight and insurance costs relative to payments for imports, African LLDCs compared with world regions, 1994

Country/region	Imports of goods, f.o.b. (US $ b)	Freight payments (US $ b)	Freight as % of imports
Industrial countries	2,762.9	86.9	3.2
Europe, non-industrial	189.2	7.2	3.8
Developing countries	1,351.0	76.9	5.7
Africa:	80.5	7.3	9.1
11 least developed LLDCs	5.2	1.1	21.0
2 other LLDCs (Botswana, Zimbabwe)	3.4	0.4	12.0
19 other least developed countries	9.6	1.3	13.1
All other Africa	62.3	4.5	7.3

Source: UNCTAD (1998) derived from IMF balance of payments data

Table 22.5 Relative costs of land and sea portions of transport between selected African cities and north-western Europe, 1995[a]

Inland town (country)	Inland mode of transport	Inland transport cost (US $/TEU)	Associated port	Ocean transport cost (US $/TEU)	Share of inland cost as % of total
Bamako (Mali)	Rail	800	Dakar	1,100	42
Bangui (CAR)	Road	2,560	Douala	1,500	63
Kigali (Rwanda)	Road	2,500	Dar es Salaam	1,300	66
Bujumbura (Burundi)	Road	3,100	Dar es Salaam	1,300	70
Lusaka (Zambia)	Road	1,900	Dar es Salaam	1,300	59
Lilongwe (Malawi)	Road	1,600	Dar es Salaam	1,300	55
Lilongwe (Malawi)	Road	1,600	Durban	1,000	62
Lilongwe (Malawi)	Road	1,050	Nacala	1,100	49

Source: UNCTAD (1998)

Note
a Costs in US $ of 20-foot equivalent unit.

Europe in 1995 (Table 22.5) showed that the inland portion of costs were generally more than half the total (in one case as high as 70 per cent); this despite, of course, very much longer distances by sea. This indicates that any attempt to increase the competitiveness of exports from landlocked countries and, for that matter, to reduce the cost burden on imports needs to focus on the level and structure of costs accruing in the transit operation.

A factor that tends further to increase international transport costs for LLDC trade is the common imbalance in volumes between imports and exports, that for the former greatly exceeding the latter. This results in a prevalence of empty returns on the backhaul legs of return journeys, raising the average cost of return trips. This leads on to relatively higher freight costs on imports due to higher charges imposed in order to compensate for the empty returns and negotiated discounts on backhaul legs. The shortfall in exports in 1996 was over 19 per cent of imports for African LLDCs (32 per cent for all LLDCs), and in many cases greater than 40 per cent, compared with 5 per cent for all developing countries (Table 22.6).

Yet another factor adversely affecting transport costs is the low density of population in the LLDCs themselves. The population per square kilometre among African LLDCs in 1997 was only twenty-one, below ten in five cases. Higher costs over the domestic section of the route associated with reduced total volume and with distribution to/collection from a dispersed population add to already high external costs. Lower traffic levels will usually lead in turn to reduced infrastructural investment in road quality.

The figures given for international transport costs in the preceding tables do not give a full picture, however. First, they refer to oncosts for goods actually traded, not for *potential* exports. Countries which export relatively high-value, low-bulk commodities, such as coffee, will show up as having low payments for freight and insurance as a proportion of f.o.b. export values. In fact, coffee-exporting countries such as Ethiopia or Uganda would wish to be exporting many other goods, particularly manufactured goods, but are restricted or prevented from doing so in substantial part by the prohibitive cost of international transport: a more

Table 22.6 Trade imbalances in SSA countries, 1996

	Exports (US $ m)	Imports (US $ m)	Surplus/deficit	
			(US $ m)	% of imports
Botswana	3,231	1,735	1,496	—
Burkina Faso	424	545	−121	22
Burundi	40	127	−87	69
CAR	230	160	70	—
Chad	252	240	12	—
Ethiopia	417	1,358	−941	69
Lesotho	189	1,001	−812	81
Malawi	481	624	−143	23
Mali	439	764	−325	43
Niger	281	543	−62	18
Rwanda	60	257	−197	77
Swaziland	893	1,079	−186	17
Uganda	604	1,188	−584	49
Zambia	1,188	1,482	−294	20
Zimbabwe	2,403	2,819	−416	15
Total, African LLDCs	11,132	13,722	−2,590	19
All LLDCs	14,892	22,052	−7,160	32
All developing countries	1,534,800	1,616,234	−81,434	5

Source: *Handbook of International Trade and Development Statistics 1996–1997*, New York: UNCTAD

diversified export range would have higher transport cost mark-ups. This also means that actual percentage transport oncosts on the export side will vary quite widely among different LLDCs according to which commodities they export: less so in the case of imports, which will be mostly similar. Landlocked countries may, indeed, have lower freight cost proportions than their transit neighbours, as has been demonstrated in one research study (Livingstone, 1986: 11 and table 14), in part because the latter have been more able to diversify their export production.

It means also that percentage transport oncosts will be greater for countries which export agricultural and mining products, which are generally bulky relative to value, than for countries which export mainly manufactured products. Table 22.7 gives figures for one low-value, high-bulk commodity and one relatively high-value, low-bulk commodity imported into Malawi (fertilizer and motor car spares respectively) and exported from Malawi (sugar and tobacco respectively). While total transport oncosts are 8.5 per cent for imported motor car spares they are 21.5 per cent for imported fertilizer. For exported tobacco they are 5.4 per cent compared with 22.6 per cent for sugar.

22.4 Effects on landlocked economies

In the case of agricultural export commodities, there may be no choice but to absorb international transport costs and accept a reduced return to labour, the alternative being purely subsistence agriculture. If a high-priced export crop is involved, such as coffee, incomes may still be very satisfactory, and since there will not be the same problem of economies of scale in production – peasant farming units being entirely viable – small economic size of a country may not affect the viability of production. This may not hold in the case of manufacturing, where capital is mobile and able to seek alternative locations, not leaving labour the option of accepting a lower wage.

Platteau and Hayami (1998) have commented on the wide spatial gaps in prices received by primary producers even within a country as a result of transport costs. Thus a study of

Table 22.7 Components of transit transport costs, Malawi, selected commodities, 1982

Commodity	Low value				High value			
	Import: fertilizer		Export: sugar		Import: motor car spares		Export: tobacco	
	US $	%	US $	%	US $	%	US $	%
Value[a] (US $/ton)	200.00	100.0	160.00	100.0	3,500.00	100.0	3,500.00	100.0
Movement costs	23.08	11.5	23.18	14.5	73.87	2.1	82.96	2.4
Other inland transport costs[b]	10.76	5.4	8.86	5.5	72.09	2.1	34.46	1.0
Inventory costs[c]	9.17	4.6	4.14	2.6	151.56	4.3	70.54	2.0
Total transit transport costs	43.01	21.5	36.18	22.6	297.52	8.5	187.96	5.4
IC/TTTC[d]	—	21.3	—	11.4	—	50.9	—	37.5

Source: Derived from UNDP/UNCTAD, Malawi Country Report, Project RAF/77/017, 1983

Notes
a For imports, landed cost at port; for exports, f.o.r. in Malawi.
b Port handling charges, c&f agent fees, etc.
c Calculated as transit time in days × c.i.f. value of goods × 12% interest charges.
d Inventory costs as % of total transit transport costs.

Inventory costs are relatively more important for high-value products. Figures exclude additional inventory costs due to uncertainty.

internal marketing of agricultural products conducted in Tanzania in the early 1990s showed that prices paid to producers in the far south of the country were only half those paid to producers in markets nearer to Dar es Salaam (Santorum and Tibaijuka, 1992: 22). Comparing foodgrain marketing costs in Africa with those in Asia, Ahmed and Rustagi (1984) found that producers in African countries received only 30–60 per cent of the terminal market price compared with 75–90 per cent in the Asian countries, transport and associated costs being the main factor. If this applies within non-landlocked countries, it will apply even more strongly to the landlocked.

The following statement is equally applicable:

> When remoteness, underdeveloped infrastructure and inefficient transport services cause a doubling or tripling of the price of exportables at the African dockside . . . relative to the farm gate prices, and when a similar price rise occurs for importables between their delivery to an African port . . . and the point of consumption, a large part of the domestic economy is actually insulated from the impact of foreign trade . . . Most African economies are thus in a semi-closed state . . . [Consequently,] inasmuch as large segments of rural economies in SSA consist of nontradables, a significant share of rural primary resources such as labour and land may remain under-exploited, even if policy reform programmes are adopted that remove major macroeconomic barriers to trade . . . [Since] very large shares of household income derive from . . . items that are nontradables . . . many people in rural areas can remain underemployed for long periods of time.
>
> (Platteau and Hayami, 1998: 371–2)

Acceptance of lower returns to factors is associated with the need to maintain international competitiveness. It has been suggested (Porter, 1990: ch. 10) that international transport costs will not have the same effect on competitiveness in different kinds of economy. In particular, in advanced industrial countries, even if these are landlocked such as Austria or Switzerland, the economies are 'innovation-driven', competing in a wide range of industries producing unique products with particular or individual characteristics such that product quality, style, up-to-date nature and associated customer service are most important and transport costs less so.

On the other hand, in 'factor-driven' economies such as less-developed countries supplying internationally homogeneous crops and other primary products, any competitive advantage derives from factor conditions, such as agricultural land or low-wage labour. Here a large number of less-developed countries are often supplying the same commodity, each accounting for a small share of the world market and therefore operating as price-takers. Since all countries have to accept a given price in the world market, producers in the landlocked countries are either uncompetitive or have to absorb the difference in transport costs and accept returns which are lower than those secured by factors – capital or labour or both – in other countries.

In the case of manufacturing, with globalisation and the international mobility of capital, LLDCs are particularly affected by alternative locational costs, including international transport costs for inputs and outputs. While the LLDCs have the advantage of low-cost labour, there is no obvious reason to locate in the landlocked country rather than in the neighbouring transit coastal country where cheap labour is often in equally elastic supply. The import content of manufacturing will reinforce any tendency to locate at ports and thus in coastal countries. Moreover, the domestic market of the latter is almost invariably the larger of the two. And there will always be many coastal countries – and islands – to choose from. Landlocked countries could thus be at the end of a long queue from the point of view of internationally footloose industry, with many implications for their long-run development.

This will apply also to the location of industry geared to a local regional market as well as overseas markets. A good description is given of a West African case (Selwyn, 1993: 5):

> Thus in the trade between Upper Volta (now Burkina Faso) and the Ivory Coast (Côte d'Ivoire), 80 per cent of Upper Volta's exports to the Ivory Coast in 1969 consisted of live animals and animal products, and a further 15 per cent of vegetable products. Ivory Coast's exports to Upper Volta were far more diversified, including cement (15.6%), wood and cork products (10.3%), textiles (9.6%), chemical products (9.2%), transport products (8.6%), food, drink and tobacco products (5.6%) and base metal products (4.9%). This structure clearly shows the peripheral relation of Upper Volta to the Ivory Coast economy.

Some elements of this situation exist, for example, in the relations between Uganda and Kenya. Despite the strong efforts made to distribute industry among the East African Community countries in the 1960s, Uganda has always had a major trade deficit in manufactures with Kenya, on which it has since been unable to make an impression, quite clearly due to its landlocked disadvantage (Livingstone, 1998).

In analysing the disadvantage of LLDCs in industrial development, we need to consider two elements, in fact. There is, first, the additional transport cost factor affecting the possibilities for establishing export industry and, even for resource-based industry, the returns to labour and capital from whatever industry is established, taking account also of the additional transport cost in respect of the import content of the manufacturing concerned.

Second, there is the locational factor, related to increasing returns and economies of scale. Krugman (1991: 5) has suggested that we should 'Step back and ask, what is the most striking feature of the geography of economic activity? The short answer is surely *concentration*.' The increasing returns and external economies of scale underlying this concentration, within countries as well as between countries, are associated with the pooling of labour; the generation of a pool of skilled/experienced labour on which existing and any new firms can draw; scale economies in the supply of specialised inputs and services to a sufficiently well-established industrial sector; and, third, 'technological spillovers': the copying and development of ideas and methods among a group of firms in the same line of activity.

This factor will augment any initial transport cost factor in favour of industry established within the coastal country near the port or elsewhere within it. But it will be further reinforced by the relative sizes of landlocked and transit countries. As noted previously, many of the LLDCs are very small in size, in terms of GDP. The resultant small domestic market will make the standard import-substitution strategy that other countries have followed even more problematic; but, because domestic production can often provide a foundation for subsequent export promotion, it can also affect the development of potential for export. Exports to neighbouring countries are often developed through such a sequence. More generally it has been stated (Srinivasan, 1985: 1) that:

> even if there are not constraints on size of the market for a product because of the possibilities for export, to the extent that penetration into foreign markets depends on the experience gained in producing and selling in the domestic market, smallness of the latter may preclude export development.

If the domestic market in the transit country is many times larger than that in the landlocked, therefore, the amount of domestic market-oriented industry established will be much greater in the former, and industries which require a combined market, because of economies of scale, will add to this. External economies/increasing returns will further extend the locational advantage over time for industry as a whole.

A recent econometric study of the factors affecting PPP-adjusted GDP per capita growth in Africa over the period 1965–90 carried out by Sachs and Warner (1997) found that landlockedness was a significant negative factor, though the most important influence was of policy factors determining the degree of openness of the economy. The study found that growth among landlocked countries could be expected to be on average 0.58 percentage points below those with access to the sea. This landlockedness factor is referred to also in a recent report on African competitiveness (World Economic Forum, 1998: 27).

A problem facing the small least developed countries as a whole, it has been suggested (UNIDO, no date: 56), is that they 'are utterly lacking in the institutional structure needed for export promotion policies'; and that transnational corporations, with their extensive marketing network throughout the world and their acquired skills in this direction, might be the best means, or one means, of overcoming this disadvantage. Transnational corporations may be persuaded to locate in a country for offshore production and export, taking advantage of cheap labour or tax concessions, independently of any domestic market. But undoubtedly the existence of such a market could provide a reason for selecting one location rather than another: Nigeria, for example, rather than Niger.

Given all the above, manufacturing for export among the LLDCs, even those in Latin America, frequently amounts to no more than a few percentage points out of total exports, which are overwhelmingly of primary products. Such manufactured exports are often resource-based, such as wood products, vegetable oils, animal feeding stuffs (cattle cake from oilseeds) or manufactured tobacco, though some countries are able to export small quantities of woven fabrics (Table 22.8).

Table 22.8 Leading exports and contribution of manufactured exports for selected LLDCs, 1993–4

Country	Leading exports	Manufactured exports and percentage share of total domestic exports
Burkina Faso	Cotton, gold, vegetables, etc.	Pulp and waste paper (0.39)
Burundi	Coffee, gold, tea	Cotton fabrics, woven (0.62)
Ethiopia	Coffee, hides/skins, leather	Petroleum products, refined (1.47) dyes, tanning products (0.48)
Malawi	Tobacco, tea, sugar	Cotton fabrics, woven (2.20) outerwear knit (1.20) veneers, plywood (0.52)
Rwanda	Coffee, tea, gold, base metal ores, hides/skins	Wood shaped, sleepers (5.79)
Uganda	Coffee, hides/skins	Wood shaped, sleepers (5.79)
Zimbabwe	Tobacco, maize	Pig iron (6.60) cotton fabrics, woven (1.74) men's outerwear (1.47)
Bolivia	Base metal ores, gold, silverware, jewellery, tin	Gas, natural and manufactured (9.43) wood shaped, sleepers (6.63)
Paraguay	Oilseeds, cotton, leather, meat, animal feed	Wood shaped, sleepers (5.79)

Source: *Handbook of International Trade and Development Statistics, 1995*, New York: United Nations, table 4.3D

22.5 Manufacturing in LLDCs: some options

Two categories of response are called for in relation to the situation in which the LLDCs find themselves. The first is to accept the existence of the transport cost handicap and its effects on comparative advantage, adjusting industrial development strategy accordingly. The second is to take steps to counter the physical and non-physical factors that raise transit transport costs themselves.

In fact, the LLDCs generally suffer from a set of disadvantages: small economic size, in terms of both population and GDP, affecting the size of the domestic market; low density of population, affecting the cost of internal infrastructure provision and again the effective domestic market; and weak resource bases, in a number of cases, with extensive semi-arid areas, all affecting the country's development options, made worse by the transport cost factor. Strategies need to be tailored to optimise the development path taking into account these constraints and the opportunities which exist, taken together.

Efforts can be made, second, to expand trade regionally with neighbouring countries, exchange opportunities here being affected by transport costs to a lesser degree. In many cases this trade is nevertheless constrained by undeveloped transport networks but offer more potential if transport systems can be improved. Participation in schemes for regional economic integration will also bring about a larger market and opportunities to expand specialisation and trade regionally, though it should be observed here that manufacturing industry tends to polarise within the common markets which have been formed among developing countries, the transport cost factor continuing to favour coastal and more developed country locations.

In manufacturing, the freight factor points to the need to carry forward as far as possible the domestic processing of resource-based export products, for the obvious reason of maximising domestic value added, but also to reduce bulk and thus the freight factor. This also points more generally to the encouragement of resource-based industries. Within manufacturing, the emphasis should otherwise be on high value-to-bulk products for export. This is related also to taking advantage of cheap labour through labour-intensive production: thus craft items and leather goods are labour-intensive as well as incorporating local skills, contributing to higher value-to-bulk ratios. In certain cases these can even take advantage of air freight. Another possibility is where (relatively) high value can be combined with a natural resource advantage, as in the seasonal export of cut flowers.

In identifying possible products for export, account needs to be taken of import content, including that of capital goods used in production, again because of its relation to value added, but also because of the freight factors which increase import costs. These will be very much greater than for coastal country locations. Again, this favours labour-intensive rather than capital-intensive production, where machinery is imported. Landlockedness, by reducing the attractiveness of a country to foreign investment, increases the scarcity of capital relative to labour, also affecting the relative advantage of capital- versus labour-intensive activities and large-scale versus small-scale industry.

At the same time, maximisation of domestic value added together with minimisation of transport cost content may be secured in some situations by importing and distributing a basic input. Thus sheet metal may be imported and converted locally into basic items such as water tanks, suitcases or pipes, using appropriate scales of production as determined by the size of local market. Very often within-country costs of transport for such items are substantial and can be reduced by producing them through a dispersed local metalworking industry. A further example is the import of cloth used as the basis of a widespread informal sector tailoring industry, common to many developing countries, generating more than proportionate value-added and significant rural employment.

In contrast to goods with high import content, items should be selected for domestic production which maximise local content: thus, instead of imported enamelware, domestically produced cooking pots and containers can be used. Such production already exists on a large scale as a response to comparative advantage, the more so the more remote and distance-affected is the country or area. A major example of a commodity produced with maximum local content (as well as with labour-intensive construction techniques) is rural housing, while others are furniture, baskets and mats (serving as carpets) and other household equipment and agricultural transport (ox carts). Production of appropriate goods may be linked more generally to the promotion of small-scale and informal sector manufacturing. Two conditions that obtain in particular in small landlocked countries, a restricted domestic market and high external transport costs raising the prices of imported goods, create exceptionally favourable economic conditions for such production.

Developing and maintaining resource-based industries, industries that are capable of sustaining a whole range of basic needs, require that the resource base itself is maintained and, as appropriate, expanded. In a number of African countries, such as Malawi and Kenya, trees established in small plantations are emerging as a cash crop, producing building poles and other materials for local use.

In relation to export manufacturing industry, it may be sensible to attempt to specialise in some specific activities. One argument for this would be in a small economy to concentrate the limited amount of savings available for manufacturing investment. A more important reason, however, is that based on possible external economies at the level of the industry, either in production or in marketing and sales promotion, which favour specialisation. The fact that developing countries that have already broken into manufacturing export markets have often done so in particular lines, initially at least, offers some evidence of this. These economies, or 'collective efficiency', derived by a cluster of small firms engaged in one line of activity or one 'industrial district', have been described (Schmitz, 1989) as deriving from 'flexible specialisation'. Specialisation of this sort in an identified market niche appears particularly suitable for small landlocked countries, since (a) local entrepreneurs are more likely to be able to establish relatively small firms; (b) the activity as a whole benefits from divisibility, allowing progressive development from any initial scale; and (c) the products themselves are generally low-bulk and labour- and, in some cases, skill-intensive.

It should be mentioned that some LLDCs are themselves transit countries for other countries further inland and may be in a good position to develop a range of manufactured exports directed to those other countries, particularly products which are resource-based. An example is sugar production in Uganda for export to Sudan and Burundi. A number of other countries in Africa which are landlocked, including Zimbabwe and Botswana, have made a certain amount of progress in producing manufactured goods for export, and this has been particularly to neighbouring countries, reflective also of transport costs to places further afield (Riddell, 1990).

22.6 Reducing the size of the international transport cost handicap

22.6.1 *Physical barriers*

It is not merely the length of the distance from landlocked country to coast that makes transportation costly, but the state of the infrastructure along the way, on the one hand, and the organisation – or disorganisation – of cargo flow on the other. Here there have been promising developments, in Asia and Latin America also, through a focus, with UNCTAD encouragement, on 'transport corridors'. In Eastern Africa, for example, a 'Northern Corridor' inland from Mombasa, has been identified, together with a 'Central Corridor' west from Dar es Salaam.

The poor state of transport route infrastructure is related in the first place to the low level of per capita income among the countries involved. Because of the high capital costs of regional infrastructural projects, and what will quite likely be longer- rather than shorter-term returns, these are particularly deserving of, and have received, attention from multilateral donors. While most of the ambitious regional integration schemes which have been initiated among developing countries have promised much more than they have achieved, particularly in the generation and wider distribution of manufacturing industry, improvements in transport infrastructure have produced more concrete and reliable benefits in the form of increased local and regional trade among partner states, including the participating LLDCs.

While both transit and landlocked countries have an interest in maintaining an efficient transport corridor between them, it remains the case that, particularly where the volume of trade generated by the landlocked country is small, the transit country's interest in maintaining the quality of infrastructure as far as the joint frontier is often less strong. Many examples exist where sections of road or rail systems fail to be fully maintained over sections within the transit country critical to the landlocked. This gap can be closed if, with encourage-ment or persuasion from international agencies, the transit country can be persuaded to increase its interest or, recognising the divergence of interest, multilateral agencies step in with the necessary capital to fund the project required on suitably soft terms.

As we have observed above, the transport costs calculated from port to capital city in the LLDC may understate international transport costs from domestic producer to overseas market or from overseas supplier to domestic consumer, due to the undeveloped nature of the internal road network and distribution/collection systems in the LLDC, consequent upon the usually low level of income per capita in these countries. More awareness is required of this aspect and of the need to provide external assistance.

A general problem along transport corridors is that of road maintenance, despite national and international attempts to regulate axle loads, as well as maintenance of rail systems. In many cases this is associated with a lack of a national strategy for maintenance, corrective action being left too long, while costs and problems of repair mount. Since this can affect all countries using a given route, the effectiveness of maintenance programmes becomes the legitimate concern of the Corridor as a whole and its Authority, where this exists.

22.6.2 Non-physical barriers

Freight charges are only part of the total international transportation costs. Costs are swollen further by handling charges at terminals, ports, warehouses and inland depots, by storage costs at any of these points, by pilferage and damage to goods, costs of customs clearance, hold-ups at border crossings, delays due to documentation problems or generally cumber-some government regulations and procedures, inefficiency and/or monopoly among carriers, and the like. If all these problems can be tackled, the transport cost disadvantage of the LLDCs can also be reduced.

Even more than with physical factors, these call for cooperation and joint action among the countries concerned. This can be within the context of a regional grouping such as the Southern Africa Development Community (SADC) or of a recognised transit transport corridor. A transit transport corridor authority can play an important and continuing role in eliminating problems and monitoring the smooth working of agreements, reducing red tape and facilitating cargo movement along the relevant routes. Such an authority, of course, must be given strong backing by the governments involved. Here it is important that the coastal countries are as persuaded of the benefits of facilitating transit trade as are the landlocked.

Administrative obstacles can be reduced by simplifying transit documentation, standardising road transit regulations such as axle-load regulations, harmonising road charges, introducing transit bond guarantee schemes and multicountry insurance, coordinating border post arrangements and establishing dry ports. Progress has been made in these directions, particularly in Southern and East Africa, although much more needs to be achieved. Even where agreements have been signed, they are often not implemented or only partially so. Thus implementation of the ECOWAS/UDEAC agreements in West Africa has been left to the political will of individual states. Customs formalities remain laborious, even those between neighbouring landlocked states themselves. The freight market is regulated, with bilateral agreements normally reserving two-thirds of the freight for carriage by vehicles from landlocked countries. As a result goods at the maritime ports may have to wait several days for vehicles from landlocked countries, even though there are coastal country vehicles available.

Border posts in Southern Africa have been reported as still a major obstacle to international road transport within the region, producing continued delays. The same problem along the Northern Corridor in East Africa is described as follows:

> The cumbersome customs procedures at the ports are exacerbated by problems related to organisation of customs services in respect of road traffic at the border posts with high traffic levels such as Busia, Malaba, Isebani and Rusumo. These offices do not have appropriate infrastructure to serve the increasing volume of traffic and the customs personnel are inefficient due to lack of adequate training and motivation. The locations of some of the offices are inappropriate, and in many cases the working hours of adjacent customs offices vary. This results in prolonged waiting times at border posts . . . the same formalities completed at one exit post are repeated at the next entry post of the neighbouring country with all the monetary and time costs involved. These factors, which result in traffic jams at the border posts and increased costs and transit times, are exacerbated by the lack of adequate telecommunications network which would allow all the customs offices to communicate amongst themselves, as well as with their central administrations.
>
> (ECA, 1996: 25)

In addition to the above, the transport problems of importers and exporters are widely compounded by inadequate communication links between ports and inland countries and between ports and overseas markets. Problems of communications lead to enormous delays in moving cargo in and out of ports as a result of irregular information regarding likely arrivals and departures of cargo. Proper telecommunications facilities are a basic necessity for trade and for development generally. Important possibilities exist also in the application of new information technologies in tracking cargo along transport routes, thus allowing clients and clearing and forwarding agents to know precisely the location of consignments in transit, reducing the unpredictability of delivery, which is itself a handicap, by allowing port or railway managements to be more efficient and coordinated, and by reducing the need for customs authorities to hold up consignments. Adoption of these technologies has been actively promoted in different countries, with significant progress in Southern Africa.

In addition, even without heavy investment in infrastructure, significant reductions in transit costs and in expensive delays to cargo can be achieved as a result of improvements in management efficiency in the various institutions which together constitute the transport corridor. Wide variations in efficiency and labour productivity that are observable at different ports indicate the scope that exists for improving handling operations. As an example of the effect simply of improved management, organisational changes carried out in the Central Corridor in Tanzania under a railway restructuring programme, along with some additions to equipment, brought major reductions in transit times in the railway system there.

In some cases greater involvement of the private sector in supporting transit operations can increase efficiency, in the first place by supplying finance to improve or extend facilities but also directly through the introduction of more commercial approaches to the management of the related operations. Thus, rather than itself maintaining responsibility for a whole range of port activities which are in fact divisible, a port authority can hive off some of these, stevedoring for instance, to different competing companies. Where stakeholders, whether chambers of commerce, shippers' councils or other private trading or transport organisations, can be involved in managing specific service operations, these will have a direct interest in providing good service at low cost, often in response to bottlenecks which they have identified. An appropriate vehicle is a formally established transport facilitation committee, which can serve as a forum in which all transit traffic agents, public and private, can participate. Such a committee can also reconcile and balance the interests of control agents, such as customs or central banks, concerned with safeguarding the collection of tax revenue or conserving foreign exchange, on the one hand, and the business community, concerned with securing speed and low-cost delivery of cargo to its destination, on the other.

One quite different factor which raises costs, potentially for both landlocked and transit countries and for donors who are providing finance for infrastructure, is related to the lack of sovereignty which the LLDCs have over their transport route and therefore their vulnerability to political or monopolistic or inefficient action by their coastal neighbour. This has led many to wish to develop an alternative, insurance, route. From the point of view of the LLDC, a problem is that very often one particular route is the most direct and, potentially, the cheapest. Developing an alternative route may involve further new and expensive infrastructural investment. This is a problem also for multilateral donors: in considering whether to make available the very substantial capital funds that may be required, they need to gauge the net benefit to the region of investing in a second route, taking into account the loss of business sustained by one party against the gain in business accruing to a second, as well as the improved security provided to the landlocked country. While due cognisance needs to be given to reducing the vulnerability of the latter, donors would wish to avoid the proliferation of costly alternative routes, suffering from excess capacity.

The benefit to the landlocked country is not only the insurance provided but also the increased efficiency induced on the initial corridor by the competition provided. A dilemma exists here for the transit country contemplating further investment in facilities or changes in organisation to improve services in that, while such improvements might be justified if a given level of traffic were to be guaranteed, the existence of choice on the part of the LLDC makes routing decisions and the projected traffic volume less predictable.

By and large, investments in developing alternative routes have been made piecemeal and not on the basis of full analysis of costs and benefits, in part because of the perceived divergence of interest among the parties involved. Further data collection and economic analysis are clearly needed, in specific regions, relating to the level and distribution of costs and benefits from diversification and from alternative transit transport routes, based on realistic projections of traffic requirements.

Where landlocked and transit countries cooperate closely, both can benefit from an efficiently run, low-cost transport corridor. Apart from the use of the transport corridor by the transit country itself, it does secure valuable foreign exchange earnings, income and employment from the transport services provided by its citizens, which it should seek to maximise, as well as possible export opportunities with its neighbour. In addition, greater expansion of volumes of traffic on one corridor enables the countries to realise greater economies of scale in their ports, road, rail and other services. For this reason it is sensible to foster a transport corridor concept under which joint interests can most easily be pursued.

References

Ahmed, R. and Rustagi, N. (1984) 'Marketing and price incentives in African and Asian countries: a comparison', in E. Dieter (ed.), *Agricultural Marketing Strategy and Price Policy*, Washington, DC: World Bank.

Amjadi, A., Reincke, U. and Yeats, A. (1996) *Did External Barriers Cause the Marginalisation of Sub-Saharan Africa in World Trade?* Policy Research Working Paper 1586, Washington, DC: World Bank, International Economic Department, March.

Bloom, D.E. and Sachs, J.D. (1998) 'Geography, demography and economic growth in Africa', Harvard University, mimeo.

Economic Commission for Africa (ECA) (1996) *The Assessment of Implementation of International Transit Facilitation Agreements along Selected Transport/Corridors in Africa*, ECA document TRANSCOM/1103, Addis Ababa: Economic Commission for Africa.

Hayami, Y. and Aoki, M. (eds) (1998) *The Institutional Foundations of East Asian Economic Development*, Proceedings of the International Economic Association Conference, Tokyo.

Krugman, P. (1991) *Geography and Trade*, Belgium: Leuven University Press, and Cambridge, MA: MIT Press.

Landes, D. (1998) *The Wealth and Poverty of Nations: Why Some are Rich and Others so Poor*, New York: W.W. Norton.

Livingstone, I. (1986) 'International transport costs and industrial development in the least developed African countries', *Industry and Development*, 19, UNIDO.

Livingstone, I. (1998) 'Developing industry in Uganda in the 1990s', in H.B. Hansen and M. Twaddle (eds), *Developing Uganda*, Oxford: James Currey.

Platteau, J.-P., and Hayami, Y. (1998) 'Resource endowments and agricultural development: Africa versus Asia', in Y. Hayami and M. Aoki (eds), *The Institutional Foundations of East Asian Economic Development*, Proceedings of the International Economic Association Conference, Tokyo.

Porter, M. (1990) *The Competitive Advantage of Nations*, New York: Free Press.

Riddell, R.A. (1990) *Manufacturing Africa: Performance and Prospects of Seven Countries in Sub-Saharan Africa*, London: Heinemann Educational Books.

Sachs, J. (1999) 'Helping the world's poorest', *The Economist*, 14 August.

Sachs, J. and Warner, A.M. (1997) 'Sources of slow growth in African economies', *Journal of African Economies*, 6 (3).

Santorum, A. and Tibaijuka, A. (1992) 'Trading responses to food market liberalisation in Tanzania', *Food Policy*, 17.

Schmitz, H. (1989) 'Flexible specialisation: a new paradigm of small-scale industrialisation', Discussion Paper no. 26.1, University of Sussex, Institute of Development Studies.

Selwyn, P. (1993) 'Industrial development in peripheral small countries', Discussion Paper no. 14, University of Sussex, Institute of Development Studies.

Srinivasan, T. (1985) 'The costs and benefits of being a small, remote island, landlocked or mini-state economy', World Bank Discussion Paper, Development Policy Issue Series, Washington, DC: World Bank.

UNCTAD (1998) *Development in the Landlocked Developing Countries: Tackling the International Transport Cost Disadvantage*, Geneva: UNCTAD.

UNIDO (1996) *The Globalization of Industry: Implications for Developing Countries beyond 2000*, Vienna: UNIDO.

UNIDO (no date) *A Strategy of Industrial Development for the Small, Resource-poor, Least Developed Countries*, Industry and Development, no. 8, Vienna: UNIDO.

World Bank (1992) *Transit Traffic Facilitation in Sub-Saharan Africa: Review of Bank Assistance, 1960–1990*, Washington, DC: World Bank, Operations Evaluation Department.

World Economic Forum (1998) *The Africa Competitiveness Report 1998*, Geneva: World Economic Forum.

23 Creating a sustainable regional framework for development

The Southern African Development Community

Carolyn Jenkins and Lynne Thomas[1]

23.1 Introduction

Globalisation and regionalism are simultaneous phenomena of the late 1900s. To some extent regional integration can be seen as complementary to the process of globalisation. For many African countries, however, there are serious concerns about small markets, the fear of marginalisation in a world increasingly dominated by powerful trading blocs, and the fear of the costs of unilateral liberalisation – especially when the large world players are most protective of sectors in which African countries might feasibly compete. Many African countries have therefore seen regional integration as an alternative to unilateral trade liberalisation, providing some of the benefits of larger markets while limiting the extent of exposure to external competition.

Regional cooperation initiatives are not new in Africa. However, while successful examples of cross-border cooperation do exist,[2] Africa's record of creating and sustaining regional frameworks is generally poor in spite of their political appeal. A range of studies indicates why this is the case.[3] Many schemes were designed without regard for members' incentives to comply; implementation has sometimes not been feasible, as countries have overlapping and incompatible memberships of different regional arrangements; and members have frequently substituted non-tariff barriers for tariffs against each other. Domestic economic policies have also undermined the effectiveness of African trade integration schemes. Moreover, the structure of demand and production is too similar across African countries to generate substantial trade creation. This suggests that any union – except, possibly, one involving South Africa – will be too small for economies of scale to be realised.

At present, there are several initiatives being pursued across Africa aimed at promoting regional integration. The Southern African Development Community (SADC) is, arguably, more likely than others to provide the basis for successful economic cooperation, due to the participation of South Africa, the continent's largest economy – accounting for nearly one-quarter of total African GDP. Part of the problem facing most African regional groupings is the lack of a large, more developed partner to provide both a significant regional market and a source of external capital and expertise, particularly in regionally integrated production processes. The involvement of South Africa in SADC has alleviated this constraint to some extent, by improving the potential for cross-border trade and investment with a relatively large and more developed neighbour.[4]

SADC is not the only regional integration initiative in which Southern African countries currently participate (Table 23.1). Many are members of the Community of Eastern and Southern Africa; others are involved in the Cross-Border Initiative (CBI); and a small subset of members are involved in the long-standing Southern African Customs Union (SACU)

Table 23.1 Membership of regional groupings in Southern Africa

SADC	COMESA	SACU	CMA	EAC	CBI
Angola	•				
Botswana		•			
DRC					
Lesotho		•	•		
Malawi	•				•
Mauritius	•				•
Mozambique					
Namibia	•	•	•		•
Seychelles	•				
South Africa		•	•		observer
Swaziland	•	•	•		•
Tanzania				•	•
Zambia	•				•
Zimbabwe	•				•

Note
SADC = Southern African Development Community; COMESA =
Community of Eastern and Southern Africa; SACU = Southern African
Customs Union; CMA = Common (Rand) Monetary Area; EAC = East
African Community; CBI = Cross-Border Initiative.

and the Common (Rand) Monetary Area (CMA). The existence of overlapping membership of regional initiatives is common across Africa and provides a confusing picture of priorities. This is one reason why some of these initiatives have not been sustainable.

In this chapter, we focus on SADC as an example of regionalism in Africa with potential to create a sustainable framework for economic growth and development. The Community's history, current structure, membership and key economic indicators are briefly reviewed. The impact on the region of the trend towards globalisation is discussed, and the importance of regional trade integration in Southern Africa is examined, together with SADC's possible role in broader trade initiatives. Finally, some of the potential barriers to continued economic cooperation in Southern Africa are highlighted. We conclude that policy coordination and outward-orientation are key challenges facing SADC in promoting sustainable growth in the region.

23.2 An overview of SADC

SADC evolved out of the Southern African Development Coordination Conference (SADCC) established in 1980. The latter's objectives were to promote self-reliance in the region and to encourage regional cooperation in development projects. The original members of SADCC sought to reduce their dependence on the rest of the world and in particular on their then hostile neighbour, South Africa. SADCC worked rather well in bringing regional political leaders together, and as a means for procuring foreign aid, although it failed to achieve the broader objective of reduced economic dependence on South Africa.

In 1992 the SADCC broadened its concerns to facilitating regional economic integration and became the Southern African Development Community. The 1992 Windhoek Treaty laid out SADC's aspirations for regional cooperation. Integration initiatives were to be given content in a series of protocols which would provide a framework for the negotiation of specific programmes for greater cooperation. The overarching aim was to build a community that could compete globally, with regional integration yielding balanced economic growth and development for the member states.

SADC operates through a variety of institutions. The Summit, made up of heads of state or government, is the ultimate policy making body. The principal executive institution is the Secretariat located in Botswana. A Council of Ministers, comprising ministers from each member state, usually with responsibility for the economy, is responsible for monitoring the progress of SADC and the appropriate implementation of regional policies. Much of the work of SADC is carried out through sectoral Commissions and Coordinating Units, which are led by specific member states and operate through government institutions. For instance, the Finance and Investment Sector Coordinating Unit is located in South Africa's National Treasury; the SADC Industry and Trade Coordinating Division is in Tanzania's trade ministry. These institutions cover a wide range of sectors reflecting SADC's objective of widespread regional cooperation, as illustrated in Table 23.2.

The current membership of SADC is set out in Table 23.3, together with key economic indicators. The original membership of SADCC consisted of nine countries. Namibia joined in 1990; South Africa and Mauritius joined SADC in 1994; and the Democratic Republic of Congo and Seychelles joined in 1998.

Total formal-sector output in SADC amounted to US $186bn in 1998. The figures presented in Table 23.3 highlight the large disparities that exist across the national economies in the region – in terms both of the size of economies and of relative wellbeing (as measured by GNP per capita). More than two-thirds of the regional economy is accounted for by South Africa, which contains around 21 per cent of the SADC population. Angola, DR Congo, Tanzania, and Zimbabwe each represent between 3 and 4 per cent of total GDP. These economies contrast with Lesotho, Malawi, Mozambique and Swaziland, each of which represents around 1 per cent or less of the SADC total. Mauritius and South Africa both have GNP per capita[5] which is, in purchasing power parity terms, at least fifteen times that of the poorest economies, Malawi and Tanzania.

Table 23.2 Sectoral responsibilities in the Southern African Development Community (SADC)

Member state	Sectoral responsibilities
Angola	Energy
Botswana	Agricultural research
	Livestock production and animal disease control
Lesotho	Water
	Environment and land management
Malawi	Inland fisheries, forestry and wildlife
Mauritius	Tourism
Mozambique	Culture and information
	Transport and communications
Namibia	Marine fisheries and resources
South Africa	Finance and investment
	Health
Swaziland	Human resources development
Tanzania	Industry and trade
Zambia	Mining
	Employment and labour
Zimbabwe	Food, agriculture and natural resources
	Crop protection

Source: Web site of the SADC: http://www.sadc.int (accessed January 1999)

Table 23.3 SADC indicators, 1998

Country	GDP (in US $ m)	Population (m)	GNP per capita (US $) purchasing power parity
Angola	7,472	12	999
Botswana	4,876	2	5,796
DRC	6,964	48	733
Lesotho	792	2	2,194
Malawi	1,688	11	551
Mauritius	4,199	1	8,236
Mozambique	3,893	17	740
Namibia	3,092	2	5,280
Seychelles	535	0.1	10,185
South Africa	133,461	41	8,296
Swaziland	1,221	1	4,195
Tanzania	8,016	32	483
Zambia	3,352	10	678
Zimbabwe	6,338	12	2,489
Total	185,899	191	

Source: *World Development Indicators 2000*, Washington, DC: World Bank

During the 1990s (and earlier), the disparities in economic welfare which are evident in Table 23.3 were accompanied by a considerable divergence of macroeconomic policy frameworks and performance, with stronger growth performances apparently linked to more conservative economic policies and more liberal trade environments (see Jenkins and Thomas, 1998; 2000).

Table 23.4 contains summary information on average economic performance and macroeconomic policy indicators between 1990 and 1998, comparing economies that can be described as being relatively more stable during this period with those that have exhibited greater instability. While this choice is somewhat arbitrary, the economies classified as having experienced instability are generally those that have exhibited some combination of: volatile swings in annual growth (ranging from significantly negative to positive numbers); relatively high consumer price inflation (in excess of 20 per cent); and large budget deficit to GDP ratios (in excess of 10 per cent). Seven economies are classified as having experienced relative stability; seven have exhibited instability.

Very few Southern African countries managed to keep their deficit to GDP ratios below 5 per cent in the first half of the 1990s, although there have been notable improvements for some countries in recent years. The problem has been particularly acute in those economies exhibiting evidence of macroeconomic instability: in these countries, the external debt to GNP ratios are high and, in many cases, rising. For these economies, weak fiscal stances have undermined the goals of monetary, financial and trade liberalisation. However, these problems are not true of all countries in the region, some of which are liberalising successfully in the context of falling public-sector deficit ratios. Here, inflation rates are falling and economic growth, while not meteoric, is positive and steady.

The more stable and open economies, on average, grew faster. Average annual growth in this period was 4.3 per cent, compared to 2.7 per cent in the weaker group of economies (excluding DR Congo, which is a significant outlier). This is consistent with recent findings for developing countries generally and Africa specifically (reviewed in Collier and Gunning,

Table 23.4 National economic performance in SADC, average 1990–8

	Real GDP growth	Range of annual growth rates	Consumer price inflation	Budget deficit (excl. grants) to GDP	Exports + imports to GDP	Investment to GDP	Savings to GDP	Aid as % of imports	External debt to GNP
Economies that have experienced relative instability									
Angola	0.1	−23.8 to 11.6	994.2	−24.5	115.2	18.9	22.1	8.0	293.8
DRC	−5.1	−13.5 to 0.7	2,089.0	—	43.8	7.0	8.6	12.3	186.3
Malawi	3.8	−10.2 to 14.7	33.6	−14.5	62.7	17.7	4.8	54.2	110.4
Mozambique	5.7	−8.2 to 12.4	44.6	−14.1	62.3	27.2	−0.4	89.4	290.6
Tanzania	3.0	−8.9 to 12.2	24.2	−2.3	53.0	22.4	0.6	51.6	138.6
Zambia	1.0	−3.4 to 6.8	80.8	−11.2	74.0	14.1	8.3	49.5	223.2
Zimbabwe	2.3	−9.0 to 7.2	25.4	−8.1	68.9	20.1	17.1	15.9	59.8
Economies that have experienced relative stability									
Botswana	4.3	−0.1 to 8.7	11.3	+4.9	90.0	27.5	36.8	4.7	14.5
Lesotho	7.2	0.7 to 12.9	12.1	−4.8	149.9	76.8	−23.9	11.9	49.9
Mauritius	5.2	4.1 to 7.2	7.0	−3.7	127.6	28.3	24.2	1.9	42.2
Namibia	3.5	−1.9 to 10.4	10.0	−3.2	114.6	20.9	13.2	8.3	12.9
Seychelles	4.1	−0.8 to 9.0	1.4	−8.7	124.0	30.0	22.8	6.7	36.6
South Africa	1.9	−2.2 to 3.4	9.6	−5.6	47.7	16.3	19.1	1.1	18.3
Swaziland	3.7	1.3 to 8.9	12.3	−1.7	167.2	27.9	22.1	4.5	21.7

Source: *World Development Indicators 2000, African Development Indicators 2000*, Washington, DC: World Bank; and Jenkins and Thomas (2000)

Note

Numbers in italics have recent years missing from the average calculation.

1999). Average savings and investment rates were generally higher in the more stable group (excluding Lesotho, which is a notable outlier), although the distinction between investment rates is not particularly significant. Aid dependence has been high in almost all of the relatively weaker economies.

There appears to be a strong correlation between openness to trade and a conservative macroeconomic policy stance (as indicated by smaller budget deficits and slower money supply growth, with corresponding lower inflation). The relative importance of foreign trade is measured by the ratio of average two-way trade (exports plus imports) to GDP. For the more stable economies, this ratio is between 90 and 170 per cent, with the exception of South Africa (which is more typical of advanced economies in terms of the share of trade in GDP). Historically, South Africa has been comparatively closed, but, since 1994, significant progress has been made in liberalising the trade regime. For the countries exhibiting greater macroeconomic and policy instability, the ratio has been much lower, with the exception of Angola whose trade is high as a result of its (offshore) oil sector. The table does not imply causation in either direction, but illustrates that maintaining openness to international trade generally requires a more stable macroeconomy where trade policy is less likely to be subordinated to the imperatives of imposing internal and external balance.

23.3 SADC and globalisation

During the 1990s, globalisation was arguably the most important feature of the world economy, driven by a combination of technological advances and the liberalisation of trade and financial flows. This brought both considerable benefits in terms of increased trade, investment and growth, and new challenges for governments, especially in the developing world, through heightened vulnerability of domestic economies to external shocks.

The SADC economy represents less than 1 per cent of world GDP and the region contains around 3 per cent of the world's population. The region, along with the rest of Africa, has benefited less from the trend towards globalisation than other developing regions. This is evident in trade and, in particular, investment flows.

23.3.1 *Trade and international capital flows*

Africa's share of world exports fell between 1990 and 1999 – from 2.4 per cent in 1990 to just 1.9 per cent in 1999. SADC's share also declined during the decade from just over 1 per cent in 1990 to 0.7 per cent in 1999, with the South African-dominated Customs Union accounting for more than half of this share.[6]

It would appear that Southern African countries have been unable to compete as world markets have opened up. One reason for this has been the inward orientation of many Southern African economies over a prolonged period, coupled with economic instability. This has had inevitable consequences for the competitiveness of domestic industries. But even where these economies have pursued outward orientation, they have generally not succeeded in establishing manufacturing sectors to compete with the fast-growing economies of Asia. Moreover, even in sectors where Southern Africa has an apparent comparative advantage, producers have suffered as a result of the continued protection of vulnerable domestic producers in industrialised countries (see, for example, Kaplan and Kaplinsky, 1999, on the deciduous fruit canning industry in South Africa).

Perhaps of greater significance is SADC's lack of access to foreign capital in an era of surges in investment to the developing world. Total capital flows to developing countries increased (in nominal terms) from US $98 billion in 1990 to US $344 billion in 1997 before falling back to US $291 billion in 1999, with private capital substantially replacing aid in the composition of external capital resources.[7] This increase has been far from evenly spread across the developing regions: between 1990 and 1996,[8] East Asia and Latin America experienced a marked increase in their shares of private capital flows in the 1990s, while the share for sub-Saharan Africa showed a notable decline.

Available data indicate that the experience of SADC countries in attracting capital flows has been mixed. For instance, Angola and Mozambique have both recorded large net inflows (relative to their GDP). For Angola, this has been driven by foreign investment in the offshore oil sector, which is relatively insulated from the country's political and social instability. On the other hand, for Mozambique these inflows have been largely driven by aid. Moreover, South Africa became a major emerging market destination for portfolio investment following political reform and the reintegration of the country into global capital markets, but expectations of large-scale foreign direct investment following the transition have not been realised. In general, most countries in SADC have remained isolated from the surge in global capital flows, due in part to their underdeveloped financial systems but also to poor perceptions of the region on the part of investors.

While there are considerable benefits to be gained from attracting increased international capital flows, there are also potentially adverse consequences. The experiences of Mexico in 1994–5 and throughout Southeast Asia in 1997–8 demonstrated the vulnerability of many emerging markets to periodic exchange rate crises. Such experiences demonstrate the need for countries with increasingly open economies to develop appropriate policy responses to volatile swings in capital flows and other forms of external shocks. This will be an important challenge for SADC governments as they seek to become more integrated in the global economy and as financial markets in the region become more sophisticated.

23.3.2 *Foreign direct investment*

In 1998, direct investment flows to the developing countries (in nominal terms) were seven times the 1990 level; of this total, around 80 per cent went to countries in East Asia and Latin America. The benefits of foreign direct investment go beyond the simple notion of capital inflows supplementing domestic savings in the financing of investment. Such flows may be accompanied by the transfer of technology and skills and are often associated with expanding trade opportunities. For these reasons, they are generally considered to be the most desirable form of capital flow.

SADC has attracted on average just 2–3 per cent of the total flows of foreign direct investment to developing countries in recent years. This total is mostly (but not exclusively) accounted for by flows to South Africa. While in US dollar terms, amounts of FDI are small, because of the small size of many of these economies inflows as a per cent of GDP for many countries have been relatively high in comparison with other emerging market economies. Nevertheless, attracting greater levels of direct investment remains a key policy challenge for governments in the region.

Creating a more conducive environment for foreign direct investment and addressing the seemingly poor perceptions that foreign investors have of the region is probably the most important challenge facing SADC if it is to take advantage of globalisation. Policies to achieve these aims will be wide-ranging. Thus, sustained fiscal and monetary discipline is critical for economic stability, while greater stability of the policy environment generally is crucial for domestic as well as foreign investment. At the microeconomic level, greater flexibility in labour markets, higher levels of skills and improved infrastructure are important. And, institutionally, improvements in the bureaucratic process of investment approval and greater transparency will also be necessary (Jenkins *et al.*, 2000b).

23.4 The SADC free trade area

In order to overcome the problems of small markets, coupled with marginalisation in the world economy, SADC member states introduced a regional free trade area (FTA) in September 2000. This is perhaps the most important initiative being undertaken by the Community. Tariffs and non-tariff barriers are to be eliminated gradually within 8 years of ratification, freeing around 90 per cent of intra-regional trade, in line with the rules of the World Trade Organisation which state that 'free trade' should cover 'substantially all' trade.

As shown earlier, at least half of SADC members have historically been relatively closed to international trade. In part, restrictive trade policies have been driven by a perceived need to protect weak domestic industries. But they have also been used as an instrument for balancing otherwise unstable macroeconomic regimes. Since the beginning of the 1990s, extensive unilateral trade liberalisation has occurred across Southern Africa – either through aid-supported structural adjustment programmes or, in the case of the Southern African Customs Union (SACU), through agreement with the World Trade Organisation. More recently, attention has turned to the role of regional trade liberalisation in encouraging faster economic growth across Southern Africa.

23.4.1 *The importance of trade openness for growth*

Until recently, economic growth in Africa has compared poorly with that of other developing regions. During the 1980s, average real annual growth in sub-Saharan Africa was just 1.7 per cent, compared with an average of 3.1 per cent for all developing countries. In the first half of the 1990s, this gap narrowed slightly (to 2.0 per cent versus 2.9 per cent) and in the latter half of the decade the situation reversed (2.2 per cent versus 1.0). However, economic growth has generally been insufficient to generate substantial improvements in wellbeing across the region.[9] Regional trade liberalisation is seen as an important step in efforts to improve growth rates in Southern Africa – in the long term, liberalisation is expected to generate a significant increase in intra-regional trade and cross-border investment.

The importance of an open trade environment in explaining economic growth performance has been demonstrated in a variety of studies covering both developed and developing countries. One conclusion to emerge from the literature is that openness to trade is associated with higher rates of economic growth or, conversely, that a lack of openness to trade is correlated with poor growth performance.[10]

International trade facilitates technology transfers, the exchange of information and opportunities to realise economies of scale, and trade agreements provide greater certainty for trade policy generally, especially in countries that have a history of reversing moves towards unilateral liberalisation. As all of this reduces the risks of – or raises the returns to – investment, a greater volume of investment is one of the reasons for the observed positive relationship between openness to trade and economic growth. This has been found to be the case in several studies which are concerned with investment in Africa.[11]

Clearly more liberal trade does not explain everything about a country's economic growth rate. There is a variety of other important explanatory factors, such as human capital (education and health), the type of investment undertaken (not only the volume), political stability, the presence or absence of market distortions, diversification away from a dependence on primary (especially non-mineral) exports, and location. In Africa, the magnitude and persistence of external shocks, deficient public service provision, and political and economic instability have also been found to correlate negatively with growth. For trade liberalisation to yield higher economic growth, a broader range of policies are needed to create a conducive environment for growth (Jenkins *et al.*, 2000b).

23.4.2 *Regional integration and economic convergence*

Regional trade liberalisation enables members – especially those which are poorer – to reap some of the gains from trade via larger markets and improved efficiency, without exposure to non-regional competition. There is evidence that regional trade groups form convergence clubs, where poorer members catch up with (converge on) richer ones through the process of trade (Ben-David, 1995; Barro and Sala-i-Martin, 1991; Dowrick and Nguyen, 1989). It has been argued that all countries that are open and integrated in the world economy are, in fact, members of a convergence club (Sachs and Warner, 1995: 41).

There is evidence that, as elsewhere, trade between African countries promotes convergence. Elsewhere we consider whether convergence processes have applied in Southern Africa using the standard data set developed by Summers and Heston (1991)[12] for the period from 1960 to 1989 (see Jenkins and Thomas, 1998). The following discussion summarises our findings.

It should be noted that, *a priori*, there is no reason to expect convergence to have already taken place in SADC. Until 2000 there were no substantive moves towards facilitating intra-regional trade or factor movements. Moreover, there is substantial variation in countries' current macroeconomic and trade policy frameworks. Our hypothesis is that, if a convergence club already exists in Southern Africa, it is between the members of SACU, who have engaged in free trade since 1910, have operated a system of fiscal transfers and, except for Botswana (although the rand has considerable weight in the pula's basket of currencies), are members of a currency union.

The extent of macroeconomic convergence in SADC as a whole, and in the sub-set of SACU countries, is discussed immediately below. Real GDP per capita is used as the best available proxy for the level of development (Kuznets, 1966; Syrquin and Chenery, 1988). Convergence in real per capita GDP is therefore an indicator of real economic convergence. There are problems attached to measures of macroeconomic convergence, and no consensus as to which measure is best. Three measures are described.

The simplest measure is σ-convergence, when the dispersion of income levels across countries diminishes over time, with dispersion typically measured by the deviation of each country's per capita income from the average for the group. If countries that are initially very different are converging, it is expected that this deviation will be growing smaller. Figure 23.1 illustrates this measure of convergence over the period 1960–89 for SADC. The pattern for SADC countries is essentially flat – indicating that no convergence in per capita incomes has occurred over the 30-year period. Indeed, the degree of dispersion was marginally higher at the end of the period than at the beginning, which suggests that the countries have, if anything, *diverged* slightly. The absence of convergence among the SADC countries may be due to several factors, including different responses to the oil and exchange-rate shocks of the 1970s, and different problems with indebtedness, and there are also uniquely domestic policy issues that have promoted or slowed growth.

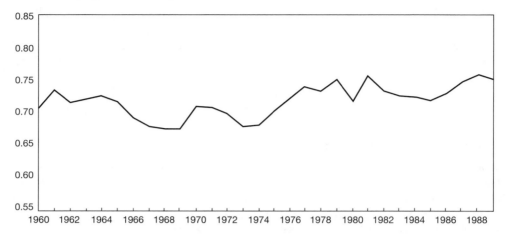

Figure 23.1 Standard deviation of log of per capita income: SADC

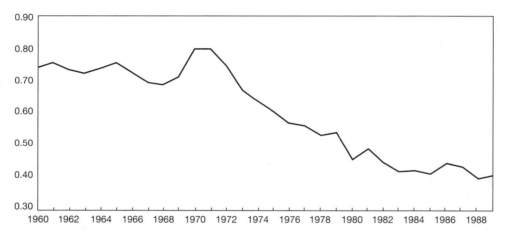

Figure 23.2 Standard deviation of log of per capita income: SACU

If, however, the sub-sample of SACU member countries is examined separately, as shown in Figure 23.2, a strikingly different pattern emerges. Although the intra-SACU dispersion of per capita incomes held roughly constant through the 1960s, it dropped steadily in the 1970s and 1980s. The result of this downward trend was that the dispersion at the end of the period was little more than half what it had been at the beginning. It is particularly interesting that neither the oil price shocks of the 1970s nor the gold price shock of 1980 – both of which would have had asymmetric effects on the SACU countries – caused any significant interruption to this pattern of convergence.

The σ-convergence measure is interesting for illustration, but does not, even in theory, explain whether poorer countries are catching up with richer ones (Quah, 1995: 15). There are alternative measures, however. For example, ß-convergence occurs when countries which are initially poorest grow faster than those which are richer, 'catching up' with richer economies. A downward-sloping plot of average growth rates on initial GDP will indicate possible ß-convergence. Using this technique, Jenkins and Thomas (1998) show that the

SACU economies have formed a convergence club, but there is no evidence of the poorer non-SACU members of SADC catching up with the richer countries.

Neither of these measures permits one to establish whether countries are diverging. However, an approach is available which allows us to estimate the probability that relatively poorer (richer) countries will raise (lower) their per capita income in the next period and thereby converge. This methodology is described in Quah (1992, 1995) and is applied to SADC countries in Jenkins and Thomas (1998). The analysis for SADC shows that there has been a tendency for countries with lower incomes to become or to remain poorer, while for countries with higher relative incomes per head, there has been a marginally higher chance of upward mobility. This suggests that cross-country incomes in Southern Africa tend to extremes at both high and low endpoints, and that that SADC countries actually *diverged* over the 30 years from 1960. This analysis was also undertaken for SACU countries: as with the σ-convergence and ß-convergence measures discussed above, a pattern of convergence is revealed in contrast to SADC as a whole.

In summary, the three measures of convergence demonstrate that SADC countries have, if anything, diverged over the 30 years between 1960 and 1989. This implies that the relatively rich have been getting richer, while the poor have been getting poorer. In contrast, there is evidence that convergence processes have applied in SACU. As noted above, there is no reason to expect that the SADC countries should have converged, as free trade in the Community is a very recent ideal. However, within the Customs Union movements of goods have been free for most of the century and smaller members have grown rapidly, particularly since the early 1970s.

The possible reasons for convergence in SACU include:

- free trade between SACU members;
- transfers from South Africa to other members under an enhanced customs revenue formula;
- the existence of a currency union (Botswana is not a member of the Common Monetary Area, but the pula largely tracks the rand);
- similar (comparatively conservative) macroeconomic policies; and
- country-specific factors which have little to do with regional arrangements (Jenkins and Thomas, 1998).

These explanations are not mutually exclusive: none rules out the importance of international trade in driving convergence in SACU. Although this evidence is not conclusive, it seems likely that access to the South African market has allowed smaller members to escape the limitations imposed by small domestic markets and this trend is at least consistent with that of other regions, both developed and developing.

23.4.3 *Implications for SADC: the potential benefits of the FTA*

The existence of one convergence club in Southern Africa provides some grounds for optimism that 'catch-up' convergence could also occur in a more closely integrated SADC. Smaller members of SADC stand to gain from regional integration in a variety of ways:

- South Africa is more than twice the size of the *sum* of the economies of the other SADC members, each of which therefore stands to increase its export markets substantially, with the potential for reaping economies of scale in domestic production.

- While exposure to South African competition will inevitably eliminate some production, more efficient firms will improve productivity and output, and diversification into products for the comparatively large South African market can be expected. Exposure to South African competition will help prepare smaller countries for greater integration into the world economy, by enhancing both quality and productivity and thereby competitiveness.
- Countries undergoing donor-funded structural adjustment programmes (SAPs) will find the credibility of their trade liberalisation enhanced, because the policy lock-in mechanism of a regional FTA should be more effective than liberalisation under SAPs has proved to be. If SADC develops an effective enforcement procedure, the costs of reversal of the SADC process could be high for a defaulting member, making it more likely that policy changes will be sustained.
- Outward investment from South Africa will both increase resources (access to savings and foreign exchange) and provide opportunities for technology transfers and better integration with South Africa's more sophisticated financial markets. Greater two-way trade together with foreign (mainly South African) investment should generate industrial development and help the diversification of production into non-traditional exports.

South Africa also stands to benefit from the regional FTA:

- There should be some market expansion, particularly for manufactured output, as the SADC countries are an important destinations for South Africa's manufactures. There are two caveats to this: first, in aggregate the SADC market is considerably smaller than South Africa's total (formal) domestic market; and second, market penetration by South African exporters will probably occur anyway, even without a SADC FTA.
- The FTA will increase opportunities for profitable cross-border investment, not least by improving the flow of information. South African investment in the region will generate additional demand for South African goods, with second-round growth effects for existing firms.
- There should be slower inward cross-border migration if the neighbours are expanding their economic – and especially industrial – capacity. Higher rates of economic growth from increased trade and greater investment should create jobs in the smaller countries, some of which are exporting labour both legally and illegally to South Africa.

However, regional trade liberalisation is not in itself a solution to creating economic growth in the region. There are three important points that need to be considered. The first is that South Africa should not be tied to the region at the expense of pursuing wider economic opportunities; second, regional integration should be viewed as a first step in the process of wider trade integration; and finally, SADC governments need to adopt macro- and microeconomic policies that are consistent with promoting trade and investment. These issues are explored below.

23.4.4 Implications for SADC: regional integration will not be enough for South Africa

One of the implications of the notion of convergence clubs is that there may be limits to the extent to which growth performance via regional arrangements can be enhanced: 'catch-up' implies that the benefits in terms of economic growth are greater the lower the initial level of income. In other words, the richest member is constrained in the extent to which economic growth can be accelerated by the forces driving catch-up. This suggests that, if South Africa is to improve its own growth performance, it will need to look beyond the region.

Securing higher future rates of economic growth in South Africa and in the region as a whole requires that South Africa expand its trade agreements beyond that with SADC. There are three reasons for this. First, South Africa does not reap significant dynamic gains from regional trade: on average it has superior technology, is the source of most of the region's investment, gains no enhanced credibility, and has limited opportunities to reap economies of scale. Second, many of the non-SACU members of SADC are instinctively protectionist, and regional integration is seen by some of them as an *alternative* to unilateral liberalisation. A frequently raised motivation for regional integration in Africa is a lowering of dependence on OECD economies. In this way, their agenda is different from that of South Africa (and the other SACU members), which has embarked on a process of closer integration into the world economy via unilateral liberalisation under its WTO agreement. Finally, South Africa needs to be in a position where it will have the opportunity to converge on both high-income and fast-growing economies. Jenkins (1997) thus argues that South Africa should look to establish a network of reciprocal FTAs with regions such as the EU, NAFTA, East Asia and possibly Australasia. It is argued that it is in the interests of SADC as a whole for the dominant partner to accelerate its growth through expanding trade with the rest of the world. The agreement between SACU and the EU is highlighted below.

23.4.5 SADC and global integration

For all SADC members, regional integration should not be perceived as an *alternative* to more general trade liberalisation, which is crucial if African economies are to grow, but rather as one step in a process of greater integration into international markets. Regional integration is complementary to global integration: it can play an important role in facilitating trade and investment through creating larger markets, which could ultimately enable SADC to compete in the global context. Continued progress in liberalising *vis-à-vis* the rest of the world is important for the entire region. Two interrelated initiatives are of particular relevance: the free trade agreement between South Africa (and the other members of SACU) and the EU; and the negotiations on the successor to the Lomé Convention, through which SADC members (with the exception of South Africa) have preferential access to the EU market.

In June 1995 negotiations commenced between South Africa and the EU over trade preferences. South Africa's request for Lomé status was turned down, but the country was offered a free trade agreement, with immediate access to Europe for most exports in return for phased exposure to European exporters. The talks with the EU were difficult, not least because of the complexity of the EU political process (Gibb, 2000). The EU's negotiating mandate was unattractive to South Africa in its early insistence that fishing rights be tied to trade negotiations and its (unreasonable) exclusion of certain agricultural products from discussion, a factor which reduced South Africa's ability to exchange concessions with the EU during talks. In addition, South African manufacturers are wary of greater exposure to foreign competition before the WTO commitments have been fully discharged, and were consequently ambivalent about the process. Finally, the country was under pressure from the smaller members of SACU/SADC, which are concerned that a free trade agreement between South Africa and the EU will expose their economies to European competition. In the end, a trade, development and cooperation agreement was signed at the end of 1999 amidst considerable rancour, driven by European interests, over the naming of fortified wines and the position of canned fruit exports.

The tariff phase-down of the EU is to occur over a maximum of 6 years (with the exception of certain agricultural and fisheries products), while South Africa has up to 10 years to eliminate tariffs on most of its imports from the EU (EU, 1999). A special safeguard is written into the agreement to cover 'disruption' of agricultural markets by exports of one party to the other; and a review within 5 years of the entire agreement is also included. The treaty also stipulates specific provision for developing and promoting cooperation not only between the parties to the agreement but also with the rest of Southern Africa. Areas specifically noted for the development of regional linkages include small enterprise development, telecommunications and information technology, energy, mining, transport and tourism. This includes, but is by no means limited to, financial assistance for regional projects by the EU.

Ultimately, the conduct of the negotiations between the EU and South Africa has sent negative signals to other developing countries as they begin to negotiate the successor to the Lomé Convention. At the time of writing, the existing Lomé arrangements had been extended for a period of 7 years, during which time negotiations on a fundamentally different form of trade and aid agreement would take place between the EU and the group of seventy African, Carribean and Pacific (ACP) countries, to be phased in over a further 12 years. The EU appears to favour the establishment of regionally negotiated trade agreements for all but the least developed of the ACP group. These agreements would be based on the principle of reciprocal but asymmetric access to markets. These proposals face considerable opposition within the ACP group.

The EU's agreement with South Africa was widely regarded as a test case for the post-Lomé arrangements. The obstacles created by the EU in concluding the agreement with South Africa has reinforced the view that the EU will not compromise on its remaining protectionism. This is likely to have increased the reluctance of the ACP countries to consider alternatives to the existing Lomé arrangements.

23.4.6 Making the FTA work

Regional trade policy is just one element of the overall economic policy framework, each component of which needs to be consistent if the goals of trade liberalisation are to be realised.[13] Although much of the work needed to make regionalism successful in Southern Africa is in the hands of domestic governments, there is scope for regional initiatives to support the FTA – for example, there is potential for improving transport links through regional networks and for developing multicountry initiatives to attract investment.[14] Moreover, the perceived need for sharing the potential gains from regional trade will inevitably require some mechanism for ensuring balanced development in the region. One alternative that is currently being explored is the creation of a regional development fund, financing investment in infrastructure, which could accompany the establishment of the FTA.

The potential role of supporting regional initiatives suggests that it will be vital for the various sectors of SADC (as illustrated in Table 23.2) to coordinate their activities. For example, while the regional FTA is the responsibility of the industry and trade sector in Tanzania, regional investment initiatives or a regional development fund would more naturally fall under the remit of the finance and investment sector in South Africa. Development corridors and spatial development initiatives could require coordination across several sectors depending on the nature of projects envisaged. Such coordination will be important to reduce duplication of effort and to ensure an appropriate targeting of limited resources. Political rivalries, however, may make such cooperation difficult in the foreseeable future.

23.5 Political barriers to regional integration

Previous sections have argued that trade liberalisation is not a panacea in itself. SADC member states must implement consistent macroeconomic policies together with micro measures to improve the environment for investment if regional integration is to deliver the long-term goal of stronger economic growth.

Compatible policy implementation is not the only barrier to successful regional integration in Southern Africa. There is also evidence of political tensions in SADC. These tensions surfaced during 1998 following the direct intervention by some member states in the war in the Democratic Republic of Congo, while other members attempted to pursue resolution through talks. This conflict, together with the continuing civil war in Angola, has had significant economic and security implications for some SADC member states. At the time of writing, there was no clear resolution in sight. New political pressures are also emerging from the economic and social crisis in Zimbabwe and there is anecdotal evidence that this instability has damaged investor perceptions of the region as a whole. While political instability remains in parts of the region, there is a continuing risk that political divisions may undermine the fragile process of regional integration in Southern Africa.

Underlying many of the political tensions in Southern Africa is a fear of the dominance of South Africa. As already noted, South Africa represents more than two-thirds of the regional economy and is expected to benefit the most, at least in the short term, from any regional trade integration initiative. There is a clear need for South Africa to be seen to be taking the concerns of its smaller neighbours seriously, although, as argued above, this should not occur at the expense of a more outward-oriented focus. South Africa has shown its willingness to make concessions – such as opening its markets to regional partners more rapidly than called for as part of the SADC FTA. It is yet to be seen whether the asymmetric phase-in of the FTA will be enough to allay fears of polarisation of industry in South Africa as a result of integration. In the longer term, the development of regionalism in Southern Africa will only prove politically sustainable if cooperation is seen to benefit each of the participating countries.

Of course, there are potential political benefits from regional integration. For example, the political and economic bargaining power of African countries might be strengthened if their voice came from regional organisations. Given the extreme difficulty of achieving an agreed pan-African view, regional organisations are more likely to be effective than continent-wide ones in presenting a collective position on globalisation issues. There is also a gain from achieving a collective view on intra-regional issues, like an anti-narcotics strategy or joint infrastructural initiatives. There is some evidence that regional groups survive, with good attendance at meetings by heads of state, so long as they achieve some diplomatic recognition and success, but that they fade away when such gains are small (Harvey, 1999: 3). In this sense, SADC has, up until now, been quite successful.

Domestic tensions may also undermine progress towards the formation of the FTA. For example, both the revival of hostilities in Angola and Zimbabwe's economic decline have the potential to disrupt those countries' compliance with any agreement, even if ratification proceeds. The optimism that was evident in the region in 1996 and 1997 has given way to a fear that the process might be difficult to get underway even though it is ratified by a majority of member states. South Africa's unilateral implementation of its own offer might be viewed as an attempt to precipitate action.

If credible sanctions for non-compliance can be established, trade integration will create a regional agency of restraint, limiting government discretion in making trade policy changes and providing more predictability and stability for importers, exporters and investors.[15] Failure to achieve a free trade area because of non-compliance by all members will mean that the SADC free trade area will suffer the fate of other FTAs in sub-Saharan Africa; and a failed attempt would set back future attempts at regional integration by many years.

23.6 Spillover effects and macroeconomic policy coordination

Within Southern Africa, there has been some debate on the need for more general economic policy coordination between member states, in particular in the form of monetary integration. International interest in policy coordination is driven in part by the spillover (cross-border) effects of macroeconomic policy, which are particularly acute when small countries are closely linked to larger economies. This is at least one motivation for the concern within SADC over this issue: it arises both because of South Africa's dominance in the region and because of the cross-border impact of uncoordinated structural adjustment programmes.[16]

Monetary cooperation usually takes the form of fixed exchange rates, where the smaller country makes a one-sided commitment to fix to the currency of the dominant country which sets policy freely. However, few fixed rate regimes have remained intact, as reserves available to support exchange rates are frequently dwarfed by potential short-term capital flows that magnify weaknesses in the commitment to the fixed exchange rate. Experiences in Europe (1992), Mexico (1994–5) and Asia (1997–8) have shown that a fixed or pegged exchange rate regime can be costly to maintain when there is a lack of credibility or similarly where the regime is perceived as unsustainable. It is worth noting, however, that the Common Monetary Area, encompassing South Africa, Lesotho, Namibia and Swaziland, has been sustained over a period of many years and has, arguably, yielded benefits for the smaller economies in terms of macroeconomic stability, in comparison to many of their Southern African neighbours.

Monetary union is optimal amongst economies that are very similar. If participants exhibit marked differences, it is imperative that one or a combination of conditions holds: there is wage flexibility; there is mobility of labour; politically acceptable transfers are possible; and countries react similarly and flexibly to shocks. On the other hand, in a group of countries marked by rigidities and immobility a degree of real economic convergence is necessary for exchange-rate/monetary policy coordination. Moreover, in practice, any group of countries contemplating a policy union will require convergence in macroeconomic stability indicators, like inflation and debt ratios, as a prerequisite for admission in order to protect other members from adverse policy spillover effects.

Many of the SADC economies are characterised by rigidities and immobility. Moreover, members react differently to external shocks; and, as shown earlier, there is divergence in policy and stability indicators. Jenkins and Thomas (1998) conclude that the apparent lack of macroeconomic convergence of the SADC countries (as illustrated above) and the significant divergence of policy and stability indicators suggests that Southern Africa is not yet in a position to establish and maintain regional monetary integration.

Premature attempts at policy coordination could have political costs that weaken the prospects for cooperation in areas such as trade and infrastructural development. Feldstein (1988) argues that an attempt to pursue coordination in a wide range of macro variables is likely to result in disagreements and disappointments that reduce the prospects for cooperation in the more limited areas of trade, defence and foreign assistance where cooperation is necessary. Moreover, the mechanisms for consultation and agreement may slow down the implementation of difficult policy decisions.

It would appear that it is premature for SADC to attempt macroeconomic policy harmonisation, even in the face of real problems of unbalanced economic power and spillover effects. Instead, by concentrating on issues that are of most immediate concern – and that are more likely to 'work' – SADC may successfully build a platform from which to embark on more ambitious regional integration initiatives in the longer term.

23.7 Concluding comments

This chapter has argued that SADC has the potential to become a sustainable regional institution promoting economic growth in Southern Africa. However, it is important that the underlying political tensions in the region are addressed so that they do not become a destabilising factor in the process of integration.

The Community as a whole needs to adopt an outward looking focus, rather than a narrow regional view. For South Africa, this is particularly important for accelerating its own economic growth. Regionalism in Southern Africa should be seen as a step towards increased participation in the global economy. In particular, the SADC FTA could become a means for pursuing trade agreements with a range of developed and developing regions.

Finally, coordination of activities within the institutional divisions of SADC will be of increasing importance as the region becomes more integrated. Regional cooperation across a wide range of sectors could contribute to creating consistent policy frameworks for increasing trade and cross-border investment. However, much of the responsibility in this area will ultimately lie in the hands of the domestic governments themselves.

Notes

1 Parts of this chapter draw on Jenkins *et al.* (2001).
2 For example, the Southern African Customs Union; the (Rand) Common Monetary Area; and the CFA franc zone. It is interesting to note that these relatively successful initiatives involve links with a more developed partner – in SACU and the CMA, the partner is South Africa; the CFA franc was previously linked to the French franc and is now linked to the euro.
3 See Berg (1988); Collier and Gunning (1996); Decaluwé *et al.* (1995); De Melo *et al.* (1993); Elbadawi (1995); Fine and Yeo (1994); Foroutan (1993).
4 It is not suggested that the relief of this constraint necessarily confers success: SADC has shown evidence of political divisions (see below), while other regional groupings, notably COMESA and the East African Community, are showing considerable political commitment to closer integration.
5 Seychelles enjoys the highest GNP per capita but as this country is so small in terms of population, it does not provide a particularly useful comparison.
6 *Direction of Trade Statistics, June 2000*, Washington, DC: International Monetary Fund.
7 *Global Development Finance 2000*, Washington, DC: World Bank
8 In the latter part of the 1990s, underlying trends in private capital flows to the developing world became less clear-cut due to financial crises in several of the major emerging markets.
9 There is significant variation in the economic growth of countries in the region. For example, in the 1980s growth rates ranged from 10.3 per cent in Botswana to 0.8 per cent in Zambia; in the first half of the 1990s, growth was 7.1 per cent in Mozambique compared to –6.6 per cent in DR Congo.
10 See, for example, Edwards (1993); Sachs and Warner (1997); Sala-i-Martin (1997). These findings are not wholly undisputed, however – see, for example, Rodriguez and Rodrik (1999) and Krishna *et al.* (1998). In a review of African growth performance, Collier and Gunning (1999) point out that Africa is less open than other regions to trade, partly due to policy and partly due to natural barriers, such as being landlocked. Almost all the studies of Africa find that impediments to trade have been detrimental to African growth performance, reducing the annual growth rate by 0.4 to 1.2 percentage points.

11 Sachs and Warner (1995); Bhattacharaya *et al.* (1996); Collier and Gunning (1996).
12 Summers and Heston's (1991) extensive cross-country data set has been used in most of the cross-country studies on economic growth; it is the most complete source of data on real GDP per capita, adjusted for purchasing power parity.
13 A recent study by Jenkins *et al.* (2000a) considers a broad range of policies that could be implemented to support the proposed SADC Free Trade Area. This study finds that in several areas current domestic policies in SADC are incompatible with the aims of regional trade liberalisation.
14 There is growing interest in the concept of cross-border development corridors in Southern Africa and, in South Africa, the government is promoting the similar concept of spatial development initiatives. Such projects generally involve government, local and foreign businesses and multilateral agencies and are aimed at promoting the development of infrastructure and investment across several sectors within a particular (cross-border) area. For example, the Maputo Corridor runs between the industrial Witwaterstand–Pretoria region of South Africa and the port of Maputo in Mozambique, passing through areas of mining, industry, agriculture, forestry and tourism (Maasdorp, 2000).
15 Investor confidence will take a long time to establish for individual countries, so the process must be accelerated. The potential benefits of a regional free trade area can only be secured if a *credible collective agency of restraint* is established by the SADC governments themselves. Other mechanisms might be able to contribute. For example, some form of partial investment guarantee to reduce the risks of South African investment in the SADC periphery could be positive. Such a scheme might attract donor support. This issue is considered below.
16 For example, Zimbabwe's devaluation was argued to have negatively affected Botswana's regional exports.

References

Barro, R. and Sala-i-Martin, X. (1991) 'Convergence across states and regions', *Brookings Papers on Economic Activity*, 1: 107–58.

Ben-David, B. (1995) 'Trade and convergence amongst countries', Discussion Paper no. 1126, London: Centre for Economic Policy Research.

Berg, E. (1988) *Regionalism and Economic Development in Sub-Saharan Africa*, Washington, DC: USAID.

Bhattacharaya, A., Montiel, P. and Sharma, S. (1996) 'Private capital flows to sub-Saharan Africa: an overview of trends and determinants', mimeo, Washington, DC: International Monetary Fund.

Centre for Research into Economics and Finance in Southern Africa (CREFSA) (1997) 'The South-East Asian crisis and implications for South Africa', *Quarterly Review*, October, London School of Economics: Centre for Research into Economics and Finance in Southern Africa.

Collier, P. and Gunning, J. (1996) 'Trade liberalisation and the composition of investment: theory and an African application', WPS/96–4, University of Oxford: Centre for the Study of African Economies.

Collier, P. and Gunning, J. (1999) 'Explaining African economic performance', *Journal of Economic Literature*, 37: 64–111.

Decaluwé, B., Njinkeu, D. and Bela, L. (1995) 'UDEAC case study', presented at the African Economic Research Consortium (AERC) workshop on 'Regional Integration and Trade Liberalisation', Harare.

De Melo, J. and Panagariya, A. (eds) (1993) *New Dimensions in Regional Integration*, Cambridge: Cambridge University Press.

De Melo, J., Panagariya, A. and Rodrik, D. (1993) 'The new regionalism: a country perspective', in J. de Melo and A. Panagariya (eds), *New Dimensions in Regional Integration*, Cambridge: Cambridge University Press.

Dollar, D. (1992) 'Outward-oriented developing economies really do grow more rapidly: evidence from 95 LDCs, 1976–1985', *Economic Development and Cultural Change*, 40 (2): 523–44.

Dowrick, S. and Nguyen, D. (1989) 'OECD comparative economic growth 1950–1985: catch-up and convergence', *American Economic Review*, 79 (5): 1010–30.

Edwards, S. (1993) 'Openness, trade liberalisation and growth in developing countries', *Journal of Economic Literature*, 31 (3): 1358–93.

Elbadawi, I. (1995) 'The impact of regional trade/monetary integration schemes on intra-sub-Saharan African trade', presented at the African Economic Research Consortium (AERC) workshop on 'Regional Integration and Trade Liberalisation', Harare.

European Union (EU) (1999) *Agreement on Trade, Development and Cooperation Between the European Community and its Member States, of the One Part, and the Republic of South Africa, of the Other Part*, 8731/99, Brussels: European Union.

Feldstein, M. (1988) 'Distinguished lecture on economics in government: thinking about international economic coordination', *Journal of Economic Perspectives*, 2 (2): 3–13.

Fine, J. and Yeo, S. (1994), 'Regional integration in sub-Saharan Africa: dead end or fresh start?', mimeo, Nairobi: African Economic Research Consortium.

Foroutan, F. (1993) 'Regional integration in sub-Saharan Africa: past experiences and future prospects', in J. de Melo and A. Panagariya (eds), *New Dimensions in Regional Integration*, Cambridge: Cambridge University Press.

Gibb, R. (2000) 'Europe's elusive foreign policy: an evaluation of the European Union's policies towards post-apartheid South Africa', mimeo, University of Plymouth.

Hall, S., Robertson, D. and Wickens, M. (1992) 'Measuring convergence of the EC economies', *Manchester School*, 55 (Supplement): 99–111.

Harvey, C. (1999) 'Macroeconomic policy and trade integration in Southern Africa', paper presented at the Industrial Strategy Project Regional Research Workshop, Irene (nr Johannesburg).

Jenkins, C. (1997) 'Regional integration is not enough', *Quarterly Review* (London School of Economics, Centre for Research into Economics and Finance in Southern Africa), April: 15–21.

Jenkins, C. and Thomas, L. (1998) 'Is Southern Africa ready for regional monetary integration?', in L. Petersson (ed.), *Post-Apartheid Southern Africa: Economic Policies and Challenges for the Future*, London: Routledge.

Jenkins, C. and Thomas, L. (2000) 'The macroeconomic policy framework', in C. Jenkins, J. Leape and L. Thomas (eds), *Gaining from Trade in Southern Africa: Complementary Policies to Underpin the SADC Free Trade Area*, Basingstoke, Hants.: Macmillan.

Jenkins, C., Leape, J. and Thomas, L. (eds) (2000a) *Gaining from Trade in Southern Africa: Complementary Policies to Underpin the SADC Free Trade Area*, Basingstoke, Hants.: Macmillan.

Jenkins, C., Leape, J. and Thomas, L. (2000b) 'Gaining from trade in Southern Africa', in C. Jenkins, J. Leape and L. Thomas (eds), *Gaining from Trade in Southern Africa: Complementary Policies to Underpin the SADC Free Trade Area*, Basingstoke, Hants.: Macmillan.

Jenkins, C., Leape, J. and Thomas, L. (2001) 'African regionalism and the SADC', in M. Telo (ed.), *The European Union and New Regionalism*, Brussels: Institut d'Etudes Européennes.

Kaplan, D. and Kaplinsky, R. (1999) 'Trade and industrial policy on an uneven playing field: the case of the deciduous fruit canning industry in South Africa', *World Development*, 27 (10): 1787–1801.

Krishna, K., Ozyildirim, A. and Swanson, N. (1998) 'Trade, investment and growth: nexus, analysis and prognosis', NBER Working Paper 6861, New York: National Bureau of Economic Research.

Kuznets, S. (1966) *Modern Economic Growth*, New Haven, CT: Yale University Press.

Maasdorp, G. (2000) 'Microeconomic policies', in C. Jenkins, J. Leape and L. Thomas (eds), *Gaining from Trade in Southern Africa: Complementary Policies to Underpin the SADC Free Trade Area*, Basingstoke, Hants.: Macmillan.

Quah, D. (1992) 'Empirical cross-section dynamics in economic growth', *European Economic Review*, 37: 426–34.

Quah, D. (1995) 'Empirics for economic growth and convergence', CEPR Discussion Paper Series 1140, London: Centre for Economic Policy Research, March.

Rodriguez, F. and Rodrik, D. (1999) 'Trade policy and economic growth: a skeptic's guide to cross-national evidence', Working Paper 7081, New York: National Bureau of Economic Research.

Sachs, J. and Warner, A. (1995) 'Economic reform and the process of global integration', Brookings Papers on Economic Activity, 1: 1–118.

Sachs, J. and Warner, A. (1997) 'Sources of slow growth in African economies', *Journal of African Economies*, 6 (3): 335–79.

Sala-i-Martin, X. (1997) 'I just ran two million regressions', *American Economic Review Papers and Proceedings*, 87 (2): 178–83.

Summers, R. and Heston, A. (1991) 'The Penn World Table (Mark 5): an expanded set of international comparisons, 1950–1988', *Quarterly Journal of Economics*, 196 (2): 327–68.

Syrquin, M. and Chenery, H. (1988) *Patterns of development, 1950–1983*, World Bank Discussion Paper no. 41, Washington, DC: World Bank.

Part VI
Gender, health and education

24 Gender and development

Policy issues in the context of globalisation

Marjorie Mbilinyi

24.1 Introduction

Gender and development policies have emerged in Africa in the context of globalisation processes which have led to increased income differentiation, impoverishment of the majority and growing gender and class inequities. Policies associated with downsizing, privatisation and liberalisation have benefited large-scale business rather than small-scale, and foreign companies more than local enterprise. State schools and hospitals have been run down, factories closed and a large number of workers thrown out of employment, both women and men. African governments lack the economic power – and will? – needed to protect the interests of domestic business and producers.

A growing number of citizens equate structural adjustment policies (SAPs) with dispossession, impoverishment and disenfranchisement. The potential for popular resistance has grown and provided an opportunity for gender activists to engender 'wider' social issues – to be illustrated below by analysis of the Gender Budget Initiative (GBI) in Tanzania.

In successive sections, this chapter examines the meaning of globalisation from a critical gender perspective; the various impacts of the restructuring policies which have been applied; responses adopted by women in Africa to their situation; the use of gendered policy analysis in securing change; and finally the example mentioned of the GBI in Tanzania.

24.2 The context of globalisation

Globalisation is a process that extends through all aspects of society: economic, political, ideological. Although it has had a long history, the present phase of globalisation – a phase marked by deepening but not widening of capitalist integration – is quite different from the earlier expansive phase of capitalism (Hoogvelt, 1997). During the colonial era, for example, third world territories were incorporated into the global capitalist system as cheap labour on plantations and mines, or as peasant producers of cheap raw materials. In the 1970s and 1980s, countries of the Far East and Latin America became major producers of electronic goods, with large segments of their population employed, often in free trade zones, by transnational corporations.

The situation today is different. Capitalism no longer has the capacity or the need to absorb new populations of people into its orbit, as a result of the transformation which has occurred in production and circulation of capital associated with information technology. Intra-product trade has replaced inter-product trade, involving export competition between producers in different countries in the same product lines (ibid.: 122). An increasing portion of global trade also consists of exchanges involving subsidiaries of one firm. Global market standards have been imposed on business everywhere – those that can cope survive, but a large number collapse or merge.

The global dimension of economic institutions such as markets is paralleled by the rapid development of global regulatory/financial institutions led by the World Bank and International Monetary Fund (IMF) and the World Trade Organisation (WTO). The global regulatory institutions have imposed a complex set of rules and regulations on nation states, transnational corporations (TNCs) and other actors. These institutions support the further expansion of big capital, with particular attention to the needs and interests of the G-7 countries and their trading blocks (Bush and Szeftel, 1998; Mohan, 1994). Macro-economic and fiscal policies are promoted which further the interests of TNCs in their search for new markets and new sources of cheap labour and reduced barriers to the free circuit of money capital and goods.

The debt crisis has empowered the core countries by providing the rationale and the means to impose economic and fiscal policies of benefit to themselves, and to create a global management system which helps to impose global market discipline around the world (Hoogvelt, 1997). In the process, the developmental state has been dismantled, leaving an impotent government apparatus beholden to its creditors. Real economic reforms that would lead to real development require a strong developmental state, capable of managing resources at all levels – as was found in the Far East in the 1980s and 1990s.

As a result of globalisation, whole segments of the world's population in the north, but especially in the south, have become 'superfluous appendages'.[1] They reportedly lack the skills necessary to compete in the global market or the market discipline to keep up with the fast pace required by the global market. Core–periphery relationships have become stronger than before, and increasingly cut across national and geographic boundaries. They include the growing number of retrenched workers in the north and the 'unemployables' who have not been employed in the formal sector for generations, including social welfare mothers.

The heightened visibility of core–periphery as a social relationship, rather than a geographic one, provides a positive outcome, however – more scope for international organising in broad social movements to challenge globalisation on its own ground, linking 'first world' and 'third world' people together. This is discussed in more detail below.

A *politics of exclusion* has developed within the global system to manage 'those disadvantaged groups and segments in all societies that can no longer perform a useful function as either producers or consumers within the global market' (ibid.: 147). Global management is 'not an economic problem but a law and order problem' (ibid.: 148). The politics of exclusion includes anti-immigration laws in Europe and North America and urban influx controls in Africa. Deepening war and military conflict in Africa is actively promoted by 'northern' states by means of massive arms exports, on the one hand, and continued donor support for defence activities on the other. HIV/AIDS is another way to manage exclusion – witness the negligible action carried out to halt the AIDS crisis in Africa by governments and donor agencies.

Official agencies also support community-based action programmes that help to contain discontent in developed and developing nations. These programmes contribute to the maintenance of globalisation by 'organising the poor and the marginalised to care for and contain and control themselves' (ibid.: 149). They respond to *neo-liberal* ideals of self-help, voluntarism and a reduced dependency/demand on the state. The majority of women-oriented projects and programmes appear to fall within this category of social management of exclusion, as will be discussed below.

This process of globalisation has been strengthened and sustained by economic reform measures, to which we now turn, drawing on recent research in Southern African countries, with an emphasis on Tanzania.

24.3 The impact of economic reforms[2]

24.3.1 Impact on poverty and employment

Economic reform measures include fiscal austerity, trade liberalisation, privatisation, reduced government support for social services, increased individual costs for social services and retrenchment of workers. The results have been mixed, in that a small minority of people has benefited in each country, while the majority have become impoverished and disenfranchised. Retrenchment in public and private sectors has led to increased job insecurity, lower pay and unemployment for many men and women, with low-income women being the most vulnerable. In cases like Mozambique, Zambia and Tanzania, there have been short-term increases in economic growth at the national level, but with ever greater regional and household disparities in income and wealth. Class polarisation has escalated and is visible in access to housing, education and health. There is a stark contrast between the luxurious consumption of the well-to-do and the deepening impoverishment of the majority.

Income disparities have increased in every African country as a result of economic reforms. The wealthy few live in large villas in exclusive suburban neighbourhoods, replete with satellite dishes and triple-car garages. Their children attend high-cost private English 'academies' starting at pre-school level, or are educated abroad. Social life is conducted in exclusive clubs and tourist hotels, where the majority of patrons are Europeans and Asians. The class/racial divide is especially notable in Tanzania, because of the contrast between the present situation and the period of the 1970s, when income differences had been successfully reduced and indigenous Tanzanians had experienced a form of cultural liberation associated with the Arusha Declaration policy and ideology.

Feminisation of poverty has occurred side by side with increased female labour force participation and increased female access to and control over cash incomes. Changes in women's economic activities, however, can only be understood if situated in the context of the dramatic reduction of male employment and/or incomes in both urban and rural areas (Mbilinyi, 1997). The majority of people employed in the formal labour market were men, and hence, in absolute terms at least, men have suffered the greatest from retrenchment policies in the public sector and downsizing in the private sector (Kaijage, 1997). At the same time, real wages in both the formal and informal sectors have declined, as have real farm incomes (see below). Men have experienced economic impotency as a result, no longer able to provide for the cash needs of their families as before, and increasingly dependent on the incomes of their female partners (Mbilinyi 1997, 2000c).

Both women and men increasingly depend on (self-)employment in the informal sector, which has no job protection, worker benefits, maternity leave, minimum wage or other worker support systems, and has been overlooked thus far by labour union organisers (Mbilinyi, 2000b). The majority of informal sector workers and operators earn extremely small incomes, which barely cover production costs, in activities which have been labelled 'survival strategies' by observers of micro-small enterprises (MSEs) (ibid.; see also Kabeer and Whitehead, 1999). Women tend to be channelled into the least remunerative occupations within the informal sector, and the same in the formal sector, working in unskilled and semi-skilled jobs often associated with 'female' tasks. Hence, women predominate in food manufacturing/processing/sale work, whether they are small-scale owners or wage employees.

One result of downsizing policies has been the movement of men into informal sector occupations once dominated by women in many countries. In Tanzania, for example, the female ratio in informal sector (self-)employment ranged from 31 per cent in rural areas to 44 per cent in most urban towns during 1991 – less than half, but much higher than the usual proportion found in the formal sector.[3] In other countries such as South Africa, Botswana and Zimbabwe, studies suggest that women outnumber men in micro-small enterprises in the informal sector.

It is important, therefore, not to over-generalise about gender patterns in employment in Southern Africa, whether in the formal or the informal sector. Men tended to predominate at the top of the occupational hierarchy in all countries, monopolising administrative and management jobs and receiving the highest wages. Women were over-represented at the bottom of the hierarchy, doing unskilled work for the lowest wages. However, in South Africa, an almost equal proportion of women and men were employed in professional and technical categories, that is, as nurses, teachers and social workers (Flood *et al.*, 1997: 22). This partly reflected the way that men had been channelled in the past into mining and manufacturing occupations by the twin systems of racial apartheid and migrant labour, leaving a gap which women filled in middle-class occupations.

Nevertheless, a gender pattern was found in *type of occupations* in all countries, whether in public or private sectors, or in the formal or informal economy. Women were over-represented in 'typically female' occupations associated with food production, preparation and sales; garment manufacture and sale; and in hair salons, bar girl service and prostitution. Male occupations were more widely spread across different kinds of jobs, especially in artisanry, where men found jobs as carpenters, welders, plumbers, tailors, electricians and mechanics, while women were mainly confined to sewing and tailoring.

The twin policies of retrenchment (downsizing) and privatisation of public parastatals have had a devastating impact on women, because of their much greater dependence on the public sector for regular employment than men (for example, in Tanzania), even though they have been in the minority within formal sector employment overall. SAPs resulted in the shrinking of less competitive sectors of manufacturing industry such as tailoring, cloth manufacturing and food processing, where, again, women employees were concentrated (Kwinjey, in SARDC, 2000). Moreover, Export Promotion Zones (EPZ) in Mauritius, for example, have been labour-intensive and highly exploitative, extracting cheap labour from young women working in clothing and electronics manufacture (Gunganah *et al.*, 1997: 23–5).

Although women had much less access to credit than men in farming, small business and small manufacture, credit opportunities have become highly restricted for both as a result of fiscal austerity measures. Most people were forced to resort to informal systems of credit, such as usury, which were more costly (see Mbilinyi, 2000b, for a detailed analysis).

Women tend to earn lower wages than men in the same occupations with the same educational credentials and work experience. The causes include the double/triple work load, gender division of labour in household work, women's 'responsibilities' in childcare and reproduction in general, and the lack of resource allocations to improve productivity in fuel, water, food preparation, local transport which continues to depend on female headloads, and laundry.

The unpaid labour of women and children, in particular, has been increased in smallholder farming and in informal sector activities so as to reduce the production costs of household economic activities; that is, unpaid labour is replacing paid wage labour, a backward step economically, socially, politically. At the same time, the unpaid labour of women and children continues to provide the bulk of reproductive needs within the household and community, in the form of cooking, procurement of water and fuel, and childcare. As women have become more active in market-oriented non-farm activities, they have been forced to withdraw their children from school to substitute for their unpaid family work. This has had an immediate negative impact on children's access to schooling, and will have long-term social as well as individual costs. Illiteracy rates have increased for both women and men in Tanzania, for example, another unfortunate legacy of economic reforms and withdrawal of public support for education and other social services (see below).

Increased poverty and rural–urban migration has also led to increased sex work among women and men and children, which is partly associated with the rise of sex tourism and expatriate workers in most countries. Young girls who migrate to town in search of a better life are recruited as they get off country trains and buses to work in brothels or on the streets. Many others turn to sex work as an escape from low pay, harsh working conditions and sexual harassment experienced in domestic service, the other main job 'opportunity' available for young rural girls in town.[4]

The effects of SAP reforms have therefore been especially harmful for women: longer work days, less access to basic resources like land and labour in some cases, reduced opportunities in formal wage employment and education, and increased financial responsibility for families and communities – too often in the absence of support from a male partner. The number of female headed households has increased in many countries, and a growing number of children are born out of wedlock and without the traditional community support systems of the past. Several writings also document the way in which male-dominant patriarchal systems have become cornerstones of the reforms. Analysis of the often contradictory impact of SAPs on gender relations, and women and youth in particular, needs to be strengthened by in-depth examination of the patriarchal construction of global capitalism.[5]

At the same time, each country has witnessed the growth of a small but notable and powerful groups of female entrepreneurs. Active in chambers of commerce and in specific women's business associations, they have expanded their holdings in such diverse sectors as tourism, transport, restaurants and retail, and mining in the formal sector. A few women are also found among the larger, more prosperous operators in the informal or MSE sector, in both urban and rural areas (Mbilinyi, 1991, 2000b). Their wealth depends necessarily on exploiting the labour of the women – and men – who work for them, usually as casual and unpaid family labour. This exemplifies the need for gender analysis to study class differentiation among women and men, in order to understand the different experiences and responses of different groups of people to the reform process (Kabeer and Whitehead, 1999).

24.3.2 *Impact on agriculture and food security*

The majority of women, and men, continue to rely on agriculture in most countries, but the viability of smallholder farming has rapidly declined (Bryceson, 1999; Malatsi, 1995; Mvula and Kakhongwa, 1997; Mbilinyi, 1997, 2000c). Economic reforms have tended to favour large-scale capitalist enterprises such as plantations and large ranches in Tanzania, for example. They have regained their earlier monopoly over support systems such as credit, extension services, and marketing channels (Mbilinyi, 1997, 2000c). Smallholder farmers have found it difficult to survive after the withdrawal of price support systems, soft loans and subsidies for farm inputs. Real returns for export and food crops have declined in many areas for smallholder farmers, and there are widening disparities between large and small farmers, and between different agro-economic zones. The majority of farm households in many areas depend heavily on off-farm activities to supplement declining farm incomes.

Women took the lead in off-farm work in Tanzania before the SAP, in the late 1970s and early 1980s, in order to access independent cash incomes over which they had some control (Mbilinyi, 1991). Both wives and young people resist having to work on farms controlled by male household heads as unpaid family labour. Their involvement in off-farm activities has grown tremendously, however, in response to the decline in smallholder agriculture in many areas, and growing class differentiation and impoverishment, resulting from the reform process.

Liberalisation has therefore had mixed outcomes for farm communities. On the one hand, many grain farmers cannot compete with cheap imported foodstuffs. They have shifted to export crops, but real returns have declined for export crops as well in many parts of the region (Mbilinyi 1997, 1998a). On the other hand, market opportunities have expanded in off-farm enterprises, which have been exploited by women and youth, in rural and urban areas. Common forms of (self-)employment include beer brewing, food processing, food preparation and sale, food trade, tailoring and casual wage employment on other small farms or on neighbouring plantations (Katapa, 2000; Mbilinyi, 2000a; Shundi, 2000: n. 6).

The contribution of female cash incomes to rural households is substantial, which has increased women's bargaining power. Gender relations have begun to change, as well as become more conflictual. Male responses vary. Some men have become more appreciative of women's contributions, and share in decision making about household incomes with their more empowered wives. Others have withdrawn their support for household needs, and used their own cash incomes for personal consumption – more booze, more women, sometimes another wife or mistress.

At the same time, the expansion of women's participation in off-farm activities should not blind us to their nature: low levels of productivity and low levels of return. Household and regional food security have been threatened by these developments, as a growing portion of rural labour and land is withdrawn from food production into export crop production or into off-farm activity (Bryceson, 1999). In Tanzania, some 6.6 million people were reportedly facing chronic food insecurity in the 1990s, especially women, children and the elderly (Mukangara and Koda, 1997).

Pressures have been increasing to open up land to market forces, as part of liberal reforms, in those countries like Tanzania which have had some form of protection for local community rights. Dualistic systems of land tenure and land rights were contradictory, in that they upheld the rights of indigenous people to their land, while also sustaining male dominant systems. Liberalisation, however, may lead to total dispossession from commons land for the entire community and an erasure of women's usufruct rights (Mbilinyi, 2000b). NGOs in Tanzania, for example, have formed a coalition, the National Land Forum, to campaign against top-down processes of decision making about land, which exclude local communities and the public at large from participation in the land reform process. They have argued for a unitary legal system, which protects the rights of local citizens, especially women and children, over land and blocks foreigners from accessing long-term rights to land (Mukangara and Koda, 1997).

Systematic forms of discrimination against women have persisted in access to and control over key resources, especially land, livestock, farm inputs, farm equipment, credit and crop incomes, especially from the major export crops. The male bias has acted as a major barrier to the further development of smallholder agriculture, given that women represent more than half the agriculture labour force in many countries but lack incentives to increase output or improve their farm practices, and thereby increase farm productivity. A vicious circle has been created, largely because of male bias in government and donor policy, which has led to ever lower levels of crop productivity in many areas and higher food insecurity at household, national and regional level.

Women farmers have not been passive victims, however. Whereas men control almost all the inherited land in most patrilineal areas of the region, women control a sizeable proportion of rented, purchased or allocated land (Mbilinyi, 1997, for Tanzania). Allocated land has usually been provided to women on an annual basis by village governments or local authorities. They have struggled to retain independent control over their own labour and those crops which they define as theirs. This includes active resistance against unpaid labour systems, thus blocking efforts by cash-crop growers to reduce production costs by shifting from wage to unpaid family labour. Stagnancy or decline in export crop output during the 1990s in Tanzania can be attributed at least partially to the resistance of women and youth against patriarchal farming systems based on unpaid family farming (ibid.).

Having raised the issue of gender discrimination in credit provision, women farmers have been quick to take advantage of donor-led projects to provide credit and/or other supports directly to women. Small women's economic groups (WEGs) have been formed throughout the region, which have enabled women to access credit, processing equipment like maize mills, and land, and to set up autonomous bank accounts in the name of the group. For all the contradictory nature of these groups, to be discussed below, they have contributed to women's economic empowerment (Mbilinyi, 1995).

24.3.3 Impact on social services

Social services in most countries of the region have been severely undermined as a result of structural adjustment policies, combined with the previous decade of under-finance and corruption. Two key aspects of SAP reforms have been cost-sharing and privatisation. Cost-sharing has shifted the burden of payment for basic education and health from the government to the people, in defiance of earlier commitments to their provision as fundamental human rights. Privatisation has contributed to the creation and/or strengthening of a two-tier system of education and health, one for the wealthy and one for the poor (Mukangara and Koda, 1997; TGNP, 1994a; SARDC, 2000).

The reform process in social services has, therefore, in practice been a *counter-reform*, a dismissal of the major steps taken in the region after independence to abolish racially divided systems of education and health, and to create a unified and integrated structure. People in the less developed rural areas, the poor, and girls and women have been especially harmed by this reversal of policy. In Tanzania, for example, the imposition of cost-sharing has led to an immediate and drastic reduction in visits to antenatal maternity clinics. Maternal mortality rates have increased since then, thus reversing the earlier downward trend.

Enrolment rates in primary education have fallen for both girls and boys in Tanzania and the drop-out rate has risen. The major reason for school drop-outs was economic: parents could not afford to pay escalating school costs; and both girls and boys were needed to work in order to contribute to household income. The opportunity costs of sending children to school outweigh the limited returns from investment in their schooling, in a context of low-quality education in most public schools. Many girls also dropped out because of pregnancy or marriage; these probably represented economic strategies as well.

Children of the most wealthy parents have been educated abroad or in neighbouring African countries. A growing percentage of upper-middle-class children have been enrolled in high-cost English-medium 'academies', beginning in pre-school; whereas the majority of children attended relatively low-cost public and community schools. Disparities in cost were matched by tremendous differences in performance. The gap has therefore grown between the skills and knowledge accessible to the wealthy and the poor, throughout the region.

Similar disparities have grown in health services as well. The well-to-do government leaders and rich entrepreneurs spent thousands of US dollars seeking treatment abroad, often at taxpayers' expense, while the poor could not afford simple treatments for malaria. Disparities were also marked at regional level, with the most up-to-date services being available in South Africa.

24.4 Women's responses

Most countries had adopted some form of affirmative action programme to increase women's access to political positions and other positions of influence at the local and national level. The most common strategy was that of preferential seats, for example, in Tanzania and Namibia. Women in Development (WID) and/or Gender and Development (GAD) departments have also been set up to mainstream gender into development policy at the sectoral level. Gender-sensitisation programmes have reached the majority of high-level civil servants in many countries. Many of these measures have depended on donor-funded programmes, however, and may appear to be donor-driven, no matter how much prior popular pressure existed for their implementation. Moreover, the results have not been commensurate with the investment.

The development of multiparty politics in several countries such as Tanzania has helped to open up the political arena to more women. Even more significant, however, has been the recent expansion and deepening of organising in civil society in many countries. Civil society organisations (CSOs) have taken the lead in raising gender awareness about key development issues. They have challenged their governments to implement the international conventions that they have signed. Linkages have been made between macro and micro issues and policies, and between the local and the national level. Efforts have been made to build coalitions among different NGOs and community organisations, around the land issue in Tanzania, for example, and around democratisation of the budget making process in South Africa, and most recently, Tanzania (Flood *et al.*, 1997; Mukangara and Koda, 1997; TGNP, 1998, 1999a). A major new development is for gender-related groups to directly enter the political arena. NGOs face a major challenge to combine forces at the grassroots and the national level (Mbilinyi, 1998b). Another challenge is to strengthen their autonomy *vis-à-vis* governments, on the one hand, and foreign donors on the other.

The increased struggle over concepts and practices of custom and tradition is another realm of politics that has begun to feature more prominently in the public eye as a result of NGO activism, social change and mainstream backlash. In many countries women and gender groups have successfully lobbied for positive legal reforms which have enhanced women's rights, including the recognition that women are full adults, capable of representing their interests in their own right, thus denying the concept of minor status under the law for women which was embodied in customary law inherited from the colonial state (see note 3). In Tanzania, the recently enacted Sexual Offences Act has provided the state with major powers to act against violence against women and children, including rape (in and out of marriage), female genital mutilation and incest.

A growing number of women have defended their rights in marriage, inheritance, property and employment through the courts system, often with the support of legal aid clinics. For example, in Botswana, the children of a Botswanan woman married to a foreign man have now been recognised as entitled to citizenship, in the same way that those born of a Botswanan man married to a foreign woman are. This was the result of legal reforms enacted in response to a national campaign led by local activist NGOs. The recently enacted Village Land Reform bill, 1999, in Tanzania incorporated many changes advocated by the Gender Land Task Force in conjunction with the National Land Forum. These included the recognition of the rights to land of female spouses; equal representation of women and men on local land committees; and a general position that recognised women's rights to own land in the same way as men.

24.5 A gendered policy analysis

Gender analysis has emerged as part of an embryonic but steadily growing women's movement in Southern Africa; that is, gender analysis has been closely linked to activism within civil society organisations. This has given it a sharp edge with which to explore basic causes of specific problems such as sexual harassment in education and employment, wife beating, the feminisation of poverty, or persistent gender inequalities at all levels of society. Critical of policies associated with national governments and international development agencies, third world feminists endeavour to link the macro to the micro level, and to engage with the policy formulation process itself. Moreover, they are constantly challenged to reflect upon their own class/gender/race position in society, which pushes many to ensure that their policy perspective reflects the interests of people living in poverty.

On the other hand, gender analysis has also been 'mainstreamed' by many donor agencies into programmes and projects, and adopted by sectoral ministries in most SADC countries. A critical analysis is needed of the impact this has had – or may have – on the oppositional element within gender activism, which is beyond the scope of this chapter. However, earlier critique of Women in Development (WID) approaches remains relevant to a substantial portion of mainstreamed gender analysis. Grounds for saying so include the common assertion that there is no perceivable difference between approaches guided by 'gender' or 'women', 'gender and development' or 'women in development'.

Women in Development analysis and action emerged in the 1970s and 1980s as efforts to highlight the contribution of women to development, and the way that they were systematically overlooked in development policies and programmes. Special programmes were designed to benefit women, with an emphasis on training and credit. WID focal points were established in most African governments, with the mandate to monitor the extent to which government policy took women into consideration, along with research, documentation and advocacy.

In Tanzania, the Women and Development Policy of 1992 called for action to ensure that women had the right to own and inherit resources and implements of production and the income emanating from their work; that contributions of women and men to development be disaggregated; that barriers against women's full access to education and training be removed; that cultural norms, values and practices which subordinate women be removed; that a culture that defends social justice be promoted, along with the talents and strengths of women; that women's issues be integrated into all sectoral plans; that women's issues be mainstreamed in all development plans (see TGNP, 1999: 10–11). The Ministry of Community Development, Women's Affairs and Children (MCDWAC) was made responsible for the advancement of women and the promotion of gender equality.

However, critics pointed out that the WID approach tended to isolate women from the rest of the community or society, in terms of space as well as time. It focused on provision of practical needs such as training, credit or food relief. Important as these may be in the short run, they did not address the fundamental causes of women's oppressed and marginalised status, which centred around relations of power and ownership/control of basic resources at all levels of society (TGNP, 1993).

Gender and Development (GAD) programmes, in contrast, focused on change of gender relations, by means of historical analysis and transformative action. Gender relations were understood to be social constructs, not biological givens, which changed over time and in different locations (of geography, social class, ethnicity, and so on). They were reconstructed on a daily basis as a result of the actions and ideas of individuals and groups, and the way in which society was governed and ruled at all levels. Gender could therefore be changed as a result of specific actions by, for example, a group of committed activist civil society organisations; hence, the assertion that gender is, or can be, a transformative analysis and practice (Mbilinyi, 1998b).

Gender relations were understood to interact with, as well as partially define, other key social relations such as class, ethnicity and north–south relations. Women differed among themselves because of class/race/national relations, for example, as did men. Moreover, people in different classes experienced gender differently. This made it difficult to generalise about the needs of all women at the national or regional level, and even within a given community. Moreover, the needs of men were also taken into account. Priority has generally been given to the perspectives of women and men who are poor, marginalised, disempowered.

A macro–meso–micro framework of gender analysis is increasingly being used in Southern African countries in efforts to democratise decision making concerning economic and social policy (Elson *et al.*, 1996; SEAGA, 1997). The framework can be used to discuss the structure of the economy; the pattern of decision making and the way in which the economy responds to policy changes; the distortions and biases which hamper effective development; and the opportunities for development transformation.

Participatory development involving both women and men connotes increased participation in making decisions at each of the three levels, macro, meso and micro. What kinds of decisions are we talking about? At the macro level, they include how to allocate public revenue with respect to different sectors, such as agriculture, industry, social services, military or debt repayment. In Tanzania, for example, agriculture policy in the 1970s was based on active state intervention to support small-scale producers by means of subsidies, pricing policies and other measures. Following SAP and liberalisation which began in the mid-1980s, there has been a withdrawal of state support for agriculture in general, and small-scale producers in particular, with devastating consequences (see Mbilinyi 1997, 1998b; TGNP 1999b, and other chapters in this volume).

Decisions at the meso level are made by sectoral ministries, in interaction with civil society organisations such as business associations, trade unions and NGOs. How services ought to be delivered, to whom, and who should bear the cost will be determined by institutional systems. Gender analysis of the Ministry of Agriculture and Cooperatives, for example, has documented the male bias which pervades nearly all departments, defined in terms of institutional structure, personnel and content of programmes (TGNP, 1999b). Men constitute the majority of professional, managerial and administrative staff. Prominence is given to export crops in research and development, crops monopolised by men at household level. The theoretical models used to guide rate of return studies pertaining to crops assume a unitary household with a male household head and 'dependent' unpaid wives and children (Mbilinyi, 1997). Failure of crop support programmes has been a frequent result of such gender 'blindness', as a result of the resistance of women and children/youth to work as unpaid family labour in smallholder farms.

Decisions about how to manage assets such as land, income or expenditure are also scrutinised at the micro level, given their impact on the capacity of women to make rational economic choices. A male bias prevails in terms of access and control over key productive assets such as land, livestock (especially cattle, sheep and goats), and cash income and credit which blocks women from making basic improvements in farming and livestock-keeping. Contrary to many project assumptions, women's labour is not elastic, because of the triple workload they carry: unpaid domestic work (cooking, provision of water and fuel, production of food for household consumption), farm and livestock production for the market, and off-farm economic activities.

Gender transformation at household level has been a major focus of many African interventions in support of women's empowerment, but they often leave out the interconnection between male bias at household level and male bias within other institutions such as the community, the market and the state.[6] The Gender Budget Initiative has incorporated the linkage between macro, meso and micro policy, as will be shown below.

24.6 The Gender Budget Initiative in Tanzania

The Gender Budget Initiative in Tanzania has been organised by Feminist Activists (FemAct), a coalition of some ten or more activist organisations, led in this case by Tanzania Gender Networking Programme (TGNP). The initiative was triggered off by the recognition in 1996 that there was a need to lobby for more resources for the education and health sectors, given the drastic drop in quality of services rendered. There was no point lobbying solely for gender equity in public programmes that were rapidly disintegrating under the onslaught of SAP measures. Drawing on the example of the Women's Budget Initiative in South Africa (Budlender, 1996), FemAct began a process of study in 1997, to learn more about the structure of the policy formulation process with respect to national and district budgets. Experts and government officials were invited to make presentations on the budget process in general, and on health and education sectors in particular, during the weekly Gender and Development Seminar Series at TGNP.

During 1997 and 1998, lobbying and advocacy activities intensified, with the support of a GBI task force consisting of NGO partners and allies within the central ministries of Finance and Planning, as well as Health and Education. Sensitisation workshops were held with budget officials in all four ministries to promote gender analysis and planning and win their support for GBI. Two parallel research processes were carried out, one at national level, to study the budget process at macro and sectoral level (TGNP, 1998) and another at district level, in one rural and one urban district (Mwateba, 1998). The research was a critique not only of the budget making process, but also of specific policies in question, such as that of education and health and, more recently, agriculture and water.

The process of carrying out the research and disseminating its findings has been as important as the actual 'findings' themselves. Members of Parliament, government officials, members of civil society organisations and others have participated in GBI workshops, in order to raise their critical awareness of democratic issues with respect to policy formulation. Questions included: Who makes the major decisions concerning allocation of resources at district level? At national level? Which stakeholders have access to adequate information so as to be able to monitor the implementation of planned expenditures? Who benefits from these allocations?

The research found that the SAP had created a gender unfriendly macroeconomic environment which prioritised efficiency over equity along gender as well as class, urban–rural and other lines (TGNP, 1998, 1999a). The budgeting process itself was found to be undemocratic and male dominated, excluding most citizens from decision making. Austerity measures and budget cuts had reduced the resources available for social services, and led to retrenchment of many women as well as men in public social services. The context of globalisation in the post-cold war world had also changed the terms of aid, from altruistic 'aid' programmes to an emphasis on trade and commerce. There was more open conditionality, with strings attached to private companies coming from donor countries in several cases. Finally, the top-down model of budgeting facilitated corruption and leakage at all levels, with a decline in transparency and accountability. International development agencies were as involved as government agencies and civil society organisations in the plunder of resources, as can be seen by careful scrutiny of the tender process.

24.7 Concluding remarks

The Gender Budget Initiative in Tanzania, led by the FemAct Coalition, seeks to increase the participation of women, at one level, and 'ordinary citizens', at another, in the decision making process over a key determinant of development strategies, that is, the budget process at district and national level. Further reflection is needed on the problematic nature of 'participation' alone, without consideration of gender/class/race/national dimensions. In whose interests would ordinary citizens or women act? To what extent would theirs be an alternative vision of development, which consciously opposes the present market-oriented approach of the government, IFIs and other development agencies?

Self-determination is a key element in Elson's (1994b: 513; emphasis added) definition of people-centred development:

> the formation and use of human capabilities through social and political and economic arrangements that put people in control of the development process . . . rather than being controlled by the development process . . . There have to be some democratically organised collective agreements organised to set the social and political framework ('rules of the game') if chaos is to be avoided. Moreover, *democratic participation in decision making is not enough: these collective agreements have to constrain the power of money over people.*

Whether development will be guided along money-centred or people-centred principles is a political question, which will be determined by the balance of power between the 'owners of money-capital' and producers/consumers, nationally and globally.

Some specific steps could be taken to restructure the international financial system, beginning with a transaction tax to be extracted from speculators in money markets – a suggestion made by economists 20 years ago, and ignored by IFIs (ibid.: 21). Economists in the north and south are resurrecting the idea again. Social regulation of the money market needs to be matched by social regulation of markets for labour and goods as well, along with the processes of production and of consumption. There needs to be interaction between market opportunities, state provision and community organisation, which transforms the present terms under which people enter into economic relations at all levels.

The following questions have been designed by Elson (ibid.: 522) to guide the movement for change:

- Opportunities in what kind of markets?
- What kind of provision for whom by what kind of state?
- What kind of community organisation with what kind of objectives?

Do the labour markets, for example, recognise women's need for safe, viable child care? Or protection from sexual harassment? Or the right of all workers to a viable living wage for themselves and their families? If not, social action is needed by workers' organisations supported by community organisations to demand appropriate changes, while raising awareness among workers (including farmers and other self-employed) about their rights, and facilitating their capacity to organise themselves on their own behalf.

Notes

1 The language formerly used by the minority regime in South Africa to maintain its racial apartheid system; globalisation processes depend on 'apartheid' policies at all levels.
2 Discussion based on the information provided in *Beyond Inequalities* reports and presented to the April 1997 Gender conference of SADC/SARDC delegates in Mbabane. See SARDC (2000); Flood *et al.* (1997); Iipinge and LeBeau (1997); Letuka *et al.* (1997); Mvula and Kakhongwa (1997); see also Elson (1994b); Kanji and Jazdowska (1993); Geisler and Narrowe (1990); Munachonga (1986); TGNP (1994); Kerr (1999).
3 Mukangara and Koda (1997: 27); for formal sector employment, see Mbilinyi (1991).
4 The NGO, TAMWA, has been carrying out research with and among domestic servants and sex workers as part of their campaign against child labour, along with kuleana.
5 For additional references: general analysis, see Bakker (1994). On Tanzania, see Kaijage (1997); Mbilinyi (1997); Meena (1991); Mukangara and Koda (1997). On Malawi, see Gladwin (1991a). For Zambia, see Malatsi (1995).
6 But see recent institutional and policy analyses in Kabeer (1994); Whitehead and Lockwood (1999); TGNP (1998, 1999a, b).

References

Bakker, I. (ed.) (1994) *The Strategic Silence*, London: Zed Books.

Bryceson, D.F. (1999) 'African rural labour, income diversification and livelihood approaches', *Review of African Political Economy*, 80: 171–89.

Budlender, D. (ed.) (1996) *The Women's Budget Institute for Democracy in South Africa*, Pretoria: Institute for Democracy in South Africa.

Bush, R. and Szeftel, M. (1998) 'Editorial: "Globalization" and the regulation of Africa', *Review of African Political Economy*, 76: 173–7.

Elson, D. (1994a) 'People, development and international financial institutions', *Review of African Political Economy*, 62: 511–24.

Elson, D. (1994b) 'People, development and international financial institutions', *Review of African Political Economy*, 75: 25–46.

Elson, D., Evers, B. and Gideon, J. (1996) 'Concepts and sources', Gender Aware Country Economic Reports, Working Paper no. 1, University of Manchester: DAC/WID Task Force.

Flood, T., Hoosain, M. and Primo, N. (1997) *Beyond Inequalities: Women in South Africa*, Capetown/Harare: University of Western Cape Gender Equity Unit and SARDC.

Geisler, G. and Narrowe, E. (1990) *'Not So Good, but Quite Ambitious': Women and Men Coping with Structural Adjustment in Rural Zambia*, Lusaka: Report to Swedish International Development Authority

Gladwin, C.H. (ed.) (1991a) *Structural Adjustment and African Women Farmers*, Gainesville, FL: University of Florida Press.

Gladwin, C.H. (1991b) 'Fertilizer subsidy removal programs and their potential impacts on women farmers in Malawi and Cameroon,' in C.H. Gladwin (ed.), *Structural Adjustment and African Women Farmers*, Gainsville, FL: University of Florida Press.

Gungana, B., Ragobar, S. and Varma, O.N. (1997) *Beyond Inequalities: Women in Mauritius*, Harare: Southern Africa Development Community.

Hoogvelt, A. (1997) *Globalisation and the Postcolonial World*, London: Macmillan Press.

Iipinge, E.M. and LeBeau, D. (1997) *Beyond Inequalities: Women in Namibia*, Harare: University of Namibia, Social Sciences Division and SARDC.

Imam, A.M., Mama, A. and Sow, F. (eds) (1997) *Engendering African Social Sciences*, Dakar: CODESRIA.

Kabeer, N. (1994) *Reversed Realities*, London: Verso.

Kabeer, N. and Whitehead, A. (1999) 'From uncertainty to risk: poverty, growth and gender in the rural African context', background paper for 1999 Status of Poverty in Africa Report, University of Sussex.

Kaijage, F. (1997) *Gender Impact of Structural Adjustment Programme on Employment in the Public Sector*, Dar es Salaam: report for ILO.

Kanji, N. and Jazdowska, N. (1993) 'Structural adjustment and women in Zimbabwe', *Review of African Political Economy*, 56.

Katapa, R. (2000) 'Gender patterns in employment in the informal sector', in M. Mbilinyi (ed.), *Gender Patterns in Micro and Small Enterprises of Tanzania*, Dar es Salaam: AIDOS.

Kerr, J. (1999) 'Responding to globalization: can feminists transform development?', in M. Porter and E. Judd (eds), *Feminists Doing Development: A Practical Critique*, London and New York: Zed Books.

Letuka, P., Matashane, K. and Morolong, B.L. (1997) *Beyond Inequalities: Women in Lesotho*, Harare: South African Development Community.

Malatsi, M. (1995) 'Possibilities for cushioning the adverse effects of the structural adjustment programs on vulnerable women in Zambia', in G. Thomas-Emeagwali (ed.), *Women Pay the Price*, Trenton, NJ: Africa World Press.

Mbilinyi, M. (1991) *Big Slavery: Agribusiness and the Crisis in Women's Employment in Tanzania*, Dar es Salaam: Dar es Salaam University Press.

Mbilinyi, M. (1992) 'Research methodologies in gender issues', in R. Meena (ed.), *Gender in Southern Africa*, Harare: SAPES Books.

Mbilinyi, M. (1995) 'The role of NGOs in promoting the economic advancement of rural women', background paper to CIRDAFRICA/IFAD Workshop, Arusha.

Mbilinyi, M. (1997) *Women Workers and Self-employed in the Rural Sector*, Dar es Salaam: Report for ILO.

Mbilinyi, M. (1998a) 'The end of smallholder farming?', GDSS paper, Dar es Salaam: Tanzania Gender Networking Programme.

Mbilinyi, M. (1998b) 'Notes on building a social movement in the globalisation era', paper presented to Annual Gender Studies Conference, 24 November, Tanzania Gender Networking Programme, Dar es Salaam.

Mbilinyi, M. (1999) *Rural Food Security in Tanzania: The Challenge for Human Rights, Democracy and Development*, Dar es Salaam: IDS, Rural Food Security Policy and Development Group, report presented to Launching Workshop (July).

Mbilinyi, M. (2000a) 'Gender and micro-small enterprises in Morogoro', in M. Mbilinyi (ed.), *Gender Patterns in Micro and Small Enterprises of Tanzania*, Dar es Salaam/Rome: AIDOS.

Mbilinyi, M. (ed.) (2000b) *Gender Patterns in Micro and Small Enterprises of Tanzania*, Dar es Salaam/Rome: Association for Women in Development (AIDOS).

Mbilinyi, M. (2000c) 'Overview of the region', in Southern Africa Development Community, *Beyond Inequalities: Women in Southern Africa*, Harare: SARDC.

Meena, R. (1984) 'Foreign aid and the question of women's liberation', *African Review*, 11 (1): 1–13.

Meena, R. (1991) 'The impact of structural adjustment programs on rural women in Tanzania', in C.H. Gladwin (ed.), *Structural Adjustment and African Women Farmers*, Gainesville, FL: University of Florida Press.

Meena, R. (ed.) (1992) *Gender in Southern Africa*, Harare: SAPES Books.

Mohan, G. (1994) 'Manufacturing consensus: (geo)political knowledge and policy-based lending', *Review of African Political Economy*, 65: 525–38.

Mukangara, F. and Koda, B. (1997) *Beyond Inequalities: Women in Tanzania*, Harare: Southern Africa Development Community.

Munachonga, M. (1986) 'The impact of economic adjustment on women in Zambia', mimeo, Lusaka: Social Development Studies Department, University of Zambia.

Mvula, P.M. and Kakhongwa, P. (1997) *Beyond Inequalities: Women in Malawi*, Harare: SARDC.

Mwateba, R. (1998) *Gender Budget Initiative: Kondoa District*, Dar es Salaam: Tanzania Gender Networking Project.

Puleng, L., Matashane, K. and Morolong, B.L. (1997) *Beyond Inequalities: Women in Lesotho*, Harare: SARDC.

Sen, G. and Grown, C. (1987) *Development, Crises and Alternative Visions*, New York: DAWN and Monthly Review Press.

Shundi, F. (2000) 'Gender patterns in MSE employment: the case of Arusha Municipality and AruMeru District', in M. Mbilinyi (ed.), *Gender Patterns in Micro and Small Enterprises of Tanzania*, Dar es Salaam/Rome: Association for Women in Development (AIDOS).

Social, Economic and Gender Analysis (SEAGA) (1997) *Macro-level Handbook*, Reader no. 1, Tucson, AZ: University of Arizona.

Southern African Research and Documention Centre (SARDC) (2000) *Beyond Inequalities: Women in Southern Africa*, Harare: SARDC and Women in Development, Southern African Awareness (WIDSAA).

Tanzania Gender Networking Programme (1993) *Gender Profile of Tanzania*, Dar es Salaam: TGNP.

Tanzania Gender Networking Programme (1994) *Structural Adjustment and Gender*, Dar es Salaam, TGNP.

Tanzania Gender Networking Programme (1998) *Gender Budget Initiative (GBI)*, Dar es Salaam: TGNP.

Tanzania Gender Networking Programme (1999a) *Budgeting with a Gender Focus*, Dar es Salaam: TGNP.

Tanzania Gender Networking Programme (1999b) *Agriculture Sector Study*, GBI Reports, Dar es Salaam: TGNP.

Whitehead, A. and Lockwood, M. (1998) 'Gendering poverty: a review of six World Bank African poverty assessments', *Development and Change*.

25 Health and sickness

Towards health systems that meet the needs of the poor[1]

Gerald Bloom and Henry Lucas, with Rosalind Goodrich

25.1 Introduction

During the 1950s–1970s health indicators improved substantially in many African countries as they invested heavily in public health services. Their governments enthusiastically endorsed the general international consensus on the relationships between health, development and poverty that culminated in the acceptance of the primary health care (PHC) strategy in 1978. Support for health services subsequently waned in response to economic crisis and changing international thinking. The focus on market solutions in the 1980s led to a view that healthcare was essentially a consumption good and that encouraging economic growth was the most effective way to improve health status.

Understanding of the role of health is changing again as development efforts have targeted poverty reduction as the primary goal. It is now generally accepted that not only do the poor face serious health problems, but that serious health problems are a major cause of poverty. And in many sub-Saharan countries the rate of health improvement has slowed or reversed over the last decade. Successful poverty reduction strategies must include measures that specifically address the health needs of the poor. This chapter explores what these measures might be.

25.2 Recent trends in mortality and ill health

There are severe limitations on the quality of health-related data in sub-Saharan Africa. As a general rule, data quality is inversely related to the level of morbidity and mortality in a given area. Poorer countries and regions tend both to have higher levels of sickness and to allocate fewer resources to data collection. Most of the commonly used indicators of health status presented below are therefore derived indirectly by applying standard epidemiological or demographic models to the few relatively reliable items of data available. It should be noted that the impact of the AIDs pandemic on the robustness of such models has not yet been fully assessed and thus even generally accepted estimates of basic morbidity and mortality indicators should be treated with some caution.

25.3 The burden of disease in sub-Saharan Africa

The extent of sickness and premature death is often measured using DALYs (disability adjusted life years). The DALY rate for a country estimates the excess years of active life lost due to disability and death per 1000 of the population, above what are regarded as achievable targets based on expert judgements. Sub-Saharan Africa, with less than 10 per cent of world population, accounts for 22 per cent of total world DALYs and 25 per cent of those lost to premature death. This latter rate is almost twice that of the second worst region, India, and seven times that of the established market economies. The pattern of disease in sub-Saharan Africa is also markedly different from other regions. Levels of infectious and parasitic diseases are almost twice those in other developing country regions.

25.3.1 Mortality

In this and the following sub-sections, attention will be focused on the distribution of mortality and ill health across the fifty countries classified as constituting sub-Saharan Africa by the World Bank, i.e. the UN list of forty-eight states plus two other small countries with disputed status.

Mortality will be considered here mainly in terms of two indicators, life expectancy and the infant mortality rate (IMR), recognising that both have limitations. In 1996, the majority of countries had life expectancies in the range 45–54 years, with six exhibiting values less than 44 years and six achieving a value over 60 years (see Figure 25.1). As commonly observed (WHO, 1998), the life expectancy for women was typically higher than that for men, an average difference of just over 3 years. This average gap has changed very little since 1970, but there are indications that differences may be narrowing in countries where HIV/AIDS has had a major impact. In Uganda and Zambia, for example, male and female life expectancies are now very similar.

25.3.2 Trends in mortality

Considerable progress has been made in terms of increasing life expectancy since 1970. In that year 60 per cent of sub-Saharan countries had life expectancies below 45 years. By 1990 this had reduced to just 12 per cent. However, the picture had become less clear by 1996. The proportion below 45 years remained constant while that below 55 years increased.

As with life expectancy, infant mortality rates vary considerably across the countries of sub-Saharan Africa. The great majority are high compared to the rest of the world, twenty-nine

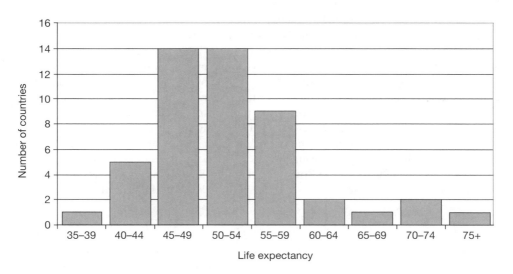

Figure 25.1 Distribution of countries by life expectancy, SSA, 1996

countries having rates above 80 per 1000 births and nine having rates above 120 per 1,000 births. The pattern described above for changes in overall mortality over time was not repeated for infant mortality. Gains over the period 1970–80 were sustained both from 1980 to 1990 and from 1990 to 1996. The main concern is that the number of countries exhibiting extremely high rates declined relatively slowly.

There is evidence that under-5 mortality also declined through the 1980s and early 1990s, possibly more rapidly in countries with high initial rates. The decline slowed in the latter part of the period, both in some lower mortality countries and in several with higher rates. Zambia stands out as the one country where child mortality increased considerably.

25.3.3 Morbidity

The pattern of sickness for sub-Saharan Africa is very different from that in other world regions, being still dominated by communicable diseases. Three such diseases – HIV/AIDS, tuberculosis and malaria – have been the focus of much recent activity, and considerable efforts have been made to produce at least reasonably reliable estimates of their incidence and prevalence.

HIV/AIDS

One of the main factors repeatedly cited to explain the relatively slow decline, and in a few cases increases in mortality, through the 1990s is the impact of acquired immune deficiency syndrome (AIDS). AIDS is the final and most serious stage of human immunodeficiency virus (HIV) infection. HIV infection attacks the immune system and leaves the body vulnerable to a variety of life-threatening illnesses. It is currently estimated that 22.5 million people in sub-Saharan Africa are living with HIV. Whereas in many parts of the world male rates exceed those of females, here it is suggested that approximately six women are infected for every five men, with consequential high rates among their children. Some 11.5 million people, including 3 million children, have died from AIDS since the onset of the epidemic. Eighty per cent of all AIDS deaths in 1998 occurred in the region, and they were estimated to account for 19 per cent of total mortality, more than twice the figure for lower respiratory infections, the second most important cause of deaths (WHO, 1999).

Timaeus (1998) argues that 'the severity of the HIV epidemic is now the dominant determinant of adult mortality' and cites the example of Zimbabwe, previously one of the lowest adult mortality countries in Africa but now one of the highest, where the probability of dying between the ages of 15 and 60 years more than doubled between 1985 and 1995. On the other hand, he cautions that this is not the case for infant and child mortality, which are largely determined by other factors. In 1997 only ten of the forty-six sub-Saharan countries with data had HIV infection rates below 2 per cent, while twelve were in excess of 10 per cent. Southern Africa had the highest rates, with estimated numbers of people between 15 and 49 years living with HIV in Botswana, Namibia, Swaziland and Zimbabwe exceeding 20 per cent, and South Africa on course to reach these levels in the near future.

Tuberculosis

There has been considerable concern at the rapid rise in reported tuberculosis in recent years, which has been closely linked to the HIV/AIDS epidemic. Total new cases in 1996 were around 1.5 million. This was projected to rise to 2.1 million by 2000, an increase of 41 per cent in 4 years. Such increases are generally attributed to co-infection with HIV that both greatly increases the risk of developing active TB and dramatically increases its progression. Failure to treat infected people and trace and treat those in contact with them has also contributed.

Malaria

Over 90 per cent of the population of sub-Saharan Africa live in areas with a significant risk of malaria and around 75 per cent live in high-risk areas. Estimation of total cases is extremely difficult, given that only a very limited proportion are officially reported, but the World Health Organisation (1997) suggests an annual occurrence of 270–480 million cases in sub-Saharan Africa, 90 per cent of all cases world-wide. Somewhat over half of cases probably occur in children under 5 years, resulting in almost 1 million deaths each year. In burden of disease terms it ranks alongside respiratory disease as the most important cause of death and disability. It also constitutes a major burden on primary healthcare services, typically being included in 20–40 per cent of all outpatient diagnoses in the region.

25.3.4 Public health

Whatever their impact, new or resurgent diseases are clearly not the only factors which may have contributed to recent mortality and morbidity trends. For example, most of the countries of sub-Saharan Africa suffered considerable economic reversals between 1980 and the mid-1990s and many introduced stabilisation and adjustment programmes specifically designed to reduce inflation and government deficits. It is clear that in the great majority of countries aggregate government spending on health and related areas has been at best static over the period, with consequent reductions in per capita expenditures.

The impact of government spending on health status is a highly contentious issue. However, it does seems clear that severe economic decline and therefore lack of expenditure leads to the deterioration of water supplies, sanitation systems and health facility buildings and equipment. Stocks of drugs are not replenished and skilled members of staff leave or become part-time, supplementing their declining salaries from alternative sources.

There also appears to be a consistent relationship between some public health and related indicators and overall mortality. Table 25.1 shows the values of six widely available indicators classified by 1996 life expectancy categories. Higher levels of life expectancy are generally positively associated with all these indicators. The main message of Table 25.1, however, is

Table 25.1 Public health indicators by life expectancy, SSA, 1989–96

Life expectancy 1996	Access per 1,000 population to:			Malnutrition: children <5(%) 1990–6	Births attended by health staff (%) 1990–6	Immunisation: children with DPT (%) 1995
	Safe water 1989–95	Sanitation 1989–95	Health services 1993			
30–49	44	30	51	32	28	45
50–54	60	44	67	29	51	63
55–59	52	39	46	21	61	61
60+	100	73	99	13	90	83
All	53	39	58	27	45	58

Source: World Bank (1998a, b)

that in a considerable number of countries in sub-Saharan Africa the general public health situation should be a major cause for concern. On average in the lowest group, even accepting the probably optimistic access measures, well under half of the population has access to safe water or sanitation, half have no access to basic health services, one-third of children are malnourished, less than one-third of births are attended and more than 50 per cent of 1-year-olds have not received DPT immunisation.

25.3.5 *Other factors influencing morbidity and mortality*

While examining the distribution of mortality and sickness data across countries may be informative, there is a risk of over-simplifying the diverse and complex range of factors bearing on health status. During the 1980s and 1990s many of the countries of sub-Saharan Africa suffered not only from the tragic impact of HIV/AIDS but from a variety of other natural and man-made shocks which led to dramatic consequences, sometimes for the whole nation and sometimes for specific regions. These included war, civil unrest, ethnic conflicts, large-scale movement of migrants and refugees, economic crisis, drought, crop failure and famine. The impact of these other shocks has been to expose people to a greater risk of ill health, whilst reducing the capacity of communities and government health systems to cope with the higher levels of sickness.

25.4 The burden of disease on the poor

Disease causes pain, disability and premature death. It also has an impact on economic development and poverty reduction. The following sub-sections explore how ill-health affects economic well-being. The aim is to indicate the wide range of factors that pro-poor health strategies must address.

25.4.1 *Loss of production by sick individuals*

A very sick or disabled person tends to be less productive than a healthy one. This is obvious when someone is seriously disabled by an acute illness. There is a wide spectrum of states of diminished capacity due to feebleness (resulting from old age or chronic illness), loss of one or more senses, reduced mobility and so forth. How much these problems reduce output depends on the degree of disability and the availability of opportunities for productive employment.

The impact of ill health is particularly evident in areas with a serious endemic health problem. For example, some parts of Africa are poor because of sleeping sickness or onchocerciasis (river blindness). Young adults with river blindness are less productive and their families become impoverished. Illness is a major block to development in affected areas. It is more difficult to measure the effect on productivity of moderately debilitating illnesses like malaria (Chima *et al.*, 1999).

Ill health can reduce returns to investment in human capital. For example, malnutrition and poor health interferes with schooling and makes expenditure on education less productive (Behrman, 1996). This underlines the painful decision households have to make when they choose between buying food, paying school fees and purchasing healthcare. It takes up to 20 years and substantial expenditure to produce a highly skilled worker. There is a heavy cost to society and to a skilled worker's extended family when he or she drops out of the labour force prematurely. This is a major cause of economic loss from AIDS. A pro-poor health strategy must take the relationship between ill health and productivity into account. This raises difficult questions regarding the relative impact of diseases of adults, management of pregnancies, child health and so forth. For example, should particular attention be given to treatment of injuries and reversible conditions that are most likely to affect household production?

25.4.2 *Financial costs*

There has been a heated debate about whether government health facilities should charge for services. There is evidence that poor households respond to increases in the cost of healthcare at government facilities by using them less. A study in Zambia, for example, found that some people turned to drug peddlers and others died without seeking medical help after user charges were increased (World Bank, 1994). Others point out that poor people already spend money on drugs, private practitioners and informal payments to government employees. They argue that it should be possible to provide individuals with better services at the same level of private expenditure, if user charges resulted in improved services and/or reduced informal payments. This point of view has been bolstered by a study in Cameroon which found that utilisation of a health system by the poor rose when increased charges were linked to quality improvements (Litvak and Bodart, 1993).

After a decade of debate, there are no simple answers regarding the role of user charges. Policy makers must assess their impact on the poor on a case-by-case basis. The fundamental problem of the poor is that they cannot afford many services from which they could benefit substantially. This problem is exacerbated when people receive little value for the money they spend on health. For example, they may travel to government health facilities that do not have drugs, or purchase doses of drugs that are unlikely to be effective. In such cases they waste resources that they can ill-afford to lose.

We need to shift our understanding of user charges from the perspective of health service managers to that of poor households. The latter often choose from a variety of public and private providers and they need to use the limited money they can spend on health services as effectively as possible. They are interested in what government can do to help them meet their needs. This reverses the question from: Should public facilities charge for some services? to What should public funding subsidise? and How can government help poor households get value for money in health care?

25.4.3 *Opportunity cost of caring for the sick*

The third element of the cost of health shocks is the time that carers, usually women, spend away from productive labour. Standing (1997) argues that prolonged financial constraints on public health services have put a higher burden of care on unpaid female household members. She warns against calls for community participation or community care that have hidden costs. Leslie (1992) describes concrete examples such as the design of child healthcare programmes that do not take into account the work patterns of mothers. There is a risk that strategies for health improvement will fail if they are based on unrealistic assumptions about the availability of unpaid household labour. This is one reason why attempts to establish national community health worker programmes have mostly been unsustainable (Walt, 1988).

The recognition that long-term care of the sick and vulnerable puts a substantial burden on households is slowly changing perceptions of the roles of outpatient and inpatient services.

There has been a tendency to view hospitals as centres of costly, high-technology care. This neglects the role of general hospitals as providers of basic nursing care for the seriously ill and of treatment that returns people to productive activity. Many hospitals face serious financial problems and cannot provide effective services. These difficulties have been exacerbated by the HIV/AIDS epidemic that has substantially increased the demand for hospital care. Loewenson and Kirkhoven (1996) report that up to 70 per cent of hospital admissions in Zimbabwe are HIV-related. These problems have increased the burden on household carers. People often must choose between feeding and providing additional nursing for hospitalised family members or caring for them at home.

The present arrangements, whereby a large proportion of a country's health workers are based in hospitals which cannot afford to provide effective services, are wasteful (Bloom, 1998). A case can be made for the establishment of alternative means of providing low-cost basic support to very sick or disabled people. For example, there have been experiments with arrangements for the care of AIDS patients in community care facilities or their own homes.

25.4.4 Households coping with an illness shock

One characteristic of healthcare is that major illness episodes are rare but costly. The serious illness of a family member is a major shock to an entire household. It is possible to conceptualise household adaptation in three phases related to the severity of the shock (Corbett, 1988 and 1989; Russell, 1996; Wilkes et al., 1997).

The household's primary adaptation consists of a reallocation of resources and depletion of reserves without substantially affecting future productivity. It may finance medical bills by running down savings, selling produce, undertaking additional paid labour and asking other family members to help. A secondary adaptation is necessary when a sickness episode (or episodes) is more severe. The household may have to choose between neglecting the sick person or compromising its capacity to withstand future shocks. The measures open to it, if it chooses the latter, include forgoing expenditure on essential inputs such as food or education, borrowing from commercial money lenders, and selling potentially productive assets. A household can recover from these measures, but its ability to withstand future shocks is reduced. The tertiary adaptation consists of survival measures including migration in search of a livelihood or the reconstitution of households into less viable entities (for example, an elderly person caring for young children). This has become a significant problem in regions where AIDS is widespread. Households are placed under enormous strain and traditional coping mechanisms may be reaching their limits (Ainsworth and Over, 1994; Barnett and Blaikie, 1992). Measures to help such households need to go beyond safety nets and be concerned with trying to restore some productive capacity to the household base.

25.4.5 The social impact of major health crises

The population of sub-Saharan Africa has faced a series of shocks including war and/or conflict-related movements of people, environmental shocks including severe crop failures, substantial economic adjustments, the HIV/AIDS epidemic, and the prolonged financial squeeze on the public sector. It is impossible to separate the impact of each shock, but together they have put enormous pressure on social support systems.

Continued stress on extended families and communities reduces their capacity to assist households and individuals that are in difficulty (Lucas and Nuwagaba, 1999). The capacity of many governments to help has also diminished. Large movements of people have created densely populated rural encampments or urban settlements where people are exposed to health threats and social coping mechanisms are eroded. Deterioration of public administration has reduced the state's capacity to protect households against shocks. All these factors increase the risk that households will be impoverished when they face a serious health problem.

Health crises follow and amplify other social stresses at a societal as well as an individual level. The major cause of death in the Darfur famine, in Sudan during the mid-1980s, was a series of epidemics resulting from the concentration of people in areas without adequate sanitation and clean water (de Waal, 1989). These outbreaks aggravated the effort to recover from the initial shock. Similar stories can be told about the social and health impact of conflict and major socioeconomic adjustment.

The AIDS epidemic illustrates the interrelationship between disease, poverty and social stress. HIV is mainly transmitted heterosexually in Africa. There are as many or more women living with HIV as there are men. This is partly for physiological reasons, but it also reflects patterns of sexual practices and of gender relations, in particular the unequal social and economic position of men and women. Being HIV-positive makes people more vulnerable to infectious diseases that are constantly present in poor communities. It amplifies the impact of existing health problems, such as tuberculosis and water-borne infections. Diseases related to HIV infection have become the major cause of death of adults aged 15–44 in more than one country. As a consequence of the high level of deaths in the 15–44 years age group, WHO estimates that as many as 10 million children have been orphaned since the beginning of the epidemic. This in turn has led to an increasing number of street children in the community and the emergence of 'child-headed' households.

There is a growing national and international response to the AIDS epidemic in Africa. This will only be effective if it tackles the problem from several angles. Measures to combat the spread of AIDS – intensive health education, for example – and to help families cope with the disease are not enough on their own. The epidemic has had a major impact on the economic well-being of affected households and communities and it has reduced the effectiveness of health systems. Resources must also be used to address these consequences of the epidemic, as part of a wider poverty reduction strategy.

25.5 Meeting the needs of the poor in a pluralist health sector

Many factors influence the health of an individual or a population. These include access to nutritious food, clean water, adequate clothing and shelter, and the means for hygienic disposal of human wastes; freedom from contamination by hazardous substances and environmental pollutants; and relevant information and skills. Many if not all such factors tend to improve with rising incomes and it is the common experience that declining levels of poverty are associated with increasing health status. Advances in preventative or curative health services may be involved, but not necessarily. None the less, most societies organise measures specifically intended to prevent and cure sickness and support the severely ill and their families. A health strategy that addresses the health and healthcare needs of the poor will have a much greater impact if it is part of a broader effort to meet all the requirements identified above.

25.5.1 The divergence between post-colonial vision and current reality

The health development strategies of most African countries were based on a shared vision of a health sector in which households and communities provided basic social support and voluntary labour for public health activities, and government provided and financed services requiring specialist knowledge, drugs and equipment. To attain this goal more rapidly, many countries trained large numbers of so-called 'medical assistants', who could provide basic services within the public sector but only under the close supervision of more highly qualified staff. In some countries civil society organisations and/or private providers also had a recognised role, but this was seen as strictly complementary to public sector facilities.

The current reality in many countries differs radically from expectations. A diverse range of providers, from drug peddlers to specialist doctors, provides a spectrum of services (frequently unsupervised and unregulated). People obtain drugs and medical items from a wide variety of public and private sources; and users pay a combination of fees and informal charges for most services, including those provided by government. The emerging pluralistic health sector resembles a weakly regulated, publicly subsidised market (Leonard, 2000; Bloom and Standing, 2001). Many national health development strategies do not acknowledge this situation adequately.

The present situation will not be transformed by simply increasing funding of government health services. It is no longer plausible to believe that the post-colonial vision can be restored. What is needed is a systematic approach that identifies plausible strategies for meeting the needs of the poor, based on a realistic assessment of the existing health sector, both public and private.

25.5.2 *The health sector after structural adjustment*

The changes in Africa's health sector have partly been due to increases in the supply of health-related goods and services. During the past 30 years many countries have created a network of health facilities and trained large numbers of health workers. Systems for distribution of goods, including health-related ones, have developed considerably. A large proportion of the population now lives within walking distance of a health worker and a supplier of health-related commodities.

The economic crisis of the 1980s, and the resulting economic and institutional changes, accelerated the creation of a market within the health sector. Public health services experienced prolonged financial constraints and many countries experienced large falls in real expenditure on health. Some governments magnified the impact of these falls by reducing expenditure on maintenance and non-salary operating costs disproportionately, maintaining staffing levels when total spending on salaries was falling, and trying to sustain all services rather than protecting the most essential. The common outcome is run-down, ill-equipped facilities, shortages of drugs and consumable inputs, and personnel who for years have worked for inadequate pay with little supervision.

The health sector's difficulties have been compounded by problems with government finance and public sector organisation. Many public services cannot pay adequate salaries. Government employees commonly have few incentives to perform well, and many have developed livelihood strategies that imply they no longer function as full-time, salaried officers. Supervisory functions have broken down because of a lack of transport and effective communications and because supervisors prefer to spend their time on other activities. The capacity of many governments to discharge their core responsibilities has been severely impaired, leading to an informal and formal market in health services. It is now accepted practice in many countries for government health workers to ask patients for informal payments.

The major barriers to access to effective healthcare for poor households are an inability to pay for goods and services, particularly during an episode of serious illness, and a lack of quality control. Some may still travel a long distance to reach any form of health facility, but many others confront a bewildering variety of public and private care providers and drug suppliers. In most localities there is no effective regulation and public health laws are widely ignored. Users find it difficult to assess the competence of different practitioners, many of whom have had relatively little training. They do not have access to reliable advice on the most cost-effective way to deal with particular health problems and frequently have to rely on their own judgement. They often waste the little money they have on inappropriate or unnecessary treatment. The result of decades of public investment in health facilities and training is often a market that does not deliver basic reliable, cost-effective services to poor people. Governments need to take this reality into account in formulating strategies for meeting the health needs of the poor.

25.6 Government and the health sector

25.6.1 What functions must government perform?

Once governments acknowledge they are not the only health service provider, they must identify their specific responsibilities more clearly. At a minimum they have to strive to achieve the following:

- enforce public health regulations;
- assist communities to dispose of human wastes and provide access to clean water;
- organise programmes to reduce exposure to malaria, tuberculosis, sexually transmitted infections and water-borne epidemic diseases;
- enforce regulation of health practitioners, drugs and other products; and
- provide information to enable people to cope with health problems more effectively and make better use of the available resources.

Many countries do not come near to meeting these objectives and any strategy must give them priority. This will be difficult to achieve unless governments are able to pay suitably skilled personnel competitive salaries and finance the other costs of providing these services. Reforms to health services, public sector employment and public finance are thus inextricably linked.

There is a considerable debate about what should happen to other government health services. One option would be to increase public finance substantially so that all employees can be paid appropriate salaries. As a variant of this, government budgets could be supplemented by substantial user charges for certain non-essential services. For example, government health facilities could earn revenue by selling over-the-counter drugs. Alternatively, the number of public employees could be substantially reduced and the range of services provided limited. People would seek other services from private sector organisations and individuals, in some cases operating under a contractual arrangement with government.

25.6.2 What health services should government finance?

Governments will have to increase their health budgets substantially and prioritise expenditure if they seriously intend to address the health problems of their population. This will require long-term commitments by governments and international donors.

Government health expenditure frequently favours the better-off. A recent study of nine African countries found that the richest 20 per cent of the population received, on average, over twice as much financial benefit from public health expenditure than the poorest 20 per cent (Castro Leal *et al.*, 1999). Active measures are needed to ensure that increases in public health budgets are linked to measures which ensure it is spent on services the poor use. There are a variety of strategies for achieving this.

Target providers of basic services

Governments could allocate resources preferentially to providers of basic services for the poor, such as primary level facilities and basic general hospitals (Noormahomed and Segall, 1994). This ensures widespread access to interventions that address common health problems. The trend towards the devolution of government functions complicates this strategy. Each authority establishes its own priorities. Their capacity to finance health services depends on their ability to raise revenue. Regional inequalities are likely to increase unless poor localities receive substantial fiscal transfers from national level. It may also be necessary to establish minimum standards for the availability of facilities, personnel and other resources.

Allocate resources to particular health problems

Another way to reach the poor is to allocate resources to particular health problems. Many governments preferentially fund preventative programmes and the treatment of illnesses, such as sexually transmitted infections, that can spread to others. They need to find ways to support potentially cost-effective interventions without creating a multiplicity of vertically organised programmes that are expensive and unsustainable (LaFond, 1995). Recent announcements by international bodies and private foundations that they intend to finance very large initiatives to tackle tuberculosis, malaria and HIV make this issue particularly important. Governments must ensure that health services will continue to be effective after external funding has ceased.

Target poor households

Countries also target poor households. For example, when they increase user charges they exempt those who cannot afford to pay. These provisions are often not translated into practice (Russell and Gilson, 1997). This is partly because health facilities have little incentive to provide services free of charge and partly because the administrative arrangements for identifying the destitute may be complex and time consuming.

Different targeting arrangements are appropriate for different kinds of services. It is simpler to subsidise certain highly cost-effective services universally on the grounds that the benefits from ensuring that few people are denied access outweighs the loss of revenue from people who could pay. Services that address particular health problems of the poor and facilities that other social groups are unlikely to use are most appropriate for this treatment.

Other services may be too costly to provide on a universal basis. It may be better to charge for some kinds of hospital care and to establish mechanisms, no matter how imperfect, which enable people to claim exemptions. This will certainly exclude some people from these services. However, this may be necessary if alternative uses of scarce government resources would provide greater benefits.

Establish a health insurance fund

Many governments provide civil servants with subsidised healthcare in public hospitals. The ministry of health is often expected to finance this benefit. This is a source of pressure for a continuation of funding of the more sophisticated hospitals. Governments could make policy making more transparent by establishing a health insurance fund to cover some or all of the cost of services in designated facilities for government employees. The government could encourage the fund to explore strategies for improving the cost-effectiveness of services. For example, beneficiaries could be discouraged from using outpatient departments of referral hospitals.

Most governments have been unwilling to address the consequences of shifting public funding away from certain services. Many hospitals have had to cope with decreasing government grants; they have found it difficult to provide effective services. Case fatality rates have risen in some cases. Governments have just begun to consider options for reconstruction and reorganisation of hospital services, including concentration on basic services for the majority of the population and either closing some expensive hospitals and/or specialist units or charging substantial fees for them. Governments will find it very difficult to charge these fees unless people can obtain hospital insurance. In choosing between alternatives, they need to take into account the needs of the poor for essential inpatient treatment and the potential benefits to people who can afford to pay for the more expensive services (Bloom, 2000).

25.6.3 What can governments do to ensure that cost-effective services are available?

The poor obtain a large proportion of their health services from personnel who were trained on the assumption they would work under close supervision. Things have not worked out that way. One of the greatest challenges the African health sector faces is to protect the population against incompetent health advice. Strategies for achieving this will combine training for health workers in the provision of effective and cost-effective treatment, improved management of clinical practice in public facilities, increasing involvement of NGOs in service provision, strengthening the capacity of local government health services to regulate service providers, and increased involvement of professional organisations and local communities in monitoring service quality.

It will take a long time to establish an effective regulatory regime. In the meantime, government can help households make better-informed decisions. People can already purchase almost any drug from local retailers. They would benefit a great deal if they knew how to treat common conditions such as coughs, fevers and diarrhoea. Governments could make this information available through a number of channels; they could also provide information on the performance of different kinds of health provider in terms of health outcomes and cost. This would enable local populations to make better informed choices and make better use of their limited resources.

25.7 How can households be helped to cope with health shocks?

There has been an increasing interest in the possibility of creating local health insurance schemes (Carrin *et al.*, 2000). Donors have supported many schemes and questions have been raised about their sustainability. There is a wide variety of local arrangements to protect individuals against serious health shocks. They vary from well-organised transfers of resources within households and extended families to funeral societies and informal understandings within communities. Governments need to find ways to build upon these initiatives to strengthen their ability to support households and the unfortunate and destitute who have no means of taking care of themselves (Table 25.2).

25.8   How can providers be made more accountable for their performance?

Health service providers are virtually unsupervised in many countries and additional funding from government or donors could simply increase their incomes. Measures are needed to make providers more accountable. Governments can play a major role in supervising and regulating health if they employ enough people with the necessary skills and provide them with sufficient resources to carry out their responsibilities. They also need to create a situation in which people responsible for these tasks perceive themselves as representatives of the public good. It will take time to establish this capacity in areas where poverty levels are high.

There is increasing interest in the potential contribution of community organisations to monitoring the use of government and donor funds and overseeing the performance of service providers (Loewenson, 1999; Cornwall *et al.*, 2000). This might be approached through the establishment of formal community structures and by making information more widely available. It will be necessary to train people to use relevant information and intervene when problems arise. Health workers will have to become accustomed to negotiating with users of their services. The success of this kind of initiative may be linked to broader democratisation efforts.

25.9   Conclusions

Governments will have to lead the adaptation of the health sector to the new economic and institutional realities. In doing so, they need to keep sight of the objectives of pro-poor development and the constraints to achieving them. The population will accept changes only if they lead to improved access to services of reasonable quality. Health workers must also feel that their aspirations will be met.

Table 25.2 Measures to reduce the impoverishing impact of health shocks

Prevention of major shocks	• Decrease risk of major illness by strengthening preventative programmes • Improve access to effective health services by subsidising essential services, reducing inefficiencies and improving quality • Strengthen the capacity of household carers to support the sick by providing relevant information and access to drugs and other inputs
Improved coping	• Improve local markets for savings and credit to enable households to manage lumpy expenditure and life-cycle expenditure patterns • Strengthen existing mechanisms to help households cope with health problems by establishing health prepayment schemes • Subsidise basic hospital inpatient services and support the establishment of low cost mechanisms to support the severely ill (such as community nursing homes)
Survival measures for highly stressed households and communities	• Safety net arrangements to meet the cost of essential health care for the very poor • Interventions to restore the productive capacity of households coping with disability or loss of family member • Emergency measures to help communities or concentrations of people cope with crop failures, war and other crises

Governments have limited control over the performance of health providers or individual health workers. They have to negotiate change. Some countries have concentrated excessively on reforming central management structures, whilst neglecting measures to meet the needs of the public or address the problems of health workers. This has led, in some cases, to a lack of popular support and a collapse of the change process.

Health sectors in sub-Saharan Africa face enormous challenges in addressing the unmet health service needs of the poor, while coping with the result of years of serious financial constraint and numerous unforeseen shocks. There are no models or prescriptions for addressing these challenges. However, there are examples of good practice in many places. For example, even though Zambia has suffered one of the worst declines in health status over the 1980s and 1990s, some districts in that country have much better health experiences than the national average (Simms *et al.*, 1998). Africa can also learn from countries in other regions where government has maintained effective public health services, despite many social and economic changes. The new vision for pro-poor health services will emerge from a process of listening to the needs of the poor, learning from mistakes and applying lessons from good practice models to address these needs.

Note

1 This chapter is based on a background paper for the 1999 Africa Poverty Status Report. The authors acknowledge useful comments on that paper by Mark Wheeler and Paul Smithson. They have also benefited from helpful suggestions by Adebiyi Edun, John Milimo and Mungai Lenneiye. The work on the background paper was financed by the Special Programme for Africa (SPA) and carried out under the auspices of the World Bank; it is reproduced with their permission. The usual disclaimer applies.

References

Ainsworth, M. and Over, M. (1994) 'AIDS and African development', *World Bank Research Observer*, 9 (2): 203–40.

Barnett, A. and Blaikie, P. (1992) *AIDS in Africa*, London: Belhaven Press.

Behrman, J. (1996) 'The impact of health and nutrition on education', *World Bank Research Observer*, 11 (1): 23–37.

Bloom, G. (1998) 'The affordability conundrum: tailoring equity to what is sustainable', in *Proceedings of a Travelling Seminar on the Attainability and Affordability of Equity in Health Care Provision*, The Philippines, 28 June–5 July 1997, Durban: Health Systems Trust.

Bloom, G. (2000) 'Equity in health in unequal societies: towards health equity during rapid social change', Working Paper no. 112, University of Sussex, Institute of Development Studies.

Bloom, G. and Standing, H. (2001) 'Pluralism and marketisation in the health sector', Working Paper No. 136, University of Sussex, Institute of Development Studies.

Carrin, G., Desmet, M. and Basaza, R. (2000) 'Social health insurance development in low-income developing countries: new roles for government and non-profit health insurance organisations', in International Social Security Association (ISSA) (ed.), *Building Social Security: The Role of Privatisation*, forthcoming

Castro Leal, F., Dayton, J., Demery, L. and Mehra, K. (1999), 'Public social spending in Africa: do the poor benefit?', *World Bank Research Observer*, 14 (1): 49–72.

Chima, R., Goodman, C. and Mills, A. (1999) 'The economic impact of Malaria in Africa: a critical review of the evidence', unpublished.

Corbett, J. (1988) 'Famine and household coping strategies', *World Development*, 16 (9): 1099–112.

Corbett, J. (1989) 'Poverty and sickness: the high costs of ill-health', *IDS Bulletin*, 2 (20): 58–62.

Cornwall, A., Lucas, H. and Pasteur, K. (2000) 'Accountability through participation', *IDS Bulletin*, 31 (1): 1–13.

de Waal, A. (1989) 'Is famine relief irrelevant to rural people?', *IDS Bulletin*, 20 (2): 63–7.

LaFond, A. (1995) *Sustaining Primary Health Care*, London: Earthscan.

Leonard, D. (2000) *Africa's Changing Markets for Health and Veterinary Services*, Basingstoke, Hants.: Macmillan Press.

Leslie, J. (1992) 'Women's time and the use of health services', *IDS Bulletin*, 23 (1): 4–7.

Litvak, J. and Bodart, C. (1993) 'User fees plus quality equals improved access to health care: results of a field experiment in Cameroon', *Social Science and Medicine*, 37 (3): 369–83.

Loewenson, R. (1999) 'Public participation in health: making people matter', IDS Working Paper no. 84, University of Sussex, Institute of Development Studies.

Loewenson, R. and Kerkhoven, R. (1996) 'The socio-economic impact of AIDS: issues and options in Zimbabwe', SAfAIDS Occasional Paper no. 1, Harare: Southern African AIDS Information Dissemination Service

Lucas, H. and Nuwagaba, A. (1999), 'Household coping strategies in response to the introduction of user charges for social services: a case study on health in Uganda', IDS Working Paper no. 86, University of Sussex, Institute of Development Studies.

Noormahomed, A. and Segall, M. (1994), 'The public health sector in Mozambique: a post-war strategy for rehabilitation and sustained development', Macroeconomics, Health and Development Series no. 14, WHO/ICO/MESD.14, Geneva: World Health Organisation.

Russell, S. (1996) 'Ability to pay for health care: concepts and evidence', *Health Policy and Planning*, 11 (3): 219–37.

Russell, S. and Gilson, L. (1997) 'User fee policies to promote health service access for the poor: a wolf in sheep's clothing?', *International Journal of Health Services*, 27 (2): 359–79.

Simms, C., Milimo, J. and Bloom, G. (1998), 'The reasons for the rise in childhood mortality during the 1980s in Zambia', IDS Working Paper no. 76, University of Sussex, Institute of Development Studies.

Standing, H. (1997) 'Gender and equity in health sector reform programmes: a review', *Health Policy and Planning*, 12 (1): 1–18.

Timaeus, I. (1998) 'Impact of the HIV epidemic on mortality in sub-Saharan Africa: evidence from national surveys and censuses', *AIDS*, 12 (Suppl. 1): S15–S27.

Walt, G. (1988) *Community Health Workers: Policy and Practice in National Programmes. A Review with Selected Annotations*, EPC Publication no. 16, London: London School of Hygiene and Tropical Medicine.

Wilkes, A., Yu, H., Bloom, G. and Gu, X. (1997) 'Coping with the costs of severe illness in rural China', IDS Working Paper no. 58, University of Sussex, Institute of Development Studies.

World Bank (1994) *Zambia Poverty Assessment*, Washington, DC: World Bank.

World Bank (1998a) *Health, Nutrition and Population Indicators*, Washington, DC: World Bank.

World Bank (1998b) *World Development Indicators 1998*, Washington, DC: World Bank.

World Health Organisation (WHO) (1997) *World Malaria Situation in 1994*, Geneva: WHO.

World Health Organisation (WHO) (1998) *Gender and Health*, Technical Paper /FRH/WHD/98.16, Geneva: WHO.

World Health Organisation (WHO) (1999) *World Health Report 1999*, Geneva: WHO.

26 Achieving schooling for all

Is gender a constraint or an opportunity?[1]

Christopher Colclough

26.1 Global targets for primary schooling

The 1990s have witnessed much greater emphasis being placed upon the need for all children to go to school. The long-established human rights case for everyone being able to have at least a basic education had been supplemented, during the 1980s, by strong evidence, from many developing countries, that primary schooling helped people to be more productive in both farm and non-farm work, and that, particularly in the case of women, it facilitated the achievement of other important development objectives, including lower fertility, better health and nutrition, reduced infant and child mortality, and a better intergenerational transmission of basic skills within households. Accordingly, the idea that there was a strong 'investment' case for the widespread provision of primary schooling became generally accepted in professional and international circles. Having been largely ignored by aid agencies, it became one of their most favoured targets for aid support – more so, even, than other levels of education.

These changed perceptions about the importance of primary schooling were nourished by a series of international conferences and, subsequently, declarations issued by the international community over the past decade. The 1989 Convention on the Rights of the Child confirmed the commitment of all signatories to 'make primary education compulsory and available free to all' (Article 29). The following year, 155 governments represented at the World Conference on Education for All committed themselves to achieving 'universal access to and completion of primary education by the year 2000', together with 'improved learning achievement, based on the attainment of defined levels of performance' (WCEFA, 1990: 3). Perhaps realising that the speed of reform required to achieve this by the turn of the century was too great, subsequent declarations from the Social Summit and the Fourth World Conference on Women envisaged a more moderate pace. The Jomtien goals were replaced by a commitment to achieve 'universal primary education by the year 2015' at the World Summit for Social Development (UN, 1995a), and the achievement of 'gender equality in primary school enrolment and completion by 2005' (Fourth World Conference on Women, Beijing, UN, 1995b). These latter targets continue to reflect the stated goals for educational development of UNICEF, UNDP, UNESCO and the World Bank. They are advocated by the Development Assistance Committee of the OECD (OECD, 1996) and are enshrined in the policy papers of a number of its members (including the Department for International Development (DfID), the Netherlands and Sweden).

This chapter briefly reviews the progress made in implementing these quantitative goals. It finds that sub-Saharan Africa, in particular, had made very little progress by mid-decade. Some of the main reasons for this circumstance are identified. The enrolment targets advocated by the international community are achievable – but they will require a willingness to undertake more comprehensive educational and financing reforms than many governments seem willing to consider. The enrolment targets will also require social reforms – dependent more upon commitment than upon resources – in order to achieve gender equity in education and society. Political and social constraints imply that the achievement of gender equity in enrolments, and of 'schooling for all', are mutually dependent. Their temporal separation by a decade, as suggested by current international targets, is unrealistic.

26.2 The enrolment record

Between 1960 and 1980, education systems in developing countries expanded rapidly. These years are sometimes described as a golden age for world economic growth, and so it was for the growth in access to education. At primary level, enrolments in Africa tripled, whilst in Asia and Latin America they more than doubled over the two decades. At secondary and tertiary levels of education, enrolments increased even more rapidly, with tenfold increases being not uncommon – albeit often starting from a very small base. This pattern of expansion, however, changed after 1980. In every major developing region, the rate of growth of school and college enrolments slowed, as compared with the previous two decades. This was particularly so for primary schooling, and especially in Africa, where primary enrolment growth fell below the rate of growth of population, and where, in a number of countries, enrolments actually declined in absolute terms.

Figure 26.1 shows the impact of these expansion patterns upon primary gross enrolment ratios (which express primary enrolments as a proportion of the school-age group) in developing regions.[2] During the period 1960–80, increases in enrolments greatly exceeded rates of population growth. Thus, gross enrolment rates (GERs) increased substantially in most countries, with the fastest growth occurring in Africa. After 1980, progress generally slowed. But in Africa there was actual retrogression, with enrolment ratios falling back by more than 5 per cent over the years to 1990.

It is clear that the reduced enrolment growth after 1980 was mainly a product of recessionary pressures during that decade. Particularly in Africa, where real per capita incomes fell substantially, children were increasingly needed at home to help (directly or indirectly) with income-earning tasks. Government budgets also fell in real per capita terms, reducing

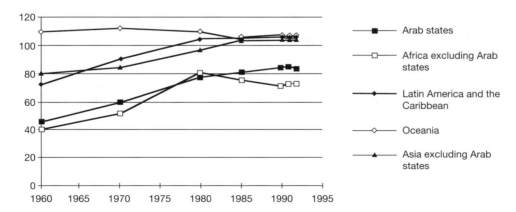

Figure 26.1 Regional gross enrolment ratios over time

the resources available for new buildings and maintenance, and causing reductions in real recurrent expenditures – usually by means of salary decline. Although some slowdown in enrolment growth in countries with average GERs of around 100 might be thought natural, owing to their having achieved 'universal' enrolment, the presence of high levels of over-age enrolment usually meant that many school-age children remain out of school. In fact, net (age-adjusted) enrolments are typically around 80 per cent of gross enrolments. Thus, GERs large enough to include all the school-aged population in these regions would have to have been around 120 – although the required size obviously differed from country to country.

After 1990, the regional statistics provide some evidence that the decline in Africa was arrested – with some increase in enrolment ratios to 1995, taking the region as a whole. However, progress remained slight, with significant gains registered in only a few countries: often, as in Malawi and Uganda, a product of the introduction of free and compulsory schooling for all. Of the thirty-three countries having data on enrolments for the 1990–5 period, sixteen achieved increases in their GERs, fourteen reported falls, with no change being reported for the remaining three countries.

26.3 The gender balance of enrolments

The dynamics of the gender balance of primary enrolments are shown in Tables 26.1 and 26.2. It can be seen that the stagnation of enrolment ratios in sub-Saharan Africa over the years 1980–95 stands in stark contrast to South Asia, where strong enrolment growth, relative to the growth of population, was maintained. Surprisingly, however, the gender balance of enrolments in SSA improved over the period, notwithstanding the disastrous picture for its total enrolments. Closer inspection of the data in Table 26.1 suggests, however, that the reason for the improvement was that, across the region as a whole, boys' enrolments fell more (relative to the size of the school-age group) than those of girls.

This is shown in more detail in Table 26.3. In twenty-eight of the thirty-nine countries in SSA for which we have time-series data, the ratio of girls to boys enrolled at primary school

Table 26.1 Trends in primary gross enrolment ratio (GER) by region and gender, 1965–95

		1965	*1980*	*1985*	*1990*	*1995*
SSA	male	52	87	84	79	81
	female	31	68	68	66	67
	total	41	77	76	73	74
Arab States	male	n.a.	92	94	93	92
	female	n.a.	67	72	75	76
	total	n.a.	80	83	84	84
LAC	male	99	106	108	109	112
	female	96	103	104	105	108
	total	98	105	106	107	110
South Asia	male	83	92	98	103	105
	female	52	61	71	77	82
	total	68	77	85	91	94
Developing countries	male	84	104	107	106	105
	female	62	85	90	92	93
	total	78	95	99	99	99

Source: UNESCO, *Statistical Yearbook 1996* and *World Education Report 1998*, Paris: UNESCO

Table 26.2 Female enrolments at primary level in developing
countries relative to male enrolments, 1980 and
1995 (%)

	1980	1995
Sub-Saharan Africa	78	83
Arab States	73	80
Latin America and Caribbean	97	94
South Asia	66	73
All developing countries	82	84

Source: UNESCO, *Statistical Yearbook 1996* and *World Education
Report 1998*, Paris: UNESCO.

Table 26.3 A comparison of total and gendered enrolment changes in SSA countries, 1980–95

No. of countries in which:	F/M increased	F/M unchanged	F/M fell	Total
GER increased	15	2	2	19
GER fell	13	5	2	20
Total	28	7	4	39

Source: UNESCO, *World Education Reports*, Paris: UNESCO, various years

Notes
GER = primary gross enrolment ratio; F/M = female primary enrolment ratio as % of male primary enrolment ratio.
Lesotho and Botswana are excluded. There, F/M fell over the period, but female enrolments exceed those of boys.

increased over the years 1980–95. However, it can be seen that in almost half (thirteen) of
these countries, the reduction of the gender gap occurred in the context of declining GERs,
implying that total enrolments in these countries increased less quickly than the school-age
group. Thus, moves towards gender equity occurred here because the proportion of boys
enrolled at school fell faster than did the proportion of girls – hardly a circumstance to be
applauded. In the remaining eleven SSA countries (almost one-third of the total), gender
ratios were either unchanged over the period, or they deteriorated further in favour of males.
Consequently, the apparent progress, as regards the gender balance of school enrolments in
SSA, turns out to be both a more complicated and a less generalised phenomenon than the
regional data might otherwise suggest.

We can conclude that the targets adopted at Jomtien and at the international meetings
that followed now appear little closer to being achieved in SSA than was the case 10 years
ago. It was clear from the outset that the Jomtien goals were very optimistic and that,
particularly in SSA and South Asia, they would not be achieved.[3] The revisions made at the
Social Summit in mid-decade, and subsequently reiterated by many agencies, now comprise
widely accepted revised targets for national and international action. The implied time-scales
are more feasible than those implied by the Jomtien targets. Nevertheless the processes at
work in the past may well conspire again to prevent their attainment. What, then, are the
main causes of recent under-performance, and what lessons for the future do they hold?

One of the more obvious potential explanations for SSA's poor enrolment record is that
governments in the region have not responded to the call to increase resources available for
primary schooling. In examining this charge, it is necessary to recognise that the financing
of any given level of school enrolments is a function of four variables: the amount of public
spending on schooling; the average cost per student; the size of the school-age population;

and the resources from private households which are spent upon schooling. Changes in any of these are capable of affecting the achievement of a given level of enrolments. Each will be briefly discussed in turn.

26.4 Public spending on primary schooling

Public spending on education in SSA, measured as a proportion of GNP, has been rising over 1985–95, and is now higher than in any major developing or developed region of the world. Table 26.4 shows that educational expenditure rose from 5.1 per cent to 5.6 per cent of GNP over those years in sub-Saharan Africa – ahead of both Latin American and South Asian countries by more than one percentage point, and also well ahead of the average for developed countries. A similar trend characterises public spending on primary schooling, which, in SSA, had risen to 2.16 per cent of GNP by the early 1990s, compared to an estimated 1.53 per cent in Latin America and 1.29 per cent in Asian countries.[4]

Table 26.4 Regional public expenditure on education, in US $ and as percentage of GNP

	Percentage of GNP				*US $ (billions)*		
	1980	*1985*	*1990*	*1995*	*1980*	*1990*	*1995*
Sub-Saharan Africa	5.1	4.8	5.1	5.6	15.8	14.8	18.8
Arab States	4.1	5.8	5.2	5.2	18.0	24.4	27.5
Latin America and the Caribbean	3.8	3.9	4.1	4.5	33.5	44.6	72.8
Eastern Asia and Oceania	2.8	3.1	3.0	3.0	16.0	32.0	59.9
Southern Asia	4.1	3.3	3.9	4.3	12.8	35.8	62.6
All developing countries	3.8	3.9	3.9	4.1	97.4	155.6	247.7
Least developed countries	2.9	3.0	2.7	2.5	3.5	4.3	5.3
Developed countries	5.2	5.0	5.0	5.1	407.8	816.4	1109.9

Source: UNESCO, *World Education Report 1998*, Paris: UNESCO, table 12

Table 26.5 shows the ten countries from sub-Saharan Africa with data for this variable for the years 1980, 1990 and 1995. Although for the intra-period movements (1980–90 and 1990–5) the total number of rises and falls were not very different, a comparison of the starting and the final years reveals that in eight countries public expenditures on primary schooling increased as a proportion of GNP, in one (Burundi) it remained unchanged, and in one (Zambia) it fell sharply. Taking these countries as a group, public expenditures on the primary system increased, on average, from 2.0 to 2.7 per cent of GNP over the years 1980–95. The table also provides some support for the view that 1990 marked some acceleration in the growth of public spending on primary schooling: its average annual growth, as a proportion of GNP, from 1990–5 was running at more than twice the level of the previous decade.[5] Thus, measured in terms of public effort, SSA is putting more emphasis on its spending priorities upon education in general, and also upon primary schooling, than other regions. It also seems that the intensity of that effort increased during the 1990s, as compared with earlier years.

However, the economic decline suffered by SSA during the 1980s and early 1990s, when real GNP per capita fell for the region as a whole by around 20 per cent, meant that the increased effort to support primary schooling was greatly outweighed by the declining resource

Table 26.5 SSA public spending on primary schooling as percentage of GNP, 1980–95

Country	1980	1990	1995
Burundi	1.2	1.6	1.2
Ethiopia	0.8	2.6	2.5
Gambia	1.5	1.6	2.5
Kenya	3.9	3.9	4.6
Lesotho	2.0	1.6	3.0
Malawi	1.2	1.4	3.4
Swaziland	2.3	2.1	3.0
Togo	1.8	1.7	2.0
Zambia	2.7	0.9	0.7
Zimbabwe	2.7	5.7	4.4
Average	2.01	2.31	2.73
Average annual change to year shown (%)		1.4	3.4

Sources: UNESCO, *World Education Report*, Paris: UNESCO, various years, and World Bank, *World Tables*, Washington, DC: World Bank

base. Table 26.4 shows that total public spending on education in SSA actually fell from US $15.8 billion to around US $14.8 billion over the 1980s, and that it managed to increase only to US $18.8 billion by 1995. This represented scarcely a 20 per cent increase, in current dollar terms, over the 15 years and, of course, it represented a very sharp real reduction in the public resources available for education. By contrast, Latin American spending almost doubled in current dollar terms, whilst that of South Asia increased five fold over the same period. These financial privations in the public sector undoubtedly provide an important part of the explanation for the modest enrolment progress made by SSA since 1990.

Although the countries of SSA can be characterised as having made greater efforts in recent years to achieve schooling for all children, there remains considerable inter-country variance. Some countries with low primary enrolments are, undoubtedly, spending far less than is both desirable, and affordable, on the primary sector. Which are they? Table 26.6 shows those SSA countries (for which data are available) with GERs less than 100 and where public spending on the primary sector amounts to less than 2 per cent of GNP. The average value for this statistic, in SSA for 1990, was 1.8 per cent, and for 1995 it had risen to 2.5 per cent. Accordingly, the countries shown were each allocating fewer resources to primary schooling than the SSA average in 1995.

Inspection of the table indicates that nine of the thirteen countries shown are Francophone, and that all of them have GERs substantially lower than 100. Furthermore, many of the countries shown have very low primary enrolments, with more than half of them having GERs less than 70. The average value for the primary public spending ratio in these seven countries is only 1.15, which indicates considerable space to increase public spending on primary schooling. The final column of the table indicates the proportion of GNP which would be required to be allocated by the government to primary schooling if GERs of 100 were to be achieved (assuming existing schooling costs and population size, and no demand constraints). It can be seen that for most of these countries the financial challenge is not great: in ten of the thirteen countries, all of the age group could be accommodated in primary schools if the level of resourcing from the public sector were to rise to the average already achieved by SSA as a whole. It can be seen from the GERs that, in fact, very little was achieved

Table 26.6 SSA 'low commitment' countries[a]

Country	GER 1990	GER 1995	Data year	Spending as % GNP	Unit costs as % p.c.y.	School-age population as % total	Required spending % of GNP for GER = 100
CAR	68	n.a.[b]	1990	1.5	11	15	2.2
Senegal	58	65	1990	1.5	15	16	2.5
Comoros	77	78	1995	1.4	8	16	1.8
Burundi	73	70	1995	1.2	13	15	1.7
Ghana	77	n.a.	1990	1.0	5	17	1.3
Chad	57	55	1995	0.9	10	16	1.7
Zambia	97	89	1995	0.7	4	21	0.8
Uganda	80	73	1990	0.6	4	19	0.7
Zaire	76	72	1990	0.5	4	18	0.6
Sierra Leone	48	n.a.	1990	0.3	3	18	0.6
Djibouti	44	38	1990	1.9	26	16	4.4
Burkina Faso	37	38	1990	1.0	17	16	2.6
Mali	24	32	1995	1.0	17	16	3.2

Source: UNESCO, *World Education Report*, various years

Notes

a 'Low commitment' is here defined as countries having GERs less than 100 in the years shown, and where public spending on primary schooling was less than 2% of GNP.

b n.a. = not available.

between 1990 and 1995. There were falls in GERs in more than half the countries shown. Where there were increases, they were usually very small.

What can be achieved, with a given aggregate expenditure, depends upon the costs per student at primary level. The three countries from Table 26.6 that would be unable to reach GERs of 100, even with an increase of public spending to its regional average value, all have high unit costs. The SSA average value for primary unit (public) costs relative to per capita income was 12 per cent in 1990 and 14 per cent in 1995. It can be seen that Djibouti, Burkina Faso and Mali each have higher unit expenditures than these values. For them it will be necessary both to increase public spending on primary schooling and to reduce unit costs if schooling for all is to be achieved in the near future.

A further group of countries comprises those in which spending on primary schooling is already high, but where the costs of school provision are likely to prevent their attainment of high enrolments. These are shown in Table 26.7. In the first three countries – Mauritania, Gambia and Tanzania – expenditure levels are high, even though costs are each close to the regional average of 12–14 per cent of GNP per capita. With little room for costs to fall, additional resources are likely to be required to sustain a move towards schooling for all. In the other countries shown – Kenya, Ethiopia, Rwanda and Mozambique – unit costs are much higher, and reforms to reduce their levels will be a necessary part of any strategy to achieve substantial enrolment expansion.[6]

26.5 The influence of population size

Clearly, for countries with given levels of income and unit costs of schooling, those with higher rates of population growth (and, thus, proportionately larger school-age populations) will face a greater relative per capita cost in achieving schooling for all than other countries.

Table 26.7 SSA cost- and resource-constrained countries[a]

Country	GER 1990	GER 1995	Data year	Spending as % GNP	Unit costs as % p.c.y.	School-age population as % total	Required spending % of GNP for GER = 100
Mauritania	51	78	1995	2.0	12	16	2.6
Gambia	64	73	1995	2.5	12	14	3.4
Tanzania	69	67	1990	2.4	11	20	3.5
Kenya	95	85	1995	4.6	15	24	5.4
Ethiopia	39	31	1995	2.5	37	15	8.1
Rwanda	71	n.a.	1990	2.8	17	20	4.0
Mozambique	64	60	1990	3.1	23	14	4.9

Source: UNESCO, *World Education Report*, Paris: UNESCO, various years

Note

a These are countries where public spending on primary schooling was 2% or more of GNP, and where unit costs of provision were high in the years shown.

Differences between countries in this respect can make an enormous difference to the cost burden of achieving universal enrolments. For example, in the developed market economies, which have zero or even negative rates of population growth, those aged 6–11 years typically comprise 5–8 per cent of the population as a whole. In Africa, by contrast, the school-age group is a much larger segment of the population. Tables 26.6 and 26.7 indicate that the range in the SSA countries shown is from 14 per cent in the Gambia to 20 per cent and more in Tanzania, Rwanda, Zambia and Kenya.[7] The figures imply that the costs of achieving a given GER in Kenya would be more than 70 per cent higher than in the Gambia, and about three times as high as in typical developed countries, assuming all other cost parameters were the same. This underlines the importance of population policy as a means of achieving and maintaining schooling for all. It is often forgotten that fertility control is potentially more powerful, in cost terms, than many direct measures of educational reform in providing all children with access to schools.

26.6 Private household demand and the problem of gender

Data on private expenditures on education in Africa are not systematically collected from official sources, so evidence as to its magnitude is rather patchy. Since almost all children attending school incur at least some associated private costs – for school books, equipment, uniforms, transport – it is likely that, were these summed, the magnitude of private expenditures would turn out to be considerable. In addition the opportunity costs of children's time are significant for poorer households. On the other hand, it is not clear how these direct and indirect costs have been changing in recent years.

Some countries charge fees for school attendance, even at primary level. Often, they were introduced as part of efforts by governments to adjust their national budgets in response to economic adversity. However, the enrolment declines which so often resulted forced a reassessment of cost-recovery policies, which still continues. Since such policies are obviously regressive for poorer households, there has been a recent trend to switch back to fee-free primary schooling. Such policy shifts have been associated with large enrolment increases in Malawi, Uganda and elsewhere, often as part of national campaigns to achieve universal primary education.

Private enrolments are reported by some observers to have been growing rapidly in Africa (Heyneman, 1999). However, the latest UNESCO statistics suggest that, at primary level, there was little change over the decade 1985–95: private enrolments increased from an average of 10 per cent to only 11 per cent of total primary enrolments in the twenty-three SSA countries with data for that period (UNESCO, 1998: table 10). There were some large increases in the Congo, Madagascar and Malawi; but these were compensated by strong falls in Cameroon, Gabon and Ghana. Elsewhere, the general picture was one of small increases in the proportion of private enrolments, by one or two percentage points.

This result is consistent with what has been happening in the public education sector. As indicated earlier, enrolments have been stagnant in SSA, relative to the age group, and falls in enrolment ratios have been common. This has been associated not only with reduced public per capita provision, but also with much reduced household incomes, which have made the direct and indirect costs of school attendance more difficult to meet – particularly amongst the poorest households. Under circumstances of downward demand pressures on public school enrolments, it is unlikely that the relatively more expensive private sector would be facing the opposite pressures. That may be happening where richer families find the services offered by increasingly under-resourced public schools to be unsatisfactory, and where the private sector offers better, if more expensive alternatives. But this is likely to be so for only a minority of households.

The ways in which the demand for primary schooling can be stimulated represent important ingredients in all strategies to achieve schooling for all. Unless demand for primary schooling can be increased in SSA, which, *inter alia*, will require some further reductions in direct costs to households, it is likely that such strategies will fail. A critical part of this endeavour concerns the question of gender. It was shown earlier that the gender gap in enrolments remains wide in most countries of SSA, and that the improvement of the female/male ratio since 1980 has been caused, in almost half of the countries concerned, by male enrolment ratios falling more quickly than those of females. Just as boys are often enrolled first in schools, so they are often withdrawn first when economic adversity sets in and as more cash incomes are needed to supplement those of adults. It cannot be assumed, therefore, that further improvements in the gender gap will follow as total enrolments begin to rise. The first response may well be a growth mainly of male enrolments, causing the gender gap again to deteriorate. This has been happening in Guinea, Ethiopia and a number of other African countries.[8]

In the longer run, obviously, as GERs rise towards 100, so too will both male and female enrolment ratios. But, as Figure 26.2 shows, the relationship is loose. Whilst the extent of gender bias in enrolments does reduce as GERs rise, it can be seen that even for African countries where GERs are close to 100, there is great variation in the gender gap. Togo, for example has 40 per cent fewer girls enrolled than boys, whereas Lesotho has about 20 per cent more boys than girls, even though the GER for these two countries is close to 100 in each case.

It is also worth noting from the scatter plot of primary GERs vs. GNP per capita (Figure 26.3) that, whilst enrolment ratios rise with income, there is considerable variability around the line. This is particularly true for countries in the US $400–500 per capita income range, where most SSA countries are grouped. A focus upon that particular segment of the scatter diagram shows that factors other than per capita income appear to determine school enrolments for the countries shown.

Figure 26.4 demonstrates a similar picture for the gender gap. Here, too, inequality in enrolments between the sexes appears to have little to do with income level. African countries

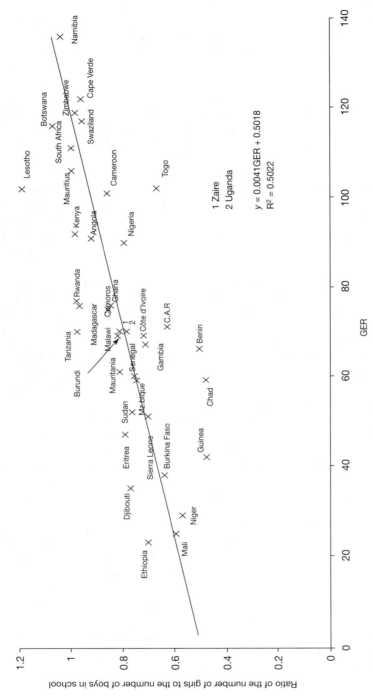

Figure 26.2 Primary gross enrolment ratios and gender gaps in SSA, 1992–3

Source: Calculated from data in UNESCO, *World Education Report 1995*, Paris: UNESCO

Note

A value of 1 on the vertical scale implies that the gross enrolment ratios for girls and boys are identical and that therefore approximately half the children in primary school are girls.

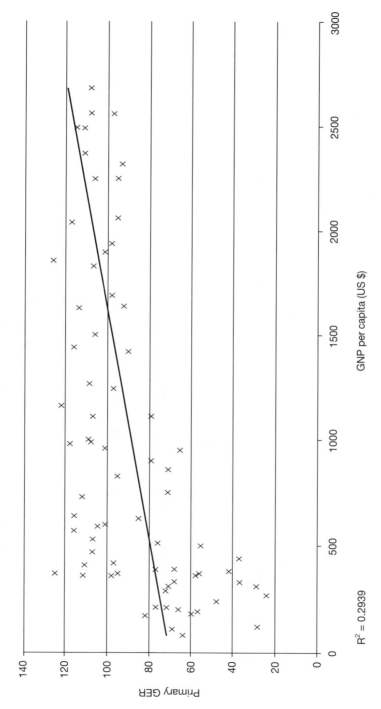

$R^2 = 0.2939$

Figure 26.3 Scatter plot of primary gross enrolment ratio against GNP per capita, 1990: low- and middle-income countries

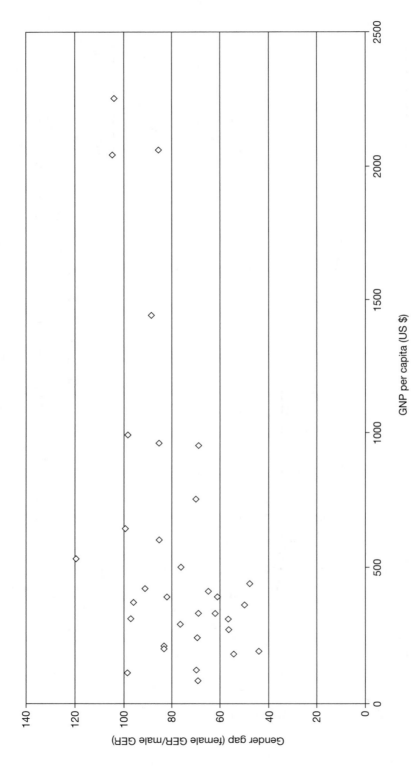

Figure 26.4 The relationship between GNP per capita and the ratio of female/male GER, Africa, 1990

in the US $300–500 per capita income range have F/M ratios of between 50 and 100, in a distribution that is unrelated to the incomes of the countries concerned.

The main point to emerge is that income growth is neither a necessary nor a sufficient condition for improvements in the gender gap in African school enrolments to be gained. As argued in a separate paper, the main cause of gender discrimination in schooling is not poverty *per se*, but adverse cultural practice, which obtains in the domains of society, the labour market, the household, and the school itself (Colclough *et al.*, 2000). Policy reform focused upon changing behaviour within each of these domains is required. Attitudes towards early marriage, changing patterns of informal job reservation, establishing equal wages and conditions of work, encouraging female role models, promoting gender training at all levels of society, improving facilities for girls in schools, changing teacher attitudes – and many more – need to become targets for policy influence. These attitudes and practices run deep, and cannot be easily changed. But some countries in Africa have demonstrated that change can occur – particularly when political leaders commit themselves to such goals. The reason why this presents more of an opportunity than a constraint is that much can be achieved in this direction without greatly enhanced resources for schooling being required. Changes to cultural practice are not easy, but neither need they be expensive. In that sense they are within everyone's grasp.

Whether they are within reach throughout SSA within the next five years, however, is an entirely different matter. Both political and social realities imply that it is unlikely that the extent of social reform required in much of Africa will be delivered so quickly. More realistic would be to link the targets for achieving gender equity and schooling for all by the same date. The former is necessary for the latter, and it is unlikely, in fact, that it will precede it to any significant extent.

26.7 Conclusions

This chapter has shown that in about half of the SSA countries for which we have data, the expansion of primary school enrolments has fallen well short of what could have been possible. In these cases, universal enrolment could have been achieved at a net GNP cost no higher than the 1995 average for SSA. Thus, whilst we have shown that, on average, SSA has increased its effort by devoting relatively more resources to primary schooling (relative to GNP) than in the 1980s, the variance of this effort is wide and the countries in the tail of the distribution have exhibited low commitment to the objectives they espoused at Jomtien.

In another group of countries commitment to achieving these goals has also been low, but high unit costs would, in any case, have prevented rapid progress towards schooling for all being made. In a further group, commitment has been high, but high costs have prevented rapid progress. A willingness to allocate greater priority to primary schooling is necessary in the first group of countries, but both groups are cost and resource constrained. Reforms to tackle both these problems will be a necessary precursor for successful expansion strategies.

In most SSA countries the difficulty of achieving universal schooling is increased by high rates of population growth. Within Africa, the difference in the cost-burden from this source alone is more than 30 per cent, and, in comparison with more developed countries, Africa's task is two to three times as great.

Most countries also face significant gender gaps, with girls' enrolments often being much lower than boys'. The evidence suggests that, in the absence of policy reforms, these are as likely to rise as to fall as the region returns to a period of enrolment progress. Gender gaps can be reduced effectively and cheaply, as some African states have shown. But it is unlikely, for a range of social and political reasons, that the achievement of gender equity will precede that of schooling for all in most countries. The target dates for meeting these objectives need to be the same.

Notes

1 Samer Al-Samarrai provided invaluable assistance in producing the data for the tables and figures.
2 Throughout the following analysis the gross enrolment ratio (GER) is used as a means of comparing children's access to schooling, even though a better measure of access to education would be given by the net enrolment ratio (NER), which excludes from the calculation those enrolled children who are outside the normal primary school age. Unfortunately, international data on net enrolments tend to be both patchy and unreliable, so the data on GERs, which are more generally available, have to be used for cross-country comparisons.
3 The argument here is not that the targets could not have been achieved; rather that the amount of domestic policy reform required presented an agenda which would be too ambitious to accomplish universally within one decade.
4 These data are calculated from UNESCO (1995).
5 This result may, of course, be sensitive to the particular selection of countries included in the comparison.
6 Further discussion of costs of, and expenditures on, primary schooling in Africa is given in Colclough and Al-Samarrai (2000).
7 These differences are also influenced by the length of primary schooling, which is 8 years in Kenya, 7 years in Tanzania, Zambia and Rwanda and 6 years in the Gambia. Thus, roughly half of the difference in eligible population size as between the Gambia and Kenya is attributable to the longer duration of primary schooling in Kenya.
8 This process is documented in Colclough *et al.* (2000).

References

Colclough, C. and Al-Samarrai, S. (2000), 'Achieving schooling for all: budgetary expenditures on education in sub-Saharan Africa and South Asia', *World Development*, 28 (11).

Colclough, C., Rose, P. and Tembon, M. (2000), 'Gender inequalities in primary schooling: the roles of poverty and adverse cultural practice', *International Journal for Educational Development*, 20: 5–27.

Heyneman, S. (1999) 'Education in sub-Saharan Africa: serious problems, significant opportunities', paper presented at Africa Summit Conference, Houston, Texas, April, mimeo.

OECD (1996) *Shaping the 21st Century: The Contribution of Development Cooperation*, Paris: Development Assistance Committee.

UNESCO (1995) *Statistical Yearbook, 1995*, Paris: UNESCO.

UNESCO (1998) *World Education Report 1998*, Paris: UNESCO.

United Nations (1995a) *World Summit for Social Development: Programme of Action*, New York: United Nations.

United Nations (1995b) *Beijing Declaration and Platform for Action*, Fourth World Conference on Women, New York: United Nations.

World Bank (1998) *World Development Indicators 1998*, Washington, DC: World Bank.

World Conference on Education for All (WCEFA) (1990) *World Declaration on Education for All and Framework for Action to Meet Basic Learning Needs*, World Conference on Education for All, Jomtien, Thailand.

Part VII

Final observations

27 Final observations on strategy and policy reform

At the beginning of the twenty-first century, it is clear that the countries of sub-Saharan Africa face a variety of severe economic, social and political problems. These range from outright armed conflict through constrained economic performance and widespread poverty; local instances of severe environmental degradation; the collapse, to varying degrees, of the post-Second World War experiment in universal primary education and health care provision; reduction in the capacity of the state to guide and lead development activity; shaky steps towards democratic, corruption-free government; to new or resurgent disease pandemics.

Recent commentators on African economic performance have stressed the importance of geography, identifying a number of different geographical factors (Krugman, 1991; Sachs, 1999; Sachs and Warner, 1997). The most basic of these would appear to be the natural resource endowment and climatic situation of various countries and zones and, for an important category of countries in SSA, the fact of being landlocked, with expensive and uncertain communication to a port and thus to overseas markets and supplies. While it is true, therefore, that broad statements can be made about the overall performance and conditions of welfare in the sub-continent as a whole, it is also the case that development strategies need to be devised for countries with different baseline conditions, in different agro-ecological zones (which may shift with climatic change), and for countries with or without the benefit of exportable mineral resources. Also, in most SSA countries the intensity of sustainable use of the natural resource base lies well below that found in comparable zones in Asia. This reflects African agriculture's significantly lower usage rates of irrigation and fertilizers, power tillers and animal power for draught and transport, high-yielding cereal varieties, and perennial and agro-forestry cropping systems.

For landlocked countries progress is being made in establishing agreements and technical mechanisms that will facilitate the transit of goods internationally, though substantial unnecessary barriers continue to exist and need to be addressed. The effect of international transport costs on comparative advantages means also that more consideration needs to be given to appropriate strategies for manufacturing, especially in small-scale industry, in industries making the most of local resources, and in industries that can take advantage of the natural protection afforded by transport costs. Opportunities for regional trade are evident and can be more fully exploited, particularly through the regional integration arrangements that are being actively pursued.

Geographical factors are behind a wide variety of environmental situations in different parts of SSA, with different causes, calling for different solutions. The environmental dimension in SSA is a major subject in itself, which has only been addressed here by Mortimore in respect of the Sahelian dryland zone, where he makes a range of recommendations. Climatic change has been most significantly demonstrated in the onset of a '30-year' dry phase in Sahelian countries from the early 1970s and in the negative impact of the 'El Niño' effect, emanating from the changes in the Pacific Ocean, in Eastern and Southern Africa from the late 1980s. Climatic change in the equatorial zone of Africa is not yet discernible (Hulme *et al.*, 2001). Nevertheless, the plausibility of rapid and dramatic global climate change, accelerated by global warming resulting from fossil fuel burning and deforestation, has been strengthened recently by evidence gained from analysis of Arctic ice-core samples (Allee, 2000). This provides historical evidence for a pattern of very rapid onsets of glaciation in Northern Europe, with a southward drift of the Mediterranean zone, with wet winters, and the arid Saharan zone moving southward into the present savanna zone.

At the same time, the need to keep pace with high population growth rates poses problems of food supply in almost every country and in particular locations of land supply and, with this, of rural employment provision. In Southern Africa, especially, but not exclusively, environmental problems are associated in part with inequitable land distribution, with land degradation due to overcrowding and/or policy neglect. Land reform in favour of smaller units is called for in recognition of their capacity to achieve more efficient land utilisation and higher total factor productivity, as is widely supported by economic analysis.

However, even within the same geographical zone and economic sector, even in the production of the same commodity, performance has varied widely between countries, indicating the need to examine the policies that have been pursued in each case. The potential advantage of mineral resources in attracting foreign investment and in providing foreign exchange has been badly misused in a great many African countries, while in a number mineral riches have actually been the source of conflict leading to poverty as well as to loss of life. The thesis put forward by Auty (2000) that resource-rich developing countries are more prone than resource-poor countries to wasteful and misdirected development policies is well supported by SSA examples. In manufacturing industry it has become very clear that sound macroeconomic policies are a precondition for progress, whatever the particular set of industries being pursued. In the past, and often still in the present, failure to follow these has produced serious shortages of foreign exchange, shortage of imported materials and huge excess capacity in the enterprises set up, of whatever nature.

In the area of economic policy the impact of structural adjustment programmes (SAPs) has been the subject of greatest debate, as reflected in a number of chapters in this volume. In agriculture it is evident that SAPs have not secured the degree of production response that had been anticipated by some of the international agencies promoting them and that other factors and constraints, as well as price, need to be examined more closely in each country. In many countries, livestock production of all kinds, particularly by poorer livestock keepers, has been neglected and the supply of meat, milk and eggs reaching poor and better-off consumers is far below its potential and below the level attained in other developing countries. Agricultural research institutions in SSA have lacked resources and direction over a long period: agricultural research policy calls for fundamental reappraisal.

A major part of the solution to both food security and environmental degradation problems requires well-focused and well-implemented action research that demonstrates the feasible quantum jumps in total factor productivity which make land and water resources sufficiently valuable that investment in them is worthwhile, both to protect and enhance their productivity. At the food security end, major gains in yield require seed and planting material possessing greatly enhanced potential that can utilise other components of a high-yield but low-input cost package. Subsistence agriculture cannot sustain high levels of purchased inputs, so the initial breakthrough to reliable marketable food surpluses must come from low-cost, high-yielding seed with other 'low external input and sustainable agriculture' (LEISA) and practices. Large food output increases can be obtained, for example, by seed that is more insect-resistant, drought-tolerant or salt-tolerant and/or that can supply key plant nutrient either for itself or other food crops.

Usually, the most rapid and effective way to secure these traits is through biogenetic engineering or genetic modification (GM), using public–private partnership (PPP) arrangements to commission research by private GM firms in the North that meet the product specifications of African governments funded by western donor agencies. Once food security and cash from a reliable food surplus is achieved, further productivity gains can be obtained by building more profitable cash crops and other enterprises within increasingly diversified farming systems. Given such incentives, farmers are prepared to invest without compulsion in further conservation measures on their plots of land (see, for example, the Machakos, Kenya, study carried out by Tiffen *et al.*, 1994).

Appropriate strategies for manufacturing post-2000 have also been the subject of lively debate. It is widely recognised by most economists, in Africa and outside, that the import-substituting policies as pursued in the 1960s and 1970s, given small domestic markets in most cases, were mistaken. At the same time the role of the state in East Asian countries in promoting key industries with the aid of protection has been documented and understood. While the concepts of 'infant industries' and 'learning by doing' have attracted renewed attention (see Tribe, 2000; Lall, 1992), it is not clear how these might be implemented in detail, in what particular context and on what scale. Better possibilities appear to exist within the small enterprise sector, taking advantage of 'flexible specialisation' within industrial 'clusters' (Schmitz, 1989) and of lower-cost small-town sites closer to rural purchasing power..

The question has been raised, in the context of the SAPs, of whether Africa is in process of actual 'de-industrialisation', either on a more temporary basis as inefficient industries are opened up to competition or, more fundamentally, in the longer term, due to a deep-seated lack of competitiveness *vis-à-vis* rising manufacturing sectors in Asia that are making fresh inroads into African markets themselves, as well as overseas markets which African manufacturers might have hoped to enter. In addition to the weak competitiveness of existing industry, there is a related problem of attracting foreign direct investment into Africa in competition with other cheap-labour locations in the context of globalisation with mobile international capital. Certainly, with SSA now very much the 'latecomer' in the field, there is need (a) to establish conditions which are more attractive to foreign investment and private capital, as Oshikoya and Mlambo state in their chapter; and (b) to expand industry meanwhile in areas in which African countries have comparative advantage. In addition to medium and small-scale industries directed towards local and regional markets as mentioned above, more systematic identification and encouragement of natural resource-based, including agricultural resource-based, industry is needed. This in turn calls for a strong emphasis on the complementary expansion and diversification of the agricultural sector that can serve as the foundation for generating further value added in industrial processes.

It should be noted that an increasing proportion of manufacturing and commercial activity in many smaller African countries is owned by transnational companies, including some originating from South Africa. While these undoubtedly make significant contributions in terms of employment, incomes and tax revenue, enforceable international corporate codes are needed in order to curb possible irresponsible or exploitative actions on their part.

Undoubtedly, the perceived lack of good governance, including military conflict, coups and mis-government by the military for extended periods in many countries, and corruption right across the continent has been a prime factor in rendering it unattractive to investors. Because this perception operates on a regional or continent-wide basis, this tends to affect all countries, irrespective of their individual records. It needs to be stressed here that it is continuity of effective governance that is important: foreign investors elsewhere have shown that they are able to live with certain levels of corruption so long as these do not impact too directly on their rates of return. They nevertheless add to costs of operation and to any other perceived or actual disadvantages; it is for this reason essential that measures be taken to reduce their various manifestations in Africa. The trend towards democratisation on the continent is a positive factor in this regard, as well as being highly desirable on its own account, and needs to be reinforced by a variety of means: aid conditionality and partnership; NGO advocacy; and human rights legislation, for example.

Due to the still relatively small size of the industrial sector within SSA countries (outside South Africa) and its modest, in a few countries even negative, growth, most people will have to continue for a considerable time to seek their livelihoods within a usually impoverished rural sector or in the 'overspill' of an urban informal sector. For the same reasons, aid agencies are correct in attempting to focus in the first place on poverty eradication, though many neglect decentralised rural development. As suggested in Chapter 14 of this volume, there is a question as to how well rapidly increasing populations can be absorbed in the rural areas, with evidence there of 'de-agrarianisation' and, in the towns and cities, of 'over-urbanisation'.

Such trends can and need to be moderated, however, by measures to raise agricultural productivity, including the provision of improved market access, with other beneficial results for the relevant populations, through the construction of bottleneck-breaking rural infrastructure, particularly roads – as is currently being emphasised by donor agencies such as the World Bank, irrigation and selective electrification. In raising agricultural productivity, advantage needs to be taken of the real opportunities that exist for agricultural diversification directed towards both domestic and export markets. With improved international communication and rising incomes in the north, and in parts of the south, international demand for tropical products should increase. The range of high-value products that have a growing world market has been tapped by only a few low-income countries to date. Product types include tropical fruits, flowers and vegetables of all types in the northern winter months; nuts, spices, medicinal plants and aromatics; and livestock products of various types. In addition, changes in production techniques or the equity of production arrangements can secure access, once produce is inspected and certificated, to higher-value consumer markets such as those for organic produce and ethically traded commodities.

Expansion in the number and importance of rural non-farm enterprises will continue to be an important part of this process and can be encouraged in a variety of ways in terms of technology and training as well as, judiciously, micro-finance schemes of various types. Many rural industries depend critically, however, on local natural resources: wood for rural construction, for example, and for charcoal manufacture. Deforestation by poverty-stricken cultivators threatens not only their own livelihoods but those of these small businesses. Given its known protective and productivity-raising qualities, multiproduct agro-forestry is potentially particularly valuable. Observed constraints in needy parts of Africa, however, include uncertain land and tree ownership rights and gaps in technical knowledge across the range of agro-forestry (as distinct from social forestry and farm forestry) systems, that is, alley cropping, multi-storey cropping, taungya and silvo-pastoral systems.

Renewing development in Africa requires that moves be completed towards the provision of substantial debt relief for the most heavily indebted countries within SSA, particularly those with the most fragile economies. There is a complementarity, however, between the according of such relief and the securing of improved governance and reduction in the prevalence of conflict. At the Prague meeting of the Group of Seven (G-7), held in September 2000, eleven African countries otherwise entitled to international debt relief were excluded from consideration because conflict prevented effective use being made of the increased resources on the ground. Even when accorded debt relief, securing benefit will depend in many cases upon a substantial improvement in governance.

The elimination of poverty in all its dimensions in Africa in fact calls for more aid, in contrast to current trends, but aid more effectively used than in the past, as noted in Chapter 1. How this is to be achieved, given past failures in imposing conditionalities, remains unclear. Delays in implementing the current debt relief proposals are unpromising. In the 1990s increased official development aid (ODA) and private funding, through both international non-governmental organisations (INGOs) and increasing numbers of 'southern' NGOs (SNGOs), most of them very small, has partly offset the decline in ODA supplied directly to African governments. This is noteworthy in that it implies a direction of assistance towards urgent welfare needs, towards disadvantaged groups, especially women, and towards populations resident in more remote areas. There is a requirement, however, for donors to monitor the cost-effectiveness of this aid as NGOs, local and international, become more substantial operators. This needs to be done without unnecessary stifling of initiative in this field or the adoption of inappropriately short-term expectations, against which Wallace, in Chapter 13, warns. The substantially increased role being accorded these institutions raises a number of further issues, as is evident from the same chapter. One question is how far INGOs should favour quasi-political advocacy and legal activity over the – perhaps more mundane – area of 'service delivery', a response in part also to the availability of the resources that would be needed to operate on a sufficient scale. It neglects, on the other hand, the fact that service delivery encompasses opportunities for the demonstration of innovative approaches, appropriate technologies, sustainable institutions and income-generating activities as a whole that are able to benefit the poor specifically and directly. A general and important question is how community- and faith-based organisations can and should, on effectiveness grounds, increase their share of donor-funded activity (see Belshaw *et al.*, 2001).

There are particular areas in which there is considerable absorptive capacity for assistance, such as health: the existence of the AIDS débâcle calls for significantly increased help from those continents fortunate enough not to have been affected on a pandemic scale. Both public and local/international NGO providers are in desperate need of additional resources for medical purposes and to provide for orphans.

In the spheres of agricultural and rural development, consideration needs to be given to the appropriate division of labour between public and private sectors and between the public sector and the civil society organisations of various kinds. At one extreme, it seems likely that the public sector will be left with the responsibility for major infrastructural provision and maintenance – road and rail networks, electricity, posts and telecommunications. At the other extreme, the private sector is likely to expand its role in commercial small-scale agriculture (including animal products, horticulture and floriculture). Especially for the small-farm sector, private firms are likely to assume extended responsibility for the provision of farm inputs (including new and improved seeds and planting material), production and market finance, and marketing arrangements. Tackling food insecurity and the transition from low-productivity subsistence production will require subsidised assistance and institutions with a capacity for efficient delivery. In this middle ground, a range of options will be needed, with public sector, commercial private sector institutions and civil society institutions of different types in competition and in conjunction.

Aid funding is being directed towards education, but resources remain woefully inadequate. Human capital formation to provide for a generally educated workforce is vital if SSA is to make up ground in competing for international investment. The demand for and value placed on education throughout Africa is exemplified by Uganda, a country with a longer history than many in Africa of good quality schooling. The recent presidential decision there to provide free universal primary education led to as much as a six-fold increase in attenders in some provinces: unfortunately, a demand not met except through a drastic reduction in quality within the state sector. More and better-quality national provision with full gender coverage is urgent throughout SSA. Improved gender coverage is important for both social and economic reasons, given the strong role of women and girls in agriculture and other rural production, their increasing importance in labour-intensive manufacturing industries, particularly for export, and the statistical association of girls' education with declining fertility rates.

How far and how fast current moves towards debt relief among 'Heavily Indebted Poor Countries' under the HIPC Initiative will improve the situation among poor SSA countries is uncertain. The Initiative was launched in 1996 and, following criticisms of bureaucratic delay and policy indecision, given somewhat more impetus in September 1999 through the 'Enhanced' Initiative (EHIPC), which more than doubled the amount of aid projected to be provided: up to twenty countries were expected to benefit by the end of 2000. In these programmes countries were to achieve, first, a decision point by which time they would have established their eligibility for assistance, followed after a fixed 3-year period by a 'completion point' at which time the full amount of debt relief would be triggered. Under EHIPC, 'floating completion points' were introduced, allowing some flexibility over the length of this period. Of the first nine countries to reach their decision points under EHIPC, seven were SSA countries.

There are some very positive aspects of the proposals. The conditions required for implementation are that countries need to establish development strategies directed towards poverty eradication, with mechanisms designed to track and report on poverty-related spending. Second, efforts are being made to involve a wide range of bilateral aid sources on top of a coordinated World Bank–IMF action programme, directing these also towards poverty-eradication objectives. Questions are being raised regarding the possibility of achieving country 'ownership' of the prepared policy documents and strategies put forward, and it may be questioned also whether entrenched national interests will permit genuine implementation in many cases.

Major international shocks such as the 2000–1 oil price hikes have major impacts on both public and private sectors in the non-oil SSA countries, on national development as a whole and through changes in government revenues on national education and health sector programmes. With the phasing out of the European Union's Stabex programme for offsetting fluctuations in countries' export proceeds, there is urgent need for international dialogue regarding both stabilisation and compensation measures aimed at cushioning the hardest hit and most vulnerable countries against these impacts.

The fundamental question which remains is how African countries are to become progressively independent of aid by raising their own levels of production. For most countries, the only way that this can be done while simultaneously addressing widespread poverty and periodic supply crises must be through rural development and the adoption of a strategy of agriculture-led development, using SSA's land and rural people as key resources. This strategy can be supported by (a) mineral development in those countries that have such resources, in marketable quantities with competitive costs of exploitation, and (b) tourism earnings, which are an important source of income in both developed and less-developed countries, potentially more important in relative terms in the latter. Given the failure of past agricultural transformation schemes (usually over-mechanised collectives or state enterprises), the focus needs to be on the intensification and diversification of small-scale agriculture, including small-to-large enterprise linkages (SLE), eliminating food insecurity, and generating a wide and inclusive pattern of income distribution that can in turn expand technical and demand linkages with non-farm activity, both locally and in associated urban centres. Environmental enhancement, productivity increase and poverty eradication need to be brought together in the local/regional context in strategies of pro-poor growth that offer opportunities for more self-reliant development.

Evidence has been provided in this volume that in the face of the forces of competition imposed by globalisation, SSA needs to accede to the realities of world markets for goods and for capital; but also that policies based simply on markets for inputs and outputs are not sufficient. Nor should the focus be simply on GDP growth: the *quality* of growth, and more broadly development, matters. Development strategies need to be inclusive in terms of social categories, gender and geographical areas; to provide for livelihoods in the short and medium term and not merely in the long; and to provide for sustainability. Recent moves in the direction of economic integration and cooperation among SSA countries are positive and suggestive of a more realistic approach towards trade facilitation than some past ambitious efforts, and these need to be pursued with much more vigour. These also need to be supported by investments in infrastructure within the transport sector and by further measures to reduce bureaucratic obstacles to intra-African trade, especially those handicapping the landlocked countries, as noted above.

The recently published *World Development Report 2000/2001* (World Bank, 2001), which is sub-titled *Attacking Poverty*, redefines the concept of poverty to include a wide range of relative poverty situations and indices, including maldistribution of income, social exclusion, human rights infringements, gender inequality and so on, in addition to the usual measures of absolute income poverty. At the same time, strategy design for poverty reduction is left much wider open compared with the Bank's 1980s' strategy of structural adjustment, for resolution at the individual country level. Domestically, different types of pro-poor strategy are clearly required according to the country and even the sub-national context. Internationally also, as the UK government's White Paper (DfID, 2000) avers, actions are required by the North to reverse the marginalisation of SSA in the context of ongoing globalisation and to direct these in particular to eradicating conditions of extreme poverty in the South. The definition of what constitutes 'poverty' needs to be adjusted not only to the nature and causality of the poverty being confronted but also to the choice of policy and project interventions and the institutional networks employed for implementation. Applied and action research on the poverty-reduction effectiveness of different strategies in similar contexts warrants high priority.

References

Allee, R.B. (2000) *The Two-mile Time Machine: Ice Cores, Abrupt Climate Change and Our Future*, Trenton, NJ: Princeton University Press.

Auty, R. (2000) 'How natural resources affect economic development', *Development Policy Review*, 18 (4): 347–64.

Belshaw, D., Calderisi, R. and Sugden, C. (eds) (2001) *Faith in Development: Partnership between the World Bank and the Churches of Africa*, Washington, DC and Oxford: World Bank/Regnum Books International.

Department for International Development (DfID) (2000) *Eliminating World Poverty: Making Globalisation Work for the Poor*, (White Paper on International Development), Cm 5006, Norwich: Stationery Office.

Hulme, M., Doherty, R., Ngara, T. and New, M. (2001) 'Global warming and African climate change: a reassessment', in Pak Sum Low (ed.), *Climate Change: Science, Technology and Policy for Africa*, New York: Elsevier.

Jalilian, H., Tribe, M. and Weiss, J. (eds) (2000) *Industrial Development and Policy in Africa*, Cheltenham, Glos.: Edward Elgar.

Krugman, P. (1991) *Geography and Trade*, Leuven: Leuven University Press, and Cambridge, MA: MIT Press.

Lall, S. (1992) 'Structural problems of African industry', in F. Stewart, S. Lall and S. Wangwe (eds), *Alternative Strategies in Sub-Saharan Africa*, Basingstoke, Hants.s: Macmillan.

Pak Sum Low (ed.) (2001) *Climate Change: Science, Technology and Policy for Africa*, New York: Elsevier.

Sachs, J. (1999) 'Helping the world's poorest', *The Economist*, 14 August.

Sachs, J. and Warner, A.M. (1997) 'Sources of slow growth in African economies', *Journal of African Economies*, 6 (3): 335–76.

Schmitz, H. (1989). 'Flexible specialisation: a new paradigm of small-scale industrialisation', IDS Discussion Paper no. 26.1, University of Sussex, Institute of Development Studies.

Stewart, F., Lall, S. and Wangwe, S. (eds) (1992) *Alternative Strategies in Sub-Saharan Africa*, Basingstoke, Hants.: Macmillan.

Tiffen, M., Mortimore, M. and Gichuki, F. (1994) *More People, Less Erosion: Environmental Recovery in Kenya*, Chichester, Sussex: John Wiley.

Tribe, M. (2000) 'The concept of "infant industry" in a sub-Saharan context', in H. Jalilian, M. Tribe and J. Weiss (eds), *Industrial Development and Policy in Africa*, Cheltenham, Glos.: Edward Elgar.

World Bank (2001) *World Development Report 2000/2001: Attacking Poverty*, New York: Oxford University Press.

Index